REVIEWS in MINERALOGY

D0400959

RECEIVED
NOV 1 - 1996
UTSA Library Serials

REACTIVE TRANSPORT IN POROUS MEDIA

Edited by

Peter C. Lichtner
Center for Nuclear Waste Regulatory Analysis

Carl I. Steefel
University of South Florida

Eric H. Oelkers
Université Paul Sabatier

Cover: Contour plot of computed mineral volume fraction and fluid streamlines (white with arrows indicating direction) after an elapsed time of 7,500 years. Fluid undersaturated with respect to the mineral is injected at the left, causing the mineral to dissolve and thus increasing the porosity and permeability in this region. Zones of red indicate regions where the mineral has been dissolved completely while the blue zones indicate regions where the initial mineral volume fraction is still present. Due to the reactive infiltration instability, the front propagates as fingers rather than as a planar front.

(Lichtner, this volume, Figure 19, page 76).

Series Editor: Paul H. Ribbe
Department of Geological Sciences
Virginia Polytechnic Institute & State University
Blacksburg, Virginia 24061 U.S.A.

WITHDRAWN
UTSA Libraries

COPYRIGHT 1996

MINERALOGICAL SOCIETY OF AMERICA

The appearance of the code at the bottom of the first page of each chapter in this volume indicates the copyright owner's consent that copies of the article can be made for personal use or internal use or for the personal use or internal use of specific clients, provided the original publication is cited. The consent is given on the condition, however, that the copier pay the stated per-copy fee through the Copyright Clearance Center, Inc. for copying beyond that permitted by Sections 107 or 108 of the U.S. Copyright Law. This consent does not extend to other types of copying for general distribution, for advertising or promotional purposes, for creating new collective works, or for resale. For permission to reprint entire articles in these cases and the like, consult the Administrator of the Mineralogical Society of America as to the royalty due to the Society.

REVIEWS IN MINERALOGY

(Formerly: SHORT COURSE NOTES)

ISSN 0275-0279

Volume 34

REACTIVE TRANSPORT IN POROUS MEDIA

ISBN 0-939950-42-1

ADDITIONAL COPIES of this volume as well as those listed on page v may be obtained at moderate cost from:

THE MINERALOGICAL SOCIETY OF AMERICA
1015 EIGHTEENTH STREET, NW, SUITE 601
WASHINGTON, DC 20036 U.S.A.

Library
University of Texas
at San Antonio

FOREWORD

This volume is the 34th in the *Reviews in Mineralogy* series begun in 1974. As with most of these volumes, it was written in preparation for a Short Course held in conjunction with a major meeting of the Mineralogical Society of America (MSA), in this case MSA's Annual Meeting with the Geological Society of America in Denver, Colorado. Carl Steefel was the editor who was most heavily involved in supporting the production of the camera-ready copy, a job much complicated by the proliferation of word-processing software. I am also indebted to Margie Sentelle for her secretarial help and my wife, Elna, for sacrificing a significant portion of her summer vacation to accommodate my efforts in the editorial office. I am grateful to the Department of Geological Sciences at Virginia Polytechnic Institute and State University for an office and for a variety of material support.

Paul H. Ribbe

Blacksburg, Virginia
September 17, 1996

PREFACE

The field of reactive transport within the Earth Sciences is a highly multidisciplinary area of research. The field encompasses a number of diverse disciplines including geochemistry, geology, physics, chemistry, hydrology, and engineering. The literature on the subject is similarly spread out as can be seen by a perusal of the bibliographies at the end of the chapters in this volume. Because these distinct disciplines have evolved largely independently of one another, their respective treatments of reactive transport in the Earth Sciences are based on different terminologies, assumptions, and levels of mathematical rigor. This volume and the short course which accompanies it, is an attempt to some extent bridge the gap between these different disciplines by bringing together authors and students from different backgrounds.

A wide variety of geochemical processes including such diverse phenomena as the transport of radiogenic and toxic waste products, diagenesis, hydrothermal ore deposit formation, and metamorphism are the result of reactive transport in the subsurface. Such systems can be viewed as open bio-geochemical reactors where chemical change is driven by the interactions between migrating fluids, solid phases, and organisms. The evolution of these systems involves diverse processes including fluid flow, chemical reaction, and solute transport, each with differing characteristic time scales. This volume focuses on methods to describe the extent and consequences of reactive flow and transport in natural subsurface systems.

Our ability to quantify reactive transport in natural systems has advanced dramatically over the past decade. Much of this advance is due to the exponential increase in computer computational power over the past generation—geochemical calculations that took years to perform in 1970 can be performed in seconds in 1996. Taking advantage of this increase of computational power, numerous comprehensive reactive transport models have been developed and applied to natural phenomena. These models can be used either qualitatively or qualitatively to provide insight into natural phenomena. Quantitative models force the investigator to validate or invalidate ideas by putting real numbers into an often vague hypothesis and thereby starting the thought process along a path that may result in acceptance, rejection, or modification of the original hypothesis. Used qualitatively, models provide insight into the general features of a particular phenomenon, rather than specific details.

One of the major questions facing the use of hydrogeochemical models is whether or not they can be used with confidence to predict future evolution of groundwater systems. There is much controversy concerning the validity and uncertainties of non-reactive fluid flow systems. Adding chemical interaction to these flow models only confounds the problem. Although such models may accurately integrate the governing physical and chemical equations, many uncertainties are inherent in characterizing the natural system itself. These systems are inherently heterogeneous on a variety of scales rendering it impossible to know precisely the many details of the flow system and chemical composition of the host rock. Other properties of natural systems such as permeability and mineral surface area, to name just two, may never be known with any great precision, and in fact may be unknowable. Because of these uncertainties, it remains an open question as to what extent numerical models of groundwater flow and reactive transport will be useful in making accurate quantitative predictions. Nevertheless, reactive transport models should be able to predict the outcome for the particular representation of the porous medium used in the model.

Finally, it should be mentioned that numerical models are often our only recourse to analyze such environmental problems as safe disposal of nuclear waste where predictions must be carried out over geologic time spans. Without such models it would be impossible to analyze such systems, because they involve times too long to perform laboratory experiments. The results of model calculations may affect important political decisions that must be made. Therefore, it is all the more important that models be applied and tested in diverse environments so that confidence and understanding of the limitations and strengths of model predictions are understood before irreversible decisions are made that could adversely affect generations to come.

This volume contains the contributions presented at a short course held in Golden, Colorado, October 25-27, 1996. We thank all of the authors for their successful efforts to produce the chapters of this volume in advance of the short course. We thank Mike Hochella for his guidance in organizing the short course and getting it approved by the MSA Council and Alex Speer for handling administrative details. We are also greatly indebted to the *Reviews in Mineralogy* Series Editor, Paul Ribbe, who patiently put up with late papers, font incompatibilities, and authors typesetting their own papers and still managed to get the volume out on time. We also thank Andy Tompson and Robert Smith, who through funding from the Subsurface Science Program managed by Frank Wobber at the U.S. Department of Energy, provided the short course with additional funds used primarily for student scholarships. Finally, we thank Dr. Wendy Harrison of the Colorado School of Mines in Golden, Colorado, for agreeing to sponsor the computer demonstration section of the short course held at CSM.

Peter C. Lichtner — San Antonio, Texas

Carl I. Steefel — Tampa, Florida

Eric H. Oelkers — Toulouse, France

List of volumes currently available in the *Reviews in Mineralogy* series

Vol.	Year	Pages	Editor(s)	Title
1-7	*out*	*of*	*print*	
8	1981	398	A.C. Lasaga R.J. Kirkpatrick	KINETICS OF GEOCHEMICAL PROCESSES
9A	1981	372	D.R. Veblen	AMPHIBOLES AND OTHER HYDROUS PYRIBOLES—MINERALOGY
9B	1982	390	D.R. Veblen, P.H. Ribbe	AMPHIBOLES: PETROLOGY AND EXPERIMENTAL PHASE RELATIONS
10	1982	397	J.M. Ferry	CHARACTERIZATION OF METAMORPHISM THROUGH MINERAL EQUILIBRIA
11	1983	394	R.J. Reeder	CARBONATES: MINERALOGY AND CHEMISTRY
12	1983	644	E. Roedder	FLUID INCLUSIONS (Monograph)
13	1984	584	S.W. Bailey	MICAS
14	1985	428	S.W. Kieffer A. Navrotsky	MICROSCOPIC TO MACROSCOPIC: ATOMIC ENVIRONMENTS TO MINERAL THERMODYNAMICS
15	1990	406	M.B. Boisen, Jr. G.V. Gibbs	MATHEMATICAL CRYSTALLOGRAPHY (Revised)
16	1986	570	J.W. Valley H.P. Taylor, Jr. J.R. O'Neil	STABLE ISOTOPES IN HIGH TEMPERATURE GEOLOGICAL PROCESSES
17	1987	500	H.P. Eugster I.S.E. Carmichael	THERMODYNAMIC MODELLING OF GEOLOGICAL MATERIALS: MINERALS, FLUIDS, MELTS
18	1988	698	F.C. Hawthorne	SPECTROSCOPIC METHODS IN MINERALOGY AND GEOLOGY
19	1988	698	S.W. Bailey	HYDROUS PHYLLOSILICATES (EXCLUSIVE OF MICAS)
20	1989	369	D.L. Bish, J.E. Post	MODERN POWDER DIFFRACTION
21	1989	348	B.R. Lipin, G.A. McKay	GEOCHEMISTRY AND MINERALOGY OF RARE EARTH ELEMENTS
22	1990	406	D.M. Kerrick	THE Al_2SiO_5 POLYMORPHS (Monograph)
23	1990	603	M.F. Hochella, Jr. A.F. White	MINERAL-WATER INTERFACE GEOCHEMISTRY
24	1990	314	J. Nicholls J.K. Russell	MODERN METHODS OF IGNEOUS PETROLOGY—UNDERSTANDING MAGMATIC PROCESSES
25	1991	509	D.H. Lindsley	OXIDE MINERALS: PETROLOGIC AND MAGNETIC SIGNIFICANCE
26	1991	847	D.M. Kerrick	CONTACT METAMORPHISM
27	1992	508	P.R. Buseck	MINERALS AND REACTIONS AT THE ATOMIC SCALE: TRANSMISSION ELECTRON MICROSCOPY
28	1993	584	G.D. Guthrie B.T. Mossman	HEALTH EFFECTS OF MINERAL DUSTS
29	1994	606	P.J. Heaney, C.T. Prewitt, G.V. Gibbs	SILICA: PHYSICAL BEHAVIOR, GEOCHEMISTRY, AND MATERIALS APPLICATIONS
30	1994	517	M.R. Carroll J.R. Holloway	VOLATILES IN MAGMAS
31	1995	583	A.F. White S.L. Brantley	CHEMICAL WEATHERING RATES OF SILICATE MINERALS
32	1995	616	J. Stebbins P.F. McMillan, D.B.Dingwell	STRUCTURE, DYNAMICS AND PROPERTIES OF SILICATE MELTS
33	1996	862	E.S. Grew L.M. Anovitz	BORON: MINERALOGY, PETROLOGY AND GEOCHEMISTRY

REACTIVE TRANSPORT IN POROUS MEDIA

TABLE OF CONTENTS, RIM VOLUME 34

Chapter 1 **P. C. Lichtner**

CONTINUUM FORMULATION OF MULTICOMPONENT-MULTIPHASE REACTIVE TRANSPORT

Chapter 2 **C. I. Steefel and K. T. B. MacQuarrie**

APPROACHES TO MODELING OF REACTIVE TRANSPORT IN POROUS MEDIA

Chapter 3 **E. H. Oelkers**

PHYSICAL AND CHEMICAL PROPERTIES OF ROCKS AND FLUIDS FOR CHEMICAL MASS TRANSPORT CALCULATIONS

Chapter 4 **C. A. J. Appelo**

MULTICOMPONENT ION EXCHANGE AND CHROMATOGRAPHY IN NATURAL SYSTEMS

Chapter 5 **D. L. Suarez and J. Šimůnek**

SOLUTE TRANSPORT MODELING UNDER VARIABLY SATURATED WATER FLOW CONDITIONS

Chapter 6 **A. F. B. Tompson & K. J. Jackson**

REACTIVE TRANSPORT IN HETEROGENEOUS SYSTEMS: AN OVERVIEW

Chapter 7 **B. E. Rittmann & J. M. VanBriesen**

MICROBIOLOGICAL PROCESSES IN REACTIVE MODELING

Chapter 8 **P. Van Cappellen and J.-F. Gaillard**

BIOGEOCHEMICAL DYNAMICS IN AQUATIC SEDIMENTS

REACTIVE TRANSPORT MODELING OF ACIDIC METAL-CONTAMINATED GROUND WATER AT A SITE WITH SPARSE SPATIAL INFORMATION

Chapter 1

CONTINUUM FORMULATION OF MULTICOMPONENT–MULTIPHASE REACTIVE TRANSPORT

Peter C. Lichtner

Center for Nuclear Waste Regulatory Analyses
San Antonio, Texas 78238 U.S.A.
(lichtner@swri.edu)

INTRODUCTION

Quantitative models of physicochemical processes in the Earth's crust can be applied to a broad range of phenomena involving fluid–rock interaction and transport of aqueous, non-aqueous and gaseous species. These phenomena encompass environmental and industrial problems as well as fundamental geologic processes. Mathematically the reactive mass transport equations represent a moving boundary problem in which it is necessary to determine the regions of space the various phases in the system—solid, liquid and gas—occupy as functions of time. Computer models can provide, if not a direct quantitative description, at least a far better qualitative understanding of the geochemical and physical processes under investigation than might otherwise be possible. Further, reactive transport models can provide a useful tool to society for evaluating various environmental hazards. For example, models can be used to predict the rate of migration of a contaminant plume resulting from acid mine drainage and heavy metal mobilization. Quantitative models can also help design remedial strategies to clean up toxic waste sites, such as removal of volatile hydrocarbons by steam injection, and others. While the emphasis of this book is on modeling, it is essential that modeling be closely integrated with field and laboratory studies if the predicted results are to have anything to do with reality.

Considerable progress has been made in the past few years in modeling reactive flow and transport in natural systems. Ever faster computers, including massively parallel machines, have enabled complex multicomponent chemical systems to be included in multi-dimensional simulations. New algorithms for high Péclet number flows have made it possible to reduce the effects of numerical dispersion. And growing thermodynamic and kinetic databases have provided fundamental data needed to model realistic geochemical systems.

Nevertheless, there remain many challenges for an accurate or even semi-quantitative description of natural systems. Reaction mechanisms need to be understood first before they can be successfully incorporated into a reactive flow and transport model. A description of fluid–rock interaction requires knowledge of mineral surface area, a quantity that is difficult to measure and may change with time. Microbial systems are still poorly understood but may be an essential aspect of many geochemical systems. Initial and boundary conditions are often unknown and unknowable. And finally, a problem of fundamental importance is upscaling properties obtained from observations at the pore and molecular scales to the macro or continuum scale at which most reactive transport models are applicable. Natural porous media are undoubtedly heterogeneous at all scales. Yet they are often approximated as if they were homogeneous. Incorporating heterogeneous porous media into reactive transport models presents a uniquely challenging problem, both conceptually as well as computationally.

0275-0279/96/0034-0001$05.00

This chapter begins with a discussion of the continuum hypothesis on which conservation equations of mass and energy describing flow and transport in porous media are founded. A general conservation law is developed from which mass and energy conservation equations can be deduced. Reactive mass transport equations are derived for a multicomponent–multiphase system involving chemical reaction of minerals, aqueous and gaseous species for a system in local partial equilibrium. It is shown that by transforming the chemical reactions to canonical form, the number of unknowns in the mass transport equations can be greatly reduced by eliminating of rates corresponding to local equilibrium reactions. Concepts of local equilibrium, scaling relations, asymptotics, ghost zones and the quasi-stationary state approximation are introduced. An analytical solution for a single component system derived from the quasi-stationary state approximation is presented. The concept of equilibration length is illustrated. The question of charge balance in the aqueous phase is investigated for species-dependent aqueous diffusion coefficients and incorporating batch sorption models with transport. Finally several examples are presented using the computer code MULTIFLO, a multicomponent–multiphase reactive transport model. These include acid mine drainage, redox front migration, reactive transport in heterogeneous porous media, an example of reaction front instability, and a simple hydrothermal system.

CONTINUUM HYPOTHESIS

A rock mass consisting of aggregates of mineral grains and pore spaces or voids is referred to as a porous medium. An actual porous medium is a highly heterogeneous structure containing physical discontinuities marked by the boundaries of pore walls which separate the solid framework from the void space. Although it is possible in principle at least, to describe this system at the microscale of a pore, such a description rapidly becomes a hopeless task as the size of the system increases and many pore volumes become involved. It is therefore necessary to approximate the system by a more manageable one. One quantitative description of the transport of fluids and their interaction with rocks is based on a mathematical idealization of the real physical system referred to as a continuum. In this theory the actual discrete physical system, consisting of aggregates of mineral grains, interstitial pore spaces, and fractures, is replaced by a continuous system in which physical variables describing the system vary continuously in space. Allowance is made for the possibility of a discrete set of surfaces across which discontinuous changes in physical properties may occur. In this fictitious representation of the real physical system solids and fluids coexist simultaneously at each point in space.

The continuum hypothesis asserts that a real physical system may be approximated by one in which the properties of the system vary sufficiently smoothly enabling the use of differential calculus to describe processes taking place in the system. The utility of the continuum hypothesis is that it enables changes in a system to be formulated in terms of partial differential equations. Real rocks, however, are far from continuous, and it must be understood at the outset that invoking the continuum hypothesis requires a major leap of faith that cannot be justified a posteriori. Certainly, the concept of a continuum applied to porous media is on a much less firm foundation compared to its use in other areas of physics. By way of contrast, in fluid mechanics a real fluid consisting of discrete molecules is replaced by a continuum in which fluid properties vary continuously throughout space. This is justified on the basis that any finite volume of fluid which characterizes changes in its macroscopic properties such as temperature and pressure, contains many molecules. For example, at standard conditions of 25°C and 1 bar, approximately 10^{22} water molecules occupy one cubic centimeter of liquid water.

In the continuum representation of a porous medium, the physical variables describing the system, which are discontinuous at the microscale or pore scale as a consequence of the granular nature of rocks, are replaced by functions which are continuous at the macroscale. The

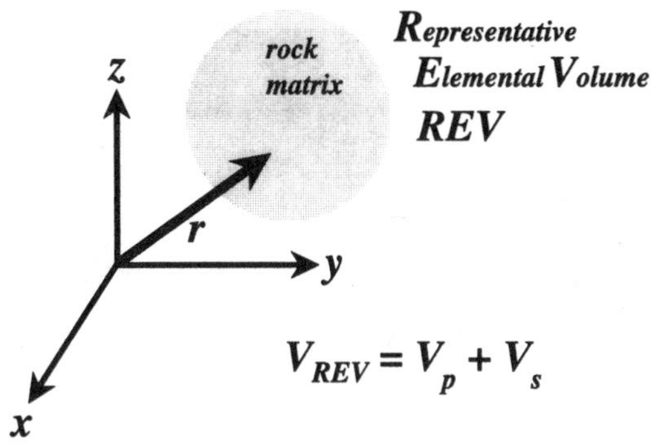

Figure 1. Schematic diagram of continuum space.

value of each physical variable assigned to a point in continuum space is obtained by locally averaging the actual physical property over some representative elemental volume, referred to simply as a REV (Bear, 1972). The essential feature of a REV is that it characterizes the properties of the system locally. The dimensions of a REV are usually large compared to the grain size, but small compared to the characteristic length scale over which the quantities of interest change. The quantities describing a system at the macroscale include the temperature, pressure, fluid density, fluid saturation, average fluid flow velocities, concentration of minerals, gaseous and aqueous species, porosity, and permeability among others. Functions defined at each in point in space and time are referred to mathematically as fields. Thus, for example, the temperature T is represented by the scalar field $T(\boldsymbol{r}, t)$ where the vector $\boldsymbol{r}$ refers to a point in continuum space with spatial coordinates (x, y, z) (see Fig. 1), and t represents the time. Bold face type is used throughout to denote vectors in 3-space. The value of the temperature at the point $\boldsymbol{r}$ is obtained by averaging the temperature over a REV centered at $\boldsymbol{r}$. Similarly the porosity, liquid saturation, mineral, aqueous and gaseous concentrations are represented by the scalar fields $\phi(\boldsymbol{r}, t)$, $s_l(\boldsymbol{r}, t)$, $\phi_m(\boldsymbol{r}, t)$, $C_i^l(\boldsymbol{r}, t)$, and $C_i^g(\boldsymbol{r}, t)$, respectively, where superscripts l and g refer to liquid and gas phases, and subscripts i and m refer to the ith solute and mth mineral, respectively. The velocity field $\boldsymbol{v}(\boldsymbol{r}, t) = (v_x(\boldsymbol{r}, t),\ v_y(\boldsymbol{r}, t),\ v_z(\boldsymbol{r}, t))$ is an example of a vector field.

Covering the entire system with a connected set of REVs provides a global description of the system. For some systems a typical size of a REV may be a hand sample collected by a geologist in the field. But a REV may also be much smaller than that, perhaps on the order of tens of mineral grains. For rocks exhibiting patterns such as a reaction halo surrounding a fracture, care must be taken to choose the size of the REV smaller than the structure being observed. A REV may not be taken too small, however, because then it no longer provides an average property of the rock–fluid system. There is no guarantee that a single set of REVs belonging to a single continuum is sufficient to characterize a rock. This is especially true of fractured rocks for which primary and secondary porosities corresponding to the matrix and fracture network can be defined. For such rocks at least two sets of REVs are needed, one for the fracture network and the other for the rock matrix. More generally a hierarchical porous medium may be required characterized by many sets of REVs, referred to as multiple interacting continua.

It is important to realize that continuum theory can provide only a macroscale description of the properties of rocks and interstitial fluids and not a microscale, or pore scale, description. This is not to say that microscale properties are not important. In fact it is the microscale properties averaged over a REV which define the macroscale properties. However, usually such averages are too difficult to perform mathematically and we must let nature carry out the averaging process for us. We may therefore attempt to measure directly the macroscale properties of a rock, such as its porosity and permeability, thereby providing a phenomenological or empirical description. Even if it is not feasible to predict the values of the various parameters entering the continuum theory from fundamental principles, a phenomenological description can provide a first attempt to model such systems.

Given an actual sample of a representative elemental volume for some porous medium, values of the various physical variables which describe the system may, in principle at least, be determined. The volume of the REV, V_{REV} or simply V, is equal to the sum of the solid volume V_s and void, or pore volume, V_p at position $\boldsymbol{r}$ in continuum space and time t:

$$V_{REV}(\boldsymbol{r},\, t) \;=\; V_s(\boldsymbol{r},\, t) + V_p(\boldsymbol{r},\, t). \tag{1}$$

As indicated, the REV volume may vary with position and even time. The total porosity ϕ is defined as the fraction of volume of the rock made up of pore space or voids

$$\phi(\boldsymbol{r},\, t) \;=\; \frac{V_p(\boldsymbol{r},\, t)}{V_{REV}(\boldsymbol{r},\, t)}. \tag{2}$$

To have any meaning in the context of the continuum hypothesis, the porosity must be independent of the choice of the REV volume V_{REV}. The pore volume may change with time as a result of chemical reactions as well as compaction of the porous medium. The total porosity consists, in general, of both connected pathways and disconnected isolated pores.

The pore and solid volumes may be further expressed in terms of the individual phases which occupy them. The solid volume is assumed to consist of an aggregate of minerals each with volume V_m. The pore or void space may be occupied by a number of fluid phases each with volume V_π. Thus the pore volume can be expressed as the sum of the fluid phase volumes

$$V_p \;=\; \sum_{\pi=1}^{N_f} V_\pi, \tag{3}$$

for N_f fluid phases, and the solid volume as the sum of mineral volumes contained in the REV:

$$V_s \;=\; \sum_{m=1}^{N_{\min}} V_m, \tag{4}$$

for $N_{\min}$ minerals, where the subscripts π and m denote the πth fluid phase and mth mineral, respectively.

Darcy's law

Fundamental to describing flow of liquid and gas in the continuum representation of a porous medium is Darcy's law. In 1856, Henry Darcy in Dijon, France carried out experiments on the flow of water through a packed sand column. For horizontal flow he found that the volumetric flow of water Q through the sand column was directly proportional to the cross sectional surface area A and the applied pressure gradient dp/dx across the sand column, and inversely proportional to the viscosity of water μ. In symbols

$$Q \;=\; -A\frac{k}{\mu}\frac{dp}{dx}, \tag{5}$$

where p denotes the fluid pressure and x the distance measured along the column. The proportionality factor k is referred to as the intrinsic permeability of the sand column. This equation is known as Darcy's law. The minus sign indicates that, in the absence of gravity, fluid always flows from a higher fluid pressure to a lower pressure. Darcy's law represents a balance between viscous forces and the pressure gradient. Inertial effects are neglected and hence Darcy's law holds only for sufficiently low velocities.

The permeability k characterizes the resistance of the sand column to the flow of water. The more permeable the sand column, the faster water flows for a given pressure gradient. The units of k are m^2, recalling that the units of pressure are force per unit area (1 Pascal = Newton m^{-2} = kg m^{-1} sec^{-2}), the units of viscosity are kg m^{-1} sec^{-1}, and Q has units of cubic meters per second (m^3 sec^{-1}). Using Darcy's law, the value of the permeability may be determined experimentally if the values of all other quantities can be measured. Typical values of permeability range as high as $10^{-7}m^2$ for gravel, to values less than $10^{-20}m^2$ for granite. Sand has a permeability of approximately $10^{-12}m^2$, or approximately 1 darcy (1 darcy = 9.8697×10^{-13} m^2) (See Oelkers, this volume). The quantity Q/A has units of velocity and is referred to as the Darcy velocity or Darcy flux

$$q = \frac{Q}{A}. \tag{6}$$

If the Darcy velocity is 1 m sec^{-1} for a pressure gradient of 1 bar m^{-1}, the permeability of the porous medium is 1 darcy.

Because flow only takes place within the pore spaces of the porous medium, for a given volumetric flow rate Q, or Darcy velocity q, the average pore velocity v is increased compared to the Darcy velocity by the reciprocal of the porosity

$$v = \frac{q}{\phi}. \tag{7}$$

In writing this relation it is assumed that the areal porosity is the same as the volumetric porosity ϕ and that all the pore space is connected. The equality of areal and volumetric porosity is generally not a bad assumption because of the random structure of most rocks at the pore scale. For metamorphic rocks which have a very low porosity on the order of 1% or less, the average fluid velocity may therefore be hundreds of times faster than the Darcy velocity. For a sand column with a porosity of around 50%, they differ by only a factor of two. It should be emphasized that the average pore velocity is quite different from the microscale fluid velocity within individual pore spaces. According to Darcy's law, the fluid velocity depends only on macroscale properties of the porous medium and not on the detailed pore geometry. The microscale properties of the medium are conveniently buried in the permeability.

Darcy's relation for the flow of water through a porous medium is empirical. It is very difficult to construct mathematical models of real porous media which enable the value of the permeability to be predicted from first principles. It is clear that the permeability must depend on microscale properties of the porous medium, namely the shapes of pore volumes and the throats connecting them. The permeability represents the average effect of such microscale properties of the pore geometry averaged over a REV and thus is a macroscale quantity. The validity of Darcy's law requires that the pore size be much smaller than other characteristic dimensions governing the flow field, such as the thickness of the sediment through which flow is taking place. In addition the flow must be laminar, that is, turbulence must not occur.

GENERAL CONSERVATION LAW

This book is concerned primarily with open systems. A system is open if it can exchange both matter and energy with its surroundings. In addition the system may undergo internal changes, for example, through chemical reactions. By contrast, a closed system is one in which the mass of the system is fixed and only energy may be exchanged with its surroundings.

A conservation equation for any of the extensive variables describing an open system (e.g., mass, energy, moles of some species, entropy, etc.) can be derived by considering the overall balance of the quantity over a representative elemental volume V. In words, the balance law for some solute species indexed by i can be expressed as

$$\begin{bmatrix} \text{rate of accumulation} \\ \text{of species } i \text{ in } V_{\text{REV}} \end{bmatrix} = \begin{bmatrix} \text{net flux of } i \text{ across} \\ \text{boundary of } V_{\text{REV}} \end{bmatrix} + \begin{bmatrix} \text{supply or removal} \\ \text{of } i \text{ in } V_{\text{REV}} \end{bmatrix}. \tag{8}$$

Supply and removal of the species may occur through chemical reactions or other source or sink processes.

Let the volume V be simply connected but otherwise have an arbitrary shape. Consider the transport of some quantity Z across the boundary of V, denoted by the symbol Γ. The amount of the quantity contained in the volume V is equal to the integral of the quantity over the volume, expressed mathematically by the integral $\int_V Z dV$. The amount escaping or entering through the boundaries of V per unit time is equal to the integral of the flux over the surface enclosing the volume, or in symbols $\int_\Gamma \boldsymbol{J}_Z \cdot \boldsymbol{dS}$, where $\boldsymbol{J}_Z$ is the flux of the quantity Z across the surface Γ, and $\boldsymbol{dS}$ points in a direction normal to the surface. The flux is taken to be positive if it is directed along the outward normal $\boldsymbol{dS}$ to the surface Γ. Finally the total amount of Z consumed or produced at the rate R_Z by chemical reactions taking place within the volume or from other sources, is equal to $\int_V R_Z dV$. Then the rate of change with time of the amount of the quantity contained within the volume V can be expressed as

$$\frac{d}{dt} \int_V Z\, dV = -\int_\Gamma \boldsymbol{J}_Z \cdot \boldsymbol{dS} + \int R_Z\, dV. \tag{9}$$

This equation can put in differential form by first transforming the surface integral using Gauss's law to yield

$$\int_\Gamma \boldsymbol{J}_Z \cdot \boldsymbol{dS} = \int_V \nabla \cdot \boldsymbol{J}_Z\, dV. \tag{10}$$

This mathematical identity is known as the divergence theorem. It applies to any vector field and states that the vector field integrated over a closed surface is equal to the divergence of the vector field integrated over the volume enclosed by the surface. Physically the divergence of a vector field represents the escaping tendency of the field from an enclosed volume. Putting this relation into the basic conservation relation, Equation (9), yields an equation in which all terms involve integrals over the volume V:

$$\frac{d}{dt} \int_V Z\, dV = -\int_V \nabla \cdot \boldsymbol{J}_Z\, dV + \int R_Z\, dV. \tag{11}$$

Rearranging this equation by placing the time derivative inside the integral, noting that

$$\frac{d}{dt} \int Z dV = \int \frac{\partial Z}{\partial t} dV, \tag{12}$$

and bringing all terms to the left hand side inside the volume integral results in the equation

$$\int_V \left\{ \frac{\partial Z}{\partial t} + \nabla \cdot \boldsymbol{J}_Z - R_Z \right\} dV = 0. \tag{13}$$

Because the volume V is arbitrary the integrand itself must vanish leading to the differential equation

$$\frac{\partial Z}{\partial t} + \nabla \cdot \boldsymbol{J}_Z = R_Z. \tag{14}$$

This equation holds at each point within the volume V and expresses the desired conservation law for the quantity Z. The quantity Z is said to be conserved if the source/sink term R_Z appearing on the right hand side vanishes and the flux only involves derivatives of Z and no other quantities. The main difficulty in applying the general conservation equation to any particular process is determining the appropriate form for the flux $\boldsymbol{J}_Z$, the source/sink term R_Z, and any constitutive relations than are needed to define material properties. By associating the quantity Z with mass, concentration of some species, and energy, the transport equations for these quantities follow.

Conservation of mass

The total mass M of fluid contained in a representative elemental volume V of a fully saturated porous medium with porosity ϕ is equal to

$$M = \frac{M}{V_p}\frac{V_p}{V}V = \phi\rho V, \tag{15}$$

where ρ refers to the intrinsic density of the fluid. The accumulation term Z is identified with the bulk mass density defined as

$$Z = \frac{M}{V} = \phi\rho. \tag{16}$$

The velocity of the fluid at each point in space is represented by the time-dependent vector field $\boldsymbol{v}(x, y, z, t)$. To determine the flux $\boldsymbol{J}_Z$ note that the total mass ΔM crossing a surface of area A in time Δt is equal to

$$\Delta M = A\rho v \Delta t, \tag{17}$$

where v denotes the fluid flow velocity normal to the surface. The flux is defined as the mass crossing a surface of unit area per unit time, or

$$J_Z = \frac{\Delta M}{A\Delta t} = \rho v, \tag{18}$$

where J_Z and v denote the magnitude of the flux and flow velocity normal to the surface. Thus taking $Z = \phi\rho, \boldsymbol{J}_Z = \rho\boldsymbol{v}$ in Equation (14), we arrive at the continuity equation

$$\frac{\partial}{\partial t}(\phi\rho) + \nabla \cdot (\rho\boldsymbol{v}) = 0. \tag{19}$$

The right hand side is zero because chemical reactions neither create nor destroy (for all practical purposes) mass. For an incompressible fluid (ρ = constant), and constant porosity, this equation reduces to

$$\nabla \cdot \boldsymbol{v} = 0. \tag{20}$$

The conservation equations for transport of solute species in a multicomponent aqueous fluid is considered next. Applying the general conservation expression Equation (14) yields the mass conservation equation

$$\frac{\partial}{\partial t}(\phi C_i) + \nabla \cdot \boldsymbol{J}_i = R_i. \tag{21}$$

This result is obtained by identifying the accumulation term with

$$Z_i = \frac{N_i}{V} = \frac{V_p}{V}\frac{N_i}{V_p} = \phi C_i, \tag{22}$$

where N_i represents the number of moles of the ith species in the REV, and the concentration of the ith species is denoted by C_i. The flux term is identified with $\boldsymbol{J}_Z = \boldsymbol{J}_i$, where $\boldsymbol{J}_i$ represents the solute flux of the ith species consisting on contributions from advection and diffusion, with the form

$$\boldsymbol{J}_i = -\phi\tau D_i \nabla C_i + \boldsymbol{v} C_i, \tag{23}$$

where D_i denotes the diffusion/dispersion coefficient and τ the tortuosity of the porous medium. In the absence of a source/sink term this equation is referred to as the continuity equation. In general this equation does not conserve charge within the aqueous phase for species-dependent diffusion coefficients. This deficiency is remedied in the section entitled Special Topics: Charge conservation. Effects of hydrodynamic dispersion resulting from fluid mixing due to contorted flow paths may be included approximately in the diffusion term (see Oelkers, this volume).

Partially saturated porous medium. In a partially saturated porous medium conservation equations for the ith species in fluid phase π_i with concentration C_i, may be obtained by identifying

$$Z_i = \frac{N_i}{V} = \frac{V_p}{V}\frac{V_{\pi_i}}{V_p}\frac{N_i}{V_{\pi_i}} = \phi s_{\pi_i} C_i, \tag{24}$$

where s_{π_i} denotes the saturation state of fluid phase π_i defined as the fraction of pore volume occupied by the fluid. Examples of different fluid phases are liquid water, water vapor, and nonaqueous liquids (NAPL) such as oil, to mention a few. The form of the conservation equation now becomes

$$\frac{\partial}{\partial t}\left(\phi s_{\pi_i} C_i\right) + \nabla\cdot\boldsymbol{J}_i = R_i. \tag{25}$$

in which the saturation state of the fluid phase appears. For multi-phase fluid flow a more complicated expression results for the fluid flux compared to the fully saturated case, involving relative permeabilities associated with each fluid phase and capillary effects between the different fluid phases [see Eqns. (162) and (163)].

Solids. For minerals the transport equations simplify because the flux term vanishes. Transport of solids as colloids or suspended sediment is not considered. The mineral concentration referenced to a REV volume is given by

$$Z_m = \frac{N_m}{V} = C_m, \tag{26}$$

for N_m moles of the mth mineral contained in V. The mineral mass transfer equation can be expressed as

$$\frac{\partial C_m}{\partial t} = R_m. \tag{27}$$

In terms of the mineral volume fraction ϕ_m, defined by

$$\phi_m = \frac{V_m}{V} = \frac{V_m}{N_m}\frac{N_m}{V} = \overline{V}_m C_m, \tag{28}$$

where $\overline{V}_m$ denotes the mineral molar volume, the mineral mass transfer equation becomes

$$\frac{\partial \phi_m}{\partial t} = \overline{V}_m R_m, \tag{29}$$

where R_m represents the source/sink term resulting from chemical reactions. To obtain this latter result it is necessary to assume that the molar volume is constant. The mineral volume fractions are related to the porosity of the porous medium by the equation

$$\phi = 1 - \sum_{m=1}^{M_{\rm min}} \phi_m. \tag{30}$$

It should be kept in mind that the porosity defined in this manner corresponds to the total porosity. This may be different from the connected porosity which appears in the transport equations.

Conservation of energy

The energy conservation equation is more complex in certain aspects than the mass conservation equation. This is because energy is contained in both fluid and solid phases. The following treatment represents a simplification of the general case in which it is assumed that thermodynamic equilibrium is achieved between the fluid and solid phases. Consider a two-phase fluid composed of liquid and gas phases. The total energy U contained in fluid and solid phases is equal to

$$U = \phi\left(s_l\rho_l U_l + s_g\rho_g U_g\right) + (1-\phi)\,\rho_{\rm rock} C_p^{\rm rock} T, \tag{31}$$

where T denotes the temperature, U_π the total internal energy, and ρ_π the density of the liquid and gas fluid phases ($\pi = l,\ g$), and $C_p^{\rm rock}$ the heat capacity of the rock matrix with density $\rho_{\rm rock}$. The energy flux has the form

$$\boldsymbol{J}_U = \boldsymbol{q}_l\rho_l H_l + \boldsymbol{q}_g\rho_g H_g - \kappa\nabla T, \tag{32}$$

where H_π denotes the enthalpy of phase π, and κ denotes the thermal conductivity of the rock matrix. The thermal conductivity is in general a complicated function of the porosity of the porous medium. The flux term takes into account both transport of energy and work done by the fluid phases. For this reason the enthalpy rather than the internal energy appears in the flux. The enthalpy and internal energy of fluid phase π differ by the pV work term (Bird et al., 1960)

$$H_\pi = U_\pi + \frac{p_\pi}{\rho_\pi}, \tag{33}$$

where p_π denotes the pressure of phase π. Putting these terms together, the energy balance equation is given by

$$\begin{aligned}&\frac{\partial}{\partial t}\left[\phi\left(s_l n_l U_l + s_g n_g U_g\right) + (1-\phi)\,\rho_{\rm rock} C_p^{\rm rock} T\right]\\ &\quad + \nabla\cdot\left(\boldsymbol{q}_l n_l H_l + \boldsymbol{q}_g n_g H_g - \kappa\nabla T\right) = Q_e,\end{aligned} \tag{34}$$

where Q_e is a source/sink term accounting for heat produced by chemical reactions, for example.

CHEMICAL REACTIONS

Chemical reactions involved in a geochemical system may be classified either as homogeneous involving species in a single phase, or heterogeneous involving species in two or more phases. It is useful to further divide the reactions into fast and slow. The terms "fast" and "slow" refer to the intrinsic rate and not to the actual rate of the reaction. By "intrinsic" rate is meant the rate the reaction would have under far-from-equilibrium conditions without taking into account the rate of supply of reactants and removal of products by advection and diffusion. Thus in effect it refers to the magnitude of the rate constant used in the kinetic rate law (see Eqn. (67a)). Fast reactions are reversible in the thermodynamic sense of a reversible process and are locally in thermodynamic equilibrium. Slow reactions represent irreversible processes in the language of thermodynamics and require a kinetic rate law to determine the reaction rate. Customarily the terms reversible and irreversible are used with a different meaning when referring to kinetic reactions. An irreversible kinetic reaction is one which proceeds in only one direction, whereas a reversible kinetic reaction can proceed in the forward and backward directions. In what follows, fast reactions are referred to as local equilibrium reactions, their rates controlled by the rate of transport of the reacting species by diffusion and advection. Slow reactions are referred to as kinetic reactions, their rates controlled by surface reactions in the case of solids, and more generally by a kinetic rate law. Local equilibrium reactions can always proceed in the forward and backward directions.

Homogeneous reactions occurring within the aqueous phase including dissociation of water, ion-pairing, and complexing reactions are generally fast, their rates determined by the rate of solute transport by advection and diffusion. Such reactions are thus said to be transport controlled. Oxidation/reduction reactions are often slow, involving the transfer of one or more electrons and must be treated kinetically (Lindberg and Runnells, 1984). Heterogeneous reactions involving minerals are generally slow compared to aqueous homogeneous reactions. Whether a particular heterogeneous reaction is fast or slow depends on various physical properties of the system. These include the magnitude of the kinetic rate constant and reactive surface area.

The law of definite proportions

As a chemical reaction proceeds from some starting point determined by the initial number of moles of each of the reacting constituents, it follows from the *law of definite proportions* that products and reactants are created and destroyed in proportional amounts (Prigogine and Defay, 1954). Thus, for example, if pyrite ($FeS_{2(s)}$) is oxidized to form ferrihydrite ($Fe(OH)_{3(s)}$) according to the reaction

$$FeS_{2(s)} + \frac{15}{4} O_{2(g)} + \frac{7}{2} H_2O \rightleftharpoons Fe(OH)_{3(s)} + 4\,H^+ + 2\,SO_4^{2-}, \tag{35}$$

for each mole of pyrite consumed, one mole of ferrihydrite is produced along with 2 moles of SO_4^{2-}, and 4 moles of H^+. In the process 3.5 moles of water and 3.75 moles of $O_{2(g)}$ are consumed. This reaction is extremely exothermic, releasing -1.468 kJ of heat per mole of pyrite oxidized.

Consider a chemical reaction involving at most N species A_i taking place in a closed system written in the general form

$$\emptyset \rightleftharpoons \sum_{i=1}^{N} \nu_i A_i, \tag{36}$$

where Ø refers to the null species. The quantities ν_i define the stoichiometric coefficients of the reaction. By convention, the species A_i is referred to as a product species if the stoichiometric coefficient is positive ($\nu_i > 0$), and as a reactant if negative ($\nu_i < 0$). The sum runs over all N species in the system, but some of the stoichiometric coefficients ν_i may vanish so not all species need take part in each reaction. As the reaction proceeds, the change in the numbers of moles of any particular species with time is proportional to the change in any other species according to the relation

$$\frac{n_1(t) - n_1^0}{\nu_1} = \frac{n_2(t) - n_2^0}{\nu_2} = \cdots = \frac{n_N(t) - n_N^0}{\nu_N} = \xi(t). \tag{37}$$

The quantities $n_i(t)$ and n_i^0 denote the number of moles of the ith species at time t, and its initial value at time $t = 0$, respectively. The quantity ξ is referred to as the *extent of reaction* or *reaction progress*, first introduced by De Donder in 1920 (Prigogine and Defay, 1954). If ξ is a known function of time, then the number of moles of each species in the system can be computed at any time t according to the relation

$$n_i(t) = n_i^0 + \nu_i\,\xi(t). \tag{38}$$

Differentiating this equation with respect to time yields an expression for the rate of change of the ith species given by

$$\frac{dn_i}{dt} = \nu_i\,\frac{d\xi}{dt}. \tag{39}$$

The reaction rate, denoted by the symbol I, is defined as the rate of change of the reaction progress variable with time, or, in symbols

$$I = \frac{d\xi}{dt}. \tag{40}$$

By convention, the reaction rate is positive if the reaction proceeds from left to right, and negative from right to left.

Next consider a set of N_R simultaneous chemical reactions taking place in a closed system as written in the form

$$\emptyset \rightleftharpoons \sum_{i=1}^{N} \nu_{ir} A_i, \quad (r = 1, \ldots, N_R). \tag{41}$$

Each chemical reaction must conserve mass and charge, so that

$$\sum_{i=1}^{N} W_i \nu_{ir} = 0, \tag{42}$$

and

$$\sum_{i=1}^{N} z_i \nu_{ir} = 0, \tag{43}$$

where W_i denotes the formula weight and z_i the charge of the ith species. The reaction rth given in Equation (41) is in thermodynamic equilibrium if the reacting constituents satisfy the mass action equation

$$K_r = \prod_{i=1}^{N} a_i^{\nu_{ir}}, \tag{44}$$

where K_r denotes the equilibrium constant, and a_i refers to the activity of the ith species. Replacing the activity by the product of the activity coefficient and the concentration,

$$a_i = \gamma_i C_i, \tag{45}$$

where γ_i represents the activity coefficient, leads to the alternative expression

$$K_r = \prod_{i=1}^{N} (\gamma_i C_i)^{\nu_{ir}}. \tag{46}$$

Molarity units are used in this expression, but molality units could equally have been chosen (Denbigh, 1981). The transport equations are naturally expressed in molarity units and it is necessary to convert from one set of units to the other. One form for the activity coefficients in electrolyte solutions is the extended Debye-Hückel relation

$$\ln \gamma_i = -\frac{z_i^2 A \sqrt{I}}{1 + \mathring{a}_i B \sqrt{I}} + \dot{b} I, \tag{47}$$

where the quantities A and B refer to Debye-Hückel parameters and $\mathring{a}_i$ represents the hydrated ionic radius of the ith species. The term involving $\dot{b}$ attempts to account for ion pair interactions. The quantity I denotes the ionic strength of the solution defined by

$$I = \frac{1}{2} \sum_{i=1}^{N_{\mathrm{aq}}} z_i^2 C_i, \tag{48}$$

where the sum is over all N_{aq} species in solution.

Source/sink term

From the law of definite proportions it follows that for a closed system, the change δn_{ir} in the number of moles of the ith species due to the rth reaction is proportional to its stoichiometric coefficient ν_{ir}, or in symbols

$$\delta n_{ir} = \nu_{ir} \delta \xi_r. \tag{49}$$

The proportionality factor $\delta\xi_r$ represents an increment in the reaction progress variable ξ_r. The reaction progress variable ξ_r is the same for all species taking part in the rth reaction. The total change due to all reactions occurring simultaneously is equal to the sum of the contributions from each reaction, or

$$\delta n_i = \sum_{r=1}^{N_R} \delta n_{ir} = \sum_{r=1}^{N_R} \nu_{ir} \delta \xi_r. \tag{50}$$

The rate of the rth reaction, represented as I_r, is defined as

$$I_r = \frac{1}{\nu_{ir}} \frac{\delta n_{ir}}{\delta t} = \frac{\delta \xi_r}{\delta t}, \tag{51}$$

where the change $\delta\xi_r$ is presumed to occur over the time interval δt. Consequently, the time rate of change of the number of moles of the ith species can be expressed as the sum of the reaction rates weighted by the stoichiometric coefficients ν_{ir}

$$\frac{\delta n_i}{\delta t} = \sum_{r=1}^{N_R} \nu_{ir} I_r, \tag{52}$$

as follows by dividing both sides of Equation (50) by δt.

In an open, continuous, system the rate I_r is referenced to a REV and represents the volumetric average rate in units of moles per unit volume per unit time. The reaction progress variable becomes the reaction progress density defined at each point in space. The change in the number of moles of some species contained in a REV can occur not only as a result of chemical reactions taking place in the system, but also because of the transport of matter into and out of the REV. In addition there may be other source/sink terms present. Denoting the contribution from external changes resulting from transport processes across the boundaries of V by $\delta_e n_i$, and that caused by internal changes resulting from chemical reactions by $\delta_i n_i$, the total change in the ith species during the time interval δt may be expressed as the sum

$$\delta n_i = \delta_e n_i + \delta_i n_i, \tag{53}$$

with

$$\delta_i n_i = \delta t \sum_{r=1}^{N_R} \nu_{ir} I_r, \tag{54}$$

and

$$\delta_e n_i = -\delta t \nabla \cdot \boldsymbol{J}_i. \tag{55}$$

The minus sign occurs because, by convention, the flux $\boldsymbol{J}_i$ is taken as positive flowing out of the representative volume. The total change in the number of moles of the ith species can thus be expressed as

$$\delta n_i = -\delta t \nabla \cdot \boldsymbol{J}_i + \delta t \sum_{r=1}^{N_R} \nu_{ir} I_r. \tag{56}$$

Dividing by δt and taking the limit $\delta t \rightarrow 0$, yields the mass conservation equations for aqueous species

$$\frac{\partial}{\partial t}\left(\phi C_i\right) + \nabla \cdot \boldsymbol{J}_i = \sum_{r=1}^{N_R} \nu_{ir} I_r. \tag{57}$$

This result is obtained by noting that

$$\lim_{\delta t \rightarrow 0} \frac{\delta n_i}{\delta t} = \frac{\partial}{\partial t}\left(\phi C_i\right). \tag{58}$$

Comparing this equation with Equation (21), the source/sink term R_i in the mass conservation equation is thus equal to

$$R_i = \sum_{r=1}^{N_R} \nu_{ir} I_r. \tag{59}$$

For minerals the transport equations simplify to

$$\frac{\partial C_m}{\partial t} = \overline{V}_m^{-1} \frac{\partial \phi_m}{\partial t} = \sum_{r=1}^{N_R} \nu_{mr} I_r, \tag{60}$$

in which the flux term is absent.

Reaction rates

To solve the transport equations requires knowledge of the reaction rates for kinetically controlled reactions. Different forms of the rate law are required depending on the type of reaction, that is whether it is homogeneous or heterogeneous, and whether it is an elementary reaction or whether it represents an overall reaction or some specific reaction mechanism.

Homogeneous reactions. The rate of an elementary kinetic homogeneous reaction of the general form

$$\emptyset \rightleftharpoons \sum_{i=1}^{N} \nu_{ir} A_i, \tag{61}$$

is given by the difference between forward and backward reaction rates

$$I_r = k_r^f \prod_{\nu_{ir}<0} (\gamma_i C_i)^{-\nu_{ir}} - k_r^b \prod_{\nu_{ir}>0} (\gamma_i C_i)^{\nu_{ir}}, \tag{62}$$

where $k_r^{f,b}$ denote the forward and backward rate constants. At equilibrium the net rate vanishes, the forward and backward rates being equal, resulting in a relation between the equilibrium constant and the kinetic rate constants

$$K_r = \frac{k_r^f}{k_r^b} = Q_r, \tag{63}$$

where the ion activity product Q_r is defined by

$$Q_r = \prod_{i=1}^{N} a_i^{\nu_{ir}}. \tag{64}$$

Heterogeneous reactions—moving boundary problem. The reaction of an aqueous solution with minerals is referred to as a heterogeneous reaction. The treatment of heterogeneous reactions is much more complicated than that of homogeneous reactions. The expression for the reaction rate that enters the mass transport equations must take into account the distribution in space of the various minerals in the system. This distribution changes continually with time as mineral alteration zones form and advance with time, dissolving and re-precipitating. Dissolution can clearly occur at some point in space only if the particular mineral in question is actually present. For precipitation, on the other hand, it is only necessary that the mineral be supersaturated with respect to the aqueous solution. In this sense the reactive transport equations define a moving boundary problem and part of the problem is to determine the regions in space each mineral occupies. This can be accomplished with the following form of the rate

$$I_m = \begin{cases} \widehat{I}_m, & \text{if } \phi_m > 0, \text{ or if } \phi_m = 0 \text{ and } A_m < 0, \\ 0, & \text{otherwise,} \end{cases} \tag{65}$$

with $\widehat{I}_m$ the kinetic rate law for the mineral as determined by the assumed reaction mechanism.

For simplicity and because of lack of knowledge of detailed reaction mechanisms, mineral reactions are often represented by an overall reaction between the mineral and aqueous solution. The form of the reaction rate is based on transition state theory (Lasaga, 1981; Aagaard and

Helgeson, 1982). Consider a heterogeneous reaction involving a solid phase reacting with an aqueous solution according to

$$\sum_{j=1}^{N_c} \widetilde{\nu}_{jm} A_j \rightleftharpoons A_m, \tag{66}$$

for solid A_m and aqueous species A_j, where $\widetilde{\nu}_{jm}$ denotes the stoichiometric reaction matrix. The reaction rate is assume to have the form

$$\widehat{I}_m = -k_m s_m \left[\prod_i a_i^{n_i} \right] \left(1 - \mathrm{e}^{-A_m/RT}\right), \tag{67a}$$

$$= -k_m s_m \left[\prod_i a_i^{n_i} \right] (1 - K_m Q_m), \tag{67b}$$

where k_m denotes the kinetic rate constant, s_m denotes the mineral surface area participating in the reaction, a_i represents the activity of the ith species, and n_i is a constant. The affinity A_m of the reaction is defined by

$$A_m = -RT \ln K_m Q_m, \tag{68}$$

with R the gas constant, and T the temperature, and where K_m denotes the corresponding equilibrium constant for the mineral reaction as written in Equation (66) with the mineral on the right-hand side, and Q_m denotes the ion activity product defined by

$$Q_m = \prod_{j=1}^{N_c} (\gamma_j C_j)^{\widetilde{\nu}_{jm}}. \tag{69}$$

Note that because of writing reaction products on the right-hand side as in Equation (66), the product $K_m Q_m$ occurs in the formulation for the rate, rather than the often used form Q_m/K_m. The quantity in round brackets in Equation (67a) is referred to as the affinity factor and provides a measure of how far the reaction is from equilibrium. At equilibrium the affinity vanishes and the affinity factor is zero, ensuring that the overall reaction rate is zero. The quantity in square brackets accounts for the dependence of the rate on the concentration of dissolved species, such as the proton activity, in addition to the affinity factor. Precipitation ($\widehat{I}_m > 0$) occurs if $A_m < 0$, and dissolution ($\widehat{I}_m < 0$) if $A_m > 0$. The reaction rate has units of moles per unit time per unit volume of bulk porous medium. Thus it represents an average rate taken over a REV.

The rate law given by Equation (67a) should really be referred to as a pseudo-kinetic rate law. This is because it refers to the overall mineral precipitation/dissolution reaction, and generally does not describe the actual kinetic mechanism by which the mineral reacts. Nevertheless, it provides a useful form to describe departures from equilibrium and is certainly no worse than the assumption of local equilibrium.

Far from equilibrium the expression for the reaction rate reduces to the forms

$$\widehat{I}_m = -k_m s_m \left[\prod_i a_i^{n_i} \right], \tag{70}$$

for $A_m \gg 0$ corresponding to dissolution, and

$$\widehat{I}_m = k_m s_m \, \mathrm{e}^{|A_m|/RT} \left[\prod_i a_i^{n_i} \right], \tag{71}$$

for $A_m \ll 0$ corresponding to precipitation. Close to equilibrium the rate becomes proportional to the chemical affinity according to the expression

$$\hat{I} = k_m s_m \left[\prod_i a_i^{n_i} \right] \frac{A_m}{RT}, \tag{72}$$

valid for $|A_m/RT| \ll 1$. There is an inherent asymmetry in the rate law regarding precipitation and dissolution that should be noted. According to Equation (71), the precipitation rate grows indefinitely as $A_m \to -\infty$, whereas according to Equation (70) the dissolution rate tends to a finite constant times the factor in square brackets as $A_m \to \infty$. Of course physically the reaction rate cannot grow indefinitely, but must be limited by the rate of transport of reactants and products to the site where the reaction takes place. Note that under such far-from-equilibrium conditions, although the rate is transport limited, the reaction is not in local chemical equilibrium.

The temperature dependence of the kinetic rate constant may be calculated from the Arrhenius equation (Lasaga, 1981)

$$k_m(T) = k_m^0 \frac{A(T)}{A(T_0)} \exp\left[-\frac{1}{R} \left(\frac{1}{T} - \frac{1}{T_0} \right) \Delta E_m \right], \tag{73}$$

where k_m^0 denotes the rate constant at T_0, $A(T)$ represents a pre-exponential factor, and ΔE_m denotes the activation energy.

Surface area. One recurring difficulty in describing heterogeneous reactions involving minerals is characterizing the reacting mineral surface area. Consider, to be specific, an aggregate of cubical grains of dimension d. Different grain geometries result in different geometric factors. If the grains are assumed to be arranged in a evenly spaced, three-dimensional array with N grains contained in V_{REV}, a total volume $b^3 = V_{\text{REV}}/N$ is associated with each grain including pore space, and the grain number density η is equal to

$$\eta = \frac{N}{V_{\text{REV}}} = \frac{1}{b^3}. \tag{74}$$

The pore volume is equal to

$$\phi = \frac{V_p}{V_{\text{REV}}} = \frac{b^3 - d^3}{b^3} = 1 - \left(\frac{d}{b} \right)^3 = 1 - \eta d^3. \tag{75}$$

The specific surface area s of the grains can be expressed in terms of the porosity either as

$$s = \frac{S}{V} = \frac{6d^2}{b^3} = \frac{6}{d} \left(\frac{d}{b} \right)^3 = \frac{6}{d}(1 - \phi), \tag{76a}$$

for fixed grain size, or, alternatively, in terms of a fixed number density as

$$s = \frac{6}{b} \left(\frac{d}{b} \right)^2 = 6\eta^{1/3}(1 - \phi)^{2/3}. \tag{76b}$$

Accordingly, a different dependence of the surface area on porosity is obtained depending on whether the grain number density η is considered constant, or the grain size itself is fixed. For $\phi \ll 1$, the surface area is approximately independent of the porosity of the porous medium for both formulations. For natural systems the grain size is given by a distribution rather than a single value, complicating the estimation of surface area.

An added complication results from the change in surface area as minerals dissolve and precipitate. One approach is to assume a two-thirds power relationship of the form

$$s_m = s_m^0 \left(\frac{\phi_m}{\phi_m^0}\right)^{2/3}, \tag{77}$$

where s_m^0 and ϕ_m^0 represent the initial mineral surface area and volume fraction, respectively. However, this formulation does not provide for the possibility of an increase in surface area with dissolution as might occur due to the formation of etch pits, for example. In a partially saturated medium the reacting mineral surface is proportional to the wetted surface area. One possibility is to assume that the pore surface is completely wetted by the aqueous solution for saturations above the residual saturation, the threshold for Darcy flow of liquid.

Boundary layer. In the case of flow past a boundary layer, a stagnant region of fluid bordering the mineral surface, solute species must diffuse from the bulk fluid across the boundary layer to the mineral surface. The reaction rate at the mineral surface must be balanced by the diffusive flux of species across the boundary layer to the surface. In symbols

$$I_{\text{surf}} = k's(C_{\text{surf}} - C_{\text{eq}}) = s'D\left(\frac{\partial C}{\partial x}\right)_{\text{surf}} = s'D\frac{C_{\text{bulk}} - C_{\text{surf}}}{\Delta l}, \tag{78}$$

where Δl denotes the boundary layer thickness, C_{bulk} denotes the bulk fluid composition, C_{surf} the concentration at the mineral surface, s denotes the effective reactive surface area of the mineral, and s' refers to the geometric area. Solving this relation for the surface concentration yields

$$C_{\text{surf}} = (1-\beta)\, C_{\text{eq}} + \beta\, C_{\text{bulk}}, \tag{79}$$

where

$$\beta = \frac{1}{1+\text{Da}}. \tag{80}$$

and

$$\text{Da} = \frac{k's\,\Delta l}{s'D}, \tag{81}$$

represents the Damköhler number (Damköhler, 1936). Substituting this result into the equation for the reaction rate, yields the following expression for the effective rate constant

$$k_{\text{eff}} = \frac{k's}{1+\text{Da}}. \tag{82}$$

The following limiting values are obtained for the effective rate constant

$$k_{\text{eff}} = \begin{cases} k's, & (\Delta l \to 0) \\ \dfrac{s'D}{\Delta l}, & (\Delta l \gg s'D/k's) \end{cases}. \tag{83}$$

This result indicates the complex relation the effective rate constant can have to measurable quantities. It may not be enough to know the rate constant and surface area alone. For intrinsically fast reactions, Da $\gg$ 1, geometry can also be an important factor.

Local chemical equilibrium. Local equilibrium is a completely different concept from that of equilibrium in a closed system. For the latter system the net reaction rate of a reaction in equilibrium is identically zero, the forward and backward rates being equal and opposite. However, for a reaction in local equilibrium, its net rate need not vanish by virtue that the system is open. The rate of a reaction in local equilibrium is thus determined by the rate of transport of matter across the boundaries of the system. The system, in this case, corresponds to a REV.

Formally, local equilibrium is obtained as the limit in which the kinetic rate constant approaches infinity. In this limit the chemical affinity approaches zero. Consequently the expression for the rate results in an indeterminate expression:

$$\lim_{k_m \to \infty} I_m = \infty \cdot 0. \tag{84}$$

To actually determine the rate for conditions of local equilibrium it is necessary to solve the transport equations.

It is important to distinguish between micro- and macroscales when referring to the concept of local equilibrium. Here microscale refers to the pore scale and macroscale to the continuum scale which is an average over many pore volumes. The limit as the rate constant approaches infinity in the continuum formulation refers to macroscale local equilibrium. At this scale each REV attains thermodynamic equilibrium; however, any particular REV is generally in disequilibrium with its neighbors. This is quite different from local equilibrium on a microscale for which the detailed pore geometry plays a role. It must be kept in mind that it is meaningless within the continuum framework to use a kinetic rate constant which gives an equilibration length that is smaller than the characteristic length associated with a REV. The characteristic equilibration length is defined as the distance required for the reaction to come to equilibrium with the fluid [see Eqn. (257)]. For this reason it may be much simpler to use a kinetic formulation with sufficiently large, but not too large, rate constants to represent macroscale local equilibrium within the continuum framework. For both microscale local equilibrium and very rapid far-from-equilibrium reaction rates, the pore geometry must be taken into account in the calculations. This becomes a very difficult task especially if the computation domain is at the macroscale.

MULTICOMPONENT REACTIVE TRANSPORT EQUATIONS

The main purpose of this section is to derive reactive transport equations for multicomponent–multiphase systems incorporating conditions of local partial equilibrium in which a subset of reactions are described through conditions of local chemical equilibrium. The remaining reactions are presumed to have kinetically controlled reaction rates. Local equilibrium reactions are described through mass action equations. Their reaction rates are eliminated from the mass transport equations by transforming them to a special form, referred to as the canonical form. The treatment given here is based on the actual physical species in solution. This approach has a number of advantages over one based on conservation of the atomic elements, especially regarding the treatment of redox reactions, although the two approaches lead to equivalent results.

The mass conservation equations for simultaneous transport of aqueous and gaseous species take the form [see Eqn. (25)]

$$\frac{\partial}{\partial t}\left(\phi s_{\pi_i} C_i\right) + \nabla \cdot \boldsymbol{J}_i = \sum_{r=1}^{N_R} \nu_{ir} I_r, \quad (i = 1, \ldots, N), \tag{85}$$

where each species, labeled by the subscript i, is associated with a particular fluid phase π_i with saturation state s_{π_i}. For example, $CO_{2(g)}$ and $CO_{2(l)}$ have different species indices associated with them since they belong to different phases.

In what follows it is useful to combine the mineral, gaseous and aqueous species transport equations into a common form by introducing the operator $Ł_i$ defined by

$$Ł_i C_i = \begin{cases} \dfrac{\partial}{\partial t}\left(\phi s_{\pi_i} C_i\right) + \nabla \cdot \boldsymbol{J}_i, & i = \text{aqueous or gaseous species}, \\[2ex] \dfrac{\partial C_i}{\partial t}, & i = \text{mineral species}. \end{cases} \tag{86}$$

With this definition the transport equations can be written concisely as

$$Ł_i C_i = \sum_{r=1}^{N_R} \nu_{ir} I_r, \quad (i = 1, \ldots, N), \tag{87}$$

for aqueous, gaseous and mineral species. For a single aqueous fluid phase with flow velocity $\boldsymbol{v}$, and species-independent diffusion/dispersion coefficient D, the differential operator $Ł_i$ is replaced by

$$Ł = \frac{\partial}{\partial t}\phi - \nabla \cdot \tau \phi D \nabla + \nabla \cdot \boldsymbol{v}. \tag{88}$$

For species-dependent diffusion coefficients the more general form of the differential operator $Ł_i$ given by Equation (86) is required.

Canonical form

The reactive mass conservation equations, Equations (57) and (60) or equivalently Equation (87), are not in a very useful form for computational purposes. To solve these equations would require knowledge of all the reaction rates I_r as functions of the species concentrations through kinetic rate laws. For many reactions, however, their intrinsic rates are sufficiently rapid that the reactions may be assumed to be in instantaneous equilibrium. Their actual rates are then controlled by the rate of transport of species to and from the site of reaction. For these reactions it would be desirable if the conservation equations could be formulated in such a way that rates corresponding to these fast reactions could be replaced by conditions of local equilibrium in the form of appropriate mass action equations. Such a system in which both local equilibrium and kinetic reactions take place simultaneously, is referred to as being in a state of local partial equilibrium.

The task of transforming the conservation equations to a form which accommodates the mass action relations can be accomplished by using the techniques of linear algebra. The basic idea is to rewrite the chemical reactions in terms of a suitable set of species, referred to variously as primary species, basis species, or components, which simplify the problem. For any linearly independent set of reactions of the form

$$\emptyset \rightleftharpoons \sum_{l=1}^{N} \nu_{lr} A_l, \quad (r = 1, \ldots, N_R), \tag{89}$$

it is possible to transform the reactions to the canonical form

$$\sum_{j=1}^{N_c} \widetilde{\nu}_{ji} A_j \rightleftharpoons A_i, \quad (i = N_c + 1, \ldots, N), \tag{90}$$

in which a single species, referred to as a secondary species, appears on the right-hand side with unit stoichiometric coefficient. The N_c species appearing on the left-hand side are primary species. The term primary species is used throughout for a system in local partial equilibrium, rather than the term component which is reserved for a closed system in thermodynamic equilibrium. There are an equal number of secondary species as there are reactions. The secondary species A_i serves to identify the ith reaction. The number of primary species is equal to the difference between the total number of species N involved in the reactions and the number of reactions N_R:

$$N_c = N - N_R. \tag{91}$$

The stoichiometric coefficients $\widetilde{\nu}_{ji}$ are related to the original stoichiometric coefficients ν_{ir} by the equation

$$\widetilde{\nu}_{ji} = -\sum_{r=1}^{N_R} \nu_{jr}(\nu^{-1})_{ri}, \quad \begin{array}{c}(j = 1, \ldots, N_c)\\ (i = N_c + 1, \ldots, N),\end{array} \tag{92}$$

where $(\nu^{-1})_{ri}$ represents the inverse of the submatrix ν_{ir} where i runs over the secondary species. The submatrix ν_{ir} is a square, nonsingular matrix. This can always be accomplished by virtue of the fact that the reactions are presumed to form a linearly independent set. The canonical form defined by Equation (90) may be derived by noting that the chemical reactions given in the general form of Equation (89) can be rewritten as follows

$$\emptyset \rightleftharpoons \sum_{j=1}^{N_c} \nu_{jr} A_j + \sum_{i=N_c+1}^{N} \nu_{ir} A_i, \quad (r = 1, \ldots, N_R), \tag{93}$$

where the first sum on the right-hand side is over primary species and the second sum over secondary species. The partitioning the reacting species into primary and secondary species is done selectively so that the submatrix ν_{ir} involved in the sum over secondary species is nonsingular. Hence it is possible to "solve" Equation (93) for the secondary species to give

$$A_i \rightleftharpoons -\sum_{r=1}^{N_R} \nu_{jr}(\nu^{-1})_{ri} A_j, \tag{94}$$

from which Equation (90) follows with $\widetilde{\nu}_{ji}$ defined by Equation (92). Because of its special form, reactions written in the form of Equation (90) are referred to as being in canonical form (Lichtner, 1985; Smith and Missen, 1982; Van Zeggeren and Storey, 1970). This terminology appears to have been first introduced by Aris and Mah (1963). The canonical form is characterized by associating a single secondary species with each reaction, with the remaining species belonging to a common subset of species which appear on the left-hand side of each reaction. Constructing the canonical form is equivalent to the Gauss-Jordan procedure used in linear algebra in which a matrix is transformed to an upper nonzero matrix and a lower unit matrix (see Steefel and MacQuarrie, this volume). It should be especially noted that the designation of a species as primary or secondary refers to the canonical form of the reaction in which the species occurs.

Reactions written in canonical form are expressed as association reactions. Several examples may help clarify the the meaning of the canonical representation. Consider the reaction describing the formation of the aqueous species $CaSO_4^\circ$. With primary species chosen as Ca^{2+} and SO_4^{2-}, and secondary species $CaSO_4^\circ$, the reaction becomes

$$Ca^{2+} + SO_4^{2-} \rightleftharpoons CaSO_4^\circ. \tag{95a}$$

Equally possible is to chose as primary species Ca^{2+} and the complex $CaSO_4^\circ$. In this case SO_4^{2-} becomes the secondary species. The reaction now has the form

$$CaSO_4^\circ - Ca^{2+} \rightleftharpoons SO_4^{2-}. \tag{95b}$$

Both forms lead to equivalent results. As a more complicated example, consider the precipitation or dissolution of goethite and iron speciation in the aqueous phase. Taking as primary species $\{Fe^{2+}, H^+, O_{2(g)}, H_2O\}$, and secondary species $\{Fe^{3+}, Fe(OH)_4^-, FeOOH\}$, the reactions have the form

$$Fe^{2+} + H^+ - \frac{1}{2}H_2O + \frac{1}{4}O_{2(g)} \rightleftharpoons Fe^{3+}, \tag{96a}$$

$$Fe^{2+} - 3H^+ + \frac{7}{2}H_2O + \frac{1}{4}O_{2(g)} \rightleftharpoons Fe(OH)_4^-, \tag{96b}$$

$$Fe^{2+} - 2H^+ + \frac{3}{2}H_2O + \frac{1}{4}O_{2(g)} \rightleftharpoons FeOOH. \tag{96c}$$

An alternative set of primary species are $\{Fe^{2+}, Fe^{3+}, H^+, H_2O\}$, and secondary species $\{O_{2(g)}$, $Fe(OH)_4^-, FeOOH\}$. This choice leads to the set of reactions

$$4Fe^{2+} - 4Fe^{3+} + 4H^+ - 2H_2O \rightleftharpoons O_{2(g)}, \tag{97a}$$

$$Fe^{3+} - 4H^+ + 4H_2O \rightleftharpoons Fe(OH)_4^-, \tag{97b}$$

$$Fe^{3+} - 3H^+ + 2H_2O \rightleftharpoons FeOOH. \tag{97c}$$

Provided equilibrium can be assumed to hold between the redox couples Fe^{2+}–Fe^{3+}–$O_{2(g)}$, the two sets of reactions are equivalent.

In the canonical representation, the charge conservation equation for a chemical reaction becomes a statement expressing the charge of the ith secondary species in terms of the charges of the primary species. Thus

$$z_i = \sum_{j=1}^{N_c} \widetilde{\nu}_{ji} z_j, \tag{98}$$

with $\widetilde{\nu}_{ji}$ defined by Equation (92). Likewise mass conservation reads

$$M_i = \sum_{j=1}^{N_c} \widetilde{\nu}_{ji} M_j. \tag{99}$$

These relations should be compared with the general forms, Equations (43) and (42).

The mass action equation corresponding to the ith canonical reaction is given by

$$\widetilde{K}_i = \frac{a_i}{\prod_{j=1}^{N_c} a_j^{\widetilde{\nu}_{ji}}}, \tag{100}$$

with equilibrium constant $\widetilde{K}_i$. The advantage of the canonical form is that this equation may be formally solved for the concentration of the secondary species C_i in terms of the concentrations of primary species. For aqueous and gaseous species it follows that

$$C_i = (\gamma_i)^{-1} \widetilde{K}_i \prod_{j=1}^{N_c} \left(\gamma_j C_j\right)^{\widetilde{\nu}_{ji}}. \tag{101}$$

To obtain this expression, the activity has been replaced by the product of the activity coefficient and the molar concentration. Actually, because the activity coefficients also depend on the concentrations of the solute species, this relation only provides an implicit expression for the concentration. For stoichiometric mineral species, replacing the index i by m and noting that $a_m = 1$, the mass action equations become an inequality:

$$\widetilde{K}_m \prod_{j=1}^{N_c} (\gamma_j C_j)^{\widetilde{\nu}_{jm}} \leq 1. \tag{102}$$

This relation provides a constraint among the primary species concentrations. Equality holds if the aqueous solution is saturated with respect to the mineral and thus the mineral is in local equilibrium with the solution. The inequality holds if the solution is undersaturated with respect to the mineral. The presence of an inequality for heterogeneous reactions reflects the moving boundary problem nature of the reactive mass transport equations.

A simple relation exists between the equilibrium constants of the original reactions K_r and the reactions written in canonical form denoted by $\widetilde{K}_i$. Taking logarithms of both sides of the mass action equation, Equation (100), yields

$$\begin{aligned} \ln \widetilde{K}_i &= \ln a_i - \sum_{j=1}^{N_c} \widetilde{\nu}_{ji} \ln a_j, \\ &= \ln a_i + \sum_{r=1}^{N_R} \sum_{j=1}^{N_c} \nu_{jr} (\nu^{-1})_{ri} \ln a_j, \end{aligned} \tag{103}$$

where to obtain the second equality, $\widetilde{\nu}_{ji}$ is replaced by its definition given in Equation (92). Taking the logarithm of Equation (44) and separating terms for primary and secondary species results in the expression

$$\ln K_r = \sum_{j=1}^{N_c} \nu_{jr} \ln a_j + \sum_{i=N_c+1}^{N} \nu_{ir} \ln a_i. \tag{104}$$

Multiplying this equation through by $(\nu^{-1})_{ri}$ and summing over all reactions yields

$$\sum_r (\nu^{-1})_{ri} \ln K_r = \ln a_i + \sum_{r=1}^{N_R} \sum_{j=1}^{N_c} \nu_{jr} (\nu^{-1})_{ri} \ln a_j. \tag{105}$$

Comparing this relation with Equation (103), it follows that

$$\ln \widetilde{K}_i = \sum_{r=1}^{N_R} (\nu^{-1})_{ri} \ln K_r, \tag{106}$$

or, alternatively

$$\widetilde{K}_i = \prod_{r=1}^{N_R} K_r^{(\nu^{-1})_{ri}}, \tag{107}$$

yielding the desired relation.

Thermodynamic databases. The relation between equilibrium constants obtained in Equation (106) or (107), is actually of direct practical use in transforming a particular thermodynamic database from one representation to another. Thermodynamic databases which store chemical reactions and their corresponding equilibrium constants, as well as other properties such as Debye-Hückel parameters, charge, molar volume, and gram-formula weight, store reactions in canonical form. The freedom of choice of primary species may be used to advantage by enabling the user of such a database in which a particular choice has been hardwired, to use any other set of species that may be more relevant to the problem at hand. This technique is used in several computer codes developed by the author (MPATH, MULTIFLO, GEM). These codes employ a modified form of the EQ3/6 database (Wolery et al., 1990). Actually the EQ3/6 database uses two sets of primary species resulting in two distinct canonical representations. One representation is referred to as the basis set and the other as the auxiliary basis. The EQ3/6 database employs a special canonical form using the species $O_{2(g)}$ as a primary species to describe redox reactions. There is nothing magical about using $O_{2(g)}$ and, in fact, in some cases it may be more convenient to use some other species. For example, under anaerobic conditions in which the oxygen concentration is just a few atoms per cubic centimeter, it may make little sense physically to use $O_{2(g)}$ even though formally, of course, it must give equivalent results and serves to parameterize the redox state. The user may wish to know the total iron II and III in the system, in which case using the redox couple Fe^{2+}–Fe^{3+} as primary species provides this information directly. Furthermore, using $O_{2(g)}$ as primary species does not allow for easy decoupling of different redox couples. In this case a customized database may be necessary (Lichtner, 1995). For problems involving a high pH it may be more desirable to use OH^- rather than H^+ as primary species.

Relation between source/sink terms. In the canonical representation the source/sink term for primary species becomes

$$R_j = -\sum_{i=N_c+1}^{N} \widetilde{\nu}_{ji} \widetilde{I}_i, \tag{108}$$

and for secondary species simply

$$R_i = \widetilde{I}_i, \tag{109}$$

where $\widetilde{I}_i$ represents the reaction rate for the canonical form of the reaction as given by Equation (90). Substituting Equation (109) into Equation (108) leads to the following relation among the source/sink terms

$$R_j + \sum_{i=N_c+1}^{N} \widetilde{\nu}_{ji} R_i = 0. \tag{110}$$

For kinetic reactions it may not be possible to associate $\widetilde{I}_i$ with a kinetic rate law since the canonical form of the kinetic reaction may not represent the actual reaction mechanism. In such cases I_r^{ke} represents the more fundamental quantity.

Local partial equilibrium

For a general geochemical system in a state of local partial equilibrium the chemical reactions taking place may be divided into two sets: those which may be described by conditions of local equilibrium, and those which are kinetically controlled and require kinetic rate laws. The reactions taking place in the system can be written in the general form

$$\emptyset \rightleftharpoons \sum_{i=1}^{N_{\mathrm{le}}} \nu_{ir}^{\mathrm{le}} A_i, \quad (r = 1, \ldots, N_R^{\mathrm{le}}), \tag{111}$$

for local equilibrium reactions assumed to be $N_R^{\rm le}$ in number, and

$$\emptyset \rightleftharpoons \sum_{i=1}^{N} \nu_{ir}^{\rm ke} A_i, \quad (r = 1, \ldots, N_R^{\rm ke}), \tag{112}$$

for kinetic reactions, where $N_R^{\rm ke}$ is the number of linearly independent kinetic reactions. The superscripts 'le' and 'ke' distinguish between local equilibrium and kinetic reactions, respectively. All species are involved, in general, in the kinetic reactions. It is assumed that only a subset of species, $N_{\rm le} \leq N$ in number, take part in local equilibrium reactions in addition to kinetic reactions. The remaining N_k species, with

$$N_k = N - N_{\rm le}, \tag{113}$$

therefore only participate in kinetic reactions and are referred to appropriately as kinetic species. The rates of reactions in local chemical equilibrium are described through imposing mass action constraints on the mass conservation equations. For these reactions it makes no difference how they are written and any rearrangement, provided only that the resulting set of reactions remains independent, may be used. The rates of kinetic reactions are described through kinetic rate laws. For these reactions the particular form of the reaction can be important and determines the reaction mechanism. However, lack of knowledge of the actual reaction mechanism often dictates use of an overall reaction in which case, just as for local equilibrium reactions, the particular form of the kinetic reaction may not be of importance either.

For a geochemical system in local partial equilibrium the mass transport equations have the form

$$\text{Ł}_i C_i = \sum_{r=1}^{N_R^{\rm le}} \nu_{ir} I_r^{\rm le} + \sum_{r=1}^{N_R^{\rm ke}} \nu_{ir}^{\rm ke} I_r^{\rm ke}, \quad (i = 1, \ldots, N). \tag{114}$$

These equations include transport of species in both aqueous and gaseous phases, and interaction with solid phases. They must be combined with $N_R^{\rm le}$ algebraic mass action equations corresponding to the local equilibrium reactions. Thus there are a total of $N + N_R^{\rm le}$ partial differential–algebraic equations to solve. The unknowns consist of the concentrations of the N chemical constituents and the $N_R^{\rm le}$ local equilibrium reaction rates. This system of equations can be reduced in number substantially by eliminating the rates of the local equilibrium reactions resulting in a system of N partial differential–algebraic equations. This may be accomplished by partitioning the reacting species into primary and secondary species and eliminating the local equilibrium reaction rates from the primary species transport equations. The secondary species transport equations are replaced by mass action equations. There are several different approaches presented in the literature for carrying out this procedure (Ramshaw, 1980; Sørensen and Stewart, 1980; Lichtner, 1985; Sevougian et al., 1993). The approach taken here follows most closely to that of Lichtner (1985).

The chemical constituents involved in local equilibrium reactions are partitioned into primary and secondary species. The number of primary species is equal to the difference between the total number of species involved in local equilibrium reactions $N_{\rm le}$ and the number of local equilibrium reactions $N_R^{\rm le}$:

$$N_c^{\rm le} = N_{\rm le} - N_R^{\rm le}. \tag{115}$$

The number of secondary species is equal to the number of local equilibrium reactions $N_R^{\rm le}$. Rearranging the species if necessary, they may assumed to be partitioned according to

$$\{\underbrace{A_1, \ldots, A_{N_c^{\rm le}}}_{\text{primary species}};\ \underbrace{A_{N_c^{\rm le}+1}, \ldots, A_{N_{\rm le}}}_{\text{secondary species}};\ \underbrace{A_{N_{\rm le}+1}, \ldots, A_N}_{\text{kinetic species}}\}. \tag{116}$$

The choice of primary species is rather arbitrary, the essential restrictions being that they occur in the reactions whose rates are to be eliminated, and that they are independent of each other. In the following it is assumed that they are chosen from the set of aqueous and gaseous species. Use of minerals as primary species is generally not a good idea. The reason for this is that any particular mineral is generally not present at all times over the entire computational domain. As a consequence this would necessitate using different sets of primary species in different regions of space at different times. For a two-phase system involving a partially saturated medium in which both gas and liquid fluid phases are present, use of a single set of primary species may be unavoidable even if the primary species are chosen from the set of aqueous and gaseous species. In this case it may be necessary to use different sets of primary species in different regions of space depending on the fluid phases present. Thus in a region containing a pure liquid phase or in a two-phase region of gas and liquid, a single set of primary species suffices taken from the set of aqueous species. However, in a pure gas phase region, primary species must all belong to the gas phase.

Denoting primary species with the subscript j, secondary species with the subscript i, and kinetic species with the subscript k, the chemical reactions can be expressed in the form

$$\emptyset \rightleftharpoons \sum_{j=1}^{N_c^{\rm le}} \nu_{jr}^{\rm le} A_j + \sum_{i=N_c^{\rm le}+1}^{N_{\rm le}} \nu_{ir}^{\rm le} A_i, \tag{117a}$$

for local equilibrium reactions, and

$$\emptyset \rightleftharpoons \sum_{j=1}^{N_c^{\rm le}} \nu_{jr}^{\rm ke} A_j + \sum_{i=N_c^{\rm le}+1}^{N_{\rm le}} \nu_{ir}^{\rm ke} A_i + \sum_{k=N_{\rm le}+1}^{N_k} \nu_{kr}^{\rm ke} A_k, \tag{117b}$$

for kinetic reactions. By construction the stoichiometric submatrix $\nu_{ir}^{\rm le}$ is nonsingular, and Equations (117a) may be solved formally for the secondary species in terms of the primary species to yield the canonical set of reactions in the form of Equation (90):

$$\sum_{j=1}^{N_c^{\rm le}} \tilde{\nu}_{ji} A_j \rightleftharpoons A_i, \quad (i = N_c^{\rm le}+1, \ldots, N_{\rm le}). \tag{118}$$

Eliminating the secondary species from the kinetic reactions yields the equivalent set of reactions

$$\emptyset \rightleftharpoons \sum_{j=1}^{N_c^{\rm le}} \left(\nu_{jr}^{\rm ke} + \sum_{i=N_c^{\rm le}+1}^{N_{\rm le}} \tilde{\nu}_{ji} \nu_{ir}^{\rm ke} \right) A_j + \sum_{k=N_{\rm le}+1}^{N_k} \nu_{kr}^{\rm ke} A_k. \tag{119}$$

To eliminate the reaction rates of the local equilibrium reactions from the mass transport equations, they may be formulated directly from the transformed chemical reactions. Alternatively, the same approach used in transforming the chemical reactions to canonical form may be applied directly the the transport equations. Following the latter approach, separate statements of the mass conservation equations for primary, secondary and kinetic species can be written as

$$Ł_j C_j = \sum_{r=1}^{N_R^{\rm le}} \nu_{jr}^{\rm le} I_r^{\rm le} + \sum_{r=1}^{N_R^{\rm ke}} \nu_{jr} I_r^{\rm ke}, \tag{120a}$$

$$Ł_i C_i = \sum_{r=1}^{N_R^{\rm le}} \nu_{ir}^{\rm le} I_r^{\rm le} + \sum_{r=1}^{N_R^{\rm ke}} \nu_{ir}^{\rm ke} I_r^{\rm ke}, \tag{120b}$$

and

$$Ł_k C_k = \sum_{r=1}^{N_R^{\rm ke}} \nu_{kr}^{\rm ke} I_r^{\rm ke}. \tag{120c}$$

There are a total of $N_c^{\rm le} = N_{\rm le} - N_R^{\rm le}$ primary species, $N_R^{\rm le}$ secondary species, and N_k kinetic species transport equations. The second set of equations for the secondary species may be solved for the $N_R^{\rm le}$ local equilibrium reaction rates to give

$$I_r^{\rm le} = \sum_{i=N_c^{\rm le}+1}^{N_{\rm le}} (\nu^{\rm le})_{ri}^{-1} \left\{ Ł_i C_i - \sum_{r=1}^{N_R^{\rm ke}} \nu_{ir}^{\rm ke} I_r^{\rm ke} \right\}. \tag{121}$$

These relations provide an expression for the local equilibrium reaction rates in terms of derivatives of the secondary species concentrations and the kinetic reaction rates. Eliminating the local equilibrium reaction rates from the primary species transport equations yields the alternative set of transport equations

$$Ł_j C_j + \sum_{i=N_c^{\rm le}+1}^{N_{\rm le}} \widetilde{\nu}_{ji} Ł_i C_i = \sum_{r=1}^{N_R^{\rm ke}} \nu'_{jr} I_r^{\rm ke}, \tag{122}$$

where $\widetilde{\nu}_{ji}$ is defined by Equation (92) with $N_c^{\rm le}$ replacing N_c, and

$$\nu'_{jr} = \nu_{jr}^{\rm ke} + \sum_{i=N_c^{\rm le}+1}^{N_{\rm le}} \widetilde{\nu}_{ji} \nu_{ir}^{\rm ke}. \tag{123}$$

The sum on the right-hand side of Equation (122) is now only over kinetic reactions. Equations (122) are now in the proper form for solving numerically. The transport equations for secondary species are eliminated and replaced by corresponding mass action equations given by Equation (100).

The total number of transport equations and unknowns is equal to the sum of the primary and kinetic species $N_c^{\rm le} + N_k$. Equation (122) for the $N_c^{\rm le}$ primary species and Equation (120c) for the N_k kinetic species, combined with the $N_R^{\rm ke}$ algebraic mass action equations given by Equations (100), provide an equal number of equations as there are unknowns. Solving this system of partial differential–algebraic equations for prescribed initial and boundary conditions, provides not only the species concentrations, but also the kinetic reaction rates as well, all as functions of time and space. Once the transport equations have been solved, the reaction rates for the local equilibrium reactions can be obtained from Equation (121).

Note that it is possible to further transform the transport equations for the primary and kinetic species to eliminate the kinetic reaction rates from a subset of these equations. Since there are $N_R^{\rm ke}$ kinetic reactions, there must be a total of

$$N_c = N_c^{\rm le} + N_k - N_R^{\rm ke}, \tag{124}$$

mass conservation equations in which the appearance of all reaction rates can be eliminated. Because

$$N_c^{\rm le} = N_{\rm le} - N_R^{\rm le} = N - N_k - N_R^{\rm le}, \tag{125}$$

it follows that

$$N_c = N - N_k - N_R^{\rm le} + N_k - N_R^{\rm ke} = N - N_R, \tag{126}$$

where N_R denotes the total number of linearly independent reactions

$$N_R = N_R^{\rm le} + N_R^{\rm ke}. \tag{127}$$

This result could have been deduced directly by applying the canonical transformation to all reactions, both local equilibrium and kinetic, simultaneously.

Linearly dependent reactions. In certain instances it may be necessary to consider linearly dependent reactions when describing a geochemical system. This can occur, for example, when different redox couples are involved. As an example, oxidation of pyrite may be described by two different mechanisms, one involving O_2 as the oxidant and the other Fe^{3+}. The two reactions may proceed in parallel over some limited range of solution composition, or one may dominate over the other. Adding to these reactions the homogeneous reaction describing the reduction of Fe(III) to Fe(II) leads to a linearly dependent set of reactions. Thus either one of the pyrite oxidation mechanisms can be obtained from the other by forming the appropriate linear combination with this reaction. The pyrite oxidation reaction is referred to as being degenerate. The degeneracy of a chemical reaction is defined as the number of reactions with which it is linearly dependent.

One way to express the degeneracy in the case of oxidation-reduction reactions is in terms of half-cell reactions. The half-cell reaction describing pyrite oxidation can be written, for example, as

$$\mathrm{Fe^{2+} + 2\,SO_4^{2-} + 16\,H^+ + 14\,e^- - 8\,H_2O \rightleftharpoons FeS_{2(s)}}. \tag{128}$$

In place of the electron, any number of possible redox couples may be substituted, a few of which are listed below:

$$\mathrm{e^-} \rightleftharpoons \frac{1}{2}\mathrm{H_2O} - \mathrm{H^+} - \frac{1}{4}\mathrm{O_2}, \tag{129a}$$

$$\rightleftharpoons \mathrm{Fe^{2+} - Fe^{3+}}, \tag{129b}$$

$$\rightleftharpoons \mathrm{Cu^{+} - Cu^{2+}}, \tag{129c}$$

$$\rightleftharpoons \frac{1}{2}\mathrm{NO_2^-} - \frac{1}{2}\mathrm{NO_3^-} - \mathrm{H^+} + \frac{1}{2}\mathrm{H_2O}, \tag{129d}$$

$$\rightleftharpoons \frac{1}{10}\mathrm{N_{2(g)}} - \frac{1}{5}\mathrm{NO_3^-} - \frac{6}{5}\mathrm{H^+} + \frac{3}{5}\mathrm{H_2O}, \tag{129e}$$

$$\rightleftharpoons \frac{1}{8}\mathrm{NH_4^+} - \frac{1}{8}\mathrm{NO_3^-} - \frac{5}{4}\mathrm{H^+} + \frac{3}{8}\mathrm{H_2O}, \tag{129f}$$

$$\rightleftharpoons \frac{1}{8}\mathrm{CH_{4(g)}} - \frac{1}{8}\mathrm{CO_{2(g)}} - \mathrm{H^+} + \frac{1}{4}\mathrm{H_2O}, \tag{129g}$$

$$\rightleftharpoons \frac{1}{8}\mathrm{HS^-} - \frac{1}{8}\mathrm{SO_4^{2-}} - \frac{9}{8}\mathrm{H^+} + \frac{1}{2}\mathrm{H_2O}. \tag{129h}$$

It is presumed that the redox couples themselves are linearly independent. If there are N_σ such redox couples present in the system of interest, then it is possible to construct N_σ linearly dependent reactions describing the oxidation of pyrite. The pyrite oxidation reaction has a degeneracy of $N_\sigma - 1$.

Let the independent reactions be written in the general form

$$\emptyset \rightleftharpoons \sum_{i=1}^{N} \nu_{ir}^{\rm dp} I_r^{\rm dp}, \quad (r = 1, \ldots N_r^{\rm dp}), \tag{130}$$

for $N_r^{\rm dp}$ reactions, where the superscript 'dp' is used to distinguish quantities pertaining to the linearly dependent reactions. Then taking into account both linearly independent and dependent reactions, the mass transport equations have the form

$$Ł_i C_i = \sum_{r=1}^{N_R} \nu_{ir} I_r + \sum_{r=1}^{N_R^{\rm dp}} \nu_{ir}^{\rm dp} I_r^{\rm dp}. \tag{131}$$

By definition, there must exist coefficients $C_{rr'}$ which relate the stoichiometric coefficients corresponding to the linearly dependent reaction as a linear combination of the stoichiometric coefficients corresponding to the linearly independent reactions. Thus

$$\nu_{ir'}^{\rm dp} = \sum_{r=1}^{N_R} \nu_{ir} C_{rr'}, \quad (r' = 1, \ldots, N_R^{\rm dp}). \tag{132}$$

Substituting the expression for $\nu_{ir'}^{\rm dp}$ given by Equation (132) into Equation (131) yields the equations

$$Ł_i C_i = \sum_{r=1}^{N_R} \nu_{ir} I_r^{\rm eff}, \tag{133}$$

where the effective reaction rate $I_r^{\rm eff}$ is defined by

$$I_r^{\rm eff} = I_r + \sum_{r'=1}^{N_R^{\rm dp}} C_{rr'} I_{r'}^{\rm dp}. \tag{134}$$

The sum on the right-hand side of Equation (133) is now only over the linearly independent reactions. Thus the effect of linearly dependent reactions is to modify the linearly independent reaction rates by the sum over the linearly dependent rates weighted by the coefficients $C_{rr'}$.

Multicomponent–multiphase mass conservation equations

So far the reactions as written in Equation (118) are completely general since no specification has been given for the species A_i, which could be either aqueous, gaseous or mineral species. Writing separate statements of Equation (118) for homogeneous and heterogeneous reactions yields

$$\sum_{j=1}^{N_c} \widetilde{\nu}_{ji} A_j \rightleftharpoons A_i^{\rm aq}, \quad (i = 1, \ldots, N_{\rm aq}), \tag{135a}$$

$$\sum_{j=1}^{N_c} \widetilde{\nu}_{jl} A_j \rightleftharpoons A_l^{g}, \quad (l = 1, \ldots, N_g), \tag{135b}$$

$$\sum_{j=1}^{N_c} \widetilde{\nu}_{jm} A_j \rightleftharpoons A_m, \quad (m = 1, \ldots, N_{\rm min}), \tag{135c}$$

for aqueous, gaseous and mineral secondary species, where $N_{\rm aq}$ denotes the number of aqueous secondary species, N_g the number of gaseous secondary species, and $N_{\rm min}$ the number of minerals. In writting these reactions the subscripts i, l and m are used to distinguish the stoichiometric coefficients $\widetilde{\nu}_{ji}$, $\widetilde{\nu}_{jl}$ and $\widetilde{\nu}_{jm}$ between aqueous, gaseous and mineral secondary species. The total number of secondary species is equal to

$$N_{\rm sec} = N_{\rm aq} + N_g + N_{\rm min}. \tag{136}$$

The primary species may be chosen as aqueous species for a two-phase or pure liquid fluid, and as gaseous species for a pure gas fluid phase.

Pure liquid fluid phase. The sum over secondary species in the primary species transport equations, Equation (122), includes in general aqueous, gaseous and mineral species. Consider first an aqueous fluid reacting with minerals. Writing the sum over secondary species separately for aqueous species and minerals yields

$$\frac{\partial}{\partial t}(\phi C_j) + \nabla\cdot\boldsymbol{J}_j + \sum_{i=1}^{N_{\rm aq}}\widetilde{\nu}_{ji}\left\{\frac{\partial}{\partial t}(\phi C_i) + \nabla\cdot\boldsymbol{J}_i\right\} + \sum_{m=1}^{N_{\rm min}}\widetilde{\nu}_{jm}\overline{V}_m^{-1}\frac{\partial\phi_m}{\partial t} = \sum_{r=1}^{N_R^{\rm ke}}\nu'_{jr}I_r^{\rm ke}, \tag{137}$$

as follows from the definition of Ł_i given in Equation (86). Collecting terms yields the primary species transport equations

$$\frac{\partial}{\partial t}\left[\phi\Psi_j + \sum_{m=1}^{N_{\rm min}}\widetilde{\nu}_{jm}\overline{V}_m^{-1}\phi_m\right] + \nabla\cdot\Omega_j = \sum_{r=1}^{N_R^{\rm ke}}\nu'_{jr}I_r^{\rm ke}. \tag{138}$$

The quantity Ψ_j is referred to as the generalized concentration and Ω_j as the generalized flux (Lichtner, 1985), defined by the relations

$$\Psi_j = C_j + \sum_{i=1}^{N_{\rm aq}}\widetilde{\nu}_{ji}C_i, \tag{139}$$

and

$$\Omega_j = \boldsymbol{J}_j + \sum_{i=1}^{N_{\rm aq}}\widetilde{\nu}_{ji}\boldsymbol{J}_i. \tag{140}$$

To solve these equations mass action expressions for minerals and aqueous species, Equations (101) and (102), must be added to complete the set of equations. This gives a total of $N_c^{\rm le} + N_{\rm aq} + N_{\rm min}$ algebraic–partial differential equations and an equal number of unknowns. These consist of the $N_c^{\rm le}$ concentrations of the primary species, $N_{\rm aq}$ aqueous secondary species concentrations, and $N_{\rm min}$ mineral volume fractions.

The analysis presented above applies to the general case of unequal diffusion coefficients (See the section entitled: Charge conservation with unequal diffusion coefficients.). For species-independent diffusion coefficients the solute flux Ω_j simplifies to the expression

$$\Omega_j = (-\tau\phi D\nabla + \boldsymbol{v})\,\Psi_j, \tag{141}$$

involving directly the generalized concentration Ψ_j, and not the individual species concentrations. This greatly simplifies the mass transport equations.

Although mineral reaction rates in Equation (138) are considered to be in local chemical equilibrium, in many instances it is more convenient computationally and essential for minerals like feldspars which are slow to react, to treat all mineral reactions kinetically. To carry this out, the mineral mass transfer equations must replace the mineral mass action equations increasing the number of differential equations necessary to solve. Mineral mass transfer equations in the canonical representation have the simple form

$$\frac{\partial \phi_m}{\partial t} = \overline{V}_m I_m. \tag{142}$$

If mineral reactions are treated as kinetically controlled with reaction rates described by Equation (65), then it is useful to eliminate the mineral concentrations from Equations (138) in favor of the reaction rates to give

$$\frac{\partial}{\partial t}(\phi \Psi_j) + \nabla \cdot \Omega_j = -\sum_{m=1}^{N_{\min}} \widetilde{\nu}_{jm} I_m + \sum_{r=1}^{N_R^{\mathrm{ke}}} \nu'_{jr} I_r^{\mathrm{ke}}. \tag{143}$$

These equations are a function of the primary species concentrations and, in addition, depend on the mineral volume fraction and surface area through the kinetic reaction rate I_m (see Eqn. (65)).

The added differential equations for the minerals do not place a great burden on the computational effort required to solve the transport equations. This is because they may be solved sequentially, as a consequence of the much slower time scales associated with mineral alteration compared to changes in the aqueous solution (see section entitled: QUASI-STATIONARY STATE APPROXIMATION). First the partial differential equations for the primary species are solved. The solution to these equations determines the solute concentrations in addition to the mineral reaction rates. Next the mineral mass transfer equations are integrated according to the explicit finite difference scheme

$$\phi_{mn}^{t+\Delta t} = \phi_{mn}^{t} + \Delta t \overline{V}_m I_{mn}, \tag{144}$$

where the subscript n refers to the nth node point, and I_{mn} denotes the reaction rate obtained in the first step. The size of the time step Δt that appears in this equation depends on how close the primary species concentrations are to a stationary state. When the system is close to a stationary state a large time step can be taken, limited only by the allowable change in mineral concentration tolerated. Because of this ability to take large time steps when solving the mineral mass transfer equations, the kinetic approach is often far more efficient than the local equilibrium approach.

Ion-exchange reactions. Ion-exchange reactions are described in more detail by Appelo (this volume). Consider a set of ion-exchange reactions involving N_{ex} cations of the form

$$z_k A_j^{z_j+} + z_j X_{z_k}^{\alpha} A_k \rightleftharpoons z_j A_k^{z_k+} + z_k X_{z_j}^{\alpha} A_j, \tag{145}$$

for exchange of the jth and kth cations, where X^{α} denotes a surface site of type α, and $X_{z_j}^{\alpha} A_j$ an adsorbed species. The reaction rate for the exchange of cations $j \leftrightarrow k$, I_{jk}^{α}, forms an antisymmetric matrix:

$$I_{jk}^{\alpha} = -I_{kj}^{\alpha}. \tag{146}$$

The rate of change of the concentration of adsorbed species $\overline{C}_l^\alpha$ is given by

$$\frac{\partial \overline{C}_l^\alpha}{\partial t} = \sum_{k=1}^{N_{\text{ex}}} z_k I_{kl}^\alpha. \tag{147}$$

This relation accounts for the change in adsorbed concentration by exchange only, but does not include changes resulting from mineral dissolution or precipitation, for example. The total surface site concentration is equal to

$$\omega_\alpha = \sum_{k=1}^{N_{\text{ex}}} z_k \overline{C}_k^\alpha. \tag{148}$$

The surface site concentration is related to the cation exchange capacity Q_α, defined as the ratio of number of exchange sites to mass of solid, by the equation

$$\begin{aligned} \omega_\alpha &= \frac{N_\alpha}{V_{\text{REV}}} = \frac{N_\alpha}{M_{\text{solid}}} \frac{M_{\text{solid}}}{V_{\text{solid}}} \frac{V_{\text{solid}}}{V_{\text{REV}}} \\ &= (1-\phi)\, \rho_{\text{solid}}\, Q_\alpha, \end{aligned} \tag{149}$$

where N_α denotes the number of exchange sites of type α contained in V_{REV}, and ρ_{solid} denotes the solid density. By virtue of the antisymmetric form of I_{kl}^α, and because the effect of mineral precipitation and dissolution is not included in the description, the total site concentration is conserved

$$\begin{aligned} \frac{\partial \omega_\alpha}{\partial t} &= \sum_{k=1}^{N_{\text{ex}}} z_k \frac{\partial \overline{C}_k^\alpha}{\partial t}, \\ &= \sum_{k=1}^{N_{\text{ex}}} \sum_{l=1}^{N_{\text{ex}}} z_k z_l I_{kl}^\alpha = 0. \end{aligned} \tag{150}$$

Because charge is conserved separately in both the aqueous and solid phases, ion-exchange reactions preserve electroneutrality. This is not necessarily the case, however, for other descriptions of adsorption, including batch surface complexation models when combined with transport processes.

Adding the contribution of ion-exchange reactions to the solute transport equations yields the equations

$$\frac{\partial}{\partial t}(\phi C_i) + \nabla \cdot \boldsymbol{J}_i = \sum_r \nu_{ir} I_r - \sum_\alpha \sum_l z_l I_{li}^\alpha. \tag{151}$$

The ion-exchange reaction rates may be eliminated using Equation (147) to yield the transport equations involving the sorbed species concentrations

$$\frac{\partial}{\partial t}\left(\phi C_i + \sum_\alpha \overline{C}_i^\alpha\right) + \nabla \cdot \boldsymbol{J}_i = \sum_r \nu_{ir} I_r. \tag{152}$$

Following the same procedure to obtain Equation (122) by eliminating the rates of the local equilibrium reactions yields the following primary species transport equations

$$\frac{\partial}{\partial t}\left(\phi \Psi_j + \sum_\alpha \overline{\Psi}_j^\alpha\right) + \nabla \cdot \Omega_j = \sum_r \nu'_{jr} I_r, \tag{153}$$

where

$$\overline{\Psi}_j^\alpha = \overline{C}_j^\alpha + \sum_i \widetilde{\nu}_{ji} \overline{C}_i^\alpha . \tag{154}$$

This formulation provides for exchange of primary, secondary and kinetic species.

To complete the set of equations it is necessary to express the exchange isotherm $\overline{C}_k$ in terms of the concentrations of the primary species. This follows from the mass action equations for the exchange reactions which have the form, neglecting activity coefficient corrections,

$$K_{jk}^\alpha = \left[\frac{C_k}{\overline{C}_k^\alpha}\right]^{z_j} \left[\frac{\overline{C}_j^\alpha}{C_j}\right]^{z_k}, \tag{155}$$

where the quantities K_{jk}^α represent selectivity coefficients. Combining this relation with the site conservation equation yields a single nonlinear equation for the jth sorption isotherm $\overline{C}_j^\alpha$:

$$\omega_\alpha = z_j \overline{C}_j^\alpha + \sum_{k \neq j} z_k C_k \left(K_{jk}^\alpha\right)^{1/z_k} \left[\frac{\overline{C}_j^\alpha}{C_j}\right]^{z_j/z_k} . \tag{156}$$

This equation implicitly defines the sorption isotherm as a function of the primary species concentrations.

Two-phase fluid flow. The same methods apply to describing the change in phase from gas to liquid and vice versa, as any other chemical reaction. For a two-phase system consisting of liquid (l) and gas (g) phases and composed of two species water (w) and air (a), there are two possible reactions: one the evaporation and condensation of water

$$H_2O_{(l)} \rightleftharpoons H_2O_{(g)}, \tag{157}$$

involving the reaction between liquid water and water vapor, and the other the partitioning of air between liquid and gas phases

$$A_{\mathrm{air}(l)} \rightleftharpoons A_{\mathrm{air}(g)}, \tag{158}$$

where A_{air} is taken to represent the dominant species in the Earth's atmosphere, namely nitrogen. The respective reaction rates are denoted by I_{H_2O} and I_{air}.

Voids in the porous medium may be filled with both liquid and gas phases. The number of moles of the ith species in the liquid and gas phases contained in a REV can be expressed as

$$Z_i^\pi = \frac{N_i^\pi}{V} = = \frac{V_p}{V}\left[\frac{V_p^\pi}{V_p}\frac{N_\pi}{V_p^\pi}\frac{N_i^\pi}{N_\pi}\right], \tag{159}$$

$$= \phi s_\pi \rho_\pi X_i^\pi, \tag{160}$$

where $\pi = l,\ g$, and N_i^π denote the number of moles of the ith species contained in the πth fluid phase with volume V_p^π and molar density ρ_π. The aqueous and gaseous mole fractions of the ith species are denoted by X_i^l and X_i^g, respectively. The liquid and gas phases occupy fractions s_l and s_g, respectively, with

$$s_l + s_g = 1. \tag{161}$$

The liquid flux $\boldsymbol{J}_i^l$ is defined by

$$\boldsymbol{J}_i^l = -\phi\tau s_l D_l \rho_l \nabla X_i^l + \boldsymbol{q}_l C_i^l, \tag{162}$$

and the gas flux by the equation

$$\boldsymbol{J}_i^g = -\phi\tau s_g D_g \rho_g \nabla X_i^g + \boldsymbol{q}_g C_i^g, \tag{163}$$

where D_π denotes the diffusion coefficient and $\boldsymbol{q}_\pi$ the Darcy velocity of phase π. The Darcy velocity is defined by

$$\boldsymbol{q}_\pi := -\frac{k k_{r\pi}}{\mu_\pi} \nabla \left(p_\pi - \rho_\pi g z \right). \tag{164}$$

In this equation k refers to the saturated permeability of the porous medium, $k_{r\pi}$ represents the relative permeability and μ_π the viscosity of phase π, g denotes the acceleration of gravity, and z the vertical height. The mole fractions X_w^π and X_a^π satisfy the identity

$$X_w^\pi + X_a^\pi = 1. \tag{165}$$

The difference between the liquid and gas pressures gives the capillary pressure p_c

$$p_c = p_g - p_l. \tag{166}$$

The effective binary gas diffusion coefficient is a function of temperature, pressure and saturation given by the expression (Vargaftik, 1975)

$$D_g^{\text{eff}} = \omega\, \tau\, \phi\, s_g\, D_g^0 \frac{p_{\text{ref}}}{p_g} \left[\frac{T + T_{\text{ref}}}{T_{\text{ref}}} \right]^\theta, \tag{167}$$

where T_{ref} and p_{ref} denote reference temperature and pressure, θ is an empirical constant ($\theta \simeq 1.8$), and ω denotes an enhancement factor. The enhancement factor is usually considered to be inversely proportional to the gas saturation s_g, which consequently cancels in the expression for the effective gas diffusion coefficient (Westcot and Wierenga, 1974; Jury and Letey, 1979).

The mass conservation equations for water are given by

$$\frac{\partial}{\partial t} \left(\phi s_l \rho_l X_w^l \right) + \nabla \cdot \boldsymbol{J}_w^l = -I_{\text{H}_2\text{O}}, \tag{168}$$

and

$$\frac{\partial}{\partial t} \left(\phi s_g \rho_g X_w^g \right) + \nabla \cdot \boldsymbol{J}_w^g = I_{\text{H}_2\text{O}}, \tag{169}$$

and for air by

$$\frac{\partial}{\partial t} \left(\phi s_l \rho_l X_a^l \right) + \nabla \cdot \boldsymbol{J}_a^l = -I_{\text{air}}, \tag{170}$$

and

$$\frac{\partial}{\partial t} \left(\phi s_g \rho_g X_a^g \right) + \nabla \cdot \boldsymbol{J}_a^g = I_{\text{air}}. \tag{171}$$

For conditions of local thermodynamic equilibrium between the liquid and gas phases, eliminating the reaction rates by adding the water and air equations for the liquid and gas phases gives

$$\frac{\partial}{\partial t}\left\{(\phi\left[s_l\rho_l X_w^l + s_g\rho_g X_w^g\right]\right\} + \nabla\cdot\left[\boldsymbol{J}_w^l + \boldsymbol{J}_w^g\right] = 0, \tag{172}$$

for water, and

$$\frac{\partial}{\partial t}\left\{(\phi\left[s_l\rho_l X_a^l + s_g\rho_g X_a^g\right]\right\} + \nabla\cdot\left[\boldsymbol{J}_a^l + \boldsymbol{J}_a^g\right] = 0, \tag{173}$$

for air. The mass action equation for evaporation and condensation of water results in the saturation curve

$$p = p_{\mathrm{sat}}(T), \tag{174}$$

the condition for coexisting liquid water and vapor. The ideal gas law is used for the equation of state for air, with Henry's law for partitioning between aqueous and gaseous phases

$$X_a^l = K_a X_a^g, \tag{175}$$

where K_a denotes Henry's constant. Adding the water and air mass balance equations eliminates the diffusion terms providing the total mass balance equation

$$\frac{\partial}{\partial t}\left\{\phi\left[s_l\rho_l + s_g\rho_g\right]\right\} + \nabla\cdot\left[\boldsymbol{q}_l\rho_l + \boldsymbol{q}_g\rho_g\right] = 0, \tag{176}$$

obtained by making use of the identities satisfied by the water and air mole fractions.

To complete the equations constitutive relations are required for the relative permeability and capillary pressure p_c in terms of the saturation state. One such set of relations is given by the phenomenological equations (van Genuchten, 1980; Suarez and Šimůnek, this volume)

$$s_l^{\mathrm{eff}}(p_c) = \left[1 + (\alpha|p_c|)^m\right]^{-\lambda}, \tag{177}$$

where α and m are parameters with

$$\lambda = 1 - \frac{1}{m}, \tag{178}$$

and the effective saturation is defined by

$$s_l^{\mathrm{eff}} = \frac{s_l - s_l^r}{1 - s_l^r}, \tag{179}$$

where s_l^r denotes the residual saturation. The liquid relative permeability k_{rl} is defined by

$$k_{rl} = \sqrt{s_l^{\mathrm{eff}}}\left[1 - \left(1 - \left(s_l^{\mathrm{eff}}\right)^{1/\lambda}\right)^{\lambda}\right]^2. \tag{180}$$

The gas phase relative permeability is assumed to be related to the liquid relative permeability by

$$k_{rg} = 1 - k_{rl}. \tag{181}$$

The relative permeability has the value of one for a fully saturated rock column and becomes zero as the saturation approaches the residual saturation s_r. At the residual saturation the capillary pressure is infinite according to Equation (177). A number of computer codes are available which describe two-phase flow in geothermal systems including TOUGH2 (Pruess, 1991), NUFT (Nitao, 1996), FEHMN (Zyvoloski et al., 1992), and METRA the two-phase flow part of the code MULTIFLO (Lichtner and Seth, 1996b; Seth and Lichtner, 1996).

Multicomponent systems. In contrast to two-phase fluid flow codes with two species water and air, very little work has been done incorporating multicomponent chemical reactions with multiphase fluid flow (White, 1995; Lichtner and Seth, 1996a, b). To derive the aqueous and gaseous primary species transport equations for a multicomponent system, the definition of the operator $Ł_i$ given in Equation (86) is substituted into Equation (122) to yield

$$\frac{\partial}{\partial t}\left(\phi\Psi_j + \sum_{m=1}^{N_{\min}} \widetilde{\nu}_{jm} \overline{V}_m^{-1} \phi_m\right) + \nabla\cdot\Omega_j = \sum_r \nu'_{jr} I_r^{\text{ke}}, \tag{182}$$

where the generalized concentration Ψ_j and flux Ω_j are defined by

$$\Psi_j = s_l \Psi_j^l + s_g \Psi_j^g, \tag{183}$$

and

$$\Omega_j = \Omega_j^l + \Omega_j^g, \tag{184}$$

with

$$\Psi_j^l = \delta_{\pi_j l} C_j + \sum_{i=1}^{N_{\text{aq}}} \widetilde{\nu}_{ji} C_i, \tag{185a}$$

$$\Psi_j^g = \delta_{\pi_j g} C_j + \sum_{l=1}^{N_g} \widetilde{\nu}_{jl} C_i, \tag{185b}$$

and

$$\Omega_j^l = \delta_{\pi_j l} \boldsymbol{J}_j^l + \sum_{i=1}^{N_{\text{aq}}} \widetilde{\nu}_{ji} \boldsymbol{J}_i^l, \tag{186a}$$

$$\Omega_j^g = \delta_{\pi_j g} \boldsymbol{J}_j^g + \sum_{l=1}^{N_g} \widetilde{\nu}_{jl} \boldsymbol{J}_l^g, \tag{186b}$$

where $\pi_j = l, g$ depending on whether the jth primary species belongs to the liquid or gas phase. The kronecker delta function $\delta_{\pi_j \pi}$ appears in Equations (185a-186b) because any particular primary species can belong to only one phase. The reaction rates of the kinetic reactions taking place in the system appear as source/sink terms on the right hand side. The reaction rates have been eliminated for local equilibrium homogeneous reactions within the aqueous and gaseous phases, and heterogeneous reactions between these phases.

Richards equation. An important and useful simplification of these equations is Richards equation in which the gas phase is considered to be inert (Richards, 1931; Suarez and Šimůnek, this volume). In this case the two-phase flow equations reduce to a single equation which may be written in the form

$$\frac{\partial}{\partial t}(\phi s_l) + \nabla \cdot (K k_r \nabla h) = 0, \tag{187}$$

where K denotes the hydraulic conductivity related to the permeability by

$$K = \frac{\rho g k}{\mu}, \tag{188}$$

and the hydraulic head h is defined by

$$h = \frac{p}{\rho g}. \tag{189}$$

The terms of the hydraulic head and conductivity, Darcy's law becomes

$$\boldsymbol{q} = -K \nabla h. \tag{190}$$

To obtain Richards equation it is assumed that the density of water is constant and the expression for the Darcy velocity $\boldsymbol{q}$ is substituted in terms of the hydraulic head h in Equation (176). Equation (187) can be expressed in terms of a single unknown, for example the liquid saturation s_l, noting that the hydraulic head can be expressed in terms of saturation as

$$h = \frac{1}{\alpha}\left[\left(s_l^{\text{eff}}\right)^{-1/\lambda} - 1\right]^{1/m}, \tag{191}$$

as follows from Equation (177). Likewise the relative permeability must be known as a function of saturation as well. This results in a nonlinear partial differential equation which must be integrated numerically [see discussion following Eqn. (336)].

Physical interpretation of the generalized concentration and flux. For the case that $\widetilde{\nu}_{jm}$ is non-negative ($\widetilde{\nu}_{jm} \geq 0$), the quantities Ψ_j^π and Ω_j^π represent the total analytical concentration and flux of the jth primary species in phase π. For a two-phase system the quantities Ψ_j and Ω_j represent the total concentration and flux summed over all fluid phases. The quantity appearing in the time derivative term in Equation (182) is denoted by W_j:

$$\begin{aligned} W_j &= \phi \Psi_j + \sum_{m=1}^{N_{\min}} \widetilde{\nu}_{jm} \overline{V}_m^{-1} \phi_m, \\ &= \phi\left(s_l \Psi_j^l + s_g \Psi_j^g\right) + \sum_{m=1}^{N_{\min}} \widetilde{\nu}_{jm} \overline{V}_m^{-1} \phi_m. \end{aligned} \tag{192}$$

The quantity W_j represents the total concentration of the jth primary species contained in fluid and solid phases.

To more fully appreciate the meaning of Ψ_j it may be helpful to consider several explicit examples. Take the reaction

$$\text{Na}^+ + \text{Cl}^- \rightleftharpoons \text{NaCl}^\circ, \tag{193}$$

with primary species Na^+ and Cl^-, and secondary species $NaCl^\circ$, defining the ion-pairing reaction of the neutral sodium chloride species. It follows that $\widetilde{\nu}_{\text{Na}^+,\text{NaCl}^\circ} = 1$, $\widetilde{\nu}_{\text{Cl}^-,\text{NaCl}^\circ} = 1$, and hence

$$\Psi_{\text{Na}^+} = C_{\text{Na}^+} + C_{\text{NaCl}^\circ}, \tag{194a}$$

and

$$\Psi_{\mathrm{Cl^-}} = C_{\mathrm{Cl^-}} + C_{\mathrm{NaCl^\circ}}, \tag{194b}$$

according to Equation (139), and thus both $\Psi_{\mathrm{Na^+}}$ and $\Psi_{\mathrm{Cl^-}}$ correspond to the total concentration of Na and Cl, respectively. However, consider next the dissociation of water

$$\mathrm{H_2O} - \mathrm{H^+} \rightleftharpoons \mathrm{OH^-}. \tag{195}$$

written with H^+ and H_2O as primary species and OH^- as secondary species. It follows in this case that $\tilde{\nu}_{\mathrm{H^+,OH^-}} = -1$, and thus

$$\Psi_{\mathrm{H^+}} = C_{\mathrm{H^+}} - C_{\mathrm{OH^-}}. \tag{196}$$

Thus for acidic solutions $\Psi_{\mathrm{H^+}}$ is positive, while for basic solutions it is negative, and it vanishes for neutral solutions.

A somewhat more complicated example is that of speciation of aluminum in solution. Possible aluminum complexes are $Al(OH)_4^-$, $Al(OH)_3^\circ$, $Al(OH)_2^+$ and $AlOH^{2+}$, formed according to the reactions

$$\mathrm{Al^{3+}} - 4\,\mathrm{H^+} + 4\,\mathrm{H_2O} \rightleftharpoons \mathrm{Al(OH)_4^-}, \tag{197a}$$

$$\mathrm{Al^{3+}} - 3\,\mathrm{H^+} + 3\,\mathrm{H_2O} \rightleftharpoons \mathrm{Al(OH)^\circ}, \tag{197b}$$

$$\mathrm{Al^{3+}} - 2\,\mathrm{H^+} + 4\,\mathrm{H_2O} \rightleftharpoons \mathrm{Al(OH)_2^+}, \tag{197c}$$

$$\mathrm{Al^{3+}} - \mathrm{H^+} + 4\,\mathrm{H_2O} \rightleftharpoons \mathrm{Al(OH)^{2+}}. \tag{197d}$$

With primary species chosen as Al^{3+} and H^+ it follows that

$$\Psi_{\mathrm{Al^{3+}}} = C_{\mathrm{Al^{3+}}} + C_{\mathrm{Al(OH)_4^-}} + C_{\mathrm{Al(OH)_3^\circ}} + C_{\mathrm{Al(OH)_2^+}} + C_{\mathrm{AlOH^{2+}}}, \tag{198a}$$

and

$$\Psi_{\mathrm{H^+}} = C_{\mathrm{H^+}} - C_{\mathrm{OH^-}} - 4\,C_{\mathrm{Al(OH)_4^-}} - 3\,C_{\mathrm{Al(OH)_3^\circ}} - 2\,C_{\mathrm{Al(OH)_2^-}} - C_{\mathrm{AlOH^{2+}}}. \tag{198b}$$

Clearly $\Psi_{\mathrm{Al^{3+}}}$ represents the total aluminum in solution, but $\Psi_{\mathrm{H^+}}$ may take on either positive, negative or zero values depending on the pH of the solution. Alternatively, one could choose as primary species $Al(OH)_4^-$ and OH^-. In this case

$$\Psi_{\mathrm{Al(OH)_4^-}} = \Psi_{\mathrm{Al^{3+}}}, \tag{199a}$$

is still the total aluminum concentration, and

$$\Psi_{\mathrm{OH^-}} = -\Psi_{\mathrm{H^+}}. \tag{199b}$$

For an example involving liquid and gas phases consider the two-phase system containing CO_2. With $\{HCO_3^-, H^+\}$ as primary species taken from the aqueous phase, the reactions of interest are

$$\mathrm{HCO_3^-} + \mathrm{H^+} - \mathrm{H_2O} \rightleftharpoons \mathrm{CO_{2(l)}}, \tag{200a}$$

$$\mathrm{HCO_3^-} - \mathrm{H^+} \rightleftharpoons \mathrm{CO_3^{2-}}, \tag{200b}$$

in the aqueous phase, and

$$HCO_3^- + H^+ - H_2O \rightleftharpoons CO_{2(g)}, \tag{200c}$$

between the liquid and the gas phases. Then for the liquid phase

$$\Psi^l_{HCO_3^-} = C^l_{HCO_3^-} + C^l_{CO_3^{2-}} + C^l_{CO_2}, \tag{201a}$$

$$\Psi^l_{H+} = C^l_{H+} - C^l_{OH^-} - C^l_{CO_3^{2-}} + C^l_{CO_2}, \tag{201b}$$

and for the gas phase

$$\Psi^g_{HCO_3^-} = C^g_{CO_2}, \tag{201c}$$

$$\Psi^g_{H+} = C^g_{CO_2}. \tag{201d}$$

Alternatively, if $CO_{2(g)}$ is used as primary species, the carbonate reactions become

$$CO_{2(g)} - 2\,H^+ + H_2O \rightleftharpoons CO_3^{2-}, \tag{202a}$$

$$CO_{2(g)} - H^+ + H_2O \rightleftharpoons HCO_3^-, \tag{202b}$$

$$CO_{2(g)} \rightleftharpoons CO_{2(l)}. \tag{202c}$$

Including Ca^{2+} as primary species gives two additional reactions

$$Ca^{2+} + CO_{2(g)} - 2\,H^+ + H_2O \rightleftharpoons CaCO_3^\circ, \tag{203a}$$

$$Ca^{2+} + CO_{2(g)} - H^+ + H_2O \rightleftharpoons CaHCO_3^-. \tag{203b}$$

It follows that

$$\Psi^l_{H+} = C^l_{H+} - C^l_{OH^-} - 2\,C^l_{CO_3^{2-}} - C^l_{HCO_3^-} - 2\,C^l_{CaCO_3^\circ} - C^l_{CaHCO_3^-}. \tag{204}$$

The quantity, $-\Psi^l_{H+}$, will be recognized as the alkalinity.

It should be noted that the so-called proton condition (Stumm and Morgan, 1981), is really nothing more than the mass conservation equation for Ψ_{H+}. A similar condition, the base condition, refers to the the mass conservation equation corresponding to OH^- used as primary species.

Example: partitioning between aqueous and gaseous phases. Simultaneous transport of two or more phases with chemical reactions taking place between the them can lead to retardation or accelerated transport of species within the phases. Consider a two-phase system consisting of a liquid and gas phase in which a single chemical reaction occurs between aqueous and gaseous phases of the form

$$A_l \rightleftharpoons A_g. \tag{205}$$

Assuming local chemical equilibrium between the two phases, the concentrations of aqueous and gaseous species are related by the mass action equation

$$C_g = K C_l, \tag{206}$$

with equilibrium constant K, in general a function of temperature and pressure. The transport equation is given by

$$\frac{\partial}{\partial t}\left[\phi\left(s_l C_l + s_g C_g\right)\right] + \nabla\cdot\left(\boldsymbol{J}_l + \boldsymbol{J}_g\right) = 0. \tag{207}$$

Substituting for the concentration of the gaseous species in terms of the aqueous species gives for the accumulation term:

$$s_l C_l + s_g C_g = \left(s_l + K s_g\right) C_l, \tag{208}$$

and for the flux term assuming that the equilibrium constant K is a constant:

$$\boldsymbol{J}_l + \boldsymbol{J}_g = -\phi[\left(s_l D_l + K s_g D_g\right)\nabla C_l] + \left(\boldsymbol{v}_l + K\boldsymbol{v}_g\right) C_l. \tag{209}$$

From these relations it follows that for K constant, the concentrations of both species obey the identical transport equation. For constant saturation, porosity, and equilibrium constant K, an effective velocity and diffusion coefficient may be introduced resulting in the effective advection–diffusion equation

$$\frac{\partial C_l}{\partial t} + \nabla\cdot\boldsymbol{v}_{\text{eff}} C_l - \nabla\cdot D_{\text{eff}}\nabla C_l = 0, \tag{210}$$

where the effective fluid velocity $\boldsymbol{v}_{\text{eff}}$ and diffusivity D_{eff} are defined by

$$\boldsymbol{v}_{\text{eff}} = \frac{\boldsymbol{v}_l + K\boldsymbol{v}_g}{\phi[s_l + K(1-s_l)]}, \tag{211}$$

and

$$D_{\text{eff}} = \frac{s_l D_l + K(1-s_l)D_g}{s_l + K(1-s_l)}. \tag{212}$$

Assuming $s_l = s_g = 1/2$ and $K \gg 1$, it follows that

$$\boldsymbol{v}_{\text{eff}} = \begin{cases} \dfrac{2}{K}\dfrac{\boldsymbol{v}_l}{\phi}, & (v_l \gg K v_g), \\ 2\dfrac{\boldsymbol{v}_g}{\phi}, & (K v_g \gg v_l), \end{cases} \tag{213}$$

and

$$D_{\text{eff}} = \begin{cases} \dfrac{D_l}{K}, & (D_l \gg K D_g), \\ D_g, & (K D_g \gg D_l), \end{cases}. \tag{214}$$

Thus in order to maintain local equilibrium, for liquid dominated flow the effective velocity is retarded by the factor $K/2$ from the aqueous fluid velocity. For gaseous dominated diffusion, the effective diffusivity is enhanced from its value in an aqueous solution and equal to the gaseous diffusion coefficient. The factor two arises because flow of each phase takes place in only half the pore volume.

For binary diffusion of water vapor with a diffusion coefficient of $D_g = 2.15\times10^{-5}\,\text{m}^2\,\text{s}^{-1}$ at 25 °C compared to $D_l = 10^{-9}\,\text{m}^2\,\text{s}^{-1}$, or four orders of magnitude larger, this leads to much more rapid diffusion in the aqueous phase than would be possible in the absence of the gas phase. This is demonstrated in Figure 2 based on a numerical calculation using the computer

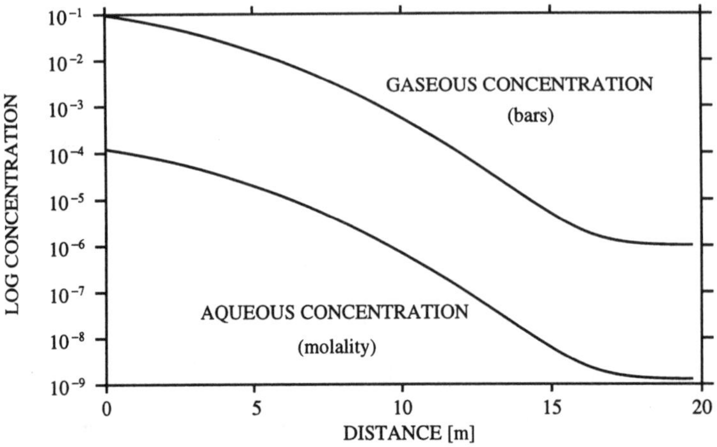

Figure 2. Aqueous and gaseous concentrations of O_2 plotted as a function of distance for pure diffusive transport.

code MULTIFLO, a two-phase, nonisothermal reactive mass transport code (Lichtner and Seth, 1996b). For a slightly soluble gas, such as O_2, $\log K_{O_2} = 2.898$. Pure diffusion is considered in a porous column 50% saturated with liquid. The porosity is taken as 10%. The concentration of dissolved oxygen and oxygen fugacity are plotted as a function of distance for an elapsed time of 3.65 days. The initial oxygen fugacity is taken as 10^{-6} bars in equilibrium with the aqueous solution. At the inlet ($x = 0$) a two-phase fluid in equilibrium with the atmosphere with a oxygen fugacity of 0.2 bars diffusives into the column with the diffusion coefficients given above. In the absence of a gas phase, there would be almost no perceptible change is the dissolved oxygen concentration during the time span of 3.65 years, in stark contrast to the case with a gas phase present as illustrated in the figure.

ASYMPTOTICS, LOCAL EQUILIBRIUM AND GHOST ZONES

A question that is continually addressed in reactive transport modeling is the relation between local equilibrium and a kinetic description of mineral reaction rates (Bahr and Rubin, 1987; Lichtner, 1993). In particular, one would like to know under what circumstances the local equilibrium description is valid. These questions can be answered by investigating the long time—or asymptotic—behavior of the system. Remarkable though it may seem, it is actually much easier in some cases to obtain the asymptotic solution to the kinetic-based mass conservation equations compared to the early time behavior. The latter requires solving a coupled set of nonlinear partial differential equations. For a system in which there is no inherent length scale, it is possible to deduce the asymptotic form of the kinetic-based reactive transport equations by merely scaling the time and space coordinates.

Scaling

As demonstrated by Lichtner (1993), scaling the x and t coordinates leads to the following identity for any field variable $F(x, t;\, v, D, \{k\})$ which is a solution to the one-dimensional reactive transport equation:

$$F(\sigma x, \sigma t;\, v, D, \{k\}) = F(x, t;\, v, \sigma^{-1} D, \sigma\{k\}), \tag{215}$$

where $\{k\}$ denotes the set of rate constants $\{k_1, k_2, \ldots\}$, and σ is a constant scale factor. This relation is predicated on the assumption that initial and boundary conditions imposed on the system are scale invariant. The scaling relation can be applied only to systems which do not contain an inherent characteristic length scale. Heterogeneities, for example, can be expected to destroy this scaling relation. In the limit $\sigma \to \infty$, the time- and space-scaled field variables approach the pure advective local equilibrium limit

$$\lim_{\sigma\to\infty} F(\sigma x, \sigma t; v, D, \{k\}) = \lim_{\sigma\to\infty} F(x, t; v, \sigma^{-1}D, \sigma\{k\}), \\ \to F_{eq}(x, t; v). \tag{216}$$

Thus the asymptotic ($t \to \infty$) limiting form of the mass transport equations may be obtained by solving the pure advective transport equations with the assumption of local chemical equilibrium. In this limit the transport equations reduce to a set of algebraic equations (Walsh et al., 1984; Lichtner, 1991). Thus by studying solutions to the pure advective mass transport equations for conditions of local chemical equilibrium, we are able to learn about the asymptotic form of solutions to the complete mass conservation equations based on a kinetic description of mineral reactions. Asymptotically the kinetic solution to the mass transport equations approaches the local equilibrium limiting solution regardless of the choice of kinetic rate constants or surface areas used in the calculation (Lichtner, 1993).

The approach to the asymptotic limit is illustrated in Figure 3 depicting the weathering of an alaskite host rock. The unaltered rock composition consists of 40% K-feldspar, 15% albite, and 40% quartz. The pure advective, step-function profile is achieved for most of the minerals with the exception of quartz for which longer time is required. Because the algebraic equations do not determine directly the reaction zone sequence which must be obtained through trial and error, they are generally tedious to solve. It is often easier to solve the differential equations representing pure advective mass transport with kinetic reactions of minerals. This approach provides a direct determination of the zone sequence. The solution can be obtained using the quasi-stationary state approximation in which the time evolution of the system is represented as a sequence of stationary states, referred to as the multiple reaction path model (Lichtner, 1988), as discussed in more detail in the section describing the quasi-stationary state approximation. The computer code MPATH has been developed to solve these equations for the case of purely advective transport, thus providing directly the asymptotic limiting solution (Lichtner, 1992).

Although local equilibrium is valid in the asymptotic limit of the kinetic transport equations, for sufficiently short time scales reaction kinetics dominate the evolution of the system when mineral alteration zones are just beginning to form. The time evolution of a geochemical system involving reactive transport of solutes can generally be divided into three overlapping regimes: an early transient period governed by kinetic mineral reactions; an intermediate transition regime; and the asymptotic regime in which the system is in local equilibrium with the exception of narrow reaction fronts. In the asymptotic regime the reaction front velocities are independent of the kinetic rate constants.

The solution to the algebraic asymptotic equations has been discussed extensively in the literature and the interested reader is referred to the works by Walsh et al. (1984), Schechter et al. (1987), Lichtner (1991), and Lichtner and Balashov (1993), for more details. The pure advective, local equilibrium limit is a highly singular limit with chemical reactions taking place at zone boundaries, their rates represented by Dirac δ-functions. The solution concentration is constant within each reaction zone with jump discontinuities across zone boundaries. Thus the system consists of a sequence of chemical shock waves propagating at retarded velocities. Replacement reactions take place at each front which separates different mineral assemblages. There are several rules for how minerals may appear and disappear across a front (Frantz and Mao, 1975, 1976, 1979; Weare et al., 1976; Walsh et al., 1984; Lichtner, 1991). The jump in

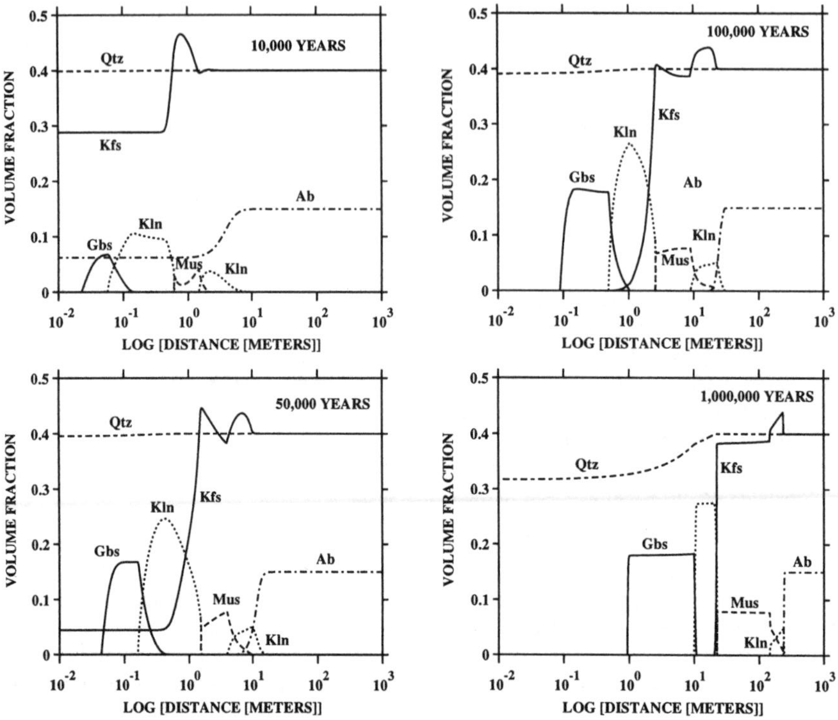

Figure 3. The evolution with time of the weathering of an alaskite host rock demonstrating the asymptotic approach to local equilibrium (Lichtner, 1992). [Used by permission of Water Resources Research.]

concentration is related to the velocity of the front v_f through mass conservation equations at the front given by

$$v_f = \frac{[\Omega_j]_f}{[\phi\Psi_j]_f + \sum_{m=1}^{N_{\min}} \widetilde{\nu}_{jm} \overline{V}_m^{-1} [\phi_m]_f}, \tag{217}$$

referred to as the generalized Rankine-Hugoniot equations (Lichtner, 1985, 1988, 1991). The square brackets $[\ \ldots\ ]_f$ denote the jump in the enclosed quantity across the front. The jump in flux is related to the jump in concentration by the relation

$$[\Omega_j]_f = v\,[\Psi_j]_f. \tag{218}$$

The mass conservation equations for each front are combined with mineral mass action equations for each zone. In order to complete the set of equations, so-called downstream equilibrium conditions must be invoked, stating that the fluid composition on the downstream side of a mineral assemblage remains in equilibrium with the assemblage until a new zone is encountered (Walsh et al., 1984; Lichtner, 1991). This set of equations provides an equal number of equations as unknowns. The main drawback to these equations is that in order to set up the problem in the first place the sequence of reaction zones must be known in advance. However, the zonation pattern is one of the properties of the system one generally wishes to predict. Therefore, a certain amount of trial and error in choosing different mineral assemblages is required to obtain a consistent solution to the problem posed.

Even so, there may not exist any physically meaningful solution to these equations as posed, even for perfectly reasonable initial and boundary conditions, for any choice of mineral zonation. This makes the problem of solving the pure advective–local equilibrium algebraic equations ever more difficult. The problem is that, although a solution to the algebraic equations can generally be found, this solution may fail to satisfy the physical requirements that all mineral concentrations be positive and that the replacement front velocities increase beginning with the front nearest to the inlet and proceeding in the downstream direction. Thus if the front positions are located at $l_{f_1} \leq \cdots \leq l_{f_n} \leq \cdots$, then the front velocities must satisfy the inequalities

$$v_{f_1} \leq \cdots \leq v_{f_n} \leq \cdots . \tag{219}$$

The reason for the nonexistence of a physical solution is a result of improperly formulating the downstream equilibrium condition. The downstream equilibrium condition asserts that the fluid composition on the downstream side of a reaction zone remain in equilibrium with the minerals in the upstream zone. However, suppose there were to exist an infinitely thin zone sandwiched between the upstream and downstream zones. Then the appropriate mineral assemblage to use in the downstream equilibrium condition is the one present in the infinitely thin zone, and not in the upstream zone. This new prescription leads to the concept of ghost zones.

Ghost zones

Ghost zones are special alteration zones that propagate with constant width, unlike normal zones which increase in size in proportion to the fluid velocity (Schechter et al., 1987; Lichtner and Balashov, 1993). A ghost zone is characterized by having equal velocities at its upstream and downstream boundaries (otherwise its width could not remain constant). A ghost zone represents a narrow zone trapped between two normal zones which propagates with constant width. As noted by Lichtner and Balashov (1993) a certain time period is required before the ghost zone attains its maximum possible width.

Ghost zones are unique to the advective transport-reaction problem. In the limit of local equilibrium and pure advective transport, ghost zones have zero width and hence are materially absent. Nevertheless, the minerals in the ghost zone still are able to buffer the downstream fluid composition, whence the term "ghost". If diffusion is included in the model, a ghost zone has a finite width governed by diffusion and advection, and finite mineral concentrations. The width of a ghost zone is directly proportional to the diffusion coefficient and inversely proportional to the fluid velocity (Lichtner and Balashov, 1993). For pure advective-kinetic mass transport, the width of the ghost zone is finite, rather than zero as in the local equilibrium limiting case. This, however, is an artifact of the kinetically controlled reaction rates, and the width is in general different if diffusion is included in the description. As a consequence, it is not possible in general to predict the width of, or mineral concentration within, the ghost zone in the pure advective-kinetic or local equilibrium limit. Nevertheless, the algebraic asymptotic solution to the mass transport equations can be useful for testing the numerical convergence and accuracy of finite difference solutions extending over geologic time spans. Clearly an algebraic solution does not suffer from round-off errors, for example, as can finite difference algorithms.

In the following example of a ghost zone shown in Figure 4, a solution with pH 4 in equilibrium with quartz infiltrates into a porous rock consisting of 60% quartz and 10% K-feldspar with 10% porosity. The system is assumed to be isothermal with a temperature of 225°C. A Darcy flow velocity of 10 m y^{-1} is used in the calculation. The effective aqueous diffusion coefficient including tortuosity has the value 3.8510^{-5} cm^2 s^{-1}. As the system evolves in time, the alteration sequence kaolinite–muscovite–K-feldspar is formed with quartz present throughout. Muscovite replaces K-feldspar and kaolinite replaces muscovite in the alteration

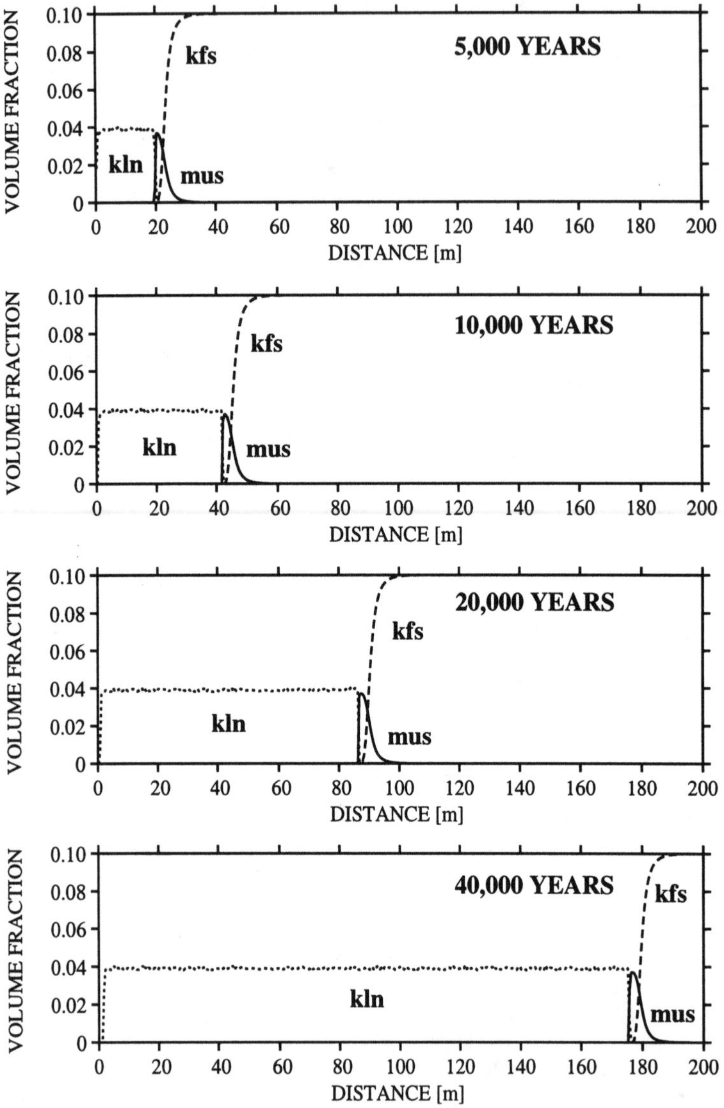

Figure 4. Formation of a muscovite ghost zone from the alteration of a K-feldspar + quartz host rock at 225°C shown for different times indicated in the figures. Note that the muscovite zone propagates without changing shape. Abbreviations used in the figure are: kln–kaolinite, mus–muscovite, and kfs–K-feldspar.

sequence as shown in Figure 4. As can be clearly seen from the figure, the muscovite zone propagates without changing shape, and thus forms a ghost zone, whereas the kaolinite zone grows continuously with time.

QUASI-STATIONARY STATE APPROXIMATION

A characteristic difference that generally holds between an aqueous solution and minerals with which it is reacting, is that any particular species is much more concentrated in minerals compared to its concentration in solution. This implies that only an extremely small change in mineral concentration is necessary to greatly affect the solution composition. As a result, the aqueous solution associated with a geochemical system may spend much of its time in a stationary state. In a stationary state the aqueous solution composition does not change with time, but can change with distance along the flow path. A stationary state is caused by the balance between transport by advection and diffusion and chemical reactions. Reaction rates may be either local equilibrium or kinetic controlled. The quasi-stationary state approximation is based on the approximation

$$\frac{\partial \Psi_j}{\partial t} \simeq 0, \tag{220}$$

the condition for existence of a stationary state. The actual time evolution of a geochemical system is represented as a sequence of stationary states.

Helgeson (1968) pioneered the application of reaction path models to geochemistry. A reaction path in a closed system is defined by the following system of ordinary differential equations

$$\frac{d\Psi_j}{dt} = -\sum_m \widetilde{\nu}_{jm} I_m, \tag{221}$$

and

$$\frac{dn_m}{dt} = I_m, \tag{222}$$

where n_m denotes the number of moles of the mth mineral. These equations are subject to the initial conditions

$$\Psi_j(0) = \Psi_j^0, \tag{223}$$

and

$$n_m(0) = n_m^0, \tag{224}$$

where Ψ_j^0 and n_m^0 designate the initial values of the solute concentrations and number of moles of each mineral, respectively. The solution to these equations defines a reaction path giving the solute and mineral concentrations, $C_j(t)$ and $n_m(t)$, parameterized by the time t. Alternatively, the reaction progress variable ξ may be used in place of the time (Helgeson, 1968).

The concept of a reaction path may also be applied to an open system in which flow of fluid is taking place. Neglecting the partial time derivative in the mass transport equations and considering purely advective transport, implies the stationary state equations

$$v\frac{d\Psi_j}{dx} = -\sum_m \widetilde{\nu}_{jm} I_m, \tag{225}$$

written for a single spatial dimension for pure advective transport. This equation is an ordinary differential equation in the space coordinate alone. Time, however, enters this equation as a parameter to specify the absolute system time state describing the state of alteration of the system.

To see the relation between this equation and the reaction path equations for a closed system, introduce the travel time t' of a fluid packet defined by

$$t' = \frac{x}{v}. \tag{226}$$

The travel time t' represents the time for a packet of fluid to travel a distance x from the inlet. In terms of the travel time, Equation (225) becomes

$$\frac{d\Psi_j}{dt'} = -\sum_m \widetilde{\nu}_{jm} I_m. \tag{227}$$

This equation is formally identical with the closed system reaction path Equation (221) with travel time identified with the absolute time of the closed system. One important difference is that in the flowing system, mineral products are left behind as the fluid packet traverses the flow path and back reaction does not occur. Thus it is not possible to compute the change in porosity at a fixed point in space using the closed system reaction path model. To do this many reaction paths must be considered.

Once a solution to the reaction path equations has been obtained for *fixed* mineral concentrations, the mineral abundances are updated using the explicit finite difference equation

$$\phi_m(x, t + \Delta t) = \phi_m(x, t) + \Delta t \overline{V}_m I_m(x, t), \tag{228}$$

where in this equation t refers to the absolute time. With the revised mineral concentrations, a new reaction path is computed and so on. In this way the aqueous species and mineral concentrations, and hence porosity, can be computed at each point in space as a function of time.

A fundamental difference exists between the closed system and the open system equations. In the closed system the time t refers to the absolute time and applies to both the solution composition and mineral concentrations. However, in the open system formulation, the travel time t' occurs in the transport equations determining the solution composition, whereas the absolute time t enters the mineral mass transfer equations which refer to the mineral concentration at a fixed point in space along the flow path. For an more extensive treatment the interested reader is referred to Lichtner (1988, 1993).

This procedure defines a multiple reaction path extending the concept of a single reaction path to many reaction paths in order to describe the evolution of a system involving time and space (Lichtner, 1988; Lichtner, 1992; Lichtner and Biino, 1992). In the pure advective approach the change in solid concentration during the transient transition period from one stationary state to the next is neglected completely. Two distinct times must be considered that must not be confused: the travel time of a fluid packet along the flow path; and the absolute system time.

In general the time evolution of a reactive transport problem can be represented by a sequence of stationary states. This representation applies to both isothermal and non-isothermal systems, and for transport by diffusion, dispersion and advection. Application to nonisothermal systems requires that the temperature change sufficiently slowly with time compared to the time required for a stationary state to form (Lichtner, 1992). To incorporate diffusion and dispersion, the transient mass conservation equations must be solved and the change in solution monitored at each time step to determine if a stationary state has been reached. The duration of each stationary state depends on the system being considered. Many such stationary states are required to describe the evolution of a system over geologic time spans. The special case of pure advective transport is worth considering separately because of the simplicity of the resulting transport equations.

One interesting result that follows immediately from the quasi-stationary state transport equations, is that for a given Darcy velocity the position of mineral reaction fronts or, equivalently, the widths of mineral alteration zones are independent of the porosity of the porous medium. At first this may seem paradoxical because, for example, if the porosity is decreased the average pore fluid velocity is proportionately increased, and one might expect that therefore reaction zones would also increase in width. However, this is cannot the case because the quasi-stationary state transport equations depend only on the Darcy velocity and not the average pore fluid velocity. The result may be understood by noting that the specific mineral surface area expressed per pore volume is inversely proportional to the porosity which exactly compensates for the increase in average pore fluid velocity as the porosity is increased.

SINGLE COMPONENT SYSTEM

It is instructive to consider a single component system with a one-dimensional flow geometry. Consider the dissolution of a solid phase comprised of a single component which occupies the region $0 \leq x < \infty$. A fluid undersaturated with respect to the solid flows into the column dissolving the solid at a dissolution front. Much of the insight into the reactive transport problem gained for this system can be applied to the general case. The governing equations represent a moving boundary problem. A constant infiltration rate is assumed. An analytical solution is given for the quasi-stationary state approximation.

Transient formulation

The transient formulation to the reactive transport problem for a single component system is given by a set of coupled partial differential equations for the concentrations of the solute species and solid phase. For pure advective transport these equations have the form for the solute concentration

$$\frac{\partial}{\partial t}\phi C + v\frac{\partial C}{\partial x} - \phi D\frac{\partial^2 C}{\partial x^2} = -k's(C - C_{\mathrm{eq}})\theta(x - l(t)), \tag{229}$$

and the solid volume fraction

$$\frac{\partial \phi_s}{\partial t} = \overline{V}_s k's(C - C_{\mathrm{eq}})\theta\,(x - l(t))\,, \tag{230}$$

where a consistent set of units are provided by the following:

C	–solute concentration [moles dm^{-3}],
C_{eq}	–equilibrium concentration [moles dm^{-3}],
v	–Darcy fluid flow velocity [m s^{-1}],
D	–diffusion coefficient [m^2 s^{-1}],
k'	–kinetic rate constant [m s^{-1}],
s	–specific surface area of solid [m^{-1}],
ϕ_s	–solid volume fraction [dimensionless],
$\overline{V}_s$	–solid molar volume [dm^3 mole^{-1}],
$l(t)$	–position of dissolution front [m],
$\theta(x)$	–Heaviside function.

The Heaviside function, defined by

$$\theta(x) = \begin{cases} 1, & x > 0, \\ 0, & \text{otherwise}, \end{cases} \tag{231}$$

occurs in the source/sink term on the right-hand side because only dissolution is considered, and therefore reaction takes place only where the solid is present. The reaction rate may be expressed alternatively as

$$\widehat{I} = -ks(1 - KC) = k's(C - C_{\rm eq}), \tag{232}$$

where K denotes the equilibrium constant for the reaction

$$A \rightleftharpoons A_s. \tag{233}$$

The rate constants k and k' are related by the equation

$$k' = kK = \frac{k}{C_{\rm eq}}. \tag{234}$$

Activity coefficient corrections are neglected.

The transport equation is subject to the initial and boundary conditions:

$$C(x, t = 0) = C_{\rm eq}, \tag{235}$$

$$\phi_s(x, t = 0) = \phi_s^0, \tag{236}$$

and

$$C(x = 0, t) = C_0. \tag{237}$$

Transient and stationary state solution. For the case in which the solid has not dissolved completely at the inlet, $l(t) = 0$, and an analytical solution exists for the transient transport equation. This is given by (Lichtner, 1988)

$$C(x, t) = C_{eq} - \frac{C_{eq} - C_0}{2}\left\{ e^{-qx}\mathrm{erfc}\left[\frac{x - Wvt/\phi}{2\sqrt{Dt}}\right] + e^{(1+W)(vx/2\phi D)}\mathrm{erfc}\left[\frac{x - Wvt/\phi}{2\sqrt{Dt}}\right]\right\}, \tag{238}$$

where

$$q = \frac{v}{2\phi D}(W - 1), \tag{239}$$

and

$$W = \left[1 + \frac{4k's\phi D}{v^2}\right]^{1/2}. \tag{240}$$

The quantity q represents the inverse of the equilibration length, the distance required for the aqueous solution to come to equilibrium with the solid phase. The corresponding stationary state solution is given by

$$C(x, t) = (C_0 - C_{eq})\, e^{-qx} + C_{eq}. \tag{241}$$

It is interesting to note that the stationary state solution is independent of diffusion for large Péclet numbers. Expressing the quantity q in terms of the Péclet number Pe yields

$$q = \frac{\mathrm{Pe}}{2\Delta l}\left\{\sqrt{1+4\frac{q_v \Delta l}{\mathrm{Pe}}}-1\right\}, \tag{242}$$

for some length scale Δl, where

$$\mathrm{Pe} = \frac{v\Delta l}{\phi D}, \tag{243}$$

and q_v represents the inverse equilibration length for pure advective transport (Lichtner, 1988)

$$q_v = \frac{k's}{v}. \tag{244}$$

For large Pe, q may be expanded in a Taylor series to yield

$$q = q_v\left(1-\frac{q_v\Delta l}{\mathrm{Pe}}+2\left(\frac{q_v\Delta l}{\mathrm{Pe}}\right)^2-\cdots\right), \tag{245}$$

and thus $q \simeq q_v$, the pure advective inverse equilibration length. This result has interesting consequences for the numerical solution of the single component transport equations. For high grid Péclet numbers a first order upwinding scheme provides the simplest approach (see Steefel and MacQuarrie, this volume). However, this method is notorious for introducing numerical diffusion into the solution (Leonard, 1979a). The magnitude of numerical diffusion can be shown to be (Fletcher, 1991)

$$D_{\mathrm{num}} = \left(1+\frac{1}{2}\mathrm{Pe}\right)D \simeq \frac{1}{2}\mathrm{Pe}\,D. \tag{246}$$

The larger the Péclet number the greater the numerical diffusion. The corresponding Péclet number becomes

$$\mathrm{Pe}_{\mathrm{num}} = \frac{v\Delta x}{\phi D_{\mathrm{num}}} \simeq 2, \tag{247}$$

where Δx denotes the grid spacing. The numerically computed inverse equilibration length taking into account numerical diffusion is thus equal to

$$q_{\mathrm{num}} = q_v\left(1-\frac{1}{2}q_v\Delta x+\cdots\right). \tag{248}$$

According to this relation the numerical solution is independent of numerical diffusion provided

$$q_v\Delta x \ll 1. \tag{249}$$

But this inequality just states that the pure advective equilibration length be greater than the grid spacing, a condition which generally must be satisfied in order for the continuum hypothesis to be valid. Thus for the single component system it appears that the effects of chemical reaction limit the impact of numerical diffusion on the stationary state solution. For the transient solution numerical diffusion can be expected to be important. However, as shown in the next section, the numerical solution can be represented as a sequence of stationary states with short duration transient time periods between each stationary state. As a result the first order upwinding scheme can be used without fear of overwhelming the solution with numerical diffusion.

Quasi-stationary state approximation

The characteristic time required for the solute concentration to reach a stationary state is of the order (Lichtner, 1991; Lasaga and Rye, 1993)

$$\tau_0 = \frac{\phi}{k's}. \tag{250}$$

By comparison, the characteristic time for substantial change to take place in the solid phase concentration is of the order

$$\tau_s = \frac{\overline{V}_s^{-1}\phi_s^0}{k's\Delta C}, \tag{251}$$

where ϕ_s^0 denotes the initial solid volume fraction, and

$$\Delta C = |C_0 - C_{eq}|. \tag{252}$$

The quantity τ_s gives the time for the solid to completely dissolve at the inlet ($x = 0$), and commencement of the dissolution front to propagate.

For pure advective transport the solute transport equation may be transformed to the dimensionless form

$$\frac{\tau_0}{\tau_s}\frac{\partial C'}{\partial t'} + \frac{\partial C'}{\partial x'} = -C'\theta(x - l(t)), \tag{253}$$

by introducing the dimensionless length and time coordinates

$$x' = \frac{x}{l_{eq}}, \tag{254}$$

$$t' = \frac{t}{\tau_s}, \tag{255}$$

and the dimensionless concentration

$$C' = \frac{C - C_{eq}}{C_0 - C_{eq}}. \tag{256}$$

The quantity l_{eq} denotes the characteristic length scale for the solute to come to equilibrium with the solid phase, equal to

$$l_{eq} = \frac{v}{k's} = \frac{v}{\phi}\tau_0. \tag{257}$$

It follows that the quasi-stationary state approximation is valid provided

$$\tau_s \gg \tau_0, \tag{258}$$

in which case the partial time derivative term may be neglected compared to the other terms in Equation (253). Substituting for τ_s and τ_0, this inequality becomes

$$\frac{\tau_0}{\tau_s} = \frac{\phi\Delta C}{\overline{V}_s^{-1}\phi_s^0} \ll 1. \tag{259}$$

Thus the condition for the quasi-stationary state to be valid is that the solid concentration be much larger than the solute concentration. Note that the rate constant and surface area cancel out and thus the applicability of the quasi-stationary state approximation does not depend on how fast the reaction occurs. For example, for dissolution of quartz at 300°C, $\Delta C \simeq 10^{-3}$ moles L^{-1}, $\overline{V}_{Qtz}=22$ cm^3 mole^{-1}, $\phi \simeq 0.1$, and $\phi_s^0 \simeq 0.9$, then $\tau_0/\tau_s \simeq 10^{-6}$.

Neglecting the partial time derivative in the solute transport equation results in the ordinary differential equation

$$v\frac{dC}{dx} = -k's(C - C_{eq})\theta(x - l(t)). \tag{260}$$

In this equation time enters as a parameter specifying the state of alteration of the solid phase through the position of the reaction front $l(t)$. This equation does not depend on the porosity of the medium, except possibly through the dependency of the surface area on porosity, and therefore, all other conditions being equal, the advance of the dissolution front is the same for different porosities.

Analytical solution. For pure advection the following analytical solution to the single component reactive transport problem is obtained using the quasi-stationary state approximation (Lichtner, 1988; 1992). The more general case is discussed in Lichtner (1988, 1993). The solute concentration is given by

$$C(x;\, t) = \begin{cases} (C_0 - C_{eq})e^{-[x - l(t)]/l_{eq}} + C_{eq}, & [x > l(t)] \\ C_0, & [x < l(t)] \end{cases}, \tag{261}$$

the solid volume fraction by

$$\phi_s(x;\, t) = \begin{cases} \phi_s^0\left(1 - e^{-[x - l(t)]/l_{eq}}\right), & [x > l(t)] \\ 0, & [x < l(t)] \end{cases}, \tag{262}$$

and the position of the reaction front by

$$l(t) = \begin{cases} v\left(\dfrac{C_{eq} - C_0}{\overline{V}_s^{-1}\phi_s^0}\right)(t - \tau_s), & (t > \tau_s) \\ 0, & (t < \tau_s) \end{cases}. \tag{263}$$

The velocity of the front is retarded compared to the average pore velocity v/ϕ by the factor

$$R = \frac{\overline{V}_s^{-1}\phi_s^0}{\phi(C_{eq} - C_0)}. \tag{264}$$

Thus

$$l(t) = \frac{v(t - \tau_s)}{\phi R}, \tag{265}$$

for $t > \tau_s$. Note that the retardation factor is independent of the kinetic rate constant k and surface area s. Thus the velocity of propagation of the dissolution front is independent of kinetics (Ortoleva et al., 1986; Lichtner, 1988; Lichtner, 1991). This is to be expected from overall mass balance considerations. Changing the kinetic rate constant only broadens or sharpens the front, but does not change its velocity which is dictated by mass conservation. These statements apply to multicomponent systems as well, and including diffusive transport (Lichtner, 1988, 1991).

The analytical solution to the single component system is illustrated in Figure 5 for the case of advective and diffusive transport. The host rock is taken as pure quartz with an initial volume fraction Of 0.5, and with a porosity of 50%. A fluid flow velocity of 10 m y^{-1} and an effective diffusion coefficient of 10^{-5} cm^2 s^{-1} is used in the example. The inlet fluid composition is equal to 10^{-4} moles L^{-1} and the equilibrium concentration of quartz is taken as 10^{-3} moles L^{-1} corresponding roughly to 300°C. As can be seen from the figure, the reaction rate is nonzero only over a narrow region equal to the equilibration length which depends on the flow velocity and rate constant. Increasing the kinetic rate constant or surface area decreases the equilibration length. For a sufficiently large rate constant the equilibration length can always be made smaller than the length of a pore. For a mineral which reacts this fast, the continuum model must fail. In such a case a concentration gradient is established in the pore volume and the bulk fluid composition is no longer an accurate measure of the fluid concentration at the mineral surface (Murphy et al., 1989).

Local chemical equilibrium is often considered to be a simplification over a kinetic description. In this approach the kinetic rate law is replaced by an algebraic equation, the mass action equilibrium constraint, reducing the problem to solving fewer partial differential equations. Attempting to describe local equilibrium conditions using a kinetic formulation with the actual experimentally determined rate constant can lead to an extremely stiff set of partial differential equations, requiring inordinately small time steps to achieve a stable solution. However, from a physical point of view it is futile to use local equilibrium in a single continuum model if the equilibration length associated with the reaction becomes smaller than the size of a pore volume. Validity of the single continuum model requires that the equilibration length be large compared to the pore size so that the reaction may sample many pore volumes. Macroscale local equilibrium can be incorporated in a kinetic description by using a sufficiently large effective rate constant such that equilibrium is achieved within a few node spaces. However, if the actual intrinsic rate of the reaction is sufficiently rapid to achieve local equilibrium on a microscale this may not give an accurate description of the system, leading to a breakdown of a single continuum representation of the porous medium. A hierarchical continuum description of porous media is required, greatly complicating the model and the data necessary to support it.

SPECIAL TOPICS

Charge conservation

One of the fundamental properties of an aqueous solution is that it is electrically neutral on a macroscopic scale. Therefore, for the mass transport equations to properly represent such systems, they must conserve charge. The analysis which follows applies to a system in which $N_k = 0$, that is there are no kinetic aqueous species. This restriction may be easily relaxed, however. With the flux defined by Equation (23), the mass conservation equation conserves electrical charge only if all species are electrically neutral, or if the diffusion coefficients are the same for all species with nonzero charges. To see this multiply Equation (21) by the charge of the ith species z_i, and sum over all species to give

$$\frac{\partial}{\partial t}\left(\phi\rho_Q\right) + \nabla \cdot \boldsymbol{i} = 0. \tag{266}$$

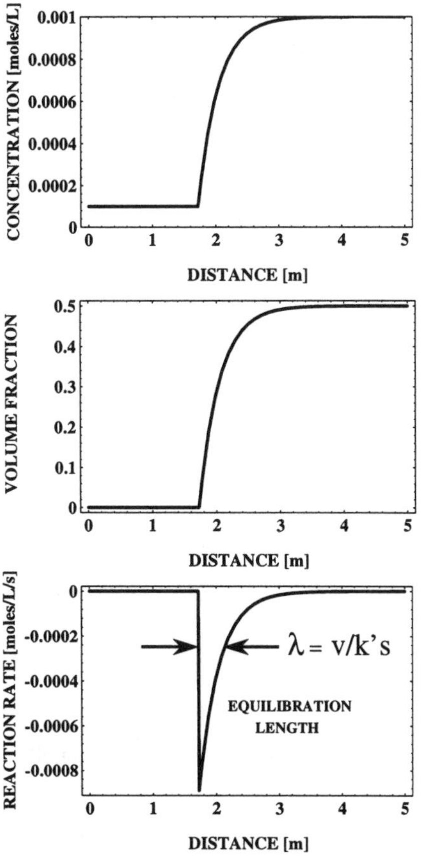

Figure 5. Solute concentration, volume fraction, and reaction rate for the quasi-stationary state approximation for a single component system plotted as a function of distance for an elapsed time of 5,000 years. See text for the values of parameters used in the calculation.

The total charge density ρ_Q is defined by

$$\rho_Q = F\sum_{i=1}^{N} z_i C_i = F\sum_{j=1}^{N_c} z_j \Psi_j, \tag{267}$$

where the Faraday constant F is introduced (F = 96487 coulombs/mole), so that ρ_Q has units of coulombs per unit volume. The current density $\boldsymbol{i}$ is defined by

$$\boldsymbol{i} = F\sum_{i=1}^{N} z_i \boldsymbol{J}_i = F\sum_{j=1}^{N_c} z_j \boldsymbol{\Omega}_j, \tag{268}$$

with units of coulombs per unit area per second, or amperes per unit area (1 ampere = 1 coulomb/second). It is assumed that the source/sink term R_i conserves charge and hence

$$\sum_{i=1}^{N} z_i R_i = 0. \tag{269}$$

If the source/sink term is a result of chemical reactions taking place in the system, this statement follows from the fact that chemical reactions conserve charge. Substituting Equation (23) for the solute flux into Equation (268) gives an expression for the current density in terms of the charge density and the gradient of the individual solute concentrations

$$\boldsymbol{i} = -F\phi\tau \sum_i z_i D_i \nabla C_i + \boldsymbol{v}\rho_Q. \tag{270}$$

As can be seen from this equation, the current density cannot be expressed as a function of the charge density ρ_Q alone unless the diffusion coefficients are the same for all charged species, $D = D_i$. If this is the case, the current density reduces to the expression

$$\boldsymbol{i} = (-\tau\phi D\nabla + \boldsymbol{v})\,\rho_Q, \tag{271}$$

and the partial differential equation for the evolution of the charge density becomes a linear function of the charge density itself

$$\frac{\partial}{\partial t}(\phi\rho_Q) + \nabla\cdot\left[(-\tau\phi D\nabla + v)\,\rho_Q\right] = 0. \tag{272}$$

Consequently, it follows that if the initial charge density vanishes,

$$\rho_Q(\boldsymbol{r},\, t=0) = 0, \tag{273}$$

and if the boundary value also vanishes

$$\rho_Q(\boldsymbol{r}_b,\, t) = 0, \tag{274}$$

where $\boldsymbol{r}_b$ refers to points on the boundary of the spatial domain, then the only possible solution satisfying these conditions is the identically zero function ($\rho_Q(\boldsymbol{r},\, t) \equiv 0$), and hence charge is conserved. Note that with unequal diffusion coefficients, the current density cannot be expressed solely as a function of the charge density and, therefore, even if the charge density is initially zero, it will not remain so as the system evolves in time. This may be demonstrated more clearly by writing the charge density equation for unequal diffusion coefficients in the form

$$\frac{\partial}{\partial t}(\phi\rho_Q) + \nabla\cdot\left[(-\tau\phi\overline{D}\nabla + v)\,\rho_Q\right] = -F\tau\phi\sum_{li}\left(\frac{1}{N} - \delta_{il}\right) D_l z_i \nabla C_i, \tag{275}$$

obtained by adding and subtracting the term $-F\tau\phi\overline{D}\nabla\rho_Q$, where

$$\overline{D} = \frac{1}{N}\sum_i D_i, \tag{276}$$

denotes the average diffusion coefficient. Thus for unequal diffusion coefficients, an effective source/sink term appears on the right-hand side. This is a consequence of species diffusing at different rates because of their different diffusion coefficients, thereby creating a charge imbalance. Physically this does not happen because electrical forces maintain charge balance. Note that the source/sink term vanishes if all diffusion coefficients are equal.

Species-dependent diffusion coefficients. As noted the mass transport equations do not conserve charge as written in the form of Equation (21) if the diffusion coefficients are different for differently charged ions. For example, the hydrated hydrogen ion has a diffusion coefficient which is an order of magnitude larger than most other ions (Newman, 1973). In order to obtain transport equations which conserve charge in this case, an additional term must be added to the solute flux providing for electrochemical migration (Haase, 1969; Newman, 1973). Electrochemical migration is caused by the presence of an electric field which is produced locally by the different rates of diffusion of differently charged ions. The solute flux $\boldsymbol{J}_i$, in general, consists of contributions from advective, diffusive and dispersive transport in addition to an electrochemical migration term. Ignoring dispersion, the flux has the form (Haase, 1969; Newman, 1973)

$$\boldsymbol{J}_i = -\tau\phi z_i \frac{D_i C_i}{RT} F\nabla\psi - \tau\phi D_i \left(\nabla C_i + C_i \nabla \ln \gamma_i\right) + \mathbf{q} C_i, \tag{277}$$

where the first term refers to electrochemical migration, the second term to aqueous diffusion, and the last term to advective transport. Here z_i, γ_i and D_i denote the charge, activity coefficient and diffusivity of the ith species, respectively. The quantity ψ represents the electrical potential, τ refers to the tortuosity of the porous medium, and $\mathbf{q}$ denotes the Darcy fluid velocity. The electrochemical migration term is proportional to the ion charge, the electric field ($\boldsymbol{E} = -\nabla\psi$), and the mobility $D_i C_i / RT$. The potential ψ is determined in such a manner as to ensure charge conservation.

With species-dependent diffusion coefficients, and taking into account corrections due to activity coefficients, the expression for the generalized flux, Equation (140), can be written in the form (Lichtner, 1995)

$$\Omega_j = -F\tau\phi\Psi_j^\epsilon \frac{\nabla\psi}{RT} - \tau\phi\left(\Gamma_j^D + \Gamma_j^\gamma\right) + \mathbf{q}\Psi_j, \tag{278}$$

obtained by inserting the expression for the flux defined in Equation (277) into equation Equation (140). The quantities Ψ_j^ϵ and Γ_j^D are defined by the expressions

$$\Psi_j^\epsilon = z_j D_j C_j + \sum_i \widetilde{\nu}_{ji} z_i D_i C_i, \tag{279}$$

and

$$\Gamma_j^D = D_j \nabla C_j + \sum_i \widetilde{\nu}_{ji} D_i \nabla C_i. \tag{280}$$

Activity coefficient corrections are accounted for by the term containing Γ_j^γ defined by

$$\Gamma_j^\gamma = D_j C_j \nabla \ln \gamma_j + \sum_i \widetilde{\nu}_{ji} D_i C_i \nabla \ln \gamma_i. \tag{281}$$

Demanding that the solution current density $\boldsymbol{i}$ defined by Equation (268) vanish, requires that the gradient in the electric potential $\nabla\psi$ be equal to

$$-\nabla\psi = \frac{1}{\kappa}\left\{\tau\phi F \sum_j z_j \left(\Gamma_j^D + \Gamma_j^\gamma\right)\right\}, \tag{282}$$

obtained by writing out Equation (268) using Equation (278), and solving for $\nabla\psi$. The inverse of the quantity κ, defined by the expression

$$\kappa = \frac{\tau\phi F^2}{RT} \sum z_j \Psi_j^\epsilon, \tag{283}$$

is referred to as the generalized Debye length. For concentrated solutions the Debye length κ^{-1} becomes small and the effects of the potential negligible. Using Equation (282), the generalized flux may be written as

$$\Omega_j = -\tau\phi \sum_l \beta_{jl} \left(\Gamma_j^D + \Gamma_j^\gamma\right) + \boldsymbol{q}\Psi_j, \tag{284}$$

by eliminating $\nabla\psi$. The matrix β_{jl} is a projection operator ($\beta^2 = \beta$) defined by

$$\beta_{jl} = \delta_{jl} - z_l \omega_j, \tag{285}$$

with

$$\omega_j = \frac{\Psi_j^\epsilon}{\sum_l z_l \Psi_l^\epsilon}. \tag{286}$$

Thus in contrast to the expression for the flux with species-independent diffusion coefficients, coupling terms now occur between the concentration gradients of the different primary species. These gradients are required to maintain electrical balance within the aqueous solution. Making use of the above choice of the electrical field so that the current density vanishes identically, it follows that the charge conservation equation for constant porosity reduces to

$$\frac{\partial}{\partial t}(\phi\rho_Q) = \phi\frac{\partial}{\partial t}(\rho_Q) = 0. \tag{287}$$

As a consequence for electrically neutral initial and boundary conditions, $\rho_Q = 0$ for all time, and charge is conserved.

Charge conservation and sorption. To investigate the ability of various sorption models to conserve charge when combined with solute transport, two distinct models are considered: ion-exchange and surface complexation. First a two-component system is considered involving the sorption of Na^+ and Ca^{2+}.

Ion-exchange. Ion-exchange of Na^+ with Ca^{2+} is described by the reaction

$$2\,XNa + Ca^{2+} \rightleftharpoons X_2Ca + 2\,Na^+. \tag{288}$$

In this reaction X^- denotes a surface site. Charge is conserved separately in both the aqueous and surface phases. Representing the rate of the exchange reaction by I_{ex}, the following transport equations hold:

$$Ł[Na^+] = 2\,I_{ex}, \tag{289a}$$

and

$$Ł[Ca^{2+}] = -I_{ex}, \tag{289b}$$

for the aqueous species Na^+ and Ca^{2+} with their corresponding concentrations denoted by the square brackets [$\cdots$], where Ł denotes the differential operator defined in Equation (88). In this simple example possible aqueous complexing reactions are omitted. This does not, however, affect the utility of the example regarding charge conservation.

It follows immediately that for the ion-exchange reaction, charge is conserved in the aqueous solution. The total charge density in solution contributed by cations is equal to

$$Q = [Na^+] + 2\,[Ca^{2+}]. \tag{290}$$

Electroneutrality is achieved by an equal and opposite contribution from anions. Because the anion transport equations are not coupled to the cation transport equations, in order for the solution to remain electrically neutral, both sets of transport equations must conserve charge individually. Therefore only the cations need be explicitly considered. It follows that

$$Ł Q \;=\; Ł[\mathrm{Na}^+] + 2Ł[\mathrm{Ca}^{2+}] \;=\; 2\,I_{\mathrm{ex}} - 2\,I_{\mathrm{ex}} \;=\; 0. \tag{291}$$

As a consequence charge is conserved and the aqueous solution will remain electrically neutral with increasing time if it is electrically neutral initially.

Surface complexation model. For sorption models with a variable number of unoccupied sites, however, charge within the aqueous solution is not in general conserved by the transport equations. This is true of surface complexation models (Dzombak and Morel, 1990). Consider the following surface complexation reactions:

$$\mathrm{X}^- + \mathrm{Na}^+ \;\rightleftharpoons\; \mathrm{XNa}, \tag{292a}$$

and

$$2\,\mathrm{X}^- + \mathrm{Ca}^{2+} \;\rightleftharpoons\; \mathrm{X_2Ca}, \tag{292b}$$

with rates I_{Na} and I_{Ca}, respectively, where X^- denotes a negatively charge surface site, and XNa and X_2Ca represent sorbed species. The mass transport equations read:

$$Ł[\mathrm{Na}^+] \;=\; -I_{\mathrm{Na}}, \tag{293a}$$

and

$$Ł[\mathrm{Ca}^{2+}] \;=\; -I_{\mathrm{Ca}}. \tag{293b}$$

The concentration of sorbed species satisfy the mass transfer equations

$$\frac{\partial}{\partial t}[\mathrm{XNa}] \;=\; I_{\mathrm{Na}}, \tag{294a}$$

$$\frac{\partial}{\partial t}[\mathrm{X_2Ca}] \;=\; I_{\mathrm{Ca}}, \tag{294b}$$

and the concentration of unoccupied sites the equation

$$\frac{\partial}{\partial t}[\mathrm{X}^-] \;=\; -I_{\mathrm{Na}} - 2\,I_{\mathrm{Ca}}. \tag{295}$$

It follows that the total number of sites, equal to

$$\omega_{\mathrm{X}} \;=\; [\mathrm{X}^-] + [\mathrm{XNa}] + 2\,[\mathrm{X_2Ca}], \tag{296}$$

is conserved:

$$\frac{\partial \omega_{\mathrm{X}}}{\partial t} \;=\; \frac{\partial}{\partial t}[\mathrm{X}^-] + I_{\mathrm{Na}} + 2\,I_{\mathrm{Ca}} \;=\; 0, \tag{297}$$

according to Equation (295), as must be the case. However, total charge within the aqueous solution is not conserved:

$$Ł Q \;=\; -I_{\mathrm{Na}} - 2\,I_{\mathrm{Ca}} \;=\; \frac{\partial}{\partial t}[\mathrm{X}^-] \neq 0. \tag{298}$$

Thus the nonconservation of charge in the surface complexation model is directly related to the rate of change with time of the unoccupied site density.

This result may seem surprising since the chemical reactions themselves conserve charge. The reason for the lack of charge conservation in the aqueous solution is because the surface sites and solute species obey different transport equations, much like the problem of maintaining charge balance if solute species diffuse with different diffusion coefficients. A flux term is not present in the site equation as the sites are fixed to the solid substrate. Thus the sites are not transported at the same rate as the solute species and a charge imbalance results.

Multicomponent system. This result can be extended easily to the general case of a multicomponent system. The generalized form of the sorption reactions given in Equations (292a) and (292b) can be written as

$$z_i \mathrm{X}^- + \sum_{j=1}^{N_c} \overline{\nu}_{ji} A_j \rightleftharpoons \mathrm{X}_{z_i} A_i, \tag{299}$$

where z_i denotes the charge of species A_i. Barred quantities refer to adsorbed species. The rate of this reaction is denoted by $\overline{I}_i$. The corresponding transport equations become, allowing for both homogeneous and heterogeneous reactions,

$$\frac{\partial}{\partial t}\left\{\phi \Psi_j + \sum_i \overline{\nu}_{ji} \overline{C}_i\right\} + \nabla \cdot \Omega_j = -\sum_m \widetilde{\nu}_{jm} I_m, \tag{300}$$

for primary species,

$$\frac{\partial \overline{C}_i}{\partial t} = \overline{I}_i, \tag{301}$$

for adsorbed species with $\overline{C}_i = [\mathrm{X}_{z_i} A_i]$, and

$$\frac{\partial [\mathrm{X}^-]}{\partial \mathrm{t}} = -\sum_i z_i \overline{I}_i, \tag{302}$$

for unoccupied surface sites. The total number of sites are equal to

$$\omega_X = [\mathrm{X}^-] + \sum_i z_i \overline{C}_i. \tag{303}$$

The total site concentration obeys the conservation law

$$\frac{\partial \omega_X}{\partial t} = \frac{\partial [\mathrm{X}^-]}{\partial \mathrm{t}} + \sum_i z_i \frac{\partial \overline{C}_i}{\partial t} = 0, \tag{304}$$

as follows from Equations (301) and (302).

The transport equation for conservation of charge can be derived by multiplying Equation (300) by the charge z_j of the jth primary species and summing over all primary species to give

$$\frac{\partial}{\partial t}\left(\phi \rho_Q + F \sum_{ij} z_j \overline{\nu}_{ji} \overline{C}_i\right) + \nabla \cdot \boldsymbol{i} = 0, \tag{305}$$

where the total charge in solution ρ_Q and current density $\boldsymbol{i}$ are defined by Equations (267) and (268). The right hand side of Equation (305) vanishes because mineral precipitation/dissolution reactions, as all chemical reactions, conserve charge. The second term in brackets on the left-hand side of Equation (305) represents the total charge on the surface. Referring to this quantity as $\overline{\rho}_Q$ it follows that

$$\overline{\rho}_Q = F\sum_{ji} z_j \overline{\nu}_{ji} \overline{C}_i = F\sum_i z_i \overline{C}_i. \tag{306}$$

With this definition, Equation (305) can be expressed as

$$\frac{\partial}{\partial t}\left(\phi \rho_Q\right) + \nabla \cdot \boldsymbol{i} = -\frac{\partial \overline{\rho}_Q}{\partial t}, \tag{307}$$

with

$$\frac{\partial \overline{\rho}_Q}{\partial t} = -F\frac{\partial [\mathrm{X}^-]}{\partial t}. \tag{308}$$

Charge is conserved only if the right-hand side of this equation vanishes identically. However, surface charge alone is *not* conserved in surface complexation models as it is in ion-exchange models. For example in the electric double layer model, surface charge plus the charge contained within the diffuse layer is conserved. This suggests that one way to correct for nonconservation of charge is to include in the mass transport equations the contribution from nonspecific adsorbed ions contained in the diffuse layer (Borkovec and Westall, 1983; Parkhurst, 1995). A difficult question to access, is what portion of the double layer nonspecific adsorbed ions are transported with the bulk solution, and what portion remain attached to the surface. The shear plane at which this occurs defines the zeta potential. At low ionic strengths the double layer may become quite large becoming a sizeable fraction of the pore diameter, possibly even resulting in overlapping double layers from opposite pore walls of very small pores. For the simpler surface complexation models which do not explicitly include the electric double layer, there does appear to be any rigorous way to conserve charge when these models are combined with solute transport.

Interpreting results of reactive transport simulations

Although the canonical representation of chemical reactions simplifies the mass conservation equations and allows easy incorporation of local equilibrium reactions, it is not a particularly useful scheme for interpreting the results of a reactive transport simulation. For this it is often useful to determine the overall reaction or reactions which summarize the coupled system of reactions used in the calculation. To analyze a set of reactions which form a subset of the total number of reactions in the system written in the canonical form

$$\sum_{j=1}^{N_c} \widetilde{\nu}_{jm} A_j \rightleftharpoons A_m, \tag{309}$$

multiply each reaction by its rate I_m, and sum over all reactions in the subset to give the overall reaction

$$\sum_{j=1}^{N_c} \alpha_j(\boldsymbol{r}, t) A_j \rightleftharpoons A_{m_0} + \sum_{m \neq m_0} \alpha_m(\boldsymbol{r}, t) A_m, \tag{310}$$

in which the mineral A_{m_0} has been chosen arbitrarily to have a unit reaction coefficient. The reaction coefficients α_j and α_m, which now depend in general on time and space in accordance with the reaction rates, are defined as

$$\alpha_j(\mathbf{r},\, t) \;=\; \sum_{m \neq m_0} \widetilde{\nu}_{jm} \frac{I_m(\mathbf{r},\, t)}{I_{m_0}(\mathbf{r},\, t)}, \tag{311}$$

and

$$\alpha_m(\mathbf{r},\, t) \;=\; \frac{I_m(\mathbf{r},\, t)}{I_{m_0}(\mathbf{r},\, t)}. \tag{312}$$

The overall reaction describes a transformation among one or more minerals and the chosen set of primary species. As indicated, the coefficients α_j and α_m for the overall reaction are generally functions of time and space, unlike the stoichiometric coefficients which are definite constants, because the reaction rates generally vary with time and space. The reaction coefficients are, however, known functions determined by solving the mass transport equations. For the overall reaction to properly describe the changes taking place in the system, it is important that a set of primary species be chosen which represent the dominant species in solution.

For example, consider the two simultaneous reactions

$$K^+ + Al^{3+} + 3SiO_2 + 2H_2O - 4H^+ \;\rightleftharpoons\; KAlSi_3O_{8(s)}, \tag{313a}$$

$$2Al^{3+} + 2SiO_2 + 5H_2O - 6H^+ \;\rightleftharpoons\; Al_2Si_2O_5(OH)_{4(s)}, \tag{313b}$$

representing dissolution of K-feldspar and precipitation of kaolinite with corresponding reaction rates I_{Kf} and I_{Ka}, respectively. Multiplying by their respective rates and adding yields the overall reaction

$$\begin{aligned} K^+ + (2\alpha+1)\, Al^{3+} + (2\alpha+3)\, SiO_2 + (5\alpha+2)\, H_2O - (6\alpha+4)\, H^+ \\ \rightleftharpoons\; KAlSi_3O_{8(s)} + \alpha\, Al_2Si_2O_5(OH)_{4(s)}, \end{aligned} \tag{314}$$

where α is defined as the ratio of the kaolinite and K-feldspar reaction rates

$$\alpha \;=\; \frac{I_{Ka}}{I_{Kf}}. \tag{315}$$

If the dominant aluminum species is $Al(OH)_4^-$, which is related to Al^{3+} by the reaction

$$Al^{3+} + 4\,H_2O - 4\,H^+ \;\rightleftharpoons\; Al(OH)_4^-, \tag{316}$$

then the overall reaction becomes

$$\begin{aligned} K^+ + (2\alpha+1)\, Al(OH)_4^- + (2\alpha+3)\, SiO_2 - (3\alpha+2)\, H_2O + 2\alpha\, H^+ \\ \rightleftharpoons\; KAlSi_3O_{8(s)} + \alpha\, Al_2Si_2O_5(OH)_{4(s)}, \end{aligned} \tag{317}$$

For example, if $\alpha = -1/2$, K-feldspar is replaced by kaolinite by conserving aluminum. Note that for this special case, both overall reactions, Equation (314) and Equation (317), give the same pH dependence.

Adding a third reaction, that of precipitation of gibbsite

$$Al^{3+} - 3\,H^+ + 3\,H_2O \;\rightleftharpoons\; Al(OH)_{3(s)}, \tag{318}$$

results in the overall reaction

$$K^+ + (2\alpha+\beta+1)\,Al^{3+} + (2\alpha+3)\,SiO_2 + (5\alpha+3\beta+2)\,H_2O - (6\alpha-3\beta+4)\,H^+ \rightleftharpoons KAlSi_3O_{8(s)} + \alpha\,Al_2Si_2O_5(OH)_{4(s)} + \beta\,Al(OH)_{3(s)}. \quad (319)$$

The parameter β is defined as the ratio of the gibbsite and K-feldspar reaction rates

$$\beta = \frac{I_{Gbs}}{I_{Kf}}. \quad (320)$$

Aluminum is now conserved provided the reaction rates satisfy the relation

$$2\alpha+\beta+1 = 0. \quad (321)$$

From the overall reaction it is possible to deduce immediately the expected changes in the pH and other species at that particular time and position in space where the reaction coefficients are evaluated. By correlating the computed changes in pH etc. with the overall reaction, one is thus able to obtain an intuitive "understanding" of the predicted system behavior. As an example of constructing overall reactions encountered in reaction path simulations of supergene enrichment see Lichtner and Biino (1992).

Inverse problem

The inverse problem is based on the integrated form of the mass conservation equations applied to a stationary state (see Glynn and Brown this volume). The purpose behind the inverse problem is to deduce the mass transfer taking place along the flow path lying between two observation points by measuring the change in concentration between the two points. The assumption of a stationary state implies that the solute concentrations and mineral reaction rates vary only with distance along a flow path, but are constant with time. In addition, it implies that the velocities of reaction fronts are sufficiently slow that they may be approximated as zero over the time interval required for fluid to flow between the observation points. Thus the positions of mineral alteration zones remain frozen during the observation period. Integrating the mass conservation equations along a flow path (streamline) from point A to point B yields the global mass conservation equation

$$\frac{d}{dt}\int_A^B \phi\left(s_l\Psi_j^l + s_g\Psi_j^g\right)dx + \Omega_j^B - \Omega_j^A = -\sum_m \int_A^B \widetilde{\nu}_{jm} I_m\, dx, \quad (322)$$

where $\Omega_j^{A,\,B}$ denotes the flux evaluated at points A and B, and x refers to the distance along the streamline considered as a straight line along the x-axis for the purposes of this discussion. Under the assumption of a stationary state the time derivative vanishes. Further, assuming pure advective transport the flux simplifies to

$$\Omega_j = q_l\Psi_j^l + q_g\Psi_j^g, \quad (323)$$

with aqueous and gaseous flow rates q_l and q_g. For a pure aqueous fluid phase, Equation (322) reduces to

$$v\left(\Psi_j^B - \Psi_j^A\right) = -\sum_m \widetilde{\nu}_{jm} \int_A^B I_m\, dx, \quad (324)$$

where Ψ_j^A, Ψ_j^B represent Ψ_j evaluated at the endpoints A and B. The remaining task is to evaluate the integral over the mineral reaction rates. Assuming that the mth mineral occupies regions $[x_m^{(i-)}, x_m^{(i+)}]$, with $i = 1, 2, \ldots$ between points A and B, then the integral over the reaction rate becomes

$$\int_A^B I_m\,dx = \sum_i \int_{x_m^{(i-)}}^{x_m^{(i+)}} I_m\,dx, \tag{325}$$

where the integral on the right-hand side extends only over those regions in which the mineral is actually present. In the inverse problem, the flow velocity v and the solution composition at the end points A and B of the flow path are presumed known. Defining the change in the total concentration of the jth primary species between points A and B, $\Delta_{AB}\Psi_j$, as

$$\Delta_{AB}\Psi_j = \Psi_j^B - \Psi_j^A, \tag{326}$$

and the mineral mass transfer coefficient α_m as

$$\alpha_m = -\frac{1}{v}\sum_i \int_{x_m^{(i-)}}^{x_m^{(i+)}} I_m\,dx, \tag{327a}$$

$$= -\frac{1}{v}\frac{d}{dt}\sum_i \int_{x_m^{(i-)}}^{x_m^{(i+)}} \overline{V}_m^{-1}\phi_m\,dx, \tag{327b}$$

yields the following set of simultaneous linear equations for the coefficients α_m

$$\sum_m \widetilde{\nu}_{jm}\alpha_m = \Delta_{AB}\Psi_j. \tag{328}$$

The transfer coefficients α_m have units of moles per unit volume. The minus sign appearing in the definition of α_m implies that α_m is positive for net dissolution or transfer of mass to the aqueous solution, and negative for precipitation or mass transfer to the solid phase. Generally there does not exist a unique solution to the inverse problem. Note that determining the mass transfer coefficients α_m does not provide information for the actual mineral reaction rates or the reaction zone widths of the primary and secondary minerals. For more details on solving Equation (328), the interested reader is referred to Glynn and Brown (this volume).

APPLICATIONS

This section presents several examples of applications of the reactive transport equations to redox problems, reaction in heterogeneous porous media, and hydrothermal systems. The calculations are carried out using the computer code MULTIFLO (Lichtner and Seth, 1996b; Seth and Lichtner, 1996). This code describes two-phase fluid flow and heat flow coupled to multicomponent reactive transport of aqueous and gaseous phases in 1, 2 or 3 spatial dimensions. The flow and reactive transport algorithms are sequentially coupled. Mineral reactions are described using kinetic rate laws. Aqueous complexing, sorption, and redox reactions are included in the code. A modified form of the EQ3/6 database (Wolery, 1990) is used in which any consistent set of primary species may be selected. The two-phase flow and heat equations are solved using a fully implicit algorithm employing upstream weighting. Mineral mass transfer equations are solved using a explicit finite difference scheme taking into account the quasi-stationary state approximation. The code has several different solution algorithms available for solving the primary species transport equations including fully implicit, operator splitting and explicit finite difference schemes. The explicit algorithm has an option for using the Leonard-TVD method for describing high Péclet number flows. The Leonard-TVD method is based on a flux limiter to ensure monotonicity of the solution combined with Leonard's third-order scheme (Van Leer, 1977; Leonard, 1979a, b; Leonard, 1984) as developed by Gupta et al. (1991). Without the flux limiter, the Leonard scheme alone can yield negative concentrations (see Steefel and MacQuarrie, this volume). A tridiagonal solver for 1D problems, direct D4 solver for small 2D problems, and the conjugate gradient solver WATSOLV (VanderKwaak et al., 1995), are available in the code.

With the exception of the Poços de Caldas redox front example which is compared qualitatively to field observations, the examples presented in this section should be viewed more as numerical experiments than predictions of actual field situations.

Application to acid mine drainage and pyrite oxidation

Application of the multicomponent mass transport equations to the problem of acid mine drainage requires two-phase flow and transport equations coupled to the oxidation of pyrite in a partially saturated porous medium. Pyrite oxidation in abandoned mines and mine tailings causes the production of acid which in turn leads to mobilization of heavy metals with consequent pollution of the environment. As oxidizing water infiltrates through a column of rock containing pyrite, oxygen is consumed by the pyrite resulting in the precipitation of iron oxide and formation of acid. A partially saturated environment provides a continuous source of oxygen limited only by its rate of diffusion through the gas phase. The net amount of acid produced depends on the neutralization capacity of the gangue minerals in the host rock, the rate of consumption of oxygen by pyrite, and the rate of diffusion of oxygen through the partially saturated zone. Pyrite may also be oxidized by ferric iron. The reduction of ferric to ferrous iron requires the presence of bacteria to catalyze the reaction.

The rate controlling step in pyrite oxidation has not been clearly identified because of the vast number of possible reaction mechanisms (Basolo and Pearson, 1967). For a review of the literature of pyrite oxidation by an aqueous solution see Nordstrom (1982). Rate laws for the oxidation of pyrite have been determined at low pH (~ 2) characteristic of acid mine drainage conditions by Singer and Stumm (1970) and Wiersma and Rimstidt (1982). In these studies it was found that under low pH conditions ferric iron is the major oxidant of pyrite. This result was recently corroborated by Moses et al. (1987) who demonstrated that under both aerobic and anaerobic environments, ferric iron was the direct oxidant of pyrite.

Stumm and Morgan (1981) describe different mechanisms for pyrite oxidation. One mechanism involves oxidation by O_2:

$$FeS_{2(s)} + \frac{7}{2} O_2 + H_2O \rightarrow Fe^{+2} + 2\,SO_4^{2-} + 2\,H^+. \tag{329}$$

Another possible mechanism for the oxidation of pyrite is reaction with Fe(III) as the oxidant according to the reaction

$$FeS_{2(s)} + 14\,Fe^{3+} + 8\,H_2O \rightarrow 15\,Fe^{2+} + 2\,SO_4^{2-} + 16\,H^+. \tag{330}$$

This latter reaction is much more rapid than oxidation by O_2, and results in a greater production of acid per mole of pyrite oxidized. Stumm and Morgan (1981) note that the oxidation of ferrous to ferric iron by the reaction

$$Fe^{2+} + \frac{1}{4} O_2 + H^+ \rightleftharpoons Fe^{3+} + \frac{1}{2} H_2O, \tag{331}$$

is extremely slow at low pH in the absence of the microorganisms *Thiobacillus* and *Ferrobacillus ferrooxidans* which act as catalyses to speed up the reaction. Hence this reaction controls the supply of Fe(III) and is consequently the rate limiting step in the oxidation of pyrite by the mechanism given in Equation (330). Ferrous iron also may be fixed by precipitation of ferrihydrite according to the reaction

$$Fe^{3+} + 3\,H_2O \rightleftharpoons Fe(OH)_{3(s)} + 3\,H^+, \tag{332}$$

releasing additional acid.

For a flow model both reaction mechanisms given in Equations (329) and (330) may occur simultaneously, but in different regions of space. To illustrate this a hypothetical example is considered in which local equilibrium is assumed for the Fe(III)–Fe(II) redox couple. The effects of bacteria *Thiobacillus* and *Ferrobacillus ferrooxidans* on pyrite oxidation are not directly considered; however, the assumption of homogeneous equilibrium of redox species provides an endmember case of their catalytic effect and allows the reduction of ferric iron to proceed at local equilibrium rates. Thus it is assumed that bacteria are present to catalyze the oxidation of Fe(II) to Fe(III) at the maximum rate possible in accordance with thermodynamics. In this example the host rock is represented by weathered material composed of the gangue minerals quartz, K-feldspar and kaolinite with minor amounts of pyrite. The rock composition used in the calculation is 4% pyrite, 20% K-feldspar and kaolinite, and 40% quartz providing only a weak neutralizing capacity of the acid produced by pyrite oxidation. The porosity is 16%. The minerals are assumed to be distributed uniformly throughout the computational domain with the exception of pyrite which is absent from the first 1 m of the profile. Gaseous diffusion coefficients in porous media are discussed by Troeh et al. (1982).

To investigate the effect of gaseous diffusion of oxygen on the formation of acid as pyrite is oxidized, a water table is presumed to be present at a depth l. The liquid saturation profile above the water table is determined using the steady state mass conservation equation for incompressible flow

$$\frac{dv(z)}{dz} = 0, \tag{333}$$

where the fluid flow velocity is given by Darcy's law written in terms of the hydraulic head h:

$$v(z) = -K_{\text{sat}}\kappa_r\left(\frac{dh}{dz} - 1\right). \tag{334}$$

Here K_{sat} denotes the saturated rock hydraulic conductivity, and κ_r the relative liquid permeability. The van Genuchten relations, Equations (177) and (180), may be used to relate relative permeability and capillary pressure to saturation. Integrating the mass conservation equation once yields

$$\int_0^z \frac{dv(z')}{dz'}\,dz' = v(z) - v(0). \tag{335}$$

Imposing a constant infiltration rate v_0 at the top of the weathering column implies $v(0) = v_0$. Substituting Equation (334) for $v(z)$ yields an ordinary differential equation for the hydraulic head:

$$\frac{dh(z)}{dz} = 1 - \frac{v_0}{K_{\text{sat}}\kappa_r[h(z)]}. \tag{336}$$

This equation, referred to as the steady state Richards equation, is subject to the boundary condition $h(l) = 0$, where l denotes the depth of the water table. To solve this equation the relative permeability must be known as a function of pressure or hydraulic head. This relation is generally nonlinear and therefore this equation must be integrated numerically. To do this it is convenient to use the program *Mathematica* (Wolfram, 1991). First define the relative permeability function `krel` by combining Equations (180) and (177) to yield

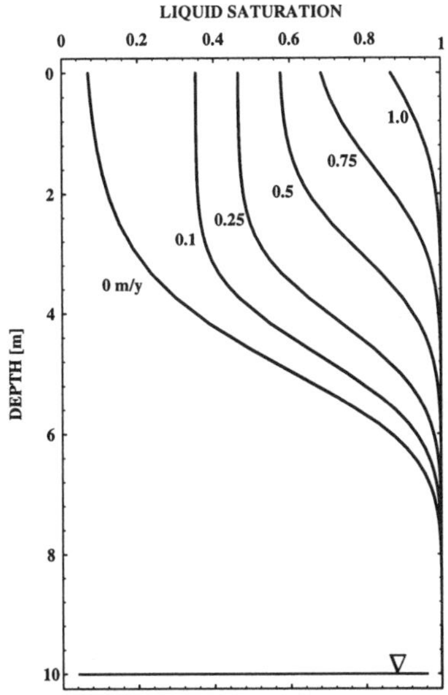

Figure 6. Liquid saturation above the water table plotted as a function of depth for different infiltration rates as calculated from Equation (336) using van Genuchten relations for relative permeability.

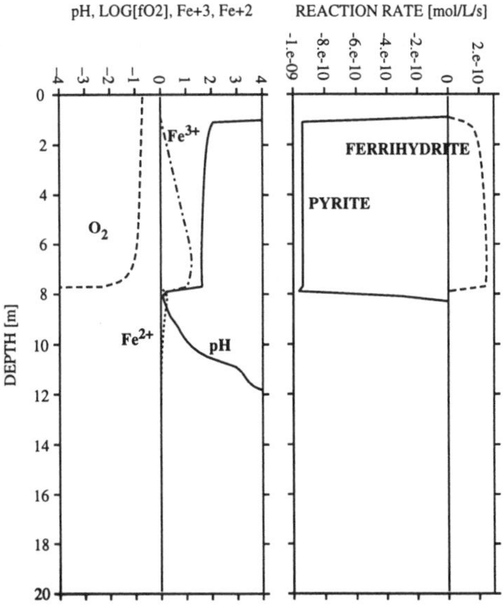

Figure 7. The pH, oxygen fugacity, and concentration of Fe^{2+} and Fe^{3+} in molal units (left panel), and reaction rates of pyrite (solid line) and ferrihydrite (dashed line).

```
krel[h_] :=
  If[
     h>=0, 1,
     (1-(-a h)^(m-1)(1+(-a h)^m)^-(1-1/m))^2 (1+(-a h)^m)^((1/m-1)/2)
   ];
```

where a is identical to α in Equation (177). The steady state Richards equation may be solved using the *Mathematica* function `NDSolve`:

```
hz = NDSolve[{h'[z]+v0/ksat/krel[h[z]]-1==0, h[10]==0}, h, {z, 10, 0}];
```

in which the water table is located at a depth of 10 m below the ground surface. The results may be easily plotted with the `Plot` function:

```
Plot[sat[Evaluate[h[z]/.hz[[1]]]], {z,0,10}, PlotRange->{0,1},
  Frame->True, FrameLabel->{"DEPTH [m]","LIQUID SATURATION"}]
```

with

```
sat[h_] := If[h>=0, 1, (1-sr) (1+(-a h)^m)^(1/m-1) + sr];
```

with `sr` the residual saturation. The results for different infiltration rates are shown in Figure 6 in which the steady-state liquid saturation profile is plotted as a function of depth. Values taken for the van Genuchten parameters used in the calculation are $m = 6.67$ and $\alpha = 0.195\ \text{m}^{-1}$, and $K_{\text{sat}} = 1.09 \times 10^{-7}\ \text{m s}^{-1}$ corresponding to a permeability of $\kappa = 10^{-14}\ \text{m}^2$.

To model the acid mine drainage problem a constant infiltration rate of 0.1 m y^{-1} is assumed. The composition of the infiltrating fluid has a pH of 5 with a partial pressure of

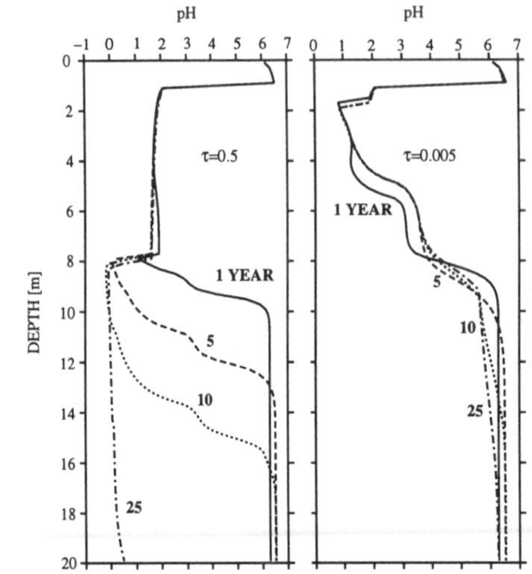

Figure 8. The pH plotted as a function of depth for the indicated times corresponding to tortuosities of 0.5 and 0.005.

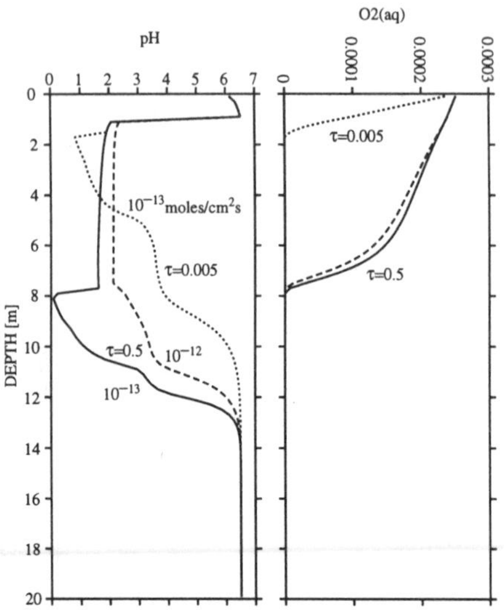

Figure 9. The pH and oxygen fugacity plotted as a function of depth for an elapsed time of 5 years.

oxygen of 0.2 bars and CO_2 of 10^{-2} bars. The initial fluid in the column is assumed to be in equilibrium with the host rock minerals quartz, kaolinite, K-feldspar, and pyrite with a pH of 6.5. Pyrite is assumed to be present one meter from the surface with a constant concentration extending down through the rest of the column. Effective rate constants of 10^{-15} moles cm^{-2} s^{-1} for quartz, 10^{-14} moles cm^{-2} s^{-1} for K-feldspar, 10^{-13} moles cm^{-2} s^{-1} for kaolinite, and 10^{-10} moles cm^{-2} s^{-1} for the secondary products ferrihydrite, siderite, and $KAlSO_4$ were used in the calculation. Shown in Figure 7 is the pH, oxygen fugacity, and concentration of Fe^{2+} and Fe^{3+}, and the reaction rates for pyrite and ferrihydrite plotted as a function of depth for an elapsed time of 5 years. A positive rate represents precipitation and a negative rate dissolution. As pyrite is oxidized above the water table the pH drops rapidly to a value of approximately 2 near the surface. Ferrihydrite precipitates throughout the oxidized zone. Siderite appears early in the calculation and then completely dissolves. The pH remains roughly constant until oxygen is completely depleted from solution, approximately 7.7 m from the surface. At this point ferrihydrite stops precipitating and the pyrite oxidation mechanism switches to Equation (330) in which Fe^{3+} is the oxidant. The pH drops further as Fe^{3+} is consumed approaching a value close to zero. It should be noted that only a portion of the pH profile is shown in Figure 7. At depths below 12 m the pH continues to rise until it reaches the initial solution pH of 6.5. It must be emphasized that the results presented here are only meant to be qualitative, at best, because of the high ionic strength solutions which are generated. It is interesting to note that negative pH values have been measured in acid mine effluents (Alpers et al., 1994).

The effect of the rate of oxygen diffusion on pyrite oxidation is illustrated in Figs. 8 and 9. The pH profiles plotted for different times corresponding to tortuosities of 0.5 and 0.005 are shown in Figure 8. For the smaller tortuosity the reduced rate of diffusion of oxygen greatly inhibits the rate of pyrite oxidation. A steady state acid plume is generated for the case of rapid oxygen diffusion with pH 0.5 at the bottom of the column. With a reduced rate of oxygen diffusion, the solution drops to a pH of approximately 1 but is quickly neutralized. Figure 9 compares the effect of oxygen diffusion and increasing the neutralizing effect of kaolinite by increasing its rate constant by an order of magnitude to 10^{-12} moles cm^{-2} s^{-1}. The solid and dashed curves correspond to a tortuosity of 0.5 and a kaolinite rate constant of 10^{-13} and 10^{-12} moles cm^{-2} s^{-1}, respectively. The dotted curves corresponds to a tortuosity of 0.005 and 10^{-13} moles cm^{-2} s^{-1} for the kaolinite rate constant. Increasing the kaolinite rate constant increases the pH throughout the upper portion of the column. The ability of oxygen to diffuse into the column, as well as the neutralizing capacity of the gangue minerals, are seen to be important factors in the generation of acid.

Poços de Caldas redox front migration–the presence of a gap

One of the characteristic features of redox fronts observed at Poços de Caldas, Brazil, is the appearance of a narrow bleached zone millimeters or less in thickness sandwiched between the iron oxide zone formed from the oxidation of pyrite and the deposition of uraninite. The bleached zone is presumably caused by oxidizing pyrite without deposition of iron oxide or uraninite. A hand sample showing the redox front and gap is presented in Figure 10. Uraninite is seen to form in nodules, with some of the nodules occurring along the iron oxide front and some further away from the front as indicated by black kidney-shaped patches. For more details of the redox fronts and mineralogy of the host rock the interested reader is referred to Waber (1990) and Waber et al. (1990).

To investigate the conditions necessary for formation of a bleached gap between the iron-hydroxide and uraninite zones, a simplified system is considered in which, with the exception of pyrite, the host rock is considered to be inert. The system is described by five components Fe–S–U–H–O. Primary species are chosen as the species: $\{Fe^{2+}, UO_2^{2+}, H^+, SO_4^{2-}, O_{2(aq)}\}$. Homogeneous equilibrium is assumed within the aqueous phase including redox sensitive species. The composition of the host rock is assumed to consist of 4% pyrite by volume and 10% porosity with the remainder of the rock considered to be inert. The initial fluid composition is assumed to have pH 6 and to be in equilibrium with pyrite and uraninite. These two minerals constrain the oxygen fugacity and total uranium concentration. Total sulfur and iron concentrations are fixed at 10^{-3} moles L^{-1}. The inlet fluid composition is assumed to be in equilibrium with the atmosphere with a pH of 4. To constrain the fluid composition further it is assumed that it is in equilibrium with ferrihydrite. The total concentration of uranium is taken as 10^{-7} moles L^{-1}. The total sulfur concentration is fixed at 10^{-6} moles L^{-1}. Note that the solution compositions are not charged balanced in the subsystem. It is assumed that additional species which are not directly involved in reactions make up the balance in charge. For example the chloride ion is one such candidate. Carbonate species are not included in the example. However, they could result in strong complexing with uranium and buffer pH changes resulting from reaction with feldspars and clay minerals present in the host rock.

Pure diffusive transport gives rise to a well-defined gap separating the iron-hydroxide and uraninite mineralization zones. This can be seen in Figure 11 where the volume fractions of pyrite, ferrihydrite and uraninite are plotted as a function of distance for times of 500 and 1,000 years. As pyrite is oxidized deposition of ferrihydrite and uraninite occur in response to the changing redox potential as it varies from oxidizing at the inlet to reducing further into the rock where the fluid has come to equilibrium with pyrite. Ferrihydrite is formed as iron released by oxidation of pyrite diffuses towards the inlet and comes in contact with oxidizing fluid. Uraninite precipitates in a narrow zone separated by a clearly defined gap from the ferrihydrite zone. The width of the gap increases with decreasing pH as shown in Figure 12. At pH 3 the gap is 5 mm wide compared to 1 mm corresponding to an inlet pH of 4.

The presence of the gap can be understood by referring to the pe-pH diagram shown in

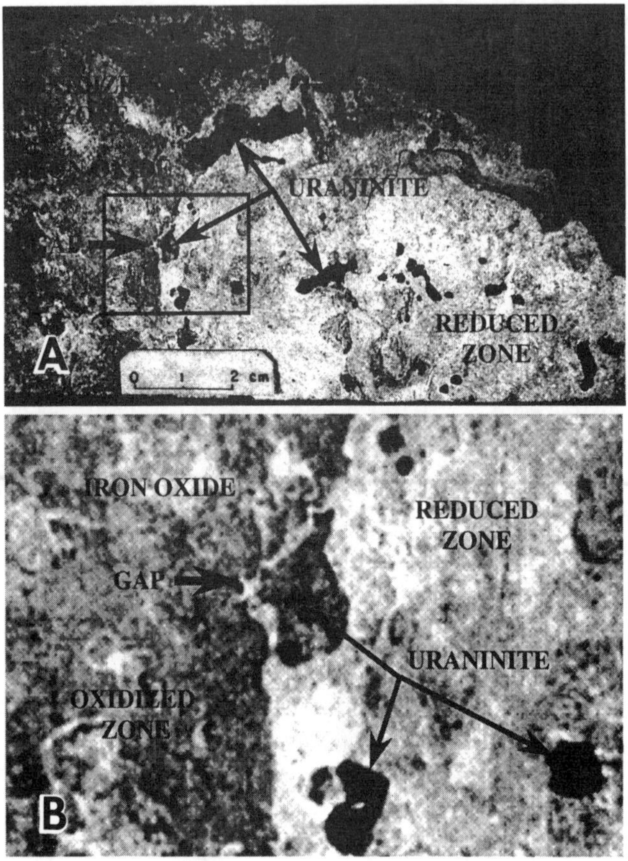

Figure 10. Redox front observed at Poços de Caldas, Brazil showing the formation of uraninite nodules and iron oxide as pyrite is oxidized (A). The narrow bleached zone is marked by arrows in the figure. Figure B is a blowup of the rectangular area in (A). Figure reproduced from the original color plate in Waber et al. (1990).

Figure 13. Starting with an oxidizing fluid, as its oxidation state is lowered by oxidation of pyrite, for pH below about 6 the fluid composition must pass through the Fe^{2+} aqueous window in which no solid phases are stable. Eventually uraninite becomes stable as the system crosses the UO_2 horizontal stability line. As expected from this analysis the width of the gap should increase with decreasing pH of the inlet solution as can be seen in Figure 12.

To investigate how the gap width varies with time the position of the reaction zone boundaries are plotted in Figure 14 as a function of time. With increasing time the width of the gap increases slightly. The width of the uraninite nodule appears to be constant as the concentration of uraninite increasing with time. The calculations are only able to explain the uraninite which is deposited at the redox front, but not the uraninite nodules which appear to form randomly further displaced from the front.

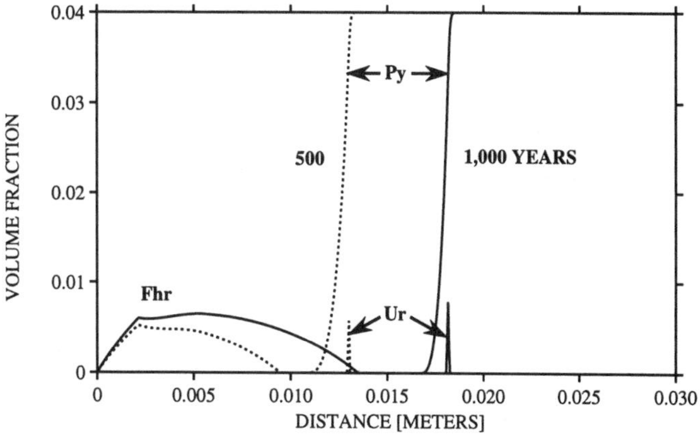

Figure 11. Volume fractions for minerals ferrihydrite (Fhr), pyrite (Py) and uraninite (Ur) plotted as a function of distance for an inlet pH 3 at elapsed times of 500 (dotted curves) and 1,000 years (solid curves).

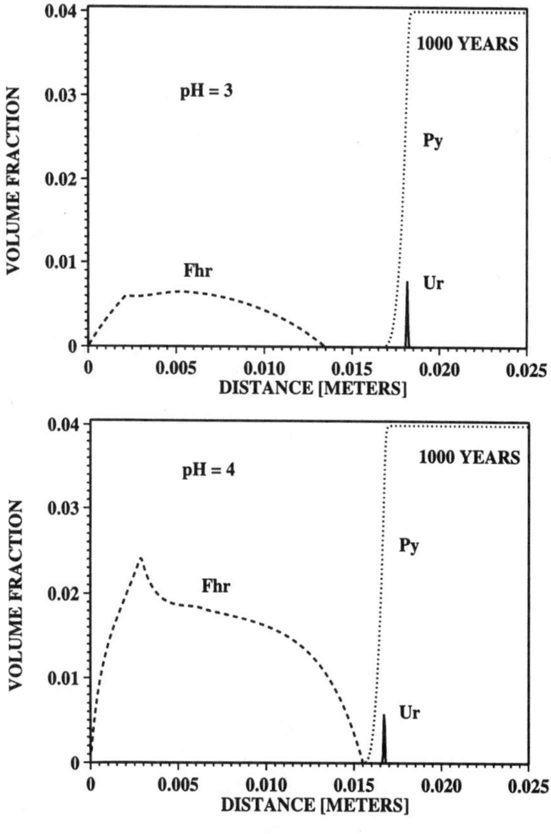

Figure 12. Volume fractions for minerals ferrihydrite (Fhr), pyrite (Py) and uraninite (Ur) plotted as a function of distance for pH 3 and 4 at an elapsed time of 1,000 years.

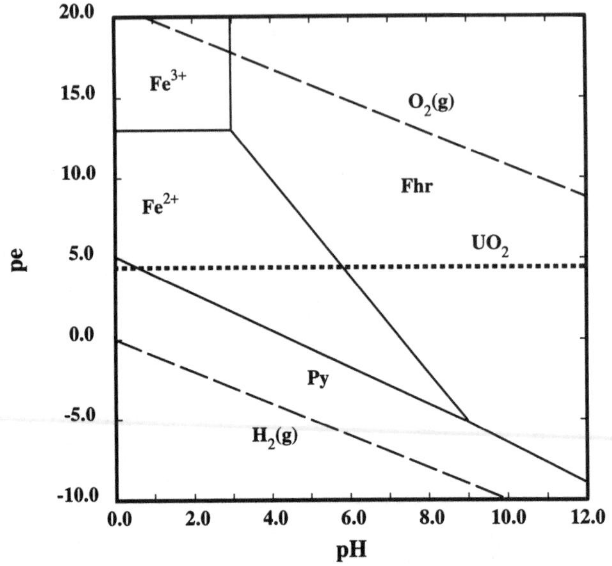

Figure 13. A pe-pH diagram for the iron system showing the saturation line for uraninite (UO_2). A gap appears between ferrihydrite (Fhr) and uraninite alteration products as pyrite (Py) is oxidized as the solution composition passes through the Fe^{2+} aqueous window in which no solid phases are stable.

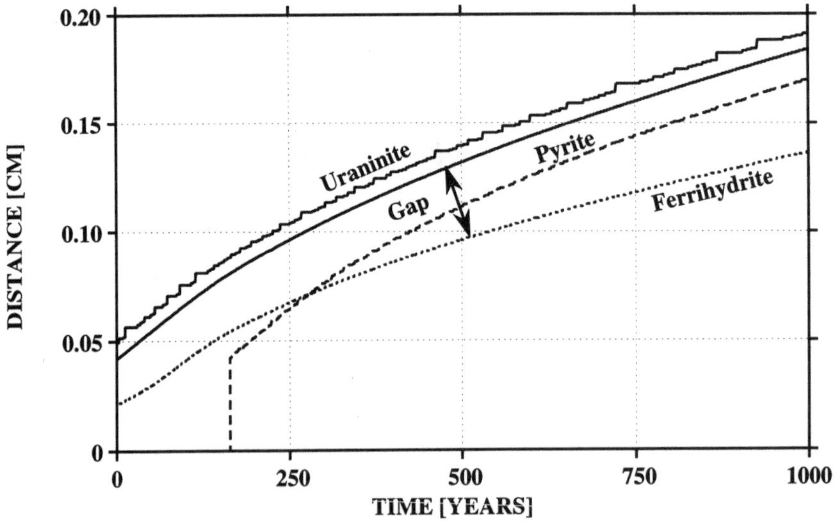

Figure 14. The position of reaction fronts showing the formation of a gap separating uraninite and ferrihydrite.

Heterogeneous porous media

Reactive transport in heterogeneous porous media presents new challenges both computationally and in our ability to verify predictions in the field and laboratory (see Tompson and Jackson, this volume). Often calculations are based on an equivalent homogeneous medium because of a lack of knowledge of the heterogeneous field and the simplicity of the calculations. The permeability field of the equivalent homogeneous medium should correspond in an average sense to the heterogeneous field. Because permeability itself is a macroscale property, even the heterogeneous permeability field refers to an approximate upscaled representation of the actual porous medium. The degree of upscaling, or smoothing, can be expected to yield different predictions of the rate of migration of contaminant plumes, for example.

There are several general statements one can make about the structure of an equivalent homogeneous permeability field. It is clear that the equivalent homogeneous field must depend on the direction of flow (Marsily, 1986; Dagan, 1989). Consider a heterogeneous field composed of N parallel units of thickness Δl_n with $\Delta l = \sum \Delta l_n$ the total thickness. Each unit has a constant but possibly different permeability κ_n. By performing flow simulations in perpendicular directions the average permeability in that direction can be determined. With boundary conditions of constant influx of fluid at one end of the computational domain, and fixed pressure at the opposing face, Darcy's law gives

$$v = -\frac{\overline{\kappa}}{\mu}\frac{\overline{\Delta p}}{\Delta l}. \tag{337}$$

The average pressure gradient across the flow field can be computed from the average pressure at each face by the equation

$$\overline{\Delta p} = \frac{1}{N}\sum_n \left(p_n^l - p_n^0\right). \tag{338}$$

The average permeability is given by

$$\overline{\kappa} = -\frac{\mu v \Delta l}{\overline{\Delta p}}. \tag{339}$$

The flow parallel to the bedding plane of the heterogeneous structure results in an equivalent average permeability equal to

$$\overline{\kappa}_{\parallel} = \frac{1}{A}\sum_n A_n \kappa_n, \tag{340}$$

where A_n denotes the cross-sectional area of the nth unit with permeability κ_n, and $A = \sum A_n$ is the total cross-sectional area. Thus in this case the equivalent homogeneous permeability is just the weighted areal average of the individual unit permeabilities. By contrast, flow perpendicular to the bedding plane yields the average permeability

$$\overline{\kappa}_{\perp}^{-1} = \frac{1}{\Delta l}\sum_n \frac{\Delta l_n}{\kappa_n}. \tag{341}$$

This is just the harmonic average weighted by the thickness of the individual unit permeabilities. A consequence of this simple analysis is that the equivalent homogeneous permeability must be represented as a tensor field.

To investigate the effect of heterogeneity on reactive flow, weathering of K-feldspar is considered as an example. Calculations are carried out using the permeability fields shown in

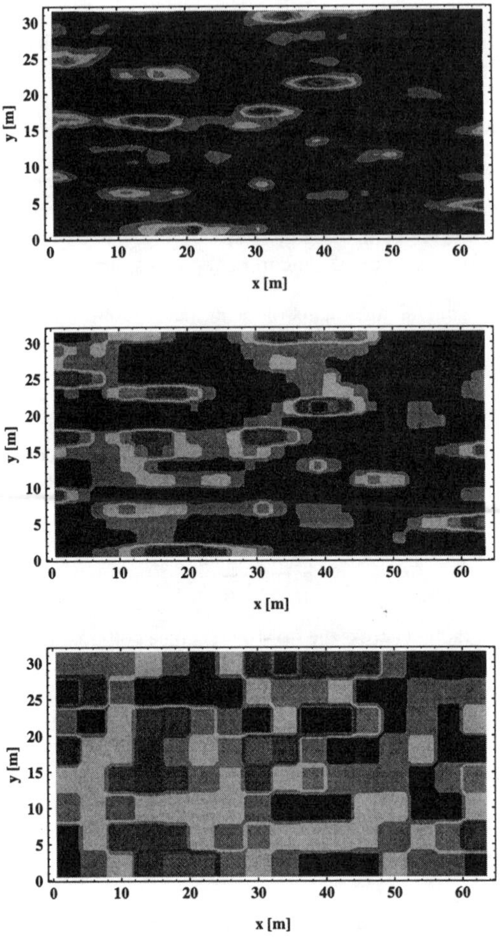

Figure 15. Upscaled heterogeneous permeability fields with resolutions of 64×32, 32×16, and 16×8. Darker regions indicate permeability highs and lighter regions lows. The average permeability is 10^{-12} m^2 with a variation of 3 orders of magnitude.

Figure 15 with scales of 64×32, 32×16, 16×8, and 1×1. An algorithm based on Real Space Renormalization Group Theory (Mohanty, 1993) was used to upscale the field beginning with a 1026×512 heterogeneous field with four orders of magnitude contrast between the smallest and largest permeability values. Successively smaller size fields were generated until a single permeability value was obtained. In each step of homogenization, the permeability contrast between the minimum and the maximum permeability values decreases and the two extreme values approach each other.

For this medium $k_{\|} = 5.829 \times 10^{-14}$ m^2, and $k_{\perp} = 3.3358 \times 10^{-14}$ m^2. The porosity is equal to 10% and the initial composition of the roc is 20% K-feldspar and 70% quartz. The porosity is considered to be constant and is not correlated with variations in the permeability. This approximation is not expected to greatly alter the qualitative features of the results. A dilute solution with pH 4.5 is allowed to infiltrate with a flow rate of 1 m y^{-1} into the porous

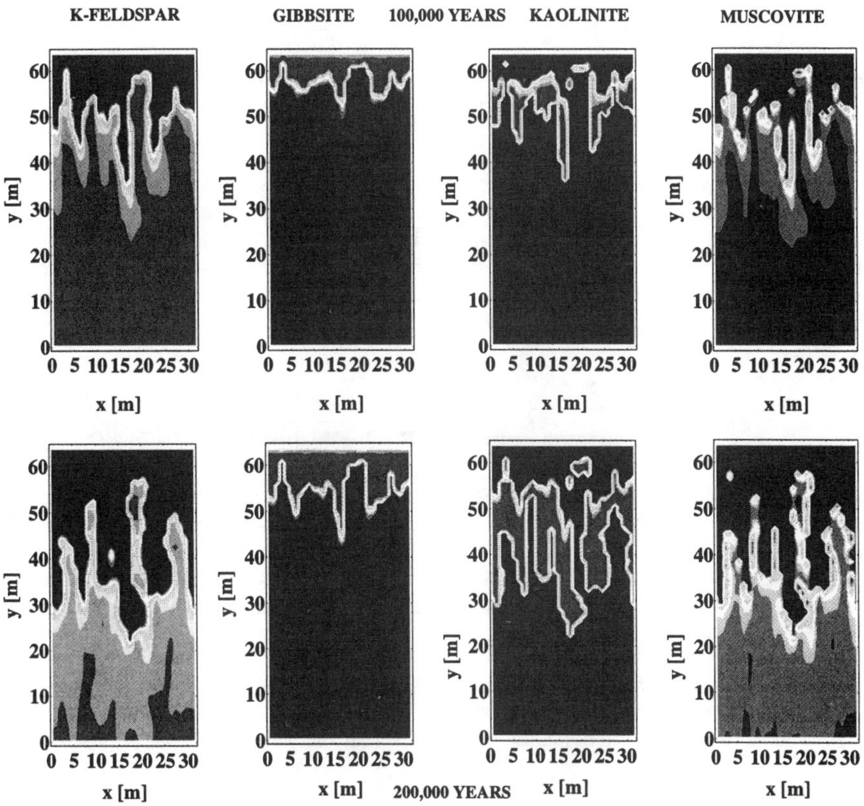

Figure 16. Volume fractions for K-feldspar, gibbsite, kaolinite and muscovite shown at 100,000 and 200,000 years for vertical infiltration of 5 m y^{-1} into a heterogeneous porous medium with resolution 64×32. For this problem the muscovite zone forms a ghost zone. Darker regions indicate zero mineral volume fraction.

rock mass. Secondary minerals allowed to form are gibbsite, kaolinite and muscovite (as a surrogate for illite). Shown in Figure 16 are the results for flow parallel, and in Figure 17 flow perpendicular to the bedding plane for the 64×32 representation. As can be seen from the figures, highly fingered reaction fronts develop in the case of flow parallel to the bedding plane, compared to that perpendicular.

The position of K-feldspar dissolution front is shown in Figure 18 along with average values at the three different scales as indicated in the figure, along with the corresponding average position of the front as given in angular brackets $< \ldots >$. The average front positions of the heterogeneous fields at the different scales correspond well with the equivalent homogeneous medium value of 13.5 m. Results obtained for flow perpendicular to the bedding gave much less fingering with similar agreement between the average front positions for the different scales and the homogeneous case.

As this example illustrates, if the only quantity of interest is the average position of the front then the equivalent homogeneous medium appears adequate. If it is also important to ascertain the degree of fingering, for example, then use of a heterogeneous medium in the calculation is necessary. However, here a fundamental difficulty arises. It can not be expected that the model calculations will ever yield the precise fingering pattern as seen in the field or even in laboratory experiments. This is because it is virtually impossible to know precisely the nature of the flow field, it being governed by an essentially random network of connected pores and fractures. At best, one might hope to predict some measure of the degree of fingering, such as the average values of higher moments of the front position, and compare this with observation.

Reaction instability. In this example flow and reactive transport for a single component system is considered in an initially statistically homogeneous, yet highly heterogeneous, permeability field. The permeability is allowed to change with reaction. Feedback between mineral dissolution and consequent changes in porosity and permeability can lead to reaction instabilities resulting in fingering phenomena (Ortoleva et al., 1987; Sherwood, 1987; Hinch and Bhatt, 1990; Steefel and Lasaga, 1990, 1994). A phenomenological power law relation of the form

$$k = k_0 \left(\frac{\phi}{\phi_0}\right)^n \left[\frac{1.001 - \phi_0^2}{1.001 - \phi^2}\right], \tag{342}$$

is used in the calculation to relate permeability and porosity, where k_0 and ϕ_0 refer to the initial permeability and porosity of the medium, and n is a constant with the value 3. The porosity is calculated from the mineral volume fraction according to Equation (30). Other forms of the permeability–porosity relation are also possible (see Oelkers this volume). However, very little experimental data is available to justify one form over another. Furthermore, such effects as pore clogging are not accounted for in this simple relation.

The initial heterogeneous porous medium is constructed by randomly assigning the permeability at each node point and normalizing the resulting field to give an average permeability of 10^{-12} m^2. Thus at the nth node, the permeability has the value

$$\kappa_n = N_G \overline{\kappa} \frac{\Re_n}{\sum_{l=1}^{N_G} \Re_l}, \tag{343}$$

where $\overline{\kappa}$ denotes the average permeability, $\Re_n$ represents a random number, and N_G denotes the total number of node points. Reaction with the solid matrix described by a first order kinetic rate law with an effective rate constant of 10^{-12} moles cm^{-2} s^{-1}. The initial porosity and solid volume fraction are both set at 50%. Initially the pore solution is assumed to be in equilibrium with respect to the solid phase with a concentration of 10^{-3} molal. The concentration of the inlet fluid is undersaturated with respect to the solid with a concentration of 10^{-6} molal. A pressure gradient of 0.005 bars across the left and right boundaries is imposed on the system with zero flux boundary conditions at the top and bottom of the computation domain. This gives an initial flow velocity of approximately 3 m y^{-1}. The grid spacing used in the calculation was 0.05 m with 100 node points. This leads to an initial grid Péclet number of 10. A first order upwinding finite difference algorithm was used in the calculation. This is not expected to greatly influence the solution according to the discussion following Equation (249).

The computed mineral volume fraction is shown in Figure 19 carried out to 7,500 years. A highly channelized flow field evolves as the solid dissolves as evidenced by the converging streamlines and dissolution pattern of the solid as shown in the figure. The arrows, spaced at equal time intervals, show the direction and relative magnitude of the flow velocity. The calculation is certainly only qualitative at best, because of the approximate relation used between permeability and porosity. Once breakthrough of the fingers occurs the calculation breaks down completely. This is because the permeability should approximate that of flow through a channel, which bears no relation to Equation (342). Before breakthrough occurs, bulk flow is governed primarily by the pressure gradient across the unreacted portion of the porous medium.

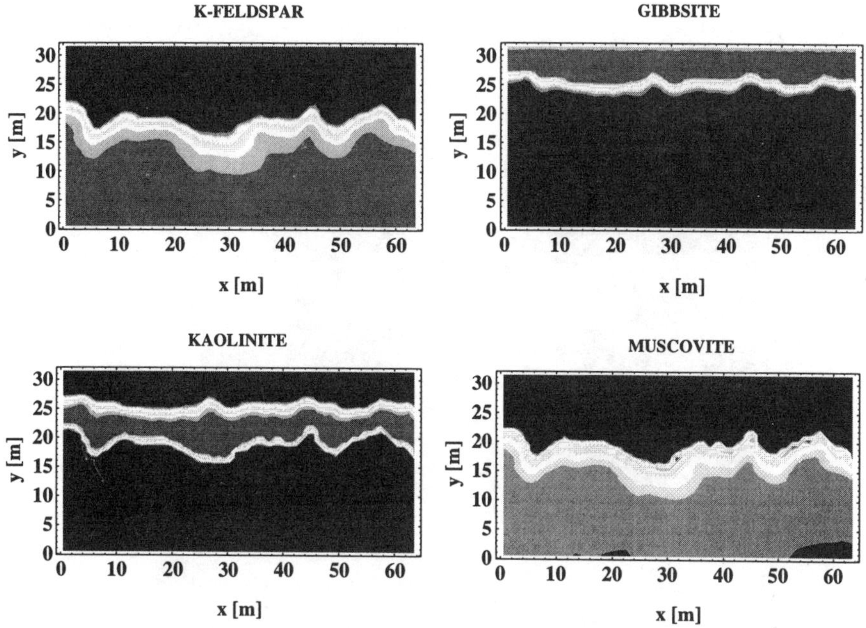

Figure 17. Volume fractions for K-feldspar, gibbsite, kaolinite and muscovite shown at 100,000 and 200,000 years for vertical infiltration of 5 m y^{-1} into a heterogeneous porous medium with resolution 64×32. For this problem the muscovite zone forms a ghost zone. Darker regions indicate zero mineral volume fraction.

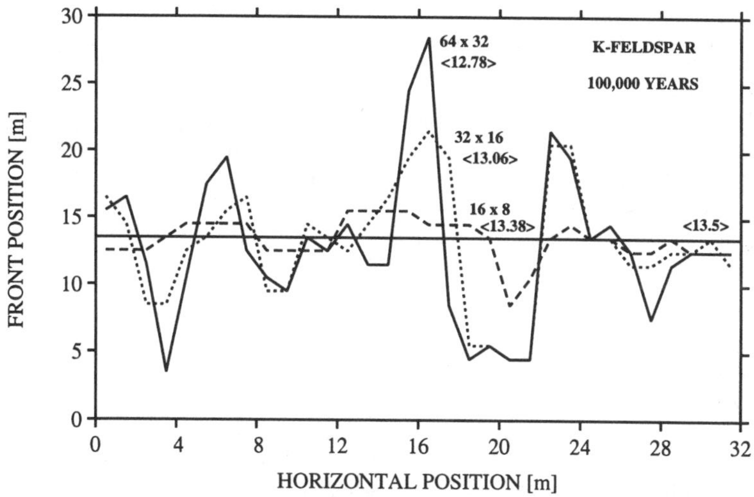

Figure 18. The average K-feldspar dissolution front for different upscaled homogenization scales of 64×32, 32×16 and 16×8 for the same realization of the heterogeneous porous medium.

This result is in stark contrast to a homogeneous porous medium with constant permeability for which the solid dissolves along a planar front. If the permeability is kept constant in the present calculation, although represented as a random field, fingering does not occur because of the initial statistical homogeneity of the field. The calculated dissolution pattern results from

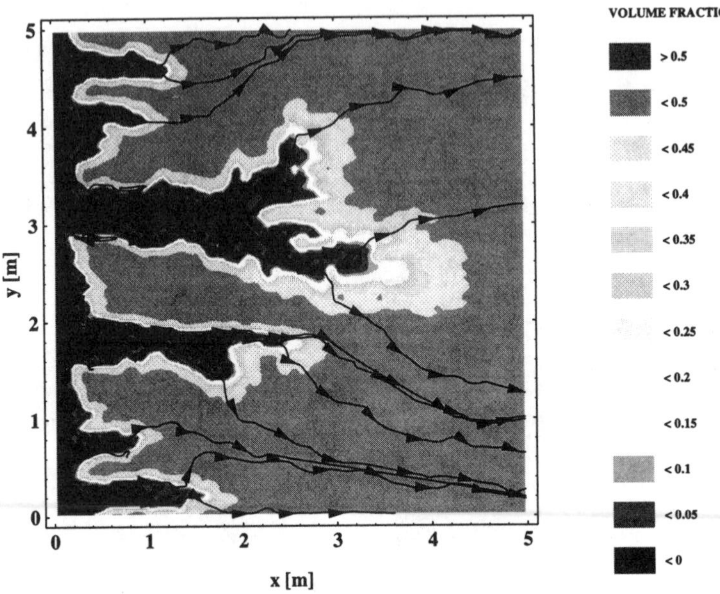

[See cover for color rendition of this figure.]

Figure 19. Contour plot of the solid volume fraction after an elapsed time of 7,500 years and streamlines plotted as a function of distance in the xy-plane. Fluid undersaturated with respect to the solid is injected at the left causing the solid to dissolve along fingers shown by the dark grey regions.

an increase in permeability as the solid matrix dissolves causing a positive feedback effect in which dissolution is enhanced along fast pathways. This behavior of frontal instability is similar to the fingering patterns produced when a less viscous fluid displaces a more viscous fluid. Such patterns have important implications, for example, in enhanced oil recovery where channelized flow could limit the usefulness of acid injection into an oil reservoir resulting in the formation of so-called wormholes (Daccord, 1987; Daccord and Lenormand, 1987; Rege and Fogler, 1989; Hoefner and Fogler, 1989; Hill et al., 1995). These results also explain the formation of karst topographies such as sinkholes, caves and underground drainage which often occur in limestone, dolomite and gypsum rock types characterized by rapid dissolution rates.

Hydrothermal system

As a final example the hydrothermal system discussed by Steefel and Lasaga (1994) is considered taking into account reaction with quartz. The flow field is calculated from the transient mass and energy conservation equations using MULTIFLO, in contrast to Steefel and Lasaga (1994) who used a steady-state solution based on the stream function. A uniform permeability of 10^{-14} m^2 with a porosity of 5% is used in the calculation. A heat source is imposed at a depth of 3,000 m which varies laterally according to a gaussian profile with a maximum temperature of 300°C at the center of the domain. The flow field is not changed throughout the calculation as quartz dissolves and reprecipitates. This is not necessarily a good approximation, but it does give an approximate account of the redistribution of quartz in the system. The quartz kinetic rate constant is taken sufficiently large that local equilibrium is approximately honored. Initially the fluid is in equilibrium with quartz with a geothermal gradient of 25°C/km assuming a constant temperature of 50°C at the surface. Zero flux boundary conditions are imposed at the sides of the computational domain with a lateral dimension of 6,000 m.

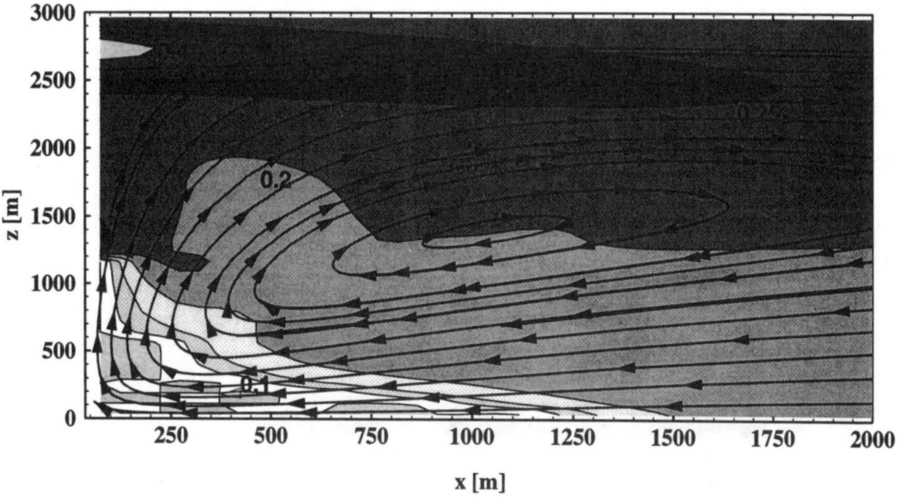

Figure 20. Flow field and distribution of quartz after an elapsed time of 10 million years resulting from hydrothermal convection. Only a portion of the flow field is shown extending to 2,000 m. The figure is symmetric about the vertical z-axis. The arrows placed along streamlines occur at equal time intervals.

Shown in Figure 20 is the flow field and the distribution of quartz after an elapsed time of 10 million years over a portion of the system. Approximately 400,000 years were required for the flow field to reach a steady state convection pattern. As can be seen from the figure, quartz dissolves in the lower hotter region, and is deposited in the cooler region by upwelling fluid forming a quartz cap across the top of the system. This behavior is a consequence of the prograde solubility of quartz. No attempt was made to take into account changes in porosity and permeability during the simulation. To what extent this result corresponds to an actual hydrothermal system is an open question. But clearly to produce a quartz cap by this mechanism would require relatively long times. Other possible mechanisms are boiling in a two-phase system, and including other sources of silica such as reaction with feldspars.

CONCLUDING REMARKS

Beginning with the continuum representation of a porous medium macroscale conservation equations of mass and energy were derived for a general multiphase-multicomponent system in local partial equilibrium. Transformation of the chemical reactions taking place in the system to canonical form allowed elimination of rates corresponding to local equilibrium reactions from the mass transport equations, thereby greatly reducing the number of unknowns. Concepts of macro- and microscale local equilibrium, scaling relations, asymptotics, ghost zones and the quasi-stationary state approximation were introduced. Microscale local equilibrium leads to a breakdown in the single continuum model with equilibration lengths becoming less than the diameter of a pore. In such cases the bulk fluid composition is no longer representative of the fluid composition at the surface of a reacting mineral. Through scaling relations, macroscale local equilibrium was found to be the limiting form of a kinetic description in the asymptotic limit of long time spans. The quasi-stationary state approxmation was introduced showing how the time evolution of a geochemical system could be represented as a sequence of stationary states. The validity of the quasi-stationary state approximation was discussed in the context of a single component system for which an analytical solution exists involving an advancing dissolution front. It was noted that the numerical solution for a single component system with linear kinetics is immune from numerical dispersion produced by a first order upwinding finite difference scheme at high Péclet number flows. It was demonstrated that incorporating sorption reactions with transport, including batch-type surface complexation models, may not preserve charge balance in the aqueous phase. Finally several examples were presented using the computer code MULTIFLO, including acid mine drainage, redox front migration, reactive transport in heterogeneous porous media, an example of reaction front instability, and a simple hydrothermal system.

Hampering a predictive capability of groundwater transport models is our inability to know the precise realization of the porous medium found in the natural system. Thus even if the physical processes described by the model were prefectly correct, the results may still not agree with observation because of differences between the representation of the porous medium used in the model compared to that found in nature. It is essential to ask the right questions, before attempting to predict the outcome of some event. Intuitively, one might expect that the concentration of some solute species at a single point will be very difficult to predict with any certainty. Yet the average concentration across an entire surface may be predicted much more accurately and confidently. Unfortunately, it is the concentration at a point—perhaps someone's well— that may needed. Nonetheless, used in a qualitative sense models may be more useful in many circumstances. The question becomes how will we know if the model predictions are even qualitatively correct if we are unable to verify them in detail because of the fundamental heterogeneities that are present in a particular flow system.

The missing link to a more realistic description of natural systems involving transport and chemical reaction is incorporating heterogeneities present in porous media. At the microscale of individual grain dimensions a porous medium consists of a highly contorted network of pores separated by aggregates of mineral grains. In fractured porous media the fracture network provides a heterogeneous flow system. At the macroscale large scale heterogeneities channel flow in different directions. All these factors can strongly influence the alteration pattern resulting from fluid/rock interaction. It is somewhat paradoxical that as more and more sophisticated representations of heterogeneous porous media are developed, it becomes increasingly difficult to compare the resulting calculations with experimental or field observations to validate the model predictions. This is because of the greater burden placed on the detail of observation required to compare with the model calculations.

One of the main challenges facing the modeler is linking field observations at the microscale with the macroscale. At the microscale, reaction mechanisms can be deduced from, for example, observations of a thin section. However, it is often difficult to make use of this information at the macroscale at which the continuum transport equations are valid. It is important to understand the cause of uncertainties in groundwater models and to be able to separate uncertainties resulting from the inherent randomness in porous media from other causes over which we have more control. Basic questions which still need to be addressed are how to upscale from the microscale of individual pore spaces to the macroscale without losing essential information.

ACKNOWLEDGMENTS

I would like to thank Tony Appelo, Carl Steefel and Eric Oelkers for reviewing the manuscript and offering valuable suggestions for its improvement. I alone, however, am responsible for any remaining errors. In addition, I would like to thank Urs Mader, Nick Waber and the Water/Rock Interaction group at the Universität Bern, and Laurant Charlet at the University of Grenoble for providing the opportunity to present parts of this material in short courses held during the Spring of 1996. Special thanks goes to Sitakanta Mohanty for providing representations of heterogeneous media for use in several examples, and to Mohan Seth for helpful discussions. This work was funded partly by an internal research and development grant (20-9872) to the author from Southwest Research Institute.

REFERENCES

Aagaard P., and Helgeson, H.C. (1982) Thermodynamic and kinetic constraints on reaction rates among minerals and aqueous solutions. I. Theoretical considerations. American Journal of Science, 282:237–285.

Alpers, C.N., Nordstrom, D.K., and Thompson, J.M. (1994) Seasonal variations of Zn/Cu ratios in acid mine water from Iron Mountain, California, *in* Alpers, C.N. and Blowes, D.W., editors, Environmental Geochemistry of Sulfide Oxidation. ACS, 550:324–344.

Aris, R. and Mah, R.H.S. (1963) Independence of chemical reactions. I & EC Fundamentals, 2:90–94.

Bahr, J.M. and Rubin, J. (1987) Direct comparison of kinetic and local equilibrium formulations for solute transport affected by surface reactions. Wat. Res. Res., 23:438–452.

Balashov, V.N., and Lichtner, P.C. (1992) Ghost zones in multicomponent systems. In preparation.

Basolo F. and Pearson R.G. (1967) Mechanisms of Inorganic Reactions: A study of metal complexes in solution, 2nd ed. Wiley, New York, 701 pp.

Bear, J. (1972) Dynamics of fluids in porous media. Elsevier. 764 pp.

Bird, R.B., Stewart, W.E., and Lightfoot, E.N. (1960) Transport phenomena. John Wiley and Sons, New York, N. Y.

Borkovec, M., and Westall, J. (1983) Solution of the Poisson-Boltzmann equation for surface excesses of ions in the diffuse layer at the oxide–electrolyte interface. J. Electroanal. Chem., 150:325–337.

Daccord, G. (1987) Chemical dissolution of a porous medium by a reactive fluid. Physical Review Letters, 58:479–482.

Daccord, G., and Lenormand, R. (1987) Fractal patterns from chemical dissolution. Nature, 325:41–43.

Dagan, G. (1989) Flow and Transport in Porous Formations. Springer-Verlag, 465 pp.

Damköhler, Von G. (1936) Einflüsse der strömung, diffusion und des wärmeüberganges auf die leistung von reactionsöfen. Zeitschrift Elektrochem. 42:846–862.

Denbigh, K. (1981) The Principles of Chemical Equilibrium. Cambridge University Press, Fourth Edition, 494 pp.

Dzombak, D.A. and Morel, F.M. (1990) Surface Complex Modeling, Hydrous Ferric Oxide. Wiley-Interscience, New York.

De Donder, T. Leçons de Thermodynamique et de Chimie Physique (new edition by F.H. van den Dungen and G. Van Lerberghe) (Paris, 1920).

Fletcher, C.A.J. (1991) Computational techniques for fluid dynamics, Volume I: Fundamental and general techniques. Springer-Verlag, 401 pp.

Frantz, J.D., and Mao, H.K. (1975) Bimetasomatism resulting from intergranular diffusion: Multimineralic zone sequences. In Carnegie Institution of Washington Year Book 74, 1974–1975, No. 1675, 417–432.

Frantz, J.D., and Mao, H.K. (1976) Bimetasomatism resulting from intergranular diffusion: I. A theoretical model for monomineralic zone sequences. Am. J. Sci., 276:817–840.

Frantz, J.D., and Mao, H.K. (1979) Bimetasomatism resulting from intergranular diffusion: II. Prediction of multimineralic zone sequences. Am. J. Sci., 297:302–323.

Gupta, A.D., Lake, L.W., Pope, G.A., Sephernoori, K., and King, M.J. (1991) High–resolution monotonic schemes for reservoir fluid flow simulation. In Situ., 15:289–317.

Jury, W.A., and Letey, J. Jr. (1979) Water movement in soil: Reconciliation of theory and experiment. Soil Sci. Soc. Am. Proc., 43:823–827.

Haase, R. (1969) Thermodynamics of irreversible processes. Dover, 513 pp.

Helgeson, H.C. (1968) Evaluation of irreversible reactions in geochemical processes involving minerals and aqueous solutions-I. Thermodynamic relations. Geochimica et Cosmochimica Acta, 32:853–877.

Helgeson, H.C., Murphy, W.M., and Aagaard P. (1984) Thermodynamic and kinetic constraints on reaction rates among minerals and aqueous solutions. II. Rate constants, effective surface area, and the hydrolysis of feldspar. Geochimica et Cosmochimica Acta, 51:3137–3153.

Hinch, E.J., and Bhatt, B. S. (1990) Stability of an acid front moving through porous rock. J. fluid Mech., 212:279–288.

Hoefner, M.L., and Fogler, H.S. (1989) Fluid-velocity and reaction-rate effects during carbonate acidizing: application of network model. SPE Production Engineering, 56–62.

Lasaga, A.C. (1981) Rate laws in chemical reactions. In Kinetics of Geochemical Processes, Reviews in Mineralogy (A.C. Lasaga and R.J. Kirkpatrick, eds.), Mineral. Soc. Am., 8:135–169.

Lasaga, A.C. and Rye, D.M. (1993) Fluid flow and chemical reactions in metamorphic systems. Am. J. Sci., 293:361–404.

Leonard, B. (1979a) A survey of finite differences of opinion on numerical muddling of the incomprehensible defective confusion equation. In: Finite Element Methods for Convection-Dominated Flows (T. J. R. Hughes, ed.): 1–17.

Leonard, B. (1979b) A stable and accurate convective modeling procedure based on quadratic upstream interpolation. Comput. Methods Applied Mech. Eng., 19:59–98.

Leonard, B. (1984) Third-order upwinding as a rational basis for computational fluid dynamics. Elsevier, New York.

Lichtner P.C. (1985) Continuum model for simultaneous chemical reactions and mass transport in hydrothermal systems, Geochimica et Cosmochimica Acta, 49, 779–800.
Lichtner, P.C. (1988) The quasi-stationary state approximation to coupled mass transport and fluid-rock interaction in a porous media. Geochimica et Cosmochimica Acta, 52:143–165.
Lichtner P.C. (1991) The quasi-stationary state approximation to fluid/rock reaction: local equilibrium revisited, *in* Ganguly, J., editor, Diffusion, atomic ordering and mass transport. Advances in Physical Geochemistry, New York, Springer Verlag, 8:454–562.
Lichtner, P.C. (1992) Time-space continuum description of fluid/rock interaction in permeable media. Water Resources Research, 28:3135–3155.
Lichtner, P.C. (1993) Scaling properties of time-space kinetic mass transport equations and the local equilibrium limit. American Journal of Science, 293:257–296.
Lichtner, P.C. (1995) Principles and practice of reactive transport modeling. Mat. Res. Soc. Symp. Proc., Kyoto, Japan, 353:117–130.
Lichtner P.C. and Balashov V. N. (1993) Metasomatic zoning: appearance of ghost zones in limit of pure advective mass transport. Geochimica et Cosmochimica Acta, 57:369–387.
Lichtner, P.C., and Biino, G. G. (1992) A first principles approach to supergene enrichment of a porphyry copper protore. I. Cu–Fe–S–H_2O subsystem. Geochimica et Cosmochimica Acta, 56:3987–4013.
Lichtner, P.C., and Seth, M. S. (1996a) Multiphase-multicomponent reactive transport nonisothermal reactive transport in partially saturated porous media. Presented at the International Conference on Deep Geological Disposal of Radioactive Waste, Canadian Nuclear Society. September 16-19, Winnipeg, Manitoba, Canada.
Lichtner, P.C., and Seth, M. S. (1996b) User's manual for MULTIFLO: Part II MULTIFLO 1.0 and GEM 1.0, Multicomponent-multiphase reactive transport model. CNWRA 96-010.
Lichtner, P.C., and Waber, N. (1992) Redox front geochemistry and weathering: Theory with application to the Osamu Utsumi uranium mine, Poços de Caldas, Brazil. Journal of Geochemical Exploration, 45:521–564.
Lindberg, R.D. and Runnells, D.D. (1984) Ground water redox reactions: An analysis of equilibrium state applied to Eh measurements and geochemical modeling. Science, 225:925-927.
Marsily, de G. (1986) Quantitative hydrology. Academic Press. 440 pp.
Mohanty, S. (1993) Experimental determination and theoretical prediction of effective thermal conductivity of porous media. Ph.D. Dissertation. University of Texas at Austin, Austin, TX.
Moses C.O., Nordstrom D.K., Herman J.S. and Mills A.L. (1987) Aqueous pyrite oxidation by dissolved oxygen and by ferric iron. Geochimica et Cosmochimica Acta, 51:1561–1571.
Murphy, W.M., Oelkers, E.H., and Lichtner, P.C. (1989) Surface reaction versus diffusion control of mineral dissolution and growth rates in geochemical processes. Chemical Geology, 78:357–380.
Newman, J.S. (1973) Electrochemical systems. Prentice Hall. 560 pp.
Nitao, J.J. (1996) Reference Manual for the NUFT Flow and Transport Code, Version 1.0. UCRL-ID-113520, Lawrence Livermore National Laboratory.
Nordstrom D.K. (1982) Aqueous pyrite oxidation and the consequent formation of secondary iron minerals. In: Acid Sulfate Weathering: Pedogeochemistry and Relationship to Manipulation of Soil Materials (eds. L.R. Hossner, J.A. Kittrick and D.F. Fanning). Soil Science Soc. Amer. Press, Madison.
Ortoleva, P. Chadam, J., Merino, E., and Sen, A. (1987) Geochemical self-organization II: The reactive-infiltration instability. American Journal of Science, 287, 1008-1040.
Ortoleva, P., Auchmuth, G., Chadam, J., Hettmer, J., Merino, E., Moore, C. H., and Ripley, E. (1986) Redox front propagation and banding modalities. Physica, 19D:334–354.
Parkhurst, D.L. (1995) User's guide to PHREEQC—a computer program for speciation, reaction-path, advective-transport, and inverse geochemical calculations. U.S. Geological Survey, Water-Resources Investigations Report 95-4227.
Prigogine, I. and Defay, R. (1954) Chemical Thermodynamics. Longmans, 543 pp.
Pruess, K. (1991) TOUGH2—A general purpose numerical simulator for multiphase fluid and heat flow. LBL-29400. Berkeley, CA, University of California, Lawrence Berkeley Laboratory.
Ramshaw, J.D. (1980) Partial chemical equilibrium in fluid dynamics. Physics of Fluids, 23:675–680.
Rege, S.D., and Fogler, H. S. (1989) Competition among flow, dissolution, and precipitation in porous media. AIChE Journal, 35:1177–1185.
Richards, L.A. (1931) Capillary conduction of liquids through porous mediums. Physics, 1:318–333.
Schechter, R.S., Bryant, S.L. and Lake, L.W. (1987) Isotherm–free chromatography: propagation of precipitation/dissolution waves. Chemical Engineering Communications, 58:353–376.
Seth, M. S., and Lichtner, P. C. (1996) User's manual for MULTIFLO: Part I Metra 1.0β, Two-phase nonisothermal flow simulator. CNWRA 96-005.
Sevougian, S.D., Schechter, R.S., and Lake, L.W. (1993) Effect of partial local equilibrium on the propagation of precipitation/dissolution waves. Ind. Eng. Chem. Res., 32:2281–2304.
Sherwood, J.D. (1987) Stability of a plane reaction front in a porous medium. Chemical Engineering Science, 42:1823–1829.

Singer, P.C. and Stumm, W. (1970) Acid mine drainage—the rate limiting step. Science, 167:1121–1123.

Smith, W.R., and Missen, R.W. (1982) Chemical reaction equilibrium analysis. Theory and algorithms. Wiley. 364 pp.

Sørensen, J.P., and Stewart, W.E. (1980) Structural analysis of multicomponent reaction models. AIChE, 26:98–111.

Stumm, W. and Morgan, J.J. (1981) Aquatic Chemistry. Second Edition, Wiley 780 pp.

Steefel, C.I., and Lasaga, A.C. (1990) Evolution of dissolution patterns: Permeability change due to coupled flow and reaction. In: Melchior, D. C. and Bassett, R. L. eds., Chemical modeling in aqueous systems II: American Chemical Society Symposium 416:212–225.

Steefel, C.I., and Van Cappellen, P. (1990) A new kinetic approach to modeling water–rock interaction. The role of nucleation, precursors, and Ostwald ripening. Geochimica et Cosmochimica Acta, 54:2657–2677.

Steefel, C.I., and Lasaga, A.C. (1994) A coupled model for transport of multiple chemical species and kinetic precipitation/dissolution reactions with application to reactive flow in single phase hydrothermal systems. American Journal of Science, 294:529–592.

Troeh, F.R., Jabero, J.D. and Kirkham, D. (1982) Gaseous diffusion equations for porous materials. Geoderma, 27:239–253.

VanderKwaak J.E., Forsyth P.A., MacQuarrie K.T.B., and Sudicky E. A. (1995) WATSOLV Sparse Matrix Iterative Solver Package Version 1.01. Unpublished report, Waterloo Centre for Groundwater Research, University of Waterloo, Waterloo, Ontario, Canada.

van Genuchten, M. (1980) A closed form equation for predicting the hydraulic conductivity of unsaturated soils. Soil Science Society of American Journal 44:892–898.

Van Leer, B. (1977) Towards the ultimate conservative difference scheme. IV. An new approach to numerical convection. J. of Comput. Physics, 23: 276.

Van Zeggeren F., and Storey, S.H. (1970) The computation of chemical equilibria. Cambridge University Press. 176 pp.

Vargaftik, N.B. (1975) Tables on the thermophysical properties of liquids and gases. Second Edition, John Wiley and Sons, New York.

Waber, N. (1990) Hydrothermal and supergene evolution of the Osamu Utsumi uranium deposit and the Morro do Ferro thorium-rare earth deposit (Minas Gerais, Brazil). Ph.D. thesis, University of Bern, 179 p.

Waber, N. Schorscher, H.D., and Peters, Tj. (1990) Mineralogy, petrology and geochemistry of the Poços de Caldas analogue study sites, Minas Gerais, Brazil, I. Osamu Utsumi uranium mine. Nagra NTB 90-20, Wettingen, Switzerland, 485 pp.

Walsh, M.P., Bryant, S.L., Schechter, R.S., and Lake, L.W. (1984) Precipitation and dissolution of solids attending flow through porous media. American Institut of Chemical Engineering Journal, 30, 317–327.

Weare, J.H., Stephens, J.R., and Eugster, H.P. (1976) Diffusion metasomatism and mineral reaction zones: General principles and applications to feldspar alteration. Am. J. Sci., 276:767–816.

Westcot, D.W., and Wierenga, P.J. (1974) Transfer of heat by conduction and vapor movement in a closed soil system. Soil Sci. Soc. Am. Proc., 38:9–14.

White, S.P. (1995) Multiphase nonisothermal transport of systems of reacting chemicals. Wat. Res. Res., 31:1761–1772.

Wiersma C.L. and Rimstidt J.D. (1984) Rates of reaction of pyrite and marcasite with ferric iron at pH 2. Geochimica et Cosmochimica Acta, 48:85–92.

Wolery, T.J., Jackson, K.J., Bourcier, W.L., Bruton, C.J., Viani, B.E., Knauss, K.G., and Delany, J.M. (1990) Current status of the EQ3/6 software package for geochemical modeling, in Melchior, D.C. and Bassett, R.L., editors, Chemical Modeling in Aqueous Systems II: American Chemical Society Symposium Series, 416:104–116.

Wolfram S. (1991) Mathematica a system for doing mathematics by computer. Second edition. Addison-Wesley, New York. 961 pp.

Zyvoloski, G., Dash, Z., and Kelkar, S. (1992) FEHMN 1.0: Finite element heat and mass transfer code. LA-12062-MS, Rev 1. Los Alamos, NM, Los Alamos National Laboratory.

Chapter 2

APPROACHES TO MODELING OF REACTIVE TRANSPORT IN POROUS MEDIA

Carl I. Steefel

Department of Geology
University of South Florida
Tampa, Florida 33620 U.S.A.

Kerry T. B. MacQuarrie[1]

Department of Earth Sciences
University of Waterloo
Waterloo, Ontario N2L 3G1 Canada

INTRODUCTION

Most of the geochemical and biogeochemical phenomena of interest to Earth scientists are the result of the coupling of some combination of fluid, heat, and solute transport with chemical reactions in the Earth's crust. Rather than being isolated systems that can be described entirely by thermodynamics, environments in the Earth's crust act as open bio-geochemical reactors where chemical change is driven by the interactions between migrating fluids, solid phases, and organisms. These interactions may involve such diverse processes as mineral dissolution and precipitation, adsorption and desorption, microbial reactions, and redox transformations. If the interactions take place over sufficiently long periods of time, they may modify the properties of the porous media itself, for example through the dissolution of matrix minerals or the cementation of pore spaces.

Because of the complexity of reactive transport processes in porous media, modeling their dynamic behavior is a challenging task for the geoscientist. The complexity of the field is in part a natural consequence of the evolution of the scientific disciplines which contribute to the broader subject of reactive transport in porous media. For example, continuing research on multicomponent aqueous geochemical systems makes it clear that the traditional means of modeling transport and reactions in groundwater systems using single ion approaches do not capture all of the chemical complexities (e.g. pH dependence of adsorption, nonlinear kinetic reaction rates) which exist in these systems. Research carried out over the last 10 years has also called into question the time-honored local equilibrium assumption (Thompson, 1959) which has been applied almost indiscriminantly in metamorphic petrology and many other fields in the geosciences (Lindberg and Runnels, 1984; Bahr and Rubin, 1987; Steefel and Van Cappellen, 1990; Lasaga and Rye, 1993; Steefel and Lasaga; 1994; Steefel and Lichtner, 1994; McNab and Narasimhan, 1994; McNab and Narasimhan, 1995; Friedly et al., 1995; Oelkers et al., 1996; Van Cappellen and Wang, 1996). In addition, the research on geochemical kinetics has revealed a wealth of possible mechanisms and rates by which minerals, organisms, and aqueous species may interact. In most cases these interactions are nonlinear and therefore require special treatment in order to incorporate them into mathematical models. Similarly, hydrologists have highlighted the complexity of natural porous media in recent years, showing clearly that the dynamic behavior of contaminant systems (and by implication, all geological

[1]Now at *Department of Civil Engineering, University of New Brunswick, Fredericton, New Brunswick E3B 5A3 Canada*

0275-0279/96/0034-0002$05.00

systems involving reactions between migrating species and a reactive porous media) is strongly dependent on both physical and chemical heterogeneities present in the system (Dagan, 1989; Gelhar, 1993; van der Zee and van Riemsdijk, 1987; Garabedian et al., 1988; Valocchi, 1989; Kabala and Sposito, 1991; Robin et al., 1991; Tompson, 1993; Bosma et al., 1993; Brusseau, 1994; Cushman et al., 1995; Hu et al., 1995; Simmons et al., 1995; Ginn et al., 1995; Tompson et al., 1996; Tompson and Jackson, this volume). The homogeneous porous media assumed by the classic geochemical reaction path, for example, has limited applicability to natural porous media. For these reasons, and facilitated by more powerful computer hardware and advanced numerical methods, the development and application of reactive transport models has accelerated in the past decade.

It is important to recognize that the numerical methods and the potential errors associated with these methods follow to some extent from the process formulations. For example, if it is appropriate to describe flow and transport through a porous media with a single velocity (i.e. the medium is homogeneous and essentially one dimensional) and the reactions can be described with a simple first-order rate law which has no feedbacks to other chemical species in the system, then there is no need to resort to the methods described below for treating nonlinear systems. Similarly, if a distribution coefficient provides an adequate chemical description of adsorption in a particular system, then the solution of the reaction-transport equations is relatively straightforward. The most demanding problems are those in which the chemistry is nonlinear because of the coupling between the various components in the system or where physical and chemical heterogeneities in the system create constantly changing conditions throughout the domain.

In this chapter, we will present an overview of some of the methods currently used to model the dynamics of reactive transport in porous media. The focus will be on numerical methods rather than analytical solutions to the governing partial differential equations, since the numerical methods have the advantage of being able to handle both nonlinearities in the governing equations (although these can present numerical difficulties) and more importantly, they are best suited for treating multicomponent chemistry. In addition, numerical methods are also better suited for treating physical and chemical heterogeneities which may occur in a particular system. Analytical solutions to the reaction-transport equation will be used primarily to assess the magnitude of various kinds of error in the numerical solution of the system of equations.

The material which follows is divided into three main sections. In the first section we present some of the most popular mathematical descriptions for systems of nonlinear reactions; consideration is given to both kinetic and equilibrium reactions, and systems which contain both. We then briefly review reaction formulations and the numerical techniques used to solve particular reaction systems. The discussion of reaction modeling is followed by a short discussion of some of the more common numerical methods used to model transport processes in porous media. This vast field is extensively treated (and reviewed) elsewhere (e.g. Huyakorn and Pinder, 1983; Celia and Gray, 1992; Zheng and Bennett, 1995). It is important, however, to include some discussion of the transport errors themselves since these errors may be just as important as the other errors which appear in the coupled multicomponent, multi-species systems (e.g. errors associated with coupling of the reaction and transport terms). For the sake of brevity, our presentation is restricted to one-dimensional domains, however, some important higher dimensional aspects are noted. The third section of this chapter considers the most commonly used methods to couple transport and reactions and what errors may arise from these coupling methods. Several examples are presented which illustrate the performance of these methods.

REACTION ALGORITHMS FOR MULTICOMPONENT SYSTEMS

Mathematical descriptions of reaction systems

The multicomponent, multi-species systems typical of those which occur in porous media require some special treatment, both because they involve multiple unknowns and because they are usually nonlinear. The mathematical description used, however, will depend on what form the reactions in the system are assumed to take. It is instructive to derive a general approach to handle multicomponent, multi-species reactive systems. Formulations for arbitrarily complex reaction systems characterized by both equilibrium and non-equilibrium reactions have been presented by Lichtner (1985), Lichtner (this volume), Friedly and Rubin (1992), Sevougian et al. (1993), and Chilakapati (1995). A clear discussion of one possible way of doing so is given by Chilakapati (1995). His approach begins with the most general case, a set of ordinary differential equations for each species in the system and reactions between the species described by kinetic rate laws. A system containing N_{tot} species and N_r reactions can be expressed as

$$\mathbf{I} \cdot \frac{d\mathbf{C}}{dt} = \nu \cdot \mathbf{R}. \tag{1}$$

The raised dot indicates matrix multiplication, $\mathbf{I}$ is the identity matrix of dimension $N_{tot} \times N_{tot}$, $\mathbf{C}$ is the vector of solute concentrations of length N_{tot}, ν is a matrix of dimension $N_{tot} \times N_r$, and $\mathbf{R}$ is a vector of length N_r. For example, the matrix ν and the vector $\mathbf{R}$ have the form

$$\nu = \begin{bmatrix} \nu_{1,1} & \nu_{1,2} & \cdots & \nu_{1,N_r} \\ \nu_{2,1} & \nu_{2,2} & \cdots & \nu_{2,N_r} \\ \vdots & \cdots & \cdots & \vdots \\ \nu_{N_{tot},1} & \nu_{N_{tot},2} & \cdots & \nu_{N_{tot},N_r} \end{bmatrix} \qquad \mathbf{R} = \begin{bmatrix} R_1 \\ R_2 \\ \cdots \\ R_{N_r} \end{bmatrix}. \tag{2}$$

The multiplication of the identity matrix by the derivatives of the individual species concentrations results in an ODE of similar form for each of the species in the system.

As an example, consider an aqueous system consisting of Ca^{+2}, H^+, OH^-, CO_3^{-2}, HCO_3^-, H_2CO_3, and $CaCO_3$(s) (calcite). We ignore H_2O for the sake of conciseness. In addition, we assume that the following reactions occur, without yet specifying whether they are to be considered equilibrium or kinetically-controlled reactions,

$$CaCO_3 \rightleftharpoons Ca^{+2} + CO_3^{-2} \qquad R_1 \tag{3}$$

$$HCO_3^- \rightleftharpoons CO_3^{-2} + H^+ \qquad R_2 \tag{4}$$

$$H_2CO_3 \rightleftharpoons CO_3^{-2} + 2H^+ \qquad R_3 \tag{5}$$

$$H^+ + OH^- \rightleftharpoons H_2O \qquad R_4. \tag{6}$$

In the above equations R_i symbolizes the rate expression for reaction i. We also make no assumptions at this stage about whether the set of reactions included are linearly independent (although the reactions listed above are). We have shown the reactions to be reversible here (thus the symbol $\rightleftharpoons$) but the results below apply whether the reactions are irreversible or reversible since at this stage, one can think of the reaction rates as simply time-dependent expressions of the mole balances inherent in a balanced chemical reaction. The reversibility or lack thereof only determines whether the sign of the reaction rate can change. The term *reversible* is generally used by thermodynamicists to refer to equilibrium reactions (Lichtner, this volume), although we prefer to use it to refer to reactions which are sufficiently close to equilibrium that the backward reaction is important. It is quite possible in a steady-state flow system, for example, for backward reactions to be important and yet not to be at equilibrium (e.g. Nagy et al., 1991; Nagy and Lasaga, 1992; Burch et al., 1993). According to this definition, the term *irreversible* is used for those reactions which proceed in only one direction (i.e. those that can be represented with a unidirectional arrow, $\longrightarrow$).

For our example aqueous system, the rates for the individual species can be written

$$\frac{d[H_2CO_3]}{dt} = -R_3 \tag{7}$$

$$\frac{d[HCO_3^-]}{dt} = -R_2 \tag{8}$$

$$\frac{d[CaCO_3]}{dt} = -R_1 \tag{9}$$

$$\frac{d[OH^-]}{dt} = -R_4 \tag{10}$$

$$\frac{d[H^+]}{dt} = R_2 + 2R_3 - R_4 \tag{11}$$

$$\frac{d[Ca^{+2}]}{dt} = R_1 \tag{12}$$

$$\frac{d[CO_3^{-2}]}{dt} = R_1 + R_2 + R_3. \tag{13}$$

In matrix form the system of equations becomes

$$\begin{bmatrix} 1 & 0 & 0 & 0 & 0 & 0 & 0 \\ 0 & 1 & 0 & 0 & 0 & 0 & 0 \\ 0 & 0 & 1 & 0 & 0 & 0 & 0 \\ 0 & 0 & 0 & 1 & 0 & 0 & 0 \\ 0 & 0 & 0 & 0 & 1 & 0 & 0 \\ 0 & 0 & 0 & 0 & 0 & 1 & 0 \\ 0 & 0 & 0 & 0 & 0 & 0 & 1 \end{bmatrix} \begin{bmatrix} \frac{d[H_2CO_3]}{dt} \\ \frac{d[HCO_3^-]}{dt} \\ \frac{d[CaCO_3]}{dt} \\ \frac{d[OH^-]}{dt} \\ \frac{d[H^+]}{dt} \\ \frac{d[Ca^{+2}]}{dt} \\ \frac{d[CO_3^{-2}]}{dt} \end{bmatrix} = \begin{bmatrix} 0 & 0 & -1 & 0 \\ 0 & -1 & 0 & 0 \\ -1 & 0 & 0 & 0 \\ 0 & 0 & 0 & -1 \\ 0 & 1 & 2 & -1 \\ 1 & 0 & 0 & 0 \\ 1 & 1 & 1 & 0 \end{bmatrix} \begin{bmatrix} R_1 \\ R_2 \\ R_3 \\ R_4 \end{bmatrix} \tag{14}$$

As written in Equation (14), the stoichiometric reaction matrix, ν, is referred to as being in *canonical* form (Smith and Missen, 1982; Lichtner, 1985; Lichtner, this volume). The system of equations is partitioned into the first four rows where the associated species (H_2CO_3, HCO_3^-, $CaCO_3(s)$, and OH^-) are involved in only one reaction while in the remaining three rows the species are involved in multiple reactions. The first four species are referred to as secondary or non-component species, while the last three are the primary or component species (Lichtner, this volume). These are also referred to as *basis* species because they form a basis which spans the concentration space. In this example, we have written all of the carbonate reactions using the species CO_3^{-2} precisely so as to restrict all of the other carbonate species to involvement in a single reaction. This is an essential first step in obtaining either the canonical formulation (Lichtner, 1985; Lichtner, this volume) or to writing the reactions in *tableaux* form (Morel and Hering, 1993), both of which assume that one is dealing with a set of linearly independent reactions, but it is not essential for what follows below. The procedure will also work if, for example, the formation of H_2CO_3 involved H^+ and HCO_3^- rather than 2 H^+ and CO_3^{-2}, although we will not obtain the conserved quantities (total H^+, total CO_3^{-2}, etc.) found in the tableaux method without additional manipulations.

The system of ODEs could be solved directly in the form of Equation (14) if the reactions are all described with kinetic rate laws. Alternatively, one can apply a Gauss-Jordan elimination process to the matrix ν and simultaneously to the identity matrix **I** until there are no pivots left (Chilakapati, 1995). The resulting transformed set of ODEs is now

$$\mathbf{M} \cdot \frac{d\mathbf{C}}{dt} = \nu^* \cdot \mathbf{R} \tag{15}$$

which partitions the system of equations into N_r ODEs associated with reactions and N_c conservation laws with zero right-hand sides (i.e. no associated reactions). The number of conservation laws or mole balance equations is equal to

$$N_c = N_{tot} - \text{rank of } \nu = N_{tot} - N_r. \tag{16}$$

N_r, therefore, refers to the number of *linearly independent* reactions between the species in the system. For the sake of clarity, we make the first N_r rows of the matrix **M** the ODEs with associated reactions and the next N_c rows the conservation equations, so that the left hand of Equation (15) takes the form

$$\begin{bmatrix} M_{1,1} & \cdots & M_{1,N_r+N_c} \\ \vdots & \cdots & \vdots \\ M_{N_r,1} & \cdots & M_{N_r,N_r+N_c} \\ M_{N_r+1,1} & \cdots & M_{N_r+1,N_r} \\ \vdots & \cdots & \vdots \\ M_{N_r+N_c,1} & \cdots & M_{N_r+N_c,N_r+N_c} \end{bmatrix} \cdot \begin{bmatrix} \frac{dC_1}{dt} \\ \vdots \\ \frac{dC_{N_r}}{dt} \\ \vdots \\ \frac{dC_{N_r+N_c}}{dt} \end{bmatrix} \tag{17}$$

In our example, the Gauss-Jordan elimination is carried out on the the matrix ν on the right hand side of Equation (14) and the same row transformations are applied to the identity matrix, **I**, yielding

$$\begin{bmatrix} 1 & 0 & 0 & 0 & 0 & 0 & 0 \\ 0 & 1 & 0 & 0 & 0 & 0 & 0 \\ 0 & 0 & 1 & 0 & 0 & 0 & 0 \\ 0 & 0 & 0 & 1 & 0 & 0 & 0 \\ 2 & 1 & 0 & -1 & 1 & 0 & 0 \\ 0 & 0 & 1 & 0 & 0 & 1 & 0 \\ 1 & 1 & 1 & 0 & 0 & 0 & 1 \end{bmatrix} \begin{bmatrix} \frac{d[H_2CO_3]}{dt} \\ \frac{d[HCO_3^-]}{dt} \\ \frac{d[CaCO_3]}{dt} \\ \frac{d[OH^-]}{dt} \\ \frac{d[H^+]}{dt} \\ \frac{d[Ca^{+2}]}{dt} \\ \frac{d[CO_3^{-2}]}{dt} \end{bmatrix} = \begin{bmatrix} 0 & 0 & -1 & 0 \\ 0 & -1 & 0 & 0 \\ -1 & 0 & 0 & 0 \\ 0 & 0 & 0 & -1 \\ 0 & 0 & 0 & 0 \\ 0 & 0 & 0 & 0 \\ 0 & 0 & 0 & 0 \end{bmatrix} \begin{bmatrix} R_1 \\ R_2 \\ R_3 \\ R_4 \end{bmatrix} \tag{18}$$

The stoichiometric reaction matrix, ν^*, now consists of a nonsingular 4 by 4 matrix (N_r by N_r) and three rows of zeros corresponding to the N_c conservation equations. Writing out the ODEs in Equation (18), we find

$$\frac{d[H_2CO_3]}{dt} = -R_3 \tag{19}$$

$$\frac{d[HCO_3^-]}{dt} = -R_2 \tag{20}$$

$$\frac{d[CaCO_3]}{dt} = -R_1 \tag{21}$$

$$\frac{d[OH^-]}{dt} = -R_4 \tag{22}$$

Table 1. Tableaux for carbonate system, neglecting H_2O as a species and component.

		Components		
		H^+	Ca^{+2}	CO_3^{-2}
Species	H_2CO_3	2		1
	HCO_3^-	1		1
	$CaCO_3$		1	1
	OH^-	-1		
	H^+	1		
	Ca^{+2}		1	
	CO_3^{-2}			1

and

$$\frac{d}{dt}\left([H^+] + 2[H_2CO_3] + [HCO_3^-] - [OH^-]\right) = 0 \tag{23}$$

$$\frac{d}{dt}\left([Ca^{+2}] + [CaCO_3]\right) = 0 \tag{24}$$

$$\frac{d}{dt}\left([CO_3^{-2}] + [H_2CO_3] + [HCO_3^-] + [CaCO_3]\right) = 0. \tag{25}$$

From the example, it is apparent that we have eliminated the reactions in the equations originally corresponding to the species H^+, Ca^{+2}, and CO_3^{-2} by making use of the relations in the first four equations. The last three equations are mole balances for *total* H^+, Ca^{+2}, and CO_3^{-2}

$$TOTH^+ = [H^+] - [OH^-] + [HCO_3^-] + 2[H_2CO_3] \tag{26}$$

$$TOTCa^{+2} = [Ca^{+2}] + [CaCO_3] \tag{27}$$

$$TOTCO_3^{-2} = [CO_3^{-2}] + [H_2CO_3] + [HCO_3^-] + [CaCO_3]. \tag{28}$$

Note that the canonical form of the stoichiometric reaction matrix is identical to the *tableaux* form popularized by Morel and coworkers (Morel and Hering, 1993; Dzombak and Morel, 1990). By transposing the last three rows of the matrix **M** in Equation (17), we can write the matrix in tableaux form (Table 1).

The procedure has yielded expressions for the total concentrations of the N_c *primary* or *component* species. A more general form is given by

$$TOT_j = C_j + \sum_{i=1}^{N_r} \nu_{ij} X_i \tag{29}$$

where C_j and X_i refer to the concentration of the primary and secondary species respectively. Note that the number of secondary species is equal to N_r, the number of linearly independent reactions in the system (i.e. the rank of the matrix ν). Equation (27) and Equation (28) are recognizable as the total concentrations of calcium and carbonate respectively. The total concentration of H^+ is written in exactly the same form as the other equations, although its physical meaning is less clear because it may take on negative values due to the negative stoichiometric coefficients in the expression. The mole balance equation for total H^+ is just the *proton condition* equation referred to in many aquatic chemistry textbooks. Oxidation-reduction reactions are also easily handled with this method. If the redox reactions are written as whole cell reactions, there is no need in any application not involving an electrical current (see Lichtner, this volume) to introduce the electron as an unknown. Writing the reactions as whole cell reactions allows redox reactions to be treated exactly like any other reaction.

Dependent chemical reactions. In the example above, all of the chemical reactions were linearly independent, that is, one of the reactions could not be obtained simply by combining one or more of the other reactions. The Gauss-Jordan elimination procedure used above is intended to find the number of linearly independent rows, N_r, whether we begin with a linearly independent set or not. In the case of a fully equilibrium system, the linearly dependent reactions can always be eliminated. Where the system is not completely at equilibrium, however, it is quite possible for a species to react via several different dependent pathways. Pursuing our example above, we could write three reactions involving the aqueous carbonate species

$$HCO_3^- \rightleftharpoons CO_3^{-2} + H^+ \qquad R_2 \tag{30}$$

$$H_2CO_3 \rightleftharpoons CO_3^{-2} + 2H^+ \qquad R_3 \tag{31}$$

$$H_2CO_3 \rightleftharpoons HCO_3^- + H^+ \qquad R_5 \tag{32}$$

of which only two are linearly dependent. However, one could envision a system in which the species H_2CO_3 is formed both by reaction of one proton with one bicarbonate ion and also by reaction of one carbonate ion with two protons. An extremely important biogeochemical example of linearly dependent reactions is the microbially-mediated oxidation of organic carbon by multiple electron acceptors (Rittmann and VanBriesen, this volume; Van Cappellen and Gaillard, this volume). So, for example, we could have a series of heterotrophic reactions in which organic carbon is oxidized by O_2 (aerobic respiration), NO_3^- (denitrification), manganese oxide (manganese reduction), iron oxide (iron reduction), and SO_4^{-2} (sulfate reduction). To the extent that intra-aqueous reactions are present linking the various electron acceptors and donors (e.g. nitrification), this set of microbially-mediated reactions will be partly or completely linearly dependent. Since each of these pathways is associated with a specific rate, a complete description of the system requires the inclusion of a number of linearly dependent pathways in the reaction scheme.

If we include the dependent reaction, R_5, in the reaction network and write the system of equations again as in Equation (14), we have

$$\begin{bmatrix} 1 & 0 & 0 & 0 & 0 & 0 & 0 \\ 0 & 1 & 0 & 0 & 0 & 0 & 0 \\ 0 & 0 & 1 & 0 & 0 & 0 & 0 \\ 0 & 0 & 0 & 1 & 0 & 0 & 0 \\ 0 & 0 & 0 & 0 & 1 & 0 & 0 \\ 0 & 0 & 0 & 0 & 0 & 1 & 0 \\ 0 & 0 & 0 & 0 & 0 & 0 & 1 \end{bmatrix} \begin{bmatrix} \frac{d[H_2CO_3]}{dt} \\ \frac{d[HCO_3^-]}{dt} \\ \frac{d[CaCO_3]}{dt} \\ \frac{d[OH^-]}{dt} \\ \frac{d[H^+]}{dt} \\ \frac{d[Ca^{+2}]}{dt} \\ \frac{d[CO_3^{-2}]}{dt} \end{bmatrix} = \begin{bmatrix} 0 & 0 & -1 & 0 & -1 \\ 0 & -1 & 0 & 0 & 1 \\ -1 & 0 & 0 & 0 & 0 \\ 0 & 0 & 0 & -1 & 0 \\ 0 & 1 & 2 & -1 & 1 \\ 1 & 0 & 0 & 0 & 0 \\ 1 & 1 & 1 & 0 & 0 \end{bmatrix} \begin{bmatrix} R_1 \\ R_2 \\ R_3 \\ R_4 \\ R_5 \end{bmatrix} \tag{33}$$

Applying the same Gauss-Jordan elimination procedure as above, the transformed set of ODEs becomes

$$\begin{bmatrix} 1 & 0 & 0 & 0 & 0 & 0 & 0 \\ 0 & 1 & 0 & 0 & 0 & 0 & 0 \\ 0 & 0 & 1 & 0 & 0 & 0 & 0 \\ 0 & 0 & 0 & 1 & 0 & 0 & 0 \\ 2 & 1 & 0 & -1 & 1 & 0 & 0 \\ 0 & 0 & 1 & 0 & 0 & 1 & 0 \\ 1 & 1 & 1 & 0 & 0 & 0 & 1 \end{bmatrix} \begin{bmatrix} \frac{d[H_2CO_3]}{dt} \\ \frac{d[HCO_3^-]}{dt} \\ \frac{d[CaCO_3]}{dt} \\ \frac{d[OH^-]}{dt} \\ \frac{d[H^+]}{dt} \\ \frac{d[Ca^{+2}]}{dt} \\ \frac{d[CO_3^{-2}]}{dt} \end{bmatrix} = \begin{bmatrix} 0 & 0 & -1 & 0 & -1 \\ 0 & -1 & 0 & 0 & 1 \\ -1 & 0 & 0 & 0 & 0 \\ 0 & 0 & 0 & -1 & 0 \\ 0 & 0 & 0 & 0 & 0 \\ 0 & 0 & 0 & 0 & 0 \\ 0 & 0 & 0 & 0 & 0 \end{bmatrix} \begin{bmatrix} R_1 \\ R_2 \\ R_3 \\ R_4 \\ R_5 \end{bmatrix} \tag{34}$$

Since the rank of the matrix ν is N_r, the number of linearly independent reactions, the Gauss-Jordan elimination procedure finds the same number of conservation or mole balance equations, $N_c = N_{tot} - N_r$. Note, however, that now two of the secondary or non-component species, H_2CO_3 and HCO_3^-, are involved in two reactions rather than a single reaction as in the linearly independent example above

$$\frac{d[H_2CO_3]}{dt} = -R_3 - R_5 \tag{35}$$

$$\frac{d[HCO_3^-]}{dt} = -R_2 + R_5. \tag{36}$$

Including equilibrium reactions. In the event that any of the reactions in the network are assumed to be at equilibrium, they can be replaced with algebraic expressions based on mass action expressions. This is done by ensuring that the equilibrium reaction rate only appears in one equation. One can ensure this by finding the pivots for the columns in the matrix, ν, corresponding to the equilibrium reactions first (Chilakapati, 1995). This procedure is equivalent to configuring the system so that the equilibrium reaction is not written in terms of more than one non-component species. This is not an issue in the case of linearly independent reactions, as in Equation (18), where each of the first N_r species are associated with only one reaction. But in the case of a linearly dependent set, as in Equation (33), this would mean that if R_5 were the reaction for which we were to assume equilibrium, then the system should be formulated so that it appears in only one ODE. This involves making either HCO_3^- or H_2CO_3 a component species.

Given the above proviso, the procedure for including equilbrium reactions is simple. Assume that the rate at which reaction R_2 proceeds is very fast such that the reaction can be assumed to be at equilibrium. Reaction 2 appears in only the second ODE, which involves the noncomponent species HCO_3^-

$$\frac{d[HCO_3^-]}{dt} = -R_2 + R_5. \tag{37}$$

This equation is replaced with the mass action expression (neglecting activity coefficients)

$$\frac{[HCO_3^-]}{[H^+][CO_3^{-2}]} = K_{eq} \tag{38}$$

where K_{eq} is the equilibrium constant for the reaction. The entries in the row in the matrix **M** are filled with zeros, i.e. we remove the time-dependence associated with this reaction. By replacing one or more of the N_r ODEs associated with reaction(s) with algebraic relations based on a mass action expression, we have transformed the set of ODEs into a set of differential-algebraic equations (DAEs) (Brenan et al., 1989; Chilakapati, 1995; Hindmarsh and Petzold, 1995a, b).

It should be noted that we still have a system of N_{tot} equations, including N_c conservation or mole balance equations, N_k equations associated with kinetic reactions, and N_e algebraic equations based on mass action expressions ($N_e + N_k = N_r$). The presence of linearly dependent reactions does not change the number of equations and unknowns, since these reactions are incorporated into existing ODEs. Alternatively, we can formally eliminate the non-component species associated with the equilibrium reaction, in our case the species HCO_3^-, by simple rearrangement of Equation (38)

$$[HCO_3^-] = [H^+][CO_3^{-2}]K_{eq}. \tag{39}$$

A more general form is given by

$$X_i = K_i^{-1} \gamma_i^{-1} \prod_{j=1}^{N_c} (\gamma_j C_j)^{\nu_{ij}} \tag{40}$$

where X_i and C_j are the secondary and component species respectively, γ_i and γ_j are their respective activity coefficients, K_i are the equilibrium constants, N_c is the number of components in the system, and ν_{ij} is the stoichiometric reaction coefficient. The total CO_3^{-2} and H^+ concentrations then become

$$TOTH^+ = [H^+] + 2[H_2CO_3] + \left([H^+][CO_3^{-2}]K_{eq}\right) - [OH^-], \tag{41}$$

$$TOTCO_3^{-2} = [CO_3^{-2}] + [H_2CO_3] + \left([H^+][CO_3^{-2}]K_{eq}\right) + [CaCO_3] \tag{42}$$

where the other non-component species are left as independent unknowns since they are associated with kinetic reactions. Note that this procedure has reduced the number of independent unknowns in the system of equations by one. The elimination of equilibrium reactions, therefore, is a computationally efficient way of formulating the system, particularly in the case where most of the reactions are assumed to be at equilibrium. Since in many aqueous systems there may be as many as 50 to 100 fast complexation reactions, the computational savings can be significant.

It is possible to eliminate all the ODEs in Equation (34) if all the reactions are considered to be at equilibrium. The system then collapses to a set of nonlinear algebraic equations (i.e. no temporal derivatives) which have the form of Equations (29) and (40). Such systems form the basis for the common equilibrium codes which have been widely used to model batch and reactive transport problems (e.g. Felmy et al., 1984; Walter et al., 1994).

We have now taken the original system of ODEs in Equation (1), transformed it to a differential-algebraic (DAE) system, and further to a fully-algebraic (equilibrium) system. While we have not discussed transport aspects at this point, it is worth observing that the chemical system formulation can have an impact on the physical transport computations. For example, for a system in which every reaction is treated kinetically, it is necessary to solve for the transport of every species. On the other hand for a partial equilibrium system, only the component species and non-component species associated with kinetic reactions need to be considered in the transport algorithm. Again returning to the preceeding example, we would not have to numerically solve a transport equation for HCO_3^- because its concentration is uniquely defined by Equation (39) at every spatial location in the domain.

TREATMENT OF TEMPORAL DERIVATIVES

By definition, all reactive transport problems involve temporal derivatives. Finite difference approximations are generally used to handle the time derivative, even where the spatial domain is treated with finite element methods. The idea behind finite difference methods is straightforward. The continuous differential operators which appear in differential equations are replaced by approximations written in terms of a finite number of discrete values of the independent variable (i.e. space, time, reaction progress etc.). The methods are intuitive to most people because they reverse the procedure by which calculus, which applies to continuous, exact differential operators, is derived. In first semester calculus, one usually obtains the exact derivative by beginning with a discrete approximation to the interval along a curve, for example, and letting that interval $\rightarrow 0$. Finite differences will be applied both to the time discretization and to the spatial discretization below, but we begin with a more general discussion of methods to advance the system in time using finite differences.

The most important distinction to be made between methods for advancing the solution in time is between *explicit* methods, where information from the present time level is used, and *implicit* methods, where the functional dependence of the rate of change of the field variable (e.g. concentration) depends on the values of the field variable at the future time level. This can be stated more succinctly in mathematical form. For example, an explicit finite difference representation of the rate of change of the concentration due to reaction can be written as

$$\frac{C^{n+1} - C^n}{\Delta t} = R(C^n) \tag{43}$$

where $n + 1$ and n refer to the future and present time levels respectively and $R(C^n)$ is the reaction rate which is evaluated using the concentrations at the present time level. The above scheme is referred to as the *forward Euler* method. Note that the reaction rate (which might have a nonlinear dependence on the concentration) is evaluated using *known* values of the concentration, so the method is extremely easy to implement. An implicit version of Equation (43), referred to as *backward Euler* method, is given by

$$\frac{C^{n+1} - C^n}{\Delta t} = R(C^{n+1}). \tag{44}$$

The previous two time discretizations can be represented more generally as

$$\frac{C^{n+1} - C^n}{\Delta t} = \epsilon R(C^{n+1}) + (1 - \epsilon)R(C^n) \tag{45}$$

where $\epsilon = 1$ gives implicit, $\epsilon = 0$ gives explicit, and $\epsilon = 0.5$ gives *centered-time weighting* (also referred to as *Crank-Nicolson* temporal weighting). Centered-time weighting is a second-order accurate scheme, while implicit and explicit are only first-order accurate; this implies that centered-weighting should provide a more accurate result for a given time step size. We postpone a more detailed analysis on the *order* of a discretization scheme until the section on modeling transport processes.

In multicomponent systems in which the reaction rate usually depends on a number of different species, the backward Euler method and centered method require the solution of a set of simultaneous equations. In the case where the reaction rate has a nonlinear dependence on one or more of the solute concentrations in the system, solving Equation (44) will require a number of iterations, each of which will involve the solution of a set of simultaneous equations.

Although the explicit method is much simpler to implement than the implicit or centered schemes, its stability is limited. A numerical method that is *unstable* will produce completely inaccurate answers (oscillations with increasing amplitude) unless sufficiently small time steps are taken. The problem with using explicit methods to solve sets of ordinary differential equations (ODEs) was discovered as early as the 1950s by chemists working with reactor systems in which multiple kinetically-controlled reactions occurred. In those systems where there was a wide range in reaction rates, they found that when using the explicit methods they were forced by the fastest reaction in the system to use a small time step. As a kinetically reacting system characterized by widely differing rates of reaction evolves in time, the fastest reactions reach steady-state first. Once these reactions have reached steady state, the concentrations of the reactive intermediaries involved in these reactions do not change, so based on accuracy considerations alone, one should be able to take larger time steps. However, when using explicit methods, the fast reactions still require a very small time step in order to maintain stability (Kee et al., 1985). This feature of dynamical systems, where stability constraints on a time step far exceed the requirements for accuracy, is referred to as *stiffness*. It is possible in some cases to use the explicit approach if appropriate assumptions about the form of the rate equations are made (Oran and Boris, 1987). Without these kind of *asymptotic* assumptions, workers who have reported success with the explicit approaches for solving sets of differential equations have usually been investigating special systems where all of the rate constants are nearly of the same value and the rate laws are linear or nearly so. The usual cure for stiffness in sets of differential equations is to use either implicit or semi-implicit methods (Press et al., 1986).

The temporal discretization methods represented by Equation (45) are only three of a number of methods commonly used to solve initial value problems. Another commonly used approach is the *predictor-corrector* method, a two-step approach which gives a more accurate and usually more stable result than the unmodified forward Euler method. To implement a predictor-corrector or corrected Euler on our model reaction system, we can use a two-step approach consisting of

$$C^* = C^n + (\Delta t) R(C^n) \tag{46}$$

which is just a standard forward Euler step, followed by a *corrector* step where the reaction rate is computed using the provisional value of the concentration, C^*

$$C^{n+1} = C^n + \frac{\Delta t}{2} \left[R(C^n) + R(C^*) \right]. \tag{47}$$

Other somewhat more complicated but generally more accurate schemes are available as well, including explicit and implicit *Runge-Kutta* methods (e.g. Press et al., 1986; Celia and Gray, 1992).

FORMULATING AND SOLVING THE CHEMICAL REACTION EQUATIONS

At this stage, we have shown that it is possible to transform a set of ODEs for reacting species of the form in Equation (1) to one of the form in Equation (15), thus partitioning the system into N_r ODEs associated with reactions and N_c conservation or mole balance equations. In addition, where reactions are sufficiently fast that equilibrium can be assumed, the differential equations associated with these reactions can be replaced by algebraic equations based on mass action expressions, thus creating a set of DAEs. The more common procedure in the case of the assumption of equilibrium with respect to some or all of the reactions, however, is to formally eliminate the non-component species associated with the equilibrium reactions, thus creating a smaller set of nonlinear ODEs. The choice of formulation, of course, is dictated primarily by the system itself. Below, we consider the individual formulations and the various approaches which can be used to solve each.

Fully kinetic formulations

A number of examples of reactive transport systems in porous media formulated completely in terms of kinetic reactions have appeared in the literature (MacQuarrie et al., 1990; Wood et al., 1994; Wood et al., 1995; Soetart et al., 1996). Most of these examples involve microbially-mediated redox reactions which have slow rates and therefore require a kinetic description. In many of the models used to describe low-temperature environments, no mention is made of thermodynamics or the equilibrium state (e.g. Soetart et al., 1996). In the case of redox reactions, this is true not only because of the slowness of the reactions, but also because the concentration of the electron acceptor or donor involved in a particular reaction goes effectively to zero before equilibrium is achieved. For example, in the case of the oxidation of organic carbon by molecular oxygen, the rate either goes to zero or is not measurable when oxygen drops to a low level (10^{-8} M). Other reactions like nitrification which use up oxygen also go to zero well before equilibrium is achieved. An equilibrium model would continue to reduce the concentration of molecular oxygen to a physically and chemically unreasonable value (eventually below 10^{-20} M). Even through the Gibbs free energies of the reactions provide the driving force for the reactions and presumably determine the sequence (in time or space) of the reactions, in many cases there is no need to include a backward reaction in the algorithm since equilibrium with individual redox pathways is never achieved. The system in this case is truly irreversible.

The microbiological reactions, which include nitrification, aerobic biodegradation, and other oxidation-reduction reactions, are most often handled using a multiple-Monod expression (Molz et al., 1986; MacQuarrie et al., 1990; Essaid et al., 1995, Rittmann and VanBriesen, this volume; Van Cappellen and Gaillard, this volume; Van Cappellen and Wang, 1996). Multiple-Monod expressions are used in favor of simpler first-order rate expressions because it is possible to include the effects of changes in concentration of the electron acceptors, electron donors, microbial biomass, and other species which may inhibit or enhance the reaction rate. As an example, we can consider the irreversible reaction describing the aerobic degradation of CH_2O in aqueous solution

$$CH_2O + O_{2(aq)} \rightarrow CO_{2(aq)} + H_2O. \tag{48}$$

The ordinary differential equation for the concentration of CH_2O is given by the multiple-Monod expression

$$\frac{d\,[CH_2O]}{dt} = -\,k_{max}\,X_m \left(\frac{[CH_2O]}{[CH_2O] + K_{CH_2O}}\right)\left(\frac{[O_{2(aq)}]}{[O_{2(aq)}] + K_{O_{2(aq)}}}\right) \tag{49}$$

where $[CH_2O]$ and $[O_{2(aq)}]$ indicate the concentration of CH_2O and $O_{2(aq)}$ respectively, k_{max}, K_{CH_2O}, $K_{O_{2(aq)}}$ are empirically determined coefficients, and X_m is the biomass concentration of the mediating microbial population. It is obvious that if the concentration of CH_2O, O_2, or X_m decrease toward zero, the reaction rate also tends toward zero. If the dynamics of biomass growth and decay are considered important, then an additional ODE for the biomass concentration, X_m, must be included in the reaction system.

Simulating mixed equilibrium-kinetic systems with kinetic formulations. There are few if any aquatic systems which do not include fast reactions which are at equilibrium. In some cases, the fast reactions (normally involving the carbonate system which controls the solution pH in most aquatic environments) can be ignored if they do not impact the rates of the reactions which are the primary focus of the model (e.g. hydrocarbon degradation, MacQuarrie et al., 1990). Even where fast reactions need to be included, however, it is possible to use a fully kinetic formulation. Several groups are now using schemes intended to simulate systems with both fast and slow reactions (thus a system with both reactions at equilibrium and out of equilibrium) which rely completely on kinetic formulations (Chilakapati, 1995; MacQuarrie, pers. comm.). The fully kinetic approach obviates the need to solve the set of mixed algebraic and differential equations which characterize mixed equilibrium-kinetic systems and avoids the *ad hoc* iteration schemes which are often employed in solving a DAE system. For aqueous inorganic reactions, a fully kinetic formulation is arrived at by making use of the *principle of microscopic reversibility* or *detailed balancing* (Lasaga, 1981; Lasaga, 1984; Morel and Hering, 1993). For an elementary reaction described by

$$A + B \rightleftharpoons C, \tag{50}$$

we can write the forward rate, $R_{forward}$, as

$$R_{forward} = k_f[A][B], \tag{51}$$

and the backward or reverse rate as

$$R_{\text{backward}} = k_b[\text{C}] \tag{52}$$

where [A], [B], and [C] indicate the concentrations of the species A, B, and C respectively and k_f and k_b are the forward and reverse rate constants respectively. The complete rate expression for species A is therefore

$$\frac{d[\text{A}]}{dt} = -k_f[\text{A}][\text{B}] + k_b[\text{C}]. \tag{53}$$

When the above reaction is at equilibrium, the forward and reverse rates must balance. One can view chemical equilibrium, therefore, as a dynamic state in which the forward and backward reaction rates balance rather than as one characterized by stasis. Since the forward and reverse rates *at equilibrium* must balance, we can write

$$k_f[\text{A}][\text{B}] = k_b[\text{C}], \tag{54}$$

which upon rearranging gives

$$\frac{k_f}{k_b} = \frac{[\text{C}]}{[\text{A}][\text{B}]} = K_{eq} \tag{55}$$

where K_{eq} is the equilibrium constant for the reaction. This principle holds in reality only where elementary reaction mechanisms are involved (Lasaga, 1984), or in certain cases where a single elementary reaction is the rate-limiting step within the overall reaction (Aagaard and Helgeson, 1982). Where this approach is used to approximate local equilibrium behavior, however, it is immaterial what the actual values of the forward and reverse rate constants are, as long as they are chosen in the proper ratio $k_f/k_b = K_{eq}$ and they are chosen to be sufficiently large such that local equilibrium is attained during the time scale of interest. The system of equations (a set of ordinary differential equations) is then solved with a stiff ODE solver (i.e. one using an implicit approach).

By simple rearrangement of terms, one can also use this approach to obtain a rate expression in terms of a single forward rate and an equilibrium constant. We can rewrite Equation (55) to obtain an expression for net rate of change of the species A

$$\frac{d[\text{A}]}{dt} = -[\text{A}][\text{B}]k_f\left(1 - \frac{[\text{C}]}{[\text{A}][\text{B}]}K_{eq}^{-1}\right) = -[\text{A}][\text{B}]k_f\left(1 - QK_{eq}^{-1}\right) \tag{56}$$

where Q is the ion activity product. This approach has been widely applied to mineral dissolution and precipitation (Aagaard and Helgeson, 1982; Lasaga, 1984), although the actual mechanism involved in the dissolution will tend to change the formulation. For example, one can write the dissolution of calcite as

$$\text{CaCO}_3(\text{s}) \rightleftharpoons \text{Ca}^{+2} + \text{CO}_3^{-2} \tag{57}$$

which would lead to a rate law assuming an elementary reaction mechanism (Lasaga, 1981) of the form

$$\frac{d[\text{CaCO}_3(\text{s})]}{dt} = -k_f[\text{CaCO}_3(\text{s})] + k_b[\text{Ca}^{+2}][\text{CO}_3^{-2}]. \tag{58}$$

Assuming that the concentration of the calcite in the above rate expression is folded into the surface area of the calcite (in the case of heterogeneous reactions, it is the interfacial surface area, not the actual concentration, which determines the rate), we find that

$$\frac{d[\mathrm{CaCO_3(s)}]}{dt} = -A_{cc}k_f\left(1 - [\mathrm{Ca^{+2}}][\mathrm{CO3^{-2}}]K_{eq}^{-1}\right) = -A_{cc}k_f\left(1 - Q_{cc}K_{eq}^{-1}\right). \tag{59}$$

Note that in this formulation, k_f is the far from equilibrium dissolution rate which has no dependence on any species in solution (i.e. far from equilibrium, this is a zero order rate law). However, if one writes the reaction

$$\mathrm{CaCO_3(s) + H^+ \rightleftharpoons Ca^{+2} + HCO_3^-}, \tag{60}$$

by rearrangement we get

$$\begin{aligned}\frac{d[\mathrm{CaCO_3(s)}]}{dt} &= -A_{cc}k_f[\mathrm{H^+}]\left(1 - \frac{[\mathrm{Ca^{+2}}][\mathrm{HCO3^-}]}{\mathrm{H^+}}K_{eq}'^{-1}\right)\\ &= -A_{cc}k_f[\mathrm{H^+}]\left(1 - Q'_{cc}K_{eq}'^{-1}\right)\end{aligned} \tag{61}$$

where the terms Q'_{cc} and K'_{eq} are used to indicate that both the ion activity product and the equilibrium constant for the reactions have been changed to reflect the calcite dissolution reaction as written in Equation (60). At a given solution composition, however, one will calculate the same ratio of Q_{cc}/K_{eq} whether the calcite dissolution reaction is written as in Equation (57) or Equation (60), but in the rate law given in Equation (61), we find that the dissolution rate now has a first-order dependence on the hydrogen ion activity far from equilibrium. In general, however, most of the rate laws devised for mineral dissolution and precipitation are more empirically than theoretically based, but the above example makes clear that the actual reaction mechanism determines the form of the rate law.

The approach outlined above is attractive since it allows us (in theory) to calculate both the dissolution and precipitation rates knowing only the far from equilibrium dissolution rate of the mineral and the equilibrium constant for the reaction. The above formulation, however, assumes that the same rate mechanism dominates over the entire range of undersaturation and supersaturation and that the precipitation rate mechanism is just the reverse of the dissolution mechanism. There are certainly some minerals for which there is good evidence that a single rate mechanism does not hold throughout. For example, Van Cappellen and Berner (1991) showed that the precipitation rate of fluorapatite changed from a squared dependence at relatively low supersaturations to a fifth order dependence at high supersaturations. They suggested that this change in the dependence of the rate on the extent of supersaturation reflected a change from a spiral growth mechanism of crystal growth to a nucleation-controlled process at high supersaturations. The requirement that for the formulation to work, the precipitation rate must be the same mechanism as the dissolution rate running in reverse also calls into question the applicability of the approach to low-temperature silicate systems where rates of precipitation are either extremely slow or non-existent (R. Wollast, pers. comm., 1995). A silicate like quartz or albite will not precipitate at all at 25°C, or if it precipitates, an amorphous phase forms which has a distinctly higher solubility than the crystalline phase which dissolves. It is therefore difficult to see how one can interpret precipitation as a simple reversal of the dissolution process in these cases.

Where the interest is in simply obtaining equilibrium with respect to a particular reaction, the details of the reaction mechanism(s) are of no interest and the principle of detailed balancing or microscopic reversibility can be used to ensure that the correct equilibrium state is obtained. It is the ratio of the forward and backward rate constants which is important, not the actual

Table 2. Packages for solving sets of ordinary differential equations (ODEs) and differential algebraic equations (DAEs)

Package	Stiff Equations	DAE Capability	Reference
TWOSTEP	No	No	Verwer, 1994
LSODA	Yes	No	Hindmarsh, 1983
VODE	Yes	No	Brown et al., 1989
DASSL	Yes	Yes	Petzold, 1983
RADAU5	Yes	Yes	Hairer and Warren, 1988
LIMEX	Yes	Yes	Deulfhard and Nowak, 1994
LSODI	Yes	Yes	Hindmarsh, 1983

magnitude of the rate constants as long as they are sufficiently large that local equilibrium is obtained. In general, however, the system will be better behaved numerically if the rate constants are chosen so that they are *sufficiently* but not *excessively* large. The other requirement for a robust numerical method in the case of a system characterized by both fast and slow reactions is that it be based on implicit methods. Since systems consisting of both fast and slow reactions are stiff, the explicit approaches will not be competitive with implicit methods unless some special treatment of the fast reactions is used.

Numerical packages for fully kinetic formulations. A number of packages are now available which can solve the stiff systems of ODEs characterizing aquatic reaction system consisting of both fast and slow reactions. Some of the packages suitable for stiff ODEs are listed in Table 2. Also included are those packages which can solve DAE systems (those consisting of both kinetic and equilibrium reactions) as well.

Mixed kinetic-equilibrium (DAE) systems

There is a growing literature on the mathematical and numerical properties of differential algebraic (DAE) systems of equations in the applied mathematics and chemical engineering literature. The properties of these systems have been discussed in Brenan et al. (1989) and in Hindmarsh and Petzold (1995a, b). In addition, a number of packages have been developed specifically for these kind of systems (e.g. DASSL, Petzold 1983). The key feature which distinguishes them from fully kinetic systems is that it is necessary to choose initial conditions which are consistent with the algebraic constraints. While these algebraic constraints may include conservation laws, the most important of them are the mass action expressions which arise when equilibrium is assumed. In order to be consistent with the algebraic constraints, therefore, the initial concentrations must be chosen so that the equilibrium constraints are satisfied. This requires a pre-equilibration of the system in order to choose the appropriate initial conditions. A speciation or equilibration routine normally precedes the time-stepping routine in most reaction path and reaction-transport codes (e.g. Steefel and Yabusaki, 1996).

Decoupled approaches for mixed kinetic-equilibrium systems. An attractive way to simulate mixed kinetic and equilibrium systems from the point of view of the computer code developer involves decoupling the solution of the equilibrium and kinetic reactions. This approach has the advantage of allowing for the maximum flexibility in the choice of individual solvers, and in many cases, allowing for the use of public domain packages suited for fully kinetic or fully equilibrium systems. The decoupled approach is another form of *operator splitting* which will be discussed more fully below when considering the methods for coupling transport and reaction. As in most of the operator splitting approaches, however, there is no clear *a priori* rule of thumb for deciding on how often each of the systems (equilibrium and kinetic) must be solved. This leads to sometimes *ad hoc* iteration schemes for the combined system of equations. For example, one cannot integrate the kinetic reactions over a significant time interval without applying the equilibrium constraints. One possible approach is to iterate between the kinetic and equilibrium reactions (i.e. *sequential iteration*) until convergence is achieved. Another approach is to use a *sequential non-iterative approach* (the equilibrium and kinetic reactions are each solved once), although this normally requires a fairly small time step. As in any operator splitting approach, the convergence of the method can be tested by reducing the time step or interval separating the equilibrium and kinetic calculations until the computed result does not change.

Some examples where decoupled approaches have been used to simulate mixed kinetic and equilibrium reaction systems in porous media can be found in Lensing et al. (1994), who coupled microbial redox reactions and aqueous inorganic chemistry, and in Marzel et al. (1994), who modeled kinetic sorption coupled with equilibrium aqueous chemistry. This approach was also used effectively by Sevougian et al. (1995) in modeling sandstone matrix acidification, an engineering practice whereby injection of acids into a subsurface formation is used to enhance the porosity and permeability.

Modeling equilibrium systems

The other limiting case in geochemical reaction modeling is the fully equilibrium model. Most of the equations which appear in equilibrium systems are nonlinear, so iterative methods appropriate for solving systems of nonlinear equations are needed. Two major approaches have been pursued in the chemical, chemical engineering, and geochemical literature (Nordstrom and Munoz, 1994; Anderson and Crerar, 1993; Bethke, 1996). The first method, which is most widely used in the geochemical community, is based on equilibrium constants embedded in mass balance equations for the various components in the system (Brinkley, 1947; Morel and Morgan, 1972; Felmy et al., 1984; Parkhurst et al., 1980; Reed, 1982; Wolery, 1979; Parkhurst, 1995). This method involves finding the root of a system of nonlinear algebraic equations. The second method employs a direct minimization of the free energy of the system (White et al., 1958; Van Zeggeren and Storey, 1970; Novak and Sevougian, 1993; Felmy, 1990; Sevougian et al., 1993; Harvie et al., 1987; DeCapitani and Brown, 1987). Although as stated by Bethke (1996), given a choice of components, the reactions and species used are the same in the two methods, the two approaches do not otherwise appear to be entirely equivalent. For one thing, the minimization procedure is not mathematically equivalent to finding the root of a set of nonlinear algebraic equations (see Press et al., 1986). Perhaps more important, however, are the differing constraints employed by the two methods. As pointed out by Felmy (1990), the equilibrium constant-based approach involves relaxing the mass balances of the various species in the system while maintaining equilibrium as a constraint. In other words, while assuming equilibrium between the various species and minerals in the system, the nonlinear iteration procedure gradually adjusts the mass of these species until the mass balance is obtained. Problems with convergence using the equilibrium constant approach can occur where the assumption of equilibrium results in large mass balance violations and the nonlinear iteration scheme is unable to converge. With appropriate guarantees that the initial mass balance is not grossly violated (see methods proposed by Wolery, 1992; Bethke, 1996), however, the equilibrium constant approach can usually be made to converge. The free energy minimization approach, in contrast, involves relaxing the equilibrium state of the overall system while maintaining the same mass throughout the iteration process. Mass is gradually transferred from one species to another until equilibrium is approached. It appears that this contrast between which parameter is relaxed and which is taken as a constraint is the major distinction between the approaches and explains why the free energy minimization approach (when properly implemented) tends to be more robust than the unmodified equilibrium constant based approach. However, as noted above, with the appropriate modifications to avoid serious mass balance violations, the equilibrium constant method can be made relatively robust as well.

Including mineral equilibria. Including equilibrium between the fluid phase and one or more minerals requires some special treatment because of the phase rule. The only limitation on the number of aqueous species which can be present in an equilibrium system is the number of possible ways the various component species can be combined to yield new secondary species. In contrast, the Gibbs phase rule,

$$f = N_c - N_p + 2, \tag{62}$$

which can be derived from the Gibbs-Duhem equation (Anderson and Crerar, 1993), states that the number of degrees of freedom in the system (i.e. the number of variables which can be freely changed), is equal to the number of components (used in the same sense as we use it here) minus the number of phases + 2. The 2 extra degrees of freedom are associated with the temperature and pressure. The maximum number of phases which can be present in an equilibrium system where pressure and temperature are externally determined corresponds to the case where there are zero degrees of freedom (i.e. the system is completely determined)

$$N_p = N_c. \tag{63}$$

This equation, taken at face value, may be slightly misleading since it might imply that if one has a three component system, one can freely choose any three minerals to be at equilibrium. Clearly the minerals must be made up of the components in the system, but also one cannot introduce a linear dependence, as would happen if we considered a system with $SiO_2(aq)$, Na^+, and Cl^- and included three different $SiO_2(aq)$-bearing phases like quartz, amorphous silica, and cristobalite. Perhaps a clearer way to see the phase rule as it applies to geochemical calculations is that one must be able to use the potential mineral phases as linearly independent component species (to "swap" them for an aqueous component species in the parlance of Bethke [1996]). In the above example involving quartz, amorphous silica, and cristobalite, of course, one cannot form a linearly independent basis set using these three minerals.

The presence of heterogeneous equilibria imposes some extra bookkeeping requirements because it is necessary to determine which out of a number of possible mineral phases are thermodynamically most favored. An algorithm is therefore necessary to choose the thermodynamically favored phases (Bethke, 1996). In addition, where minerals dissolve completely they must of course be removed completely from the tableaux or stoichiometric reaction matrix.

Changing basis sets. As noted above, the choice of a set of component species to describe a chemical system (in the terminology of linear algebra, to construct a basis) is not unique. On the one hand, as we have shown in the example involving the carbonate system, one can describe the entire system with a single basis, i.e. there is no need to reformulate the system each time one changes the ingredients added to the system (unless of course these ingredients cannot be described in terms of the available components). On the other hand, however, since the choice of a basis set is not unique, one can choose alternative bases to describe the system. Some of the geochemical (Bethke, 1996; Wolery, 1992) and geochemical transport software (Steefel and Yabusaki, 1996; Lichtner, 1992) allow for specifying alternative basis sets, that is, including component species which differ from those used to write the reactions in the thermodynamic database. Some of the software even allows for dynamic selection of bases in the course of a calculation (Bethke, 1996; Lichtner, 1992; Steefel and Yabusaki, 1996; Sevougian et al., 1993).

Although two different basis sets may be mathematically equivalent (as can be shown practically by carrying out the same calculation with the two different basis sets) the basis sets may not behave the same numerically. Since the concentrations of some species may go effectively to zero in the course of a calculation (for example, the concentration of O_2 in an organic rich sediment), basis switching may be used to "swap out" this component species in favor an alternative, more abundant one. However, one can show that if the logarithms of the concentrations are solved for, very small (if physically unrealistic) concentrations may be retained in the course of the calculation, thus eliminating the need for basis switching. There appear to be at least some instances, however, where basis switching improves the convergence of the geochemical calculations, especially in the case of equilibrium redox reactions. It is our experience that the basis switching, to be effective, has to be applied nearly every iteration within a Newton-Raphson iteration loop, since the concentration of a redox component like O_2 may go to 0 in the course of a single time step. Schemes for changing bases using linear algebra are described in Bethke (1996) and Lichtner (1992).

Solving equilibrium problems with minimization methods. As discussed above, the Gibbs free energy minimization methods are not entirely identical to the equilibrium constant-based methods, although the two should give identical results where both succeed. The minimization methods use mass balance as a constraint throughout and gradually adjust the equilibrium state of the system until global equilibrium is achieved (Anderson and Crerar, 1993)

$$\text{minimize } G = \sum_{i=1}^{N_{tot}} \mu_i n_i \tag{64}$$

where G is the Gibbs free energy of the system, μ_i are the chemical potentials of the individual species in the system, and n_i is their mole numbers. The Gibbs free energy approach can take advantage of a large number of robust minimization methods which have been developed for use in many branches of science (see Press et al., 1986 for a partial review of methods). Gibbs free energy minimization routines which have been applied to aquatic chemistry include the VCS algorithm (Smith and Missen, 1982; Novak, 1990; Sevougian et al., 1993) and GMIN (Felmy, 1990).

Solving the nonlinear equations

In most aquatic systems which include multiple components and species, the governing equations are nonlinear whether kinetic or equilibrium reactions are included. Iterative methods are therefore required in order to solve the system of equations. The equilibrium constant-based methods involve finding the root of the governing equations as do the fully kinetic (non-equilibrium) formulations. By far the most common approach for finding the root of nonlinear sets of equations is the Newton-Raphson method (Press et al., 1986; Bethke, 1996). The Gibbs free energy-based methods employ minimization methods to solve the nonlinear set of equations (DeCapitani and Brown, 1987; Harvie et al., 1987; Felmy, 1990).

Newton-Raphson method. The Newton-Raphson method can be applied to any set of nonlinear equations involving one or more unknowns. For our purposes, this includes speciation problems (i.e. purely equilibrium systems with no time dependence), time-dependent mixed equilibrium and kinetic systems, and fully kinetic systems. Using finite differences, the time-dependent systems are converted to systems of nonlinear algebraic equations and solved in a similar fashion to the speciation problems, although errors associated with the time discretization have to be controlled. As an example of using the Newton-Raphson method to solve a time-dependent batch reaction calculation (this could apply to either a closed system batch calculation or to a reaction path calculation), consider the reactions given in Equations (3) to (6). We assume in this case that all reactions except for calcite dissolution are at equilibrium and we choose H^+, CO_3^{-2}, $CaCO_3$, and Ca^{+2} as component species. In addition, we use a backward Euler method to difference the time derivative and we ignore H_2O for the sake of conciseness. Eliminating the "equilibrium" species using the methods described above, the set of functions we need to solve is given by

$$\frac{1}{\Delta t}\left(TOTH^{+^{n+1}}(H^+, CO_3^{-2}) - TOTH^{+^{n}}\right) = 0; \qquad f_1 \qquad (65)$$

$$\frac{1}{\Delta t}\left(TOTCO_3^{-2^{n+1}}(H^+, CO_3^{-2}, CaCO_3) - TOTCO_3^{-2^{n}}\right) = 0; \qquad f_2 \qquad (66)$$

$$\frac{1}{\Delta t}\left(TOTCa^{+2^{n+1}}(Ca^{+2}, CaCO_3) - TOTCa^{+2^{n}}\right) = 0; \qquad f_3 \qquad (67)$$

$$\frac{1}{\Delta t}\left(C_{CaCO_3}^{n+1} - C_{CaCO_3}^{n}\right) + R_1(Ca^{+2^{n+1}}, CO_3^{-2^{n+1}}) = 0; \qquad f_4 \qquad (68)$$

where $R_1(Ca^{+2}, CO_3^{-2})$ is the rate of dissolution of calcite considered here to be a function of the Ca^{+2} and CO_3^{-2} concentrations. The symbols f_1, f_2, f_3, and f_4 are the designations for the functions to be evaluated. In this example, we have four nonlinear algebraic equations and four independent unknowns.

In more general form, we have $N_c + N_k$ functional relations to be zeroed, involving the variables $C_i = 1, 2, \ldots N_c + N_k$

$$f(C_1, C_2, \ldots, C_{N_c+N_k}) = 0 \qquad i = 1, 2, \ldots, N_c + N_k. \qquad (69)$$

If we let $\mathbf{C}$ refer to the entire vector of concentrations C_i, then each of the functions can be expanded in a Taylor series

$$f_i(\mathbf{C} + \delta\mathbf{C}) = f_i(\mathbf{C}) + \sum_{j=1}^{N_c+N_k} \frac{\partial f_i}{\partial C_j}\delta C_j + O(\delta\mathbf{C}^2). \qquad (70)$$

By neglecting the terms of order $\delta\mathbf{C}^2$ and higher, we obtain a set of linear equations which can be solved for the corrections $\delta\mathbf{C}$

$$\sum_{j=1}^{N_c+N_k} \frac{\partial f_i}{\partial C_j}\delta C_j = -f_i. \qquad (71)$$

The derivative terms in Equation (71) make up the entries in the *Jacobian* matrix while the right hand side is just the values of the functions (those in Equation (65) through (68) above) at the present iteration level. Note that in the example above, the Jacobian matrix is 4 by 4, although not all the entries of the matrix are non-zero (i.e. not all of the functions depend on all of the unknown species concentrations). Standard linear algebra methods are used to solve the set of equations for the concentration corrections and then the concentrations are updated according to

$$C_i^{new} = C_i^{old} + \delta C_i \quad i = 1, \ldots, N_c + N_k. \qquad (72)$$

These new values of the concentrations are then used to reevaluate the functions, f_i and optionally the Jacobian matrix is updated as well. This process is continued until convergence is achieved.

In matrix form, the Newton step for the example above becomes

$$\begin{bmatrix} \partial f_1/\partial C_{H^+} & \partial f_1/\partial C_{CO_3^{-2}} & 0 & 0 \\ \partial f_2/\partial C_{H^+} & \partial f_2/\partial C_{CO_3^{-2}} & 0 & \partial f_2/\partial C_{CaCO_3} \\ 0 & 0 & \partial f_3/\partial C_{Ca^{+2}} & \partial f_3/\partial C_{CaCO_3} \\ 0 & \partial f_4/\partial C_{CO_3^{-2}} & \partial f_4/\partial C_{Ca^{+2}} & \partial f_4/\partial C_{CaCO_3} \end{bmatrix} \begin{bmatrix} \delta C_{H^+} \\ \delta C_{CO_3^{-2}} \\ \delta C_{Ca^{+2}} \\ \delta C_{CaCO_3} \end{bmatrix} = - \begin{bmatrix} f_1 \\ f_2 \\ f_3 \\ f_4 \end{bmatrix} \qquad (73)$$

followed by

$$C^{new}_{H^+} = C^{old}_{H^+} + \delta C_{H^+} \tag{74}$$

$$C^{new}_{CO_3^{-2}} = C^{old}_{CO_3^{-2}} + \delta C_{CO_3^{-2}} \tag{75}$$

$$C^{new}_{Ca^{+2}} = C^{old}_{Ca^{+2}} + \delta C_{Ca^{+2}} \tag{76}$$

$$C^{new}_{CaCO_3} = C^{old}_{CaCO_3} + \delta C_{CaCO_3}. \tag{77}$$

Computing the Jacobian matrix. Generally the most time consuming part of solving systems of nonlinear equations is the calculation of the entries in the Jacobian matrix. Many ODE and DAE solvers will calculate the derivatives making up the Jacobian matrix for the user who provides only the functions for which the roots of the equations are sought. These packages use divided differences to approximate the function with respect to the jth component species concentration at a concentration C_0

$$\left.\frac{\partial f(C)}{\partial C_j}\right|_{C=C_0} \approx \frac{f(C_0 + \epsilon_j) - f(C_0)}{\epsilon_j} \tag{78}$$

where ϵ_j refers to the perturbation of the jth component species. This approach has the advantage that user/programmer need only provide the function. If there are N_c unknowns in the system, this procedure requires $N_c \times N_c$ additional function evaluations. There are several problems involved in using this numerical approximation to the derivatives however: (1) it is difficult to determine if the difference approximation is accurate and (2), usually the divided differences require considerably more CPU time to compute than do the analytical derivatives. If the divided difference approximation is inaccurate, the Newton-Raphson scheme may either fail to converge or may do so slowly. The perturbation ϵ must be chosen so that it is not so large that truncation error results, but also not so small that the functions in the numerator of Equation (78) cancel each other (Bischof et al., 1995).

Symbolic algebra routines like Maple and Mathematica can calculate analytical derivatives given a function, but they cannot begin with or generate more complicated FORTRAN constructs like do loops, subroutines, and branches. More recently software has been developed to calculate analytical derivatives automatically from FORTRAN 77 functions. The ADIFOR package (Bischof et al., 1995) calculates first-order derivatives using the chain rule of differential calculus while retaining standard FORTRAN syntax. Automatic differentiation offers the advantage of producing derivatives of potentially complicated FORTRAN functions which are accurate up to machine precision and also the convenience of updating the derivatives easily if the original functions are changed.

MODELING TRANSPORT PROCESSES

To this point, we have not discussed numerical aspects related to solving the transport portion of a reactive transport problem. There is a vast literature on numerical methods used to model transport processes in porous media, although the field is still an active one since many of the issues have not been satisfactorily resolved. This is particularly true for systems dominated by advective transport, i.e. high Peclet number systems where avoiding spurious numerical effects can be difficult. Normally, partial differential equations for transport are solved by *Eulerian* methods, for example finite difference or finite element methods which balance mass fluxes in stationary subregions of the domain. Although we will briefly outline the basis for the finite element method, we will explore the finite difference approach in greater detail so as to emphasize the basic methods by which a set of partial differential equations involving solute transport are discretized both in space and time. For the sake of conciseness, we will restrict our discussion and examples mostly to single-solute transport in single-phase (e.g. saturated porous media) systems. Obviously multicomponent systems require solving transport for many species, however, all species in the system are generally simulated using the same numerical method. The advection-dispersion equation we require a solution for can be written in its simplest form as

$$\frac{\partial\left(\theta_w C\right)}{\partial t} + q\,\frac{\partial C}{\partial x} - \frac{\partial}{\partial x}\left(\theta_w D\,\frac{\partial C}{\partial x}\right) = 0 \tag{79}$$

where C is the aqueous phase concentration, q is the volumetric fluid flux (*Darcy velocity* = $\theta_w v$, where v is the average linear velocity), θ_w is the volumetric water content, and D is the hydrodynamic dispersion coefficient. The fluid flux is obtained from a solution to a fluid continuity equation. Mathematical formulations for the dispersion tensor for general three-dimensional transport are given, for example, by Bear (1972) and Burnett and Frind (1987). In Equation (79) we have not included any reaction term; we will introduce such terms in the last section of the chapter.

Before proceeding, it is helpful to consider what are desirable properties of a numerical solution to the transport equation. The three main properties we seek are: *accuracy* in both space and time, which for example implies that numerical (artificial) diffusion and mass conservation errors are minimized, *monotonicity*, meaning that non-physical solutions (e.g. negative concentrations) are not produced, and computational efficiency. As discussed previously, *stability* is also a necessary property for a numerical scheme, however monotonicity is a more stringent condition than stability (Unger et al., 1996).

Finite difference methods for spatial discretization

Finite difference methods have the advantage that they are relatively easy to visualize and understand (to some extent, in contrast to finite element methods). Where the interest is in the solute transport equation, of course, the discretization usually involves both space and time as the two independent variables, although there are special methods applicable to purely advective transport where a single variable, travel time, is used. The most common and straightforward procedure is to set up a grid or mesh representing the spatial domain and then to step through discrete time steps to advance the solution, thus approximating both the space and time evolution of the system. Figure 1 shows typical one and two-dimensional *stencils* (3 point and 5 point respectively) used to discretize a spatial domain. The discretization of the spatial domain may be fixed for the course of a simulation or it may be allowed to dynamically evolve. Each grid cell represents an representative elementary volume (REV) (Lichtner, this volume). Simple finite difference schemes which are block-centered (Fig. 1) assume that the property at the node point itself is representative of the entire cell. The block-centered finite difference schemes are very closely related to *control volume* or *integrated finite difference* discretizations (e.g. Narasimhan and Witherspoon, 1976).

Finite difference approximations. Standard finite difference approximations for derivatives can be obtained from Taylor series expansions. This method is also useful in highlighting the magnitude of the error associated with each of the approximations (Celia and Gray, 1992). Consider a one-dimensional system as in Figure 1 where a grid cell, i, is linked to 2 nearest neighbors at grid points $i-1$ and $i+1$. The grid spacing, Δx, is given by $x_{i+1}-x_i$. A Taylor series expansion of the concentration about the grid cell i can be written as

$$C_{i+1} = C_i + \left.\frac{\partial C}{\partial x}\right|_{x_i} \frac{x_{i+1}-x_i}{1!} + \left.\frac{\partial^2 C}{\partial x^2}\right|_{x_i} \frac{(x_{i+1}-x_i)^2}{2!} + \ldots \tag{80}$$

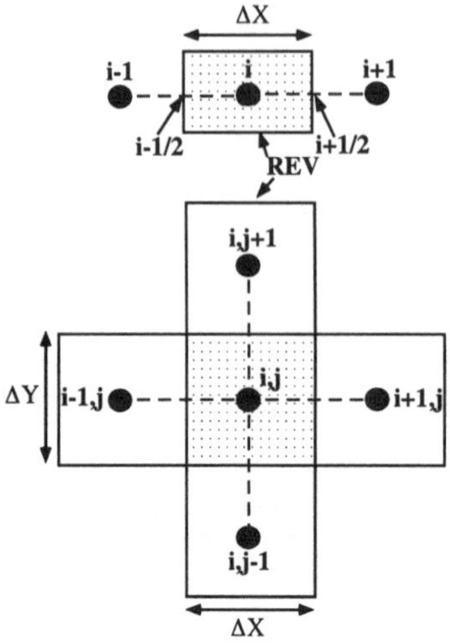

Figure 1. Block-centered finite difference discretization scheme in one and two dimensions using a 3 point and 5 point stencil respectively. Grid spacing need not be constant as in this figure. In the block-centered scheme, material properties and dynamic terms like the velocities are computed at the cell interfaces.

which would become exact if an infinite number of terms were retained. Rearranging Equation (80) to obtain an expression for the first derivative, we find that

$$\left.\frac{\partial C}{\partial x}\right|_{x_i} = \frac{C_{i+1} - C_i}{x_{i+1} - x_i} - \frac{x_{i+1} - x_i}{2}\left.\frac{\partial^2 C}{\partial x^2}\right|_{x_i} - \cdots \tag{81}$$

The difference between the true derivative and the finite difference approximation is therefore

$$\left.\frac{\partial C}{\partial x}\right|_{x_i} - \frac{C_{i+1} - C_i}{x_{i+1} - x_i} = -\frac{x_{i+1} - x_i}{2}\left.\frac{\partial^2 C}{\partial x^2}\right|_{x_i} - \cdots \tag{82}$$

which is referred to as the *truncation error.*

It is possible to get a higher order (i.e. more accurate) approximation of the derivative $\partial C/\partial x$ by using a central difference approximation. If we use a Taylor expansion of the concentration written with respect to the points x_i and x_{i-1}

$$C_{i-1} = C_i - \left.\frac{\partial C}{\partial x}\right|_{x_i}\frac{x_i - x_{i-1}}{1!} + \left.\frac{\partial^2 C}{\partial x^2}\right|_{x_i}\frac{(x_i - x_{i-1})^2}{2!} - \cdots \tag{83}$$

and subtract this equation from Equation (80), we find that the truncation error is given by

$$\frac{\partial C}{\partial x} - \frac{C_{i+1} - C_{i-1}}{2\Delta x_i} = -\frac{(\Delta x_i)^2}{6}\left.\frac{\partial^3 C}{\partial x^3}\right|_{x_i} - \cdots \tag{84}$$

In this expression, the second derivative terms have canceled so that only terms of the order $(\Delta x)^2$ are neglected. The lowest order term which is neglected is said to be *the order of the approximation.* The central difference approximation of the first derivative is therefore said to be a second-order approximation while the forward difference scheme is a first-order

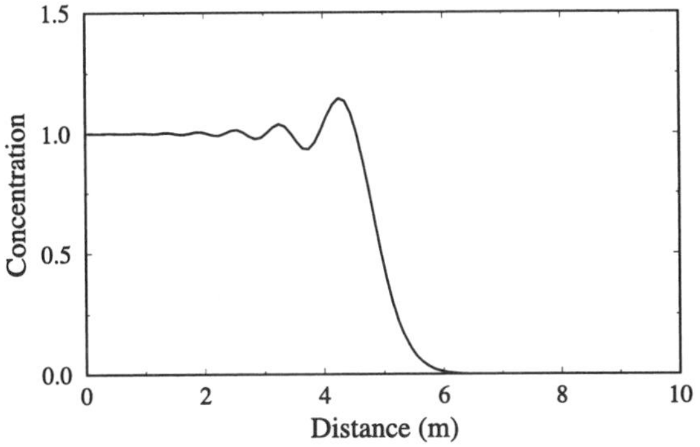

Figure 2. Oscillations occur in the vicinity of a concentration front using the central difference approximation at grid Peclet numbers greater than 2. This simulation uses a grid Peclet number of 20.

approximation. As is readily apparent from the Taylor series expansion, the central difference approximation is second-order only when the grid spacing is constant (otherwise the second derivative terms do not completely cancel). Based on this criteria alone, the central difference approximation would be preferred because it includes less error than the first-order forward difference approximation. Higher order approximations are also possible. This kind of simple analysis of the error in various finite difference approximations would suggest that it is always preferable to use the higher order approximation since it is more accurate. Things are not quite so simple, however, since as we shall see below, the higher order approximations may introduce spurious oscillations (i.e. they may not be monotone). This is particularly true in the case of high Peclet number transport.

Grid Peclet number. The Peclet number is a non-dimensional term which compares the characteristic time for dispersion and diffusion given a length scale with the characteristic time for advection. In numerical analysis, one normally refers to a grid Peclet number

$$Pe_{\text{grid}} = \frac{v\Delta x}{D} \tag{85}$$

where $v = q/\theta_w$ is the average linear (pore water) velocity and the characteristic length scale is given by the grid spacing Δx. The problem with the central difference approximation for the first derivative term appears at grid Peclet numbers > 2 (Unger et al., 1996). At grid Peclet numbers below 2 (i.e. dispersion/diffusion and advection are either of approximately the same importance or the system is dominated by dispersion and/or diffusion), the central difference approximation is monotone. Figure 2 shows a numerically calculated concentration profile using a central difference approximation for the first derivative at a grid Peclet number of 20. Note that while the central difference scheme captures the overall topology and position of the front (reflecting its higher order), it introduces spurious oscillations in the vicinity of the front. This problem, which is a feature of all the unmodified higher order schemes, becomes more severe as the grid Peclet number is increased. While in certain cases it may be possible to simply refine the grid (i.e. reduce Δx) so that the grid Peclet number is less than 2, this often leads to an intractable problem because of the excessive number of grid points needed.

In contrast to the second-order accurate central difference approximation, the first order scheme representation of the first derivative can be made monotone if an *upwind* or *upstream* scheme (also referred to as a *donor cell* scheme) is used. This consists of finite differencing the first derivative using the grid cell upstream (i.e. the cell from which the advective flux is derived) and the grid cell itself

$$\frac{\partial C}{\partial x} \approx \frac{C_i - C_{i-1}}{x_i - x_{i-1}} \tag{86}$$

where the $i - 1$ node is assumed to be upstream. The price for this monotone scheme, however, is that it introduces *numerical dispersion* because of the truncation of the terms greater than first-order (of order Δx). The result is shown for grid Peclet numbers of 1, 10, and 100 in Figure 3, comparing the analytical solution in each case with the numerical.

Courant number. Another important non-dimensional parameter in numerical analysis is the *Courant* or *Courant-Friedrichs-Lewy* (CFL) number. This parameter gives the fractional distance relative to the grid spacing traveled due to advection in a single time step

$$\text{CFL} = \frac{v\Delta t}{\Delta x}. \tag{87}$$

It is possible to show using a Fourier error analysis that for a forward difference in time approximation (i.e. explicit), no matter what approximation is used for the spatial derivatives, that the transport equation is stable for values of the CFL < 1. This stability constraint for explicit transport equations states that one cannot advect the concentration more than one grid cell in a single time step. Unger et al. (1996) derive the monotonicity constraints for a variety of spatial and temporal weighting schemes, and show that only fully-implicit transport methods are not subject to the CFL constraint. A similar expression may be derived for systems characterized by purely diffusive transport, giving rise to a *diffusion number*

$$N_D = \frac{D(\Delta t)}{(\Delta x)^2} \tag{88}$$

where again the stability constraint for an explicit formulation is that N_D be < 1.

Amplitude and phase errors. A Fourier analysis of the advection-dispersion equation can be carried out to show that there are different kinds of error associated with the upwind and the central difference approximations of the first derivative (Celia and Gray, 1992). One class of errors are *amplitude* errors which result in excessive smearing of a concentration front. The upwind formulation is characterized by significant amplitude errors except at a CFL number of 1 and using a forward in time discretization. In this case, the upwind scheme duplicates the analytical solution to within the resolution of the grid. The central difference approximation, in contrast, shows no amplitude errors throughout the range of CFL numbers. The central difference approximation of the first derivative, however, shows phase errors which result in the non-physical oscillation in the vicinity of a sharp front observed in Figure 2. The problems associated with the various schemes when applied to high Peclet number transport, therefore, are derived from more than one kind of error. Ideally, one would like to find a higher order scheme which is accurate because of its avoidance of amplitude errors while avoiding the phase errors associated with these methods (i.e. they are monotone).

Finite element methods for spatial discretization

Finite element techniques are very popular for solving transport problems, including transport in porous media. In one dimension, finite difference and linear finite element solutions obtained with the same nodal spacing have the same accuracy. The above discussion related to oscillations and numerical dispersion apply equally to finite element techniques, that is, grid Peclet and CFL criteria must be satisfied in order to obtain accurate and monotone solutions. There are, however, some important differences between standard finite difference and finite

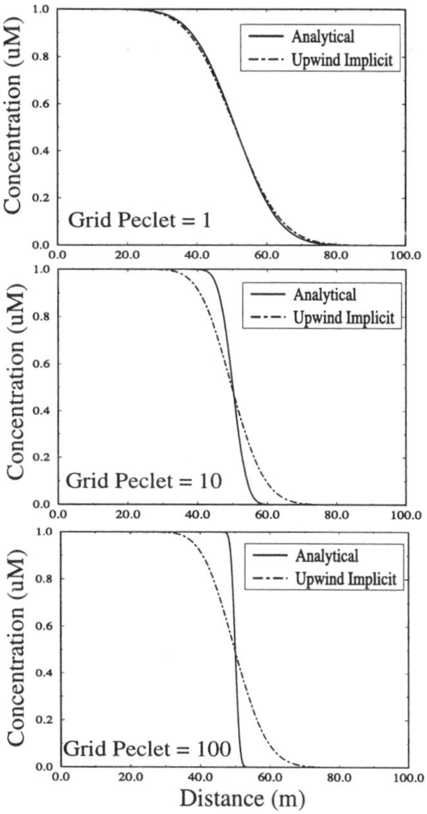

Figure 3. Effect of grid Peclet number on calculated concentration profiles using upwind, fully implicit formulation.

element methods in two or three dimensional problems. Thus we present a very brief introduction to the concepts which form the basis for the finite element method. Complete treatment of the subject can be found in the literature, for example, Huyakorn and Pinder (1983) and Istok (1989) present detailed coverage for subsurface applications.

The finite element method arrives at a solution in three steps. First, the domain is discretized and a trial solution consisting of an interpolation function (usually linear), which passes through the unknown nodal values of concentration, is constructed over the entire grid. Then, by demanding that the trial solution minimize the residual error, on average, a system of ODEs is obtained (one for each node in the grid). Finally, this system is converted into a set of algebraic equations by applying finite differences to the temporal derivative.

If we consider a one-dimensional domain shown in Figure 4, we can approximate the concentration profile, $C(x)$, in any element using linear interpolation functions

$$\hat{C}(x) = C_1\left(1 - \frac{x}{L_e}\right) + C_2\left(\frac{x}{L_e}\right) \tag{89}$$

where $\hat{C}(x)$ is the trial solution, C_1 an C_2 are the concentrations at node 1 and 2, respectively,

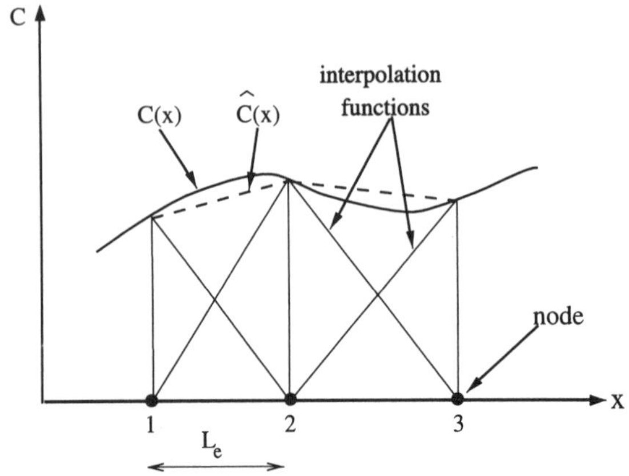

Figure 4. Schematic of the trial solution to a continuous function. The linear interpolation functions commonly used in the finite element method are indicated

and L_e is the element length. This can also be written as

$$\hat{C}(x) = \sum_{j=1}^{2} C_j N_j(x) \tag{90}$$

where j is the nodal index for the element and $N_j(x)$ are the *basis functions* for the linear element, that is

$$N_1(x) = 1 - \frac{x}{L_e}, \quad N_2(x) = \frac{x}{L_e}. \tag{91}$$

The *method of weighted residuals* (Huyakorn and Pinder, 1983) is usually used to minimize residual errors; often the weighting functions are chosen to be the same as the basis functions, which yields the *Galerkin method.* As an example, we present the integral equation for a steady-state diffusion problem

$$\int_R \left[\frac{\partial}{\partial x} \left(D \frac{\partial \hat{C}(x)}{\partial x} \right) N_j(x) dx \right] = 0 \tag{92}$$

where R is the domain region, and j=1 to NN, where NN is the total number of nodes in the domain. The trial solution, Equation (90), is next substituted into Equation (92) and the spatial derivative is reduced using integration by parts (Huyakorn and Pinder, 1983).

The finite element method has the well known advantage that it can be applied with a variety of element shapes (in two and three dimensions) and is thus well suited for non-rectangular geometries. In two-dimensions, triangular elements with linear basis functions can be used to discretize almost any conceivable domain. Arbitrary discretization of a domain, however, can lead to nonphysical results, for example mass diffusion from a node with a low concentration to a node with a higher concentration, and specific discretization criteria must be satisified (e.g. Cordes and Kinzelbach, 1996). An additional difference between finite element discretizations and standard finite difference methods in two or more dimensions is the ability of finite elements to easily handle off-diagonal terms in the disperson tensor. For example, if we apply finite differences in two dimensions, concentration gradients are calculated along a line perpendicular to a surface (the interface between the nodes) when in fact tensor quantities present non-zero tangential components at this interface. Ferraresi and Marinelli (1996) have suggested a correction to the integrated finite difference method to allow for transport deriving from tensor quantities; the proposed method borrows from the finite element concept of two-dimensional interpolation on discrete elements.

It should be noted that *control volume* discretization can be applied using nonrectangular elements, and thus these more physically-intuitive methods can also be employed to discretize complex geometries (e.g. Forsyth, 1991; Di Giammarco et al., 1996).

High-resolution spatial schemes.

Because of the great interest in numerically simulating high Peclet number transport systems (the problem is not restricted to porous medium, but appears in all dynamical systems involving advection, including oceanography, atmospheric science etc.), a large number of alternative schemes have been proposed. These include the method of characteristics (or its modified versions) (e.g. Konikow and Bredehoeft, 1978; Arbogast et al., 1992; Chilakapati, 1993; Zheng and Bennett, 1995) and adjoint Eulerian-Lagrangian methods (Celia et al., 1990). Both of these approaches are based on the treatment of the advective part of the transport equation using a Lagrangian scheme (a reference frame in which one follows the advective displacement of the fluid packet). An Eulerian (fixed) reference frame is then used to simulate dispersive/diffusive transport. The approach reduces numerical dispersion by decreasing the effective grid Peclet number for the fixed Eulerian grid. These schemes can be used with finite element approaches as well. While somewhat more complicated than the simple approaches outlined above, they are capable in many systems of virtually eliminating numerical dispersion. For example, in the modified method of characteristics as developed by Chilakapati (1993), one computes the changes in concentration at a grid point by tracking back along a fluid streamline to the starting point of a fluid packet at the beginning of a time step. This elegant approach, however, can be difficult to implement in physically heterogeneous two- and three-dimensional systems where the streamline mapping can become very complicated. In addition, to obtain good results it may be necessary to track a large number of fluid packets, thus quickly leading to excessive CPU time and memory requirements. Nonetheless, the method of characteristics and related approaches are still widely used when it is critical that numerical dispersion be avoided.

Another class of high resolution Eulerian methods uses higher-order approximations for the first derivatives, but hybridizes these with low-order schemes in an attempt to obtain monotone solutions. The solutions have the higher-order accuracy in smooth regions and the low-order accuracy near discontinuities (e.g. near plume fronts). The price to be paid for these schemes is that they are nonlinear, even when applied to initially linear problems such as Equation (79). In this class are the flux-corrected transport (FCT) methods (Boris and Book, 1973; Oran and Boris, 1987; Zalesak, 1987; Hills et al., 1994) which normally give excellent results when applied to non-reactive solute transport (Hills et al., 1994; Yabusaki et al., in prep.). However, as in some of the other methods discussed here, very low level oscillations still exist which can give spurious results when nonlinear reactions are coupled into the solution. In an example given below we solve for reactive transport in a physically heterogeneous domain using a 4th order FCT scheme.

We have found that another scheme, referred to as the *total variation diminishing* or TVD scheme, gives more nearly oscillation-free behavior (Harten, 1983; Sweby, 1984; Yee, 1987; Gupta et al., 1991). The TVD method is one of a class of methods which use limiters to ensure monotonicity of the solution (Van Leer, 1977a,b; Leonard, 1979a,b; Leonard, 1984). In the example below, we use the implementation of the TVD suggested by Gupta et al. (1991) which is combined with a third-order Leonard scheme (Leonard, 1979b).

Example of reactive transport in a physically heterogeneous porous media. Although a thorough review of the high resolution monotone methods is beyond the scope of this work, it is useful to compare two of the schemes with which we have extensive experience. As an example problem, consider the influx of a groundwater containing initially 0.1 M dissolved organic carbon (DOC) through the left boundary of a physically heterogeneous domain. Organic carbon degrades by a first-order rate law via sulfate reduction according to the reaction

$$CH_2O + \frac{1}{2}SO_4^{2-} \longrightarrow HCO_3^- + \frac{1}{2}HS^- + \frac{1}{2}H^+. \tag{93}$$

SO_4^{2-} is replenished in solution by the equilibrium dissolution of gypsum ($CaSO_4^{2-} \cdot 2H_2O$) while the carbonate produced is largely consumed by the equilibrium precipitation of calcite ($CaCO_3$). Flow is from left to right across the domain. Downstream of the front, the groundwater is in equilibrium with calcite and gypsum and should be unaffected by the front upstream. Since the medium is assumed to be chemically homogeneous, the initial aqueous concentrations computed from the equilibrium relations are homogeneous as well. The simulation was run as a pure advective problem using the FCT (Hills et al., 1994) and TVD (Gupta et al., 1991) algorithms and using a 200 by 88 grid. The calculation was carried out with MCTRACKER, a massively parallel version of the code OS3D (Steefel and Yabusaki, 1996). The simulations were run on a CM-5 massively parallel computer using 32 to 128 processors. The $TOTCa^{+2}$ concentration is shown after 0.5 days in Figure 5. The TVD method shows increased Ca^{+2} concentrations about 1/3 of the way across the domain, with the leading edge of the concentration front matching the position of a tracer released at the same time as the organic carbon plume. Downstream of the front, essentially nothing is happening in the TVD simulation, which is the expected behavior. In contrast, the FCT method shows low-level oscillations both within the organic carbon plume and downstream of the front. It should be noted that these oscillations are not evident when the tracer concentration is plotted at the same time; the low-level oscillations (or non-monotonicity) characteristic of the 4th order FCT become amplified when nonlinear chemistry is involved. We have also carried out other calculations on an equilibrium redox problem using the 4th order FCT where the method failed due to the low-level oscillations interacting with the highly nonlinear chemistry (Regnier et al., in press). A transport scheme based on the well-known Lax-Wendroff scheme (Press et al., 1986) also failed on this problem. In contrast, the problem was solved with both upwind and 3rd order TVD formulations. This points out that care must be taken in evaluating transport algorithms. Many, like the well-known Lax-Wendroff scheme and the FCT scheme evaluated here, are not sufficiently monotone that they can be used on the highly nonlinear reactive transport problems that often occur in hydrogeology.

METHODS FOR COUPLING REACTION AND TRANSPORT

In recent years much of the discussion about numerical approaches to reactive transport has centered on the question of how to couple the reaction and transport terms. The reaction term affects the local concentrations which in turn determine the fluxes of aqueous species. In the same way, the magnitude of the fluxes can affect the reaction rates. An accurate solution of the overall problem, therefore, requires that the terms be coupled at some level. Several methods have been proposed to solve the coupled set of equations. The most straightforward way conceptually, but the most demanding from a computational efficiency point of view, is to solve the governing equations, including both reaction and transport terms, simultaneously. This approach is referred to as a *one-step* or *global implicit* method (Kee et al., 1987; Oran and Boris, 1987; Steefel and Lasaga, 1994; Steefel and Yabusaki, 1996). Alternatively, it is possible to use operator splitting techniques to decouple the reaction and transport calculations. The classic time-splitting approach to solving the coupled set of equations (referred to by T. Xu [pers. comm., 1996] as the *sequential non-iterative approach* or SNIA) consists of solving the

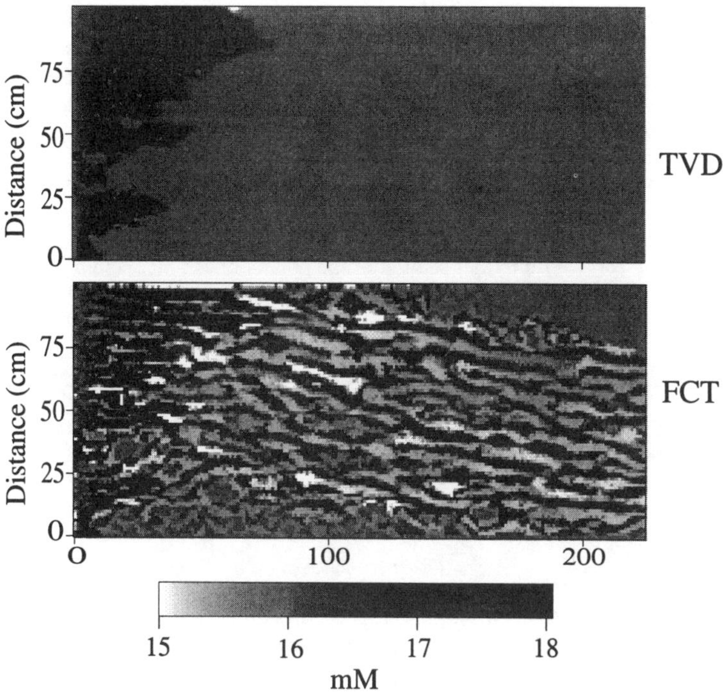

Figure 5. Comparison of the 4th order FCT method (Hills et al., 1994) and the 3rd order TVD method (Gupta et al., 1991) on a problem involving organic carbon degradation via sulfate reduction coupled to two equilibrium dissolution-precipitation reactions. The non-monotone character of the FCT algorithm, although not detectable in a plot of non-reactive solute concentrations, results in unacceptable oscillations when amplified by nonlinear chemical reactions. In contrast, the TVD method performs much better on this reactive transport problem.

reaction and transport equations within a single timestep in sequence, with no iteration between the two. A method championed by Yeh and Tripathi (1989; 1991) is the *sequential iteration approach* or SIA where the reaction and transport are solved separately but iteration between the two calculations is carried out until a converged solution is attained. There are variants on these approaches, including a predictor-corrector approach (Raffensperger and Garven, 1995) and the Strang splitting approach (Zysset and Stauffer, 1992). The formulations are discussed below and their performance is compared on several test problems.

One-step or global implicit approach

Since the publication of the influential paper by Yeh and Tripathi (1989) which rejected the use of the one-step or global implicit approach as being too CPU time and memory intensive, relatively little discussion of one-step approaches for the reaction-transport problem has appeared in the hydrogeochemical literature. This is despite the fact that the global implicit approach has been used sucessfully for some time in the chemical and mechanical engineering fields (Kee et al., 1987), although even in these fields the relative merits of the decoupled or operator splitting approaches versus the one-step approach still rages (Oran and Boris, 1987; Kee et al., 1987).

In one component systems (for example, modeling of the system quartz-SiO_2(aq) or stable isotope exchange) or in systems in which there is no coupling via reactions of the various species and minerals, there is no particular difficulty in simply solving the entire system using standard iterative or direct linear algebra methods. In this case, the presence of a reaction term does not increase the size of the matrix over and above its size in the case of transport alone. Using as an example a one-dimensional diffusion-reaction problem where the porosity, diffusion coefficient, and grid spacing are assumed constant, the fully implicit approach in finite difference form is given by

$$\frac{C^{n+1} - C^n}{\Delta t} = D\frac{C_{i+1}^{n+1} - 2C_i^{n+1} + C_{i-1}^{n+1}}{\Delta X^2} + R(C_i^{n+1}) \tag{94}$$

where i, $i - 1$, and $i + 1$ refer to the location of the nodal points and n and $n + 1$ refer to the present and future time levels respectively. Note that the form of Equation (94) assumes that the values of the concentration which appear in the accumulation term (the left hand side), the transport term, and the reaction term are those at the future time level. If the reaction term R is linear, Equation (94) can be rewritten as a set of simultaneous linear equations of the form

$$b_1C_1 + c_1C_1 = d_1 \tag{95}$$
$$a_2C_1 + b_2C_2 + c_2C_3 = d_2 \tag{96}$$
$$\cdots \tag{97}$$
$$a_{N-1}C_{N-2} + b_{N-1}C_{N-1} + c_{N-1}C_N = d_{N-1} \tag{98}$$
$$a_NC_{N-1} + b_NC_N = d_N \tag{99}$$

where the lower case characters a_i, b_i, and c_i are the coefficients, d_i refers to the right hand side, and N is the number of spatial points. The one-dimensional system can be represented by a tridiagonal matrix of the form

$$\begin{bmatrix} b_1 & c_1 & 0 & \cdots & & & \\ a_2 & b_2 & c_2 & \cdots & & & \\ & & & \cdots & & & \\ & & & \cdots & a_{N-1} & b_{N-1} & c_{N-1} \\ & & & \cdots & 0 & a_N & b_N \end{bmatrix} \begin{bmatrix} C_1 \\ C_2 \\ \cdots \\ C_{N-1} \\ C_N \end{bmatrix} = \begin{bmatrix} d_1 \\ d_2 \\ \cdots \\ d_{N-1} \\ d_N \end{bmatrix} \tag{100}$$

which is easily and rapidly solved with the Thomas algorithm (Press et al., 1986). The reaction term, if present, is imbedded in the coefficient b_i multiplying the concentration at the nodal point i. Since the transport equations are linear themselves, no iteration is required unless the reaction terms are nonlinear.

The problem is only slightly more difficult in the case of two and three dimensional transport. In two dimensions where a 5 point stencil is used (Fig. 1), any one row of the coefficient matrix will have 5 non-zero entries. Since the system is still sparse, however, it can be readily solved in most cases using modern sparse matrix solvers (e.g. WATSOLV, VanderKwaak et al., 1995). Again, where there is no coupling between species introduced by the reactions, each species concentration field can be solved for independently of the other species in the system.

Using a global implicit approach becomes considerably more difficult in the case of a multicomponent, multi-species system both because the coupling of species via reactions enlarges the size of the coefficient matrix and because it typically results in sets of nonlinear equations which must be solved. Just as in the case of the time-dependent reaction systems, the nonlinearities resulting from the reaction terms require the use of iterative methods. The Newton-Raphson technique described above can be used, however, to solve the global set of equations, i.e. a set of equations containing both transport and reaction terms (Steefel and Lasaga, 1994). The size of the Jacobian matrix which must be constructed and solved becomes considerably larger since each function will include contributions from the concentrations both at that nodal point itself and from neighboring nodal points which are used in the finite difference or finite element discretization. For example, in the case of one-dimensional transport and N_c unknown concentrations at each nodal point, the form of the Newton equations to be solved is

$$\sum_{k=1}^{N_c} \frac{\partial f_{i,j}}{\partial C_{k,j}} \delta C_{k,j} + \sum_{k=1}^{N_c} \frac{\partial f_{i,j}}{\partial C_{k,j+1}} \delta C_{k,j+1} + \sum_{k=1}^{N_c} \frac{\partial f_{i,j}}{\partial C_{k,j-1}} \delta C_{k,j-1} = -f_{i,j} \tag{101}$$

where i refers to the component number, k is the unknown component species number, and j, $j+1$, and $j-1$ are the nodal points. The Jacobian matrix in the case of one-dimensional transport takes a *block tridiagonal* form

$$\begin{bmatrix} \mathbf{A}_{1,1} & \mathbf{A}_{1,2} & 0 & \cdots & & & \\ \mathbf{A}_{2,1} & \mathbf{A}_{2,2} & \mathbf{A}_{2,3} & \cdots & & & \\ & & & \cdots & & & \\ & & & \cdots & \mathbf{A}_{N-1,N-2} & \mathbf{A}_{N-1,N-1} & \mathbf{A}_{N-1,N} \\ & & & \cdots & 0 & \mathbf{A}_{N,N-1} & \mathbf{A}_{N,N} \end{bmatrix} \begin{bmatrix} \delta \mathbf{C_1} \\ \delta \mathbf{C_2} \\ \cdots \\ \delta \mathbf{C_{N-1}} \\ \delta \mathbf{C_N} \end{bmatrix} = - \begin{bmatrix} \mathbf{f_1} \\ \mathbf{f_2} \\ \cdots \\ \mathbf{f_{N-1}} \\ \mathbf{f_N} \end{bmatrix} \tag{102}$$

where the entries in the Jacobian matrix (the **A**) are submatrices of dimension N_c by N_c in the case where there are N_c unknowns. The entries $\delta \mathbf{C_i}$ refer here to the entire vector of unknown concentrations at any particular nodal point while the functions $\mathbf{f_i}$ include the entire vector of equations for the unknown concentrations at each nodal point. The one-dimensional problem can be solved relatively rapidly using a block tridiagonal solver (Steefel and Lichtner, 1994; Hindmarsh, 1977). In the case of multidimensional systems, the equations can be solved by block iterative methods (Steefel and Lasaga, 1994) or by using sparse matrix solvers which in certain cases require minor modifications to handle the block structure of multicomponent problems (VanderKwaak et al., 1995; Steefel and Yabusaki, 1996).

The major obstacle to the use of global implicit methods for multicomponent reactive transport systems is the need for large amounts of computer memory (Yeh and Tripathi, 1989). For example, for a two-dimensional domain discretized with a 250 by 250 five-point connection grid and having 10 component species, the Jacobian matrix stored in compressed form as a rectangular matrix of dimension 500 by 62,500 could require about 250 MB of memory. A second issue is the additional CPU time required to construct the global Jacobian matrix. If an unmodified Newton step is carried out, the entire Jacobian matrix must be constructed at each iteration. However, it is possible to take advantage of what are referred to as modified Newton methods where the Jacobian matrix is either only periodically updated or only updated locally in space where the functions have not converged. In addition, not all of the entries in the Jacobian matrix are different, thus reducing the number which must actually be computed. For example, in a one-dimensional system of the kind represented in Equation (102), most of the calculations which go in to constructing the submatrix $\mathbf{A}_{1,1}$ can be used in constructing $\mathbf{A}_{2,1}$ (since the Jacobian mostly involves the derivatives of the total concentrations with respect to the unknown component species).

Sequential non-iterative approach (SNIA)

At the other extreme from the one-step or global implicit method is the *sequential non-iterative approach* (SNIA), also called the operator splitting or time splitting approach. In this approach, a single timestep consists of a transport step followed by a reaction step using the transported concentrations. Physically, one can think of the method as equivalent to taking part of the contents of one beaker (i.e. one node) and adding it to a second beaker downstream. After this physical process is completed, reactions occur which modify the chemical composition of each of the beakers in the sequence. Mathematically, the method can be represented as a two step sequential process consisting of a transport step

$$\frac{(C_i^{transp} - C_i^n)}{\Delta t} = L(C_i)^n, \quad (i = 1, ..., N_{tot}), \tag{103}$$

followed by a reaction step

$$\frac{(C_i^{n+1} - C_i^{transp})}{\Delta t} = R_i^{n+1} \quad (i = 1, ..., N_{tot}) \tag{104}$$

where L is the spatial operator. The problem can also be formulated using the total concentrations, thus reducing the number of equations which must be solved.

Clearly the potential problem with the method is that it views the addition of fluid from one cell to another as being sufficiently rapid that the reactions only begin after the physical transport is complete. The greatest difficulties arise in kinetic systems at the physical boundaries of the system where the same amount of reaction is applied to a fluid parcel which just entered the system as is applied to a parcel which has been in the system for the entire timestep (Valocchi and Malmstead, 1992). For a decay reaction where the concentrations are continually reduced, therefore, the SNIA or operator splitting method would tend to overestimate the amount of reaction.

Strang splitting. It is possible to reduce some of the errors associated with the SNIA approach by using a time-centered method referred to as *Strang splitting* (Strang, 1968; Zysset and Stauffer, 1992). This method involves centering the reaction step between two transport steps. The scheme can also be used with multiple reaction and transport steps, but in each case the reaction step will be centered. Where only a single reaction step is used, the method takes the form

$$\frac{(C_i^{transp} - C_i^n)}{\Delta t/2} = L(C_i)^n, \quad (i = 1, ..., N_{tot}), \tag{105}$$

followed by a reaction step

$$\frac{(C_i^{reacted} - C_i^{transp})}{\Delta t} = R_i, \quad (i = 1, ..., N_{tot}), \tag{106}$$

which is in turn followed by another 1/2 transport step

$$\frac{(C_i^{n+1} - C_i^{reacted})}{\Delta t/2} = L(C_i)^{reacted}, \quad (i = 1, ..., N_{tot}). \tag{107}$$

Sequential iteration approach (SIA)

The *sequential iteration approach* (SIA) has been suggested as a method which avoids the construction and manipulation of the large matrices characteristic of the global implicit approach and which corrects at the same time the errors which may occur in the use of the sequential non-iterative or classic operator splitting approach (Yeh and Tripathi, 1989; Yeh and Tripathi, 1991; Engesgaard and Kipp, 1992; Zysset and Stauffer, 1992; Zysset et al., 1994; Walter et al., 1994; Šimůnek and Suarez, 1994). The presentation of the method, however, has been less than clear in the literature and in fact, several different implementations of the method are possible. The basic idea is to arrive at a fully coupled solution at the $n + 1$ time level for both the transport and reaction terms. But rather than solving for the entire system of equations as in the global implicit method, the SIA approach accomplishes the coupling by iterating between the reaction and transport terms. Conceptually the most straightforward method involves solving the reactive transport equation in its full form at every step of the iterative process, alternating which term (reaction or transport) is included on the right hand side as a source term from the previous iteration (Zysset and Stauffer, 1992)

$$\frac{\left(C^{n+1,m+1} - C^n\right)}{\Delta t} - L(C^{n+1,m+1}) = R^{n+1,m} \tag{108}$$

$$\frac{\left(C^{n+1,m+2} - C^n\right)}{\Delta t} - R^{n+1,m+2} = L(C^{n+1,m+1}). \tag{109}$$

In this case, the iteration continues until the concentration computed in Equation (108) agrees with the concentration computed in Equation (109) to within some tolerance.

A more commonly used approach is to react the concentrations coming out of the transport (plus reaction) step given by Equation (108). In the case of purely equilibrium systems, this means the transported concentrations are equilibrated and the correction to the reaction term is added back into the right hand side of Equation (108). It is important to point out that, in contrast to some of the descriptions of the method in the literature, the second step in this case does not involve a full reaction step but rather is a refinement of the reaction term which continues until the reaction term converges. It can be shown that this form of iteration refinement of the reaction term is actually equivalent to the formulation given in Equations (108) and (109) above. With a simple rearrangement of Equation (108) so that we solve for the transport term

$$L(C^{n+1,m+1}) = \frac{\left(C^{n+1,m+1} - C^n\right)}{\Delta t} - R^{n+1,m} \tag{110}$$

and substituting this result into Equation (109), we find that

$$\frac{\left(C^{n+1,m+2} - C^n\right)}{\Delta t} - R^{n+1,m+2} = \frac{\left(C^{n+1,m+1} - C^n\right)}{\Delta t} - R^{n+1,m} \tag{111}$$

or

$$\frac{\left(C^{n+1,m+2} - C^{n+1,m+1}\right)}{\Delta t} = R^{n+1,m+2} - R^{n+1,m} = \Delta R^{n+1,m+2}. \tag{112}$$

The reaction rate computed at the latest iteration level can then be substituted back into the Equation (108) or equivalently the change in the reaction term, $\Delta R^{n+1,m+2}$, can be used to refine the reaction term used on the right hand side of Equation (108). The process is continued until the reaction rate and concentrations at the $n + 1$ time level converge.

The SIA method can also be formulated in terms of total concentrations. If these total concentrations include both aqueous and solid phases (i.e. aqueous complexes, surface complexes, and minerals), no explicit reaction term would appear in Equation (108). A respeciation of the total concentration, or in the case of kinetic reactions, solving of a set of ODEs to get the individual species concentrations is still required however.

Potential numerical problems with the SIA method. Although the SIA method appears attractive because of its modular structure and its reduction of operator splitting error, it may occasionally show numerically unstable behavior depending on the problem considered. For problems involving decay-type reactions (e.g. Monod kinetics), the method appears to work reasonably well. When solving other systems of reactions, however, there may be difficulties in getting the iterative procedure to converge (Engesgaard and Kipp, 1992). We encountered convergence difficulties in some of the examples discussed below.

Comparison of coupling schemes

In this section we compare the performance of the coupling schemes discussed above on several example problems. In the first two examples, we examine linear, single-solute reactions in order to compare the numerical solutions to available analytical solutions. Two more realistic multicomponent examples are also investigated. To quantitatively assess the errors introduced by the various coupling methods, we employ the L_2 norm for the vector of concentration differences, defined by

$$\| x \|_2 = \sqrt{\sum_{i=1}^{N_x} x_i^2} \tag{113}$$

where x is the vector of concentration differences (e.g. determined by taking the difference between a numerical solution and an analytical solution at all nodes in the domain), and N_x is the number of concentration values in the domain. As such, the L_2 norm values are specific to each example given below.

Example involving first-order decay. In order to test the coupling methods, we use a simple example involving a first-order decay reaction for which the analytical solution is available (Bear, 1979)

$$\partial \frac{(\theta_w C)}{\partial t} = \frac{\partial}{\partial x}\left(\theta_w D \frac{\partial C}{\partial x}\right) - q\frac{\partial C}{\partial x} - \theta_w k C. \tag{114}$$

The solute concentration is assumed to be 0 at $t = 0$ within the domain. The concentration is fixed at the boundary, $x = 0$, to be 1. For our test problem, we use a volumetric water content, θ_w, of 1.0, a velocity of 100 m yr^{-1}, a dispersivity of 0.2 m, a grid spacing of 0.4 m, and a rate constant, k, of 100 yr^{-1}. This yields a grid Peclet number of 2.0. Figure 6 shows a comparison of the analytical solution for the problem at 0.5 years to the SNIA and Strang methods at a variety of CFL numbers. Note the characteristic feature of the SNIA method at larger time steps in which the concentrations close to the boundary are below the analytical solution. As discussed above, this is due to the fact that any fluid crossing the boundary in a single time step is reacted as if it was in the system that entire time step (Valocchi and Malmstead, 1992). The SNIA, therefore, overestimates the amount of reaction, although the error becomes less significant with the reduction in the timestep. Where this boundary effect is absent, as in problems involving decay of a reactive constituent that begins inside the system, the SNIA performs better. The Strang approach performs better at the boundary than the SNIA, but calculates the front to be in advance of the analytical solution. The SIA method reproduces the analytical curve almost exactly.

Example involving equilibrium adsorption. As pointed out by Barry et al. (1996), conclusions about the accuracy of a particular coupling method depend on the problem considered. They pointed out the problems with the Strang approach when applied to an equilibrium linear adsorption problem. In this case, the governing differential equation is given by

$$\frac{\partial C}{\partial t}\left(1 + \frac{\rho_D}{\theta_w} K_d\right) = D\frac{\partial^2 C}{\partial x^2} - v\frac{\partial C}{\partial x} \tag{115}$$

where ρ_D is the bulk density of the aquifer and K_d is the distribution coefficient (L/kg). For this example, we use a volumetric water content of 1.0, a bulk density of 1.0, a K_d of 10, a

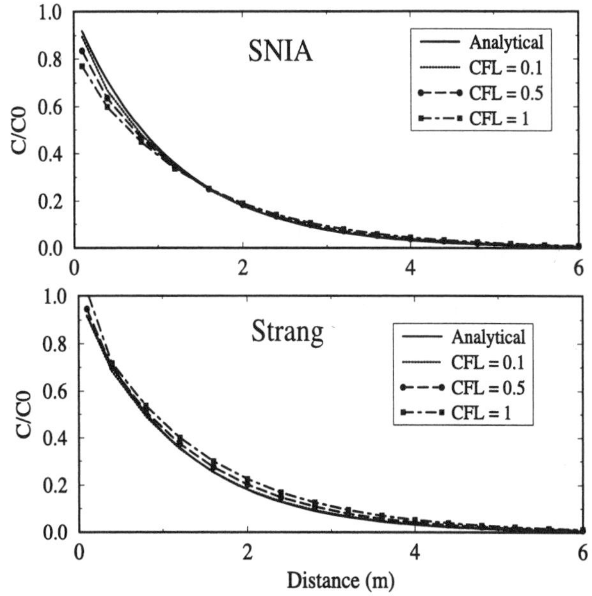

Figure 6. Comparison of the sequential non-iterative approach (SNIA) and Strang splitting approach at a variety of CFL numbers with the analytical solution on a problem involving a first-order decay reaction. The SIA approach reproduces the analytical curve. See text for parameters used in run.

velocity of 5 m yr^{-1}, a dispersivity of 0.1 m, and a grid spacing of 0.2 m yielding a grid Peclet number of 2. We compare the results from the SNIA, Strang, SIA methods at a CFL = 0.625 (Δt = 0.025 yr^{-1}) with the analytical solution (Fig. 7). The error associated with each of the methods is tabulated in Table 3. The SIA and global implicit schemes give the same answer which differs only slightly from the analytical solution due to minor transport errors. Note that the Strang scheme improves only very slightly on the SNIA method using the same time step. By reducing the time step, the error of the SNIA can be reduced, but at least in this simple problem, reducing the error in the SNIA approach to the level of the SIA scheme would require more CPU time than the SIA method does.

Example involving Monod kinetics. We now present a one-dimensional reactive transport example comparing various operator splitting schemes to a one-step method. The one-step method is based on the algorithm presented by MacQuarrie et al. (1990) and uses Galerkin finite-element spatial discretization and a Gauss-Seidel iteration scheme to solve the coupled reactive transport problem. For the operator splitting alogrithms we use a control volume finite element scheme for transport, Crank-Nicolson temporal weighting, and the implicit ODE solver VODE (Brown et al., 1989) for kinetic reactions. The problem involves two reacting solutes, an organic substrate (i.e. electron donor) and electron acceptor, and a dynamic biomass population. The electron acceptor is uniformly present at 1.0 mg/L at t = 0, when the substrate is introduced at x = 0 at a concentration of 1.0 mg/L. The nodal spacing is 0.4 m; this results in a maximum grid Peclet number (for the substrate) of 2.0. The relevant transport and Monod kinetic parameters are given by MacQuarrie (in preparation). A comparison of the results from the two models is shown in Figure 8 for Strang splitting with a time step of 0.667 days. For this time step size the grid CFL number for a nonreactive solute is 0.95, while the maximum for the reactive species is 0.4 (for the electron acceptor), thus transport-related numerical errors are minimized. As shown in Figure 8, the agreement between the solutions is very good. A

Table 3. Accuracy (based on L_2 norm) and CPU time comparison for various reaction-transport coupling schemes applied to equilibrium linear adsorption problem.

Method	Δt	L_2 norm	Normalized CPU Time
SNIA	0.25	1.911	1.00
SNIA	0.125	1.335	1.53
SNIA	0.0625	1.036	2.82
Strang	0.25	1.682	1.12
SIA	0.25	0.772	1.53
GI	0.25	0.764	2.88

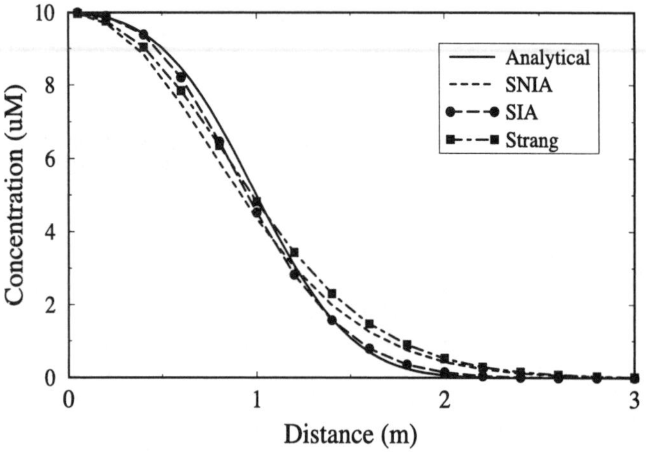

Figure 7. Comparison of the SNIA, SIA, and Strang splitting approaches at a CFL of 0.625 with the analytical solution for an equilibrium linear adsorption problem ($K_d = 10$). The SIA scheme duplicates the global implicit scheme, both of which differ from the analytical curve due to the presence of minor transport error.

significant amount of substrate and electron donor consumption occurs in the first 5 m of the domain, which results in a biomass increase in this region of about one order of magnitude.

Table 4 presents the the L_2 norms for the electron acceptor profiles at a time of 10 days; in each case the L_2 norm is computed by comparing the operator-splitting solution to the one-step solution. The electron acceptor always had the largest L_2 norm in this example. Results are given in Table 4 for Strang splitting with three different time steps, and for the SNIA and SIA methods as implemented by Zysset and Stauffer (1992). The results in Table 4 indicate that the Strang scheme quickly converges as the time step is reduced. It is interesting to note that for the smallest time step investigated, the SIA of Zysset and Stauffer (1992) and the Strang scheme have the same level of accuracy, however, the SIA requires more than twice the total CPU time. For the smallest time step, the CPU time requirements for the the Strang splitting and SNIA are about the same, however the SNIA has a error approximately one order of magnitude larger. These trends are consistent with previous numerical comparisons for more simplified

Table 4. Accuracy (based on L_2 norm) and CPU time comparison for three types of operator splitting applied to a Monod kinetics reactive-transport problem.

Splitting Method	Global Δt	L_2 norm for electron acceptor	Normalized CPU Time	
			Transport	Chemistry
Strang	1.333	0.070	1.0	1.4
Strang	0.667	0.037	1.9	2.1
Strang	0.333	0.015	4.0	3.8
SNIA	0.333	0.10	2.1	4.1
SIA	0.333	0.011	7.1	13.1

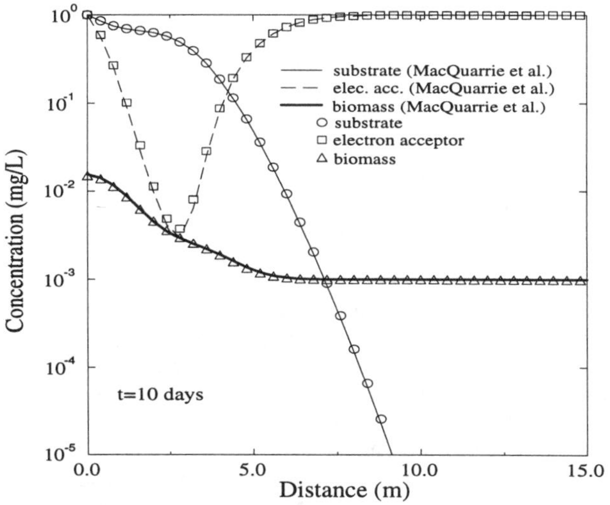

Figure 8. Comparison of Strang operator splitting solution to a one-step solution for a Monod-kinetics reactive transport problem. Lines are one-step results computed with the model of MacQuarrie et al. (1990) and symbols are the Strang results using a time step of 0.667 days.

kinetic reaction systems (Valocchi and Malmstead, 1992; Zysset et al., 1994; Kaluarachchi and Morshed, 1995) and show that operator splitting methods can be accurate and efficient for kinetic reaction problems.

Using VODE, the solution of the chemical ODE system for this example problem took about 50 to 65% of the total CPU time - this results because of the low number of species (three) and the relatively nonstiff chemistry. All the simulations presented in Table 4 were repeated using the explicit TWOSTEP solver (Verwer, 1994), which produced the same solutions but with chemistry solve times which were less than the VODE solve times by a factor of two to three. Zysset et al. (1994) also noted a similar reduction in chemistry solve time when they substituted an explicit solver for the standard GEAR package. However, we have found that when Monod kinetics are coupled to aqueous inorganic carbonate reactions using a fully-kinetic formulation, the TWOSTEP solver fails completely. This is because of the increased stiffness of the system. Therefore, as discussed previously in this chapter, it is apparent that implicit

Table 5. Tableaux for cobalt surface complexation example. Surface complexes designated with >.

	Co^{+2}	H^{+}	Fe^{+3}	Al^{+3}	$SiO_2(aq)$	Na^{+}	Cl^{-}	HCO_3^{-}	>FeOH
Co^{+2}	1								
H^{+}		1							
Fe^{+3}			1						
Al^{+3}				1					
$SiO_2(aq)$					1				
Na^{+}						1			
Cl^{-}							1		
HCO_3^{-}								1	
$Al(OH)_2^{+}$		-1		1					
$Al(OH)^{+2}$		-2		1					
$Al(OH)_3(aq)$		-3		1					
$Al(OH)_4^{-}$		-4		1					
$CO_2(aq)$		1						1	
CO_3^{-}		-1						1	
H_3SiO4^{-}		-1			1				
>FeOH									1
$>FeO^{-}$		-1							1
$>FeOH_2^{+}$		1							1
$>FeOH\text{-}Co^{+2}$	1	-1							1

ODE algorithms will be the more robust solvers for general application to subsurface reactive transport.

Example of multicomponent aqueous and surface complexation. In this section, we compare several of the methods for coupling reaction and transport discussed above by applying them to a problem involving both aqueous and surface complexation. The problem includes pH-dependent, equilibrium adsorption on an Fe-hydroxide surface using a non-electrostatic surface complexation model to describe the adsorption process. The Co^{+2} is weakly adsorbed at low pH and is strongly adsorbed at higher pH (Fig. 9), so if a fluid packet experiences a significant shift in pH, the retardation of the Co^{+2} can change dramatically. Aqueous cobalt is introduced at the left boundary of the system at a constant flow rate of 5 m/year and with a dispersivity of 0.1 meters. A uniform grid spacing of 0.2 meters results in a grid Peclet number of 2. At this value of the grid Peclet number, the central difference representation of the gradient appearing in the advective term is stable and second order accurate. We use a Crank-Nicolson temporal weighting scheme for the transport and a backward difference (fully implicit) method for the reaction step. The tableaux for the surface and aqueous complexation example is given in Table 5. The initial and boundary conditions and the equilibrium constants for the problem are given in Table 6. The calculations are carried out with the codes OS3D and GIMRT (Steefel and Yabusaki, 1996).

The strong pH dependence of the adsorption results in more complex behavior than is observed in the case of a linear equilibrium adsorption problem such as that which was described above. Figure 10 shows the pH evolution of the one-dimensional system as a function of time. The initial pH within the domain is 8.22 which results in strong adsorption of any cobalt infiltrating the system at this stage (see Fig. 9). As the lower pH fluids infiltrate the system, however, the previously adsorbed cobalt is desorbed resulting in the concentrations locally *above* the boundary input value of 10 μM. In this particular system, adsorbed cobalt is concentrated at a transient front corresponding to the pH front in the system. Cobalt shows a retardation factor of about 3, a result which could not be predicted with a simple linear adsorption model because of the nonlinear dependence on the pH front which is itself retarded.

Table 6. Initial and boundary conditions and equilibrium constants for cobalt surface complexation example.

Species	Boundary (M)	Initial (M)	Log K_{eq}
Co^{+2}	9.8×10^{-6}	5.7×10^{-13}	0.0
H^{+}	1.0×10^{-5}	6.0×10^{-9}	0.0
Fe^{+3}	1.0×10^{-8}	7.3×10^{-25}	0.0
Al^{+3}	3.5×10^{-10}	7.7×10^{-19}	0.0
$SiO_2(aq)$	1.0×10^{-8}	1.0×10^{-4}	0.0
Na^{+}	1.0×10^{-8}	1.4×10^{-1}	0.0
Cl^{-}	4.4×10^{-4}	1.4×10^{-1}	0.0
HCO_3^{-}	4.9×10^{-7}	8.1×10^{-4}	0.0
OH^{-}	1.0×10^{-9}	1.7×10^{-6}	14.0
$Al(OH)_2^{+}$	3.4×10^{-10}	1.2×10^{-15}	5.0
$Al(OH)^{+2}$	2.8×10^{-10}	1.7×10^{-12}	10.1
$Al(OH)_3(aq)$	2.4×10^{-11}	2.5×10^{-10}	16.2
$Al(OH)_4^{-}$	2.5×10^{-12}	4.2×10^{-8}	22.2
$CO_2(aq)$	1.1×10^{-5}	1.1×10^{-5}	-6.3
CO_3^{-}	2.3×10^{-12}	6.3×10^{-6}	10.3
H_3SiO4^{-}	1.5×10^{-13}	2.6×10^{-6}	9.8
>FeOH	1.0×10^{-4}	5.1×10^{-4}	0.0
$>FeO^{-}$	2.6×10^{-11}	2.1×10^{-7}	11.6
$>FeOH_2^{+}$	4.1×10^{-4}	1.2×10^{-6}	-5.6
$>FeOH\text{-}Co^{+2}$	2.1×10^{-7}	9.9×10^{-11}	2.7

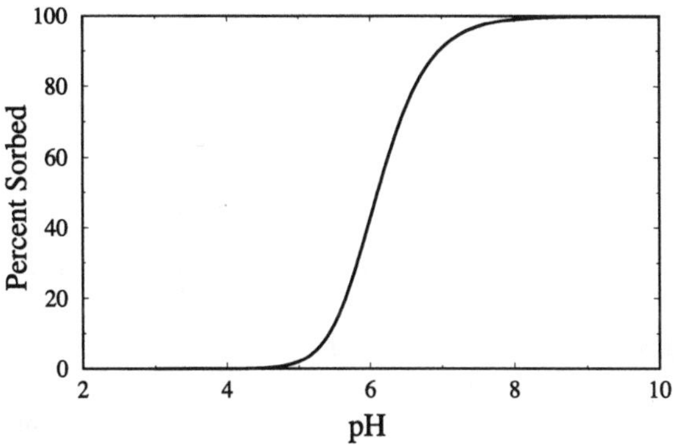

Figure 9. Co^{+2} adsorption on an Fe-hydroxide surface as a function of pH. Calculation assumes 5.12×10^{-4} total surface hydroxyl sites per liter solution.

Comparison of the various methods shows that the SIA is the most effective at reducing the operator splitting error at lowest computational cost (Table 7), although some of the SIA simulations had difficulty in converging due to the problems described above. The global implicit method, which had no difficulty at all with the problem, nontheless required significantly more time than any of the other methods.

Summary of results from method comparisons. The examples considered above, which

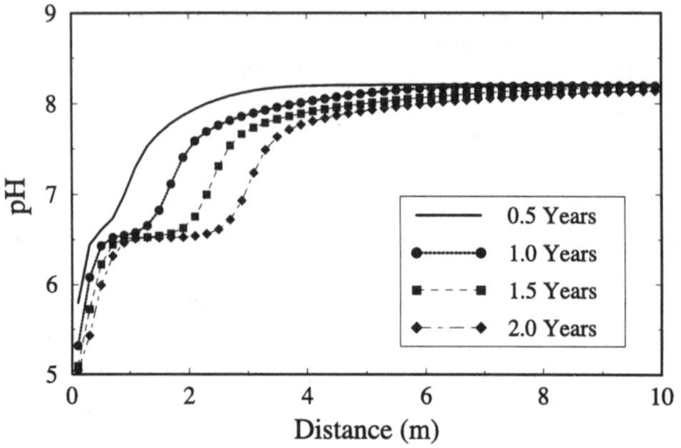

Figure 10. Time evolution of pH profile in problem involving multicomponent aqueous and surface complexation. The propagation of the pH front (which is retarded itself by adsorption) causes the desorption of Co^{+2} observed in Figure 11.

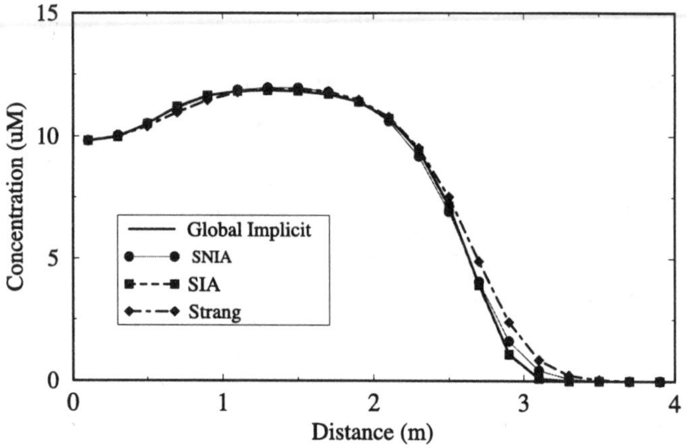

Figure 11. Comparison of the SNIA, SIA, Strang splitting, global implicit (one-step) approaches at a CFL of 0.5 on a multicomponent aqueous and surface complexation problem. Co^{+2} profile is at 2 years.

range from simple decay and equilibrium adsorption reactions to more highly nonlinear Monod kinetic and multicomponent adsorption problems, indicate that under certain circumstances, the SNIA and Strang splitting approachs may result in operator splitting error. Reducing the timestep decreases the error, but the simulations suggest it may be more computationally expensive to reduce the errors inherent in the SNIA approach than it would be using the SIA with a larger time step. In the case of Strang splitting, the conclusions appears to depend on the kind of chemistry problem considered, as suggested by Barry et al. (1996). In the case of a realistic one-dimensional problem involving Monod kinetic reactions, the Strang approach is computationally more efficient than the SIA, but for equilibrium adsorption reactions (whether linear or nonlinear) the SIA appears to be more efficient, although convergence difficulties in some problems may make it harder to use.

While the SIA method had no difficulty in converging when applied to the Monod kinetics example presented above, problems did occur when it was applied to the problem involving

Table 7. Accuracy (based on L_2 norm for Co^{+2}) and CPU time comparison for various reaction-transport coupling schemes applied to a multicomponent aqueous and surface complexation problem.

Method	Δt	L_2 norm for Co^{+2}	Normalized CPU Time
SNIA	0.2	0.770	1.00
SNIA	0.1	0.450	1.82
SNIA	0.05	0.275	3.38
Strang	0.2	1.847	1.27
Strang	0.1	0.779	2.15
Strang	0.05	0.512	4.05
SIA	0.2	0.062	4.58
GI	0.2	0.000	7.01

pH-dependent adsorption of cobalt. The difficulties arise when a significant percentage of the mass transported into a grid cell is reacted within a time step. The first iteration is generally carried out with no reaction term, i.e. the transport is non-reactive. In the case we considered, the initial concentration of cobalt in grid cells downstream of the boundary was low, so the amount transported into the downstream grid cells exceeded the initial concentration. The reaction step calculated in the second iteration (Eqn. 109), however, reduced the aqueous cobalt concentration brought into the cell via transport, partitioning the bulk of it onto the Fe-hydroxide surface. This change in concentration was then the source term which was used in the next solution of Equation (108). If the flux into the grid cell remained constant, there would be no particular problem since the flux would still tend to balance the reaction rate computed in the previous iteration. But the flux into downstream nodes will be reduced due to the adsorption occurring upstream. The SIA method, however, has no way of knowing this (unlike the global implicit method), so the same relatively large, negative source term is used despite the fact that the flux term, which would tend to increase the concentration in the cell, is lowered over previous iterations. The result can be negative concentrations. The problems can be "cured" in part by requiring that all concentrations computed in Equation (108) be non-negative. In the cobalt adsorption problem, this required a number of reductions of the source term and repeats of the iterative cycle in order to get convergence. Similar problems arise when the SIA method is applied to even a simple problem involving a linear adsorption isotherm and strong partitioning of the species onto the solid phase.

Another possible disadvantage to using the SIA approach is that it requires an implicit solution for the transport equations in order to get a fully implicit solution to the overall reaction-transport problem. This makes it impossible to use such explicit methods as the TVD scheme outlined above. There are, however, implicit TVD schemes which may give the desired higher order accuracy (see for example, Unger et al., 1996; Blunt and Rubin, 1992), although they make the transport equations nonlinear which necessitates the use of iterative methods. It is important to point out that in many cases, the errors associated with using fully implicit, lower order transport methods may be larger than the operator splitting errors associated with the SNIA or Strang schemes. This point can be further reinforced by running the cobalt surface complexation example at a higher grid Peclet number (200 instead of 2). In this case, we will have to potentially contend with both operator splitting errors and with spatial discretization errors of the kind shown in Figure 3. The global implicit method is free from operator splitting

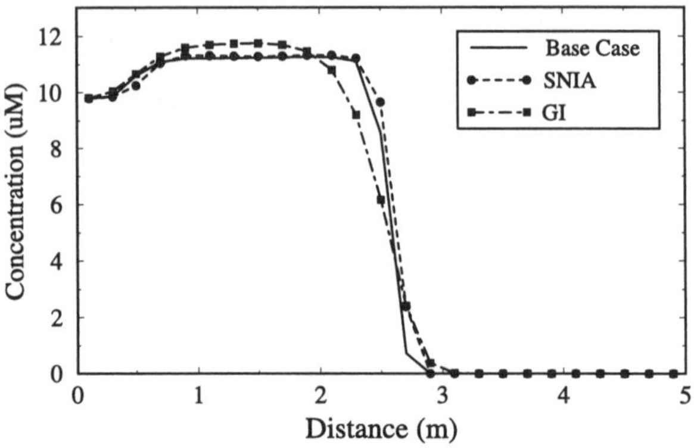

Figure 12. Cobalt surface complexation example run at a grid Peclet number of 200 comparing the SNIA approach (using the 3rd order TVD) and the global implicit (using a 1st order upwind) against a SNIA calculation using a 100 times smaller timestep (*Base Case*). The numerical dispersion in the global implicit method contributes more error than does the operating splitting error associated with using the SNIA method with a relatively large time step.

errors but both the implicit solution of the transport equation and the use of the first-order upwind scheme add significant numerical dispersion. The SNIA method in this case uses the 3rd order TVD scheme previously demonstrated in Figure 5. The Co^{+2} concentration profile for the SNIA using the TVD scheme and the global implicit using an upwind formulation (both using a timestep of 0.02 years) are compared in Figure 12 with the SNIA using a timestep 100 times smaller (the "base case", which is largely free of operator splitting error). Note that in this case, the numerical dispersion contributes more error to the solution than does the operating splitting error associated with using the SNIA method with a large time step. In any modeling effort, therefore, it is important to consider all of the varieties of error which may occur in a particular problem.

In none of the problems discussed above was the global implicit or one-step method more efficient, although it does solve the problems considered. The same caveat concerning the use of implicit transport methods (especially when combined with a low order transport scheme like the upwind) applies to the global implicit methods, however. The real use of the global implicit schemes comes in steady-state problems where the transient error associated with concentration front propagation is absent. Steady-state problems, while rare in environmental applications in hydrogeology, are common in certain environments (Lichtner, this volume; Van Cappellen and Gaillard, this volume). Where one is interested in long time scales, the global implicit method may be more efficient because of its ability to take larger time steps. The SIA method, however, could ostensibly be applied to these same problems and is worth testing more fully on a wider range of problems.

SUMMARY

In this chapter we have attempted to review some of the major issues associated with modeling multicomponent, multi-species reactive transport in porous media. By focusing initially on the reaction system, we showed how it is possible to formulate the problem to include any combination of equilibrium and non-equilibrium reactions, whether linearly independent or not. While it is possible to reformulate fully kinetic reaction schemes, there is no real computational advantage in doing so in most cases. The greatest simplifications occur with completely or partly equilibrium systems where the reformulation of the system of equations allows the number of independent unknowns to be reduced. This can result in significant reductions in the amount of CPU time required to solve the problems, particularly where the aqueous system involves many tens of equilibrium reactions as is often the case. The discussion of the formulation of the reactions was followed by a review of commonly used approaches to solve the resulting systems of equations.

Following the discussion on reaction systems, methods for discretizing and solving the transport equations were addressed. Using the finite difference methods as an example, we analyzed the sources of error associated with various formulations. Because these formulations are subject to more than one kind of error, it is not straightforward to eliminate the errors altogether, especially in the case of high Peclet number transport. Several schemes are available, however, which substantially reduce the numerical dispersion associated with lower order transport methods.

The last part of the chapter dealt with the problem of how to couple the reaction and transport terms, with a focus on what if any errors occur due to the coupling scheme itself. We reviewed the major approaches, which include the global implicit or one step approach, where the reaction and transport terms are solved simultaneously, the sequential non-iterative approach (SNIA), where reaction and transport terms are coupled in sequence without iteration, and the sequential iteration approach, where iteration between separate solves of the reaction and transport terms occurs until a fully implicit solution is achieved. We then used the various methods on a number of one-dimensional problems ranging from first-order decay and linear adsorption where an analytical solution is available to somewhat more complicated systems involving multiple-Monod kinetics and multicomponent surface complexation. Our results show that in some cases the SIA gives the smallest error to CPU time ratio, although in other cases a sequential non-iterative approach is more efficient. While the global implicit can solve all of the problems, it is the least efficient from a computational point of view. We concluded, however, with the proviso that the choice of method may also be dictated by the presence of other kinds of error like numerical dispersion which may swamp the operator splitting error.

ACKNOWLEDGMENTS

We are grateful for the reviews of the manuscript by Steve Yabusaki, Uli Mayer, and James Johnson. Partial financial support for the senior author was provided by the Environmental Molecular Science Laboratory at the Pacific Northwest National Laboratory which is operated for the U.S. Department of Energy by Battelle Memorial Institute under Contract DE-AC06-76RLO 1830.

REFERENCES

Aagaard, P. and Helgeson, H.C. (1982) Thermodynamic and kinetic constraints on reaction rates among minerals and aqueous solutions, I, Theoretical considerations. Am J Sci 282:237-285.

Anderson, G.M. and Crerar, D.A. (1993) Thermodynamics in Geochemistry: The Equilibrium Model. Oxford University Press, New York.

Arbogast, T., Chilakapati, A., and Wheeler, M.F. (1992) A characteric-mixed method for contaminant transport and miscible displacement. In: Numerical Methods in Water Resources (T.F. Russell, R.E. Ewing, C.A. Brebbia, W.G. Gray, and G.F. Pinder, eds.), Computational Mechanics Publications, Southampton, UK:77-84.

Bahr, J.M. and Rubin, J. (1987) Direct comparison of kinetic and local equilibrium formulations for solute transport affected by surface reactions. Wat Resource Research 23:438-452.

Barry, D.A., Miller, C.T., and Culligan-Hensley, P.J. (1996) Temporal discretization errors in non-iterative split-operator approaches to solving chemical reaction/groundwater transport models. J Contam Hydrol 22:1-17.

Bear J. (1972) Dynamics of Fluids in Porous Media. American Elsevier, New York..

Bear J. (1979) Hydraulics of Groundwater. McGraw-Hill, New York.

Bethke, C.M. (1996) Geochemical Reaction Modeling. Oxford University Press, New York.

Bischof, C., Carle, A., Khademi, P., and Mauer, A. (1995) The ADIFOR 2.0 system for the automatic differentiation of Fortran 77 programs. Argonne Preprint ANL-MCS-P481-1194.

Blunt, M., and Rubin, B. (1992) Implicit flux limiting schemes for petroleum reservoir simulation. J Comput Phys 102:194-210.

Boris, J.P., and Book, D.L. (1973) Flux corrected transport I: SHASTA, A fluid transport algorithm that works. J Comput Phys 11:38-69.

Bosma, W.J.P. and Van der Zee, S.E.A.T.M. (1993) Transport of reacting solutes in a one-dimensional, chemically heterogeneous porous medium. Wat Resource Research 29:117-131.

Brenan, K.E., Campbell, S.L., and Petzold, L.R. (1989) Numerical Solution of Initial-Value Problems in Differential-Algebraic Systems. North-Holland, Amsterdam.

Brinkley, S.R., Jr., (1947) Calculation of the equilibrium composition of systems of many components. J Comput Phys 15:107-110.

Brown, P.N., Byrne, G.D., and Hindmarsh, A.C. (1989) VODE, a variable-coefficient ODE solver. SIAM J Sci Stat Comp 10:1038-1051.

Brusseau, M. L. (1994) Transport of reactive contaminants in heterogeneous porous media. Rev of Geophys 32:285-313.

Burch, T.E., Nagy, K.L., and Lasaga, A.C. (1993) Free energy dependence of albite dissolution kinetics at 80°C and pH 8.8. Chem Geol 105:137-165.

Burnett, R.D. and Frind, E.O. (1987) Simulation of contaminant transport in three dimensions 2. Dimensionality effects. Wat Resource Research 23:695-705.

Celia, M.A. and Gray, W.G. (1992) Numerical Methods for Differential Equations: Fundamental Concepts for scientific and engineering applications. Prentice-Hall, Englewood Cliffs, N.J.

Celia, M.A., Russell, T.F., Herrera, I., and Ewing, R.E. (1990) An Eulerian-Lagrangian localized adjoint method for the advection-dispersion equation. Adv Wat Resource 13:187-206.

Chilakapati, A. (1993) Numerical Simulation of Reactive Flow and Transport through the Subsurface. Ph.D. dissertation, Rice University, Houston, TX.

Chilakapati, A. (1995) RAFT: A simulator for reactive flow and transport of groundwater contaminants. PNL-10636, Pacific Northwest Laboratory Report.

Cordes, C. and Kinzelbach, W. (1996) Comment on "Application of the mixed hybrid finite element approximation in a groundwater flow model: Luxury or necessity?" by R. Mosé, P. Siegel, P. Ackerer, and G. Chavent. Wat Resource Research 32:1905-1909.

Cushman, J.H., Hu, Bill X., and Deng, F. (1995) Nonlocal reactive transport with physical and chemical heterogeneity: Localization errors. Wat Resource Research 31:2219-2237.

Dagan, G. (1986) Statistical theory of groundwater flow and transport: Pore to laboratory, laboratory to formation, and formation to regional scale. Wat Resource Research 22:120-134.

Dagan, G. (1989) Flow and Transport in Porous Formations. Springer-Verlag, New York.

Dagan, G., and Cvetkovic, V. (1993) Spatial moments of a kinetically sorbing solute plume in a heterogeneous aquifer. Wat Resource Research 29:4053-4061.

DeCapitani, C. and Brown, T.H. (1987) The computation of chemical equilibrium in complex systems containing non-ideal solutions. Geochim Cosmochim Acta 51:2639-2652.

Deuflhard, P. and Nowak, U. (1985) Extrapolation integrators for quasilinear implicit ODE's. SFB 213, TR 332, Technical Report, University of Heidelberg.

Di Giammarco, P., Todini, E., and Lamberti, P. (1996) A conservative finite elements approach to overland flow: the control volume finite element formulation. J Hydrol 175:267-291.

Dzombak, D.A. and Morel, F.M.M. (1990) Surface Complex Modeling, Hydrous Ferric Oxide. Wiley-Interscience, New York.

Engesgaard, P. and Kipp, K.L. (1992) A geochemical model for redox-controlled movement of mineral fronts in ground-water flow systems: A case of nitrate removal by oxidation of pyrite. Wat Resource Research 28:2829-2843.

Essaid, H.I., Bekins, B.A., Godsy, E.M., Warren, E., Baedecker, M.J., and Cozzarelli, I.M. (1995) Simulation of aerobic and anaerobic biodegradation processes at a crude oil spill site. Wat Resource Research 31:3309-3327.

Felmy, A.R. (1990) GMIN: A computerized chemical equilibrium model using a constrained minimization of the Gibbs free energy. PNL-7281, Pacific Northwest Laboratory Report.

Felmy, A.R., Girvin, D.C., and Jenne, E.A. (1984) MINTEQ: A computer program for calculating aqueous geochemical equilibria. EPA 600/3-84-032, U.S. Environmental Protection Agency, Office of Research and Development, Athens, Georgia.

Ferraresi, M. and Marinelli, A. (1996) An extended formulation of the integral finite difference method for groundwater flow and transport. J Hydrol 175:453-471.

Forsyth, P.A. (1991) A control volume finite element approach to NAPL groundwater contamination. SIAM J Sci Stat Comp 12:1029-1057.

Friedly, J.C. and Rubin, J. (1992) Solute transport with multiple equilibrium-controlled or kinetically-controlled chemical reactions. Wat Resource Research 28:1935-1953.

Friedly, J.C., Davis, J.A., and Kent, D.B. (1995) Modeling hexavalent chromium reduction in groundwater in field-scale transport and laboratory batch experiments. Wat Resource Research 31:2783-2794.

Garabedian, S., Gelhar, L.W., and Celia, M.A. (1988). Large-scale dispersive transport in aquifers: Field experiments and reactive transport theory. Tech. Rep. 315, R.M. Parsons Lab., Dep. of Civ. Engl, Mass. Inst. Technol, Cambridge.

Gelhar, L.W. (1993) Stochastic Subsurface Hydrology. Prentice-Hall, Englewood Cliffs, N.J.

Ginn, T.R., Simmons, C.S. and Wood, B.D. (1995) Stochastic-convective transport with nonlinear reaction: Biodegradaton with microbial growth, Wat Resource Research 31:2689-2700.

Gupta, A.D., Lake, L.W., Pope, G.A., Sephernoori, K., and King, M.J. (1991) High–resolution monotonic schemes for reservoir fluid flow simulation. In Situ 15:289-317.

Hairer, E. and Wanner, G. (1988) Radau5 - an implicit Runge-Kutta code. Technical report, Dept. de mathématiques, Université de Genève.

Harten, A. (1983) High resolution schemes for hyperbolic conservation laws. J Comput Phys 49:357-393.

Harvie, C.E., Greenberg, J.P., and Weare, J.H. (1987) A chemical equilibrium algorithm for highly non-ideal multiphase systems: free energy minimization. Geochim Cosmochim Acta 51:1045-1057.

Hills, R.G., Fisher, K.A., Kirkland, M.R., and Wierenga, P.J. (1994) Application of flux-corrected transport to the Las Cruces Trench site. Wat Resource Research 30:2377-2386.

Hindmarsh, A.C. (1977) Solution of block–tridiagonal systems of linear algebraic equations. UCID-30150.

Hindmarsh, A.C. (1983) ODEPACK, A systemized collection of ODE solvers. In: Scientific Computing, (R.S. Stepleman, ed.), North-Holland, Amsterdam:55-64.

Hindmarsh, A.C., and Petzold, L.R. (1995a) Algorithms and software for ordinary differential equations and differential/algebraic equations, Part I: Euler methods and error estimation. Comput in Phys 9:34-41.

Hindmarsh, A.C., and Petzold, L.R. (1995b) Algorithms and software for ordinary differential equations and differential/algebraic equations, Part II: Higher-order methods and software packages. Comput in Phys 9:148-155.

Huyakorn, P.S. and Pinder, G.F. (1983) Computational Methods in Subsurface Flow. Academic Press, San Diego, Ca.

Istok, J. (1989) Groundwater modeling by the finite element method. Water Resources Monograph 13, American Geophysical Union.

Kabala, Z.J., and Sposito, G. (1991) A stochastic model of reactive solute transport with time-varying velocity in a heterogeneous aquifer. Wat Resource Research 27:341-350.

Kaluarachchi, J.J, and Morshed, J. (1995) Critical assessment of the operator-splitting technique in solving the advection-dispersion-reaction equation: 1. First-order reaction. Adv Wat Resource 18:89-100.

Kee, R.J., Petzold, L.R., Smooke, M.D., and Grcar, J.F. (1985) Implicit methods in combustion and chemical kinetics modeling. In: Multiple Time Scales (J.U. Brackbill and B.I. Cohen, eds.), Academic Press, New York.

Konikow, L.F. and Bredehoeft, J.D. (1978) Computer model of two-dimensional solute transport and dispersion in ground water. In : Techniques of Water-Resources Investigations, book 7, chap. C2, U.S. Geol. Survey, Reston, Va.

Lasaga, A.C. (1981) Rate laws in chemical reactions. In Kinetics of Geochemical Processes, Reviews in Mineralogy 8 (A.C. Lasaga and R.J. Kirkpatrick, eds.), Mineral Soc Am:135-169.

Lasaga, A.C. (1984) Chemical kinetics of water-rock interactions. J Geophys Res 89:4009-4025.

Lasaga, A.C. and Rye, D.M. (1993) Fluid flow and chemical reactions in metamorphic systems. Am J Sci 293:361-404.

Lensing, H.J., Vogt, M., and Herrling, B. (1994) Modeling of biologically mediated redox processes in the subsurface. J Hydrol 159:125-143.

Leonard, B. (1979a) A stable and accurate convective modeling procedure based on quadratic upstream interpolation. Comput Meth Applied Mech Eng 19:59-98.

Leonard, B. (1979b) A survey of finite differences of opinion on numerical muddling of the incomprehensible defective confusion equation. In: Finite Element Methods for Convection-Dominated Flows (T.J.R. Hughes, ed.):1-17.

Leonard, B. (1984) Third-order upwinding as a rational basis for computational fluid dynamics. Elsevier, New York.

Lichtner, P.C. (1985) Continuum model for simultaneous chemical reactions and mass transport in hydrothermal systems. Geochim Cosmochim Acta 49:779-800.

Lichtner, P.C. (1992) Time-space continuum description of fluid/rock interaction in permeable media. Wat Resource Research 28:3135-3155.

Lindberg, R.D. and Runnells, D.D. (1984) Ground water redox reactions: An analysis of equilibrium state applied to Eh measurements and geochemical modeling. Science 225:925-927.

MacQuarrie, K.T.B., Sudicky, E.A., and Frind, E.O. (1990) Simulation of biodegradable organic contaminants in groundwater 1. Numerical formulation in principal directions. Wat Resource Research 26:207-222.

Marzel, P., Seco, A., Ferrer, J., and Gambaldón, C. (1994) Modeling multiple reactive solute transport with adsorption under equilibrium and nonequilibrium conditions. Adv Wat Resource 17:363-374.

McNab, W.W., Jr. and Narasimhan, T.N. (1994) Modeling reactive transport of organic compounds in groundwater using a partial redox disequilibrium approach. Wat Resource Research 30:2619-2635.

McNab, W.W., Jr. and Narasimhan, T.N. (1995) Reactive transport of petroleum hydrocarbon constituents in a shallow aquifer: Modeling geochemical interactions between organic and inorganic species. Wat Resource Research 31:2027-2033.

Molz, F.J., Widdowson, M.A., and Benefield, L.D. (1986) Simulation of microbial growth dynamics coupled to nutrient and oxygen transport in porous media. Wat Resource Research 22:1207:1216.

Morel, F.M.M. and Hering, J.G. (1993) Principles and Applications of Aquatic Chemistry. John Wiley, New York.

Morel, F.M.M. and Morgan, J. (1972) A numerical method for computing equilibria in aqueous chemical systems. Environ Sci and Tech 6:58-67.

Nagy, K.L., Blum, A.E., and Lasaga, A.C. (1991) Dissolution and precipitation kinetics of kaolinite at 80°C and pH 3: The dependence on solution saturation state. Am J Sci 291:649-686.

Nagy, K.L. and Lasaga, A.C. (1992) Dissolution and precipitation kinetics of gibbsite at 80°C and pH 3: The dependence on solution saturation state. Geochim Cosmochim Acta 56:3093-3111.

Narasimhan, T.N. and Witherspoon, P.A. (1976). An integrated finite differences method for analyzing fluid flow in porous media. Wat Resource Research 12:57-64.

Nordstrom, D.K. and Munoz, J.L. (1994) Geochemical Thermodynamics, 2nd ed., Blackwell, Boston.

Novak, C.F. (1990) Metasomatic Patterns Produced by Infiltration or Diffusion in Permeable Media. Ph.D. Dissertation, The University of Texas at Austin, Austin, TX.

Novak, C.F. and Sevougian, S.D. (1993) Propagation of dissolution/precipitation waves in porous media. In: Migration and Fate of Pollutants in Soils and Subsoils (D. Petruzzelli, F.G. Helfferich, eds.) NATO ASI Series. Series G: Ecological Sciences, 32, Springer-Verlag, Berlin:275-307.

Oelkers, E.H., Bjorkum, P.-A., and Murphy, W.M. (1996) A petrographic and computational investigation of quartz cementation and porosity reduction in North Sea sandstones. Am J Sci 296:420-452.

Oran, E.S., and Boris, J.P. (1987) Numerical Simulation of Reactive Flow. Elsevier, New York.

Parkhurst, D.L. (1995) User's guide to PHREEQC - A comuter program for speciation, reaction-path, advective transport, and inverse geochemical calculations. U.S. Geol. Survey Water Resources Investigations:95-4227.

Parkhurst, D.L., Thorstenson, D.C., and Plummer, L.N. (1980) PHREEQE – a computer program for geochemical calculations. U.S. Geological Survey Water Resources Investigations Report 80-96.

Petzold, L.R. (1983) A description of DASSL: A differential/algebraic system solver. In: Scientific Computing (R.S. Stepleman, ed.), North-Holland, Amsterdam:65-68.

Press, W.H., Flannery, B.P., Teukolsky, S.A., and Vetterling, W.T. (1986) Numerical Recipes: The Art of Scientific Computing. Cambridge University Press, Cambridge.

Raffensperger, J.P. and Garven, G. (1995) The formation of unconformity-type uranium ore deposits. 2. Coupled hydrochemical modeling. Am J Sci 295:639-696.

Reed, M.H. (1982) Calculation of multicomponent chemical equilibria and reaction processes in systems involving minerals, gases, and an aqueous phase. Geochim Cosmochim Acta 46:513-528.

Regnier, P., Wollast, R., and Steefel, C.I. (in press) Long-term fluxes of reactive species in macrotidal estuaries. Marine Chemistry.

Robin, M.J.L., Sudicky, E., Gillham, R., and Kachanoski, R. (1991) Spatial variability of strontium distribution coefficients and their correlation with hydraulic conductivity in the Canadian Forces Base Borden aquifer. Wat Resource Research 27:2619-2632.

Sevougian, S.D., Lake, L.W., and Schechter, R.S. (1995) KGEOFLOW: A new reactive transport simulator for sandstone matrix acidizing. Soc. Petr. Eng. Production and Facilities, February:13-19.

Sevougian, S.D., Schechter, R.S., and Lake, L.W. (1993) Effect of partial local equilibrium on the propagation of precipitation/dissolution waves. Ind Eng Chem Res 32:2281-2304.

Simmons, C.S., Ginn, T.R., and Wood, B.D. (1995) Stochastic-convective transport with nonlinear reaction: Mathematical framework. Wat Resource Research 31:2675-2688.

Šimůnek, J. and Suarez, D.L. (1994) Two-dimensional transport model for variably saturated porous media with major ion chemistry. Wat Resource Research 30:1115-1133.

Smith, W.R. and Missen, R.W. (1982) Chemical Reaction Equilibrium Analysis: Theory and Algorithms. Wiley, New York.

Soetaert, K., Herman, P.M.J., and Middelburg, J.J. (1996) A model of early diagenetic processes from shelf to abyssal depths. Geochim Cosmochim Acta 60:1019-1040.

Steefel, C.I. and Lasaga, A.C. (1994) A coupled model for transport of multiple chemical species and kinetic precipitation/dissolution reactions with application to reactive flow in single phase hydrothermal systems. Am J Sci 294:529-592.

Steefel, C.I. and Lichtner, P.C. (1994) Diffusion and reaction in rock matrix bordering a hyperalkaline fluid-filled fracture. Geochim Cosmochim Acta 58:3595-3612.

Steefel, C.I. and Van Cappellen, P. (1990) A new kinetic approach to modeling water–rock interaction: The role of nucleation, precursors, and Ostwald ripening. Geochim Cosmochim Acta 54:2657-2677.

Steefel, C.I. and Yabusaki, S.B. (1996) OS3D/GIMRT, Software for Multicomponent-Multidimensional Reactive Transport, User Manual and Programmer's Guide. PNL-11166, Pacific Northwest National Laboratory, Richland, Washington.

Strang, G.. (1968) On the construction and comparison of difference schemes. SIAM J Numer Anal 5:506-517.

Sweby, P.K. (1984) High resolution schemes using flux limiters for hyperbolic conservation laws. SIAM J Numer Anal 21:995-1011.

Thompson, J.B. (1959) Local equilibrium in metasomatic processes. In: Researches in Geochemistry (P.H. Abelson, ed.), John Wiley, New York:427-457.

Tompson, A.F.B. (1993) Numerical simulation of chemical migration in physically and chemically heterogeneous porous media. Wat Resource Research 29:3709-3726.

Tompson, A.F.B., Schafer, A.L., and Smith, R.W. (1996) Impacts of physical and chemical heterogeneity on cocontaminant transport in a sandy porous medium. Wat Resource Research 32:801-818.

Unger, A.J.A., Forsyth, P.A., and Sudicky, E.A. (1995) Variable spatial and temporal weighting schemes for use in multi-phase compositional problems. Adv Wat Resource 19:1-27.

Valocchi, A.J. (1989) Spatial moment analysis of the transport of kinetically sorbing solutes through stratified aquifers. Wat Resource Research 25:273-279.

Valocchi, A.J. and Malmstead, M. (1992) Accuracy of operator splitting for advection-dispersion-reaction problems. Wat Resource Research 28:1471-1476.

Van Cappellen, P. and Berner, R.A. (1991) Fluorapatite crystal growth from modified seawater solutions. Geochim Cosmochim Acta 55:1219-1234.

Van Cappellen, P. and Wang, Y. (1996) Cycling of iron and manganese in surface sediments: a general early diagenetic model. Am J Sci 296:197-243.

VanderKwaak, J.E., Forsyth, P.A., MacQuarrie, K.T.B., and Sudicky, E.A. (1995) WATSOLV Sparse Matrix Iterative Solver Package Versions 1.01. Unpublished report, Waterloo Centre for Groundwater Research, University of Waterloo, Waterloo, Ontario, Canada.

Van Der Zee, S.E.A.T.M. and Riemsdijk, W.H.V. (1987) Transport of reactive solute in spatially variable soil systems. Wat Resource Research 23:2059-2069.

Van Leer, B. (1977a) Towards the ultimate conservative difference scheme. III. Upstream-centered finite-difference schemes for ideal compressible flow. J Comput Phys 23:263-275.

Van Leer, B. (1977b) Towards the ultimate conservative difference scheme. IV. A new approach to numerical convection. J Comput Phys 23:276.

Verwer, J.G. (1994) Gauss-Seidel iteration for stiff ODEs from chemical kinetics. SIAM J Sci Comp 15:1243-1250.

Walter, A.L., Frind, E.O., Blowes, D.W., Ptacek, C.J., and Molson, J.W. (1994) Modeling of multicomponent reactive transport in groundwater: 1. Model development and evaluation. Wat Resource Research 30:3137-3148.

White, W.B., Johnson, S.M., and Dantzig, G.B. (1958) Chemical equilibrium in complex mixtures. J Comput Phys 28:751-755.

Wolery, T.J. (1979) Calculation of chemical equilibrium between aqueous solution and minerals: the EQ3/6 software package. Lawrence Livermore Laboratory URCL-52658.

Wolery, T.J. (1992) EQ3NR, a computer program for geochemical aqueous speciation-solubility calculations: Theoretical Manual, users's guide, and related documentation (Version 7.0). Publ. UCRL-MA-110662 PT III. Lawrence Livermore National Laboratory, Livermore, Ca.

Wood, B.D., Dawson, C.N., Szecsody, J.E., and Streile, G.P. (1994) Modeling contaminant transport and biodegradation in a layered system. Wat Resource Research 30:1833-1846.

Wood, B.D., Ginn, T.R., and Dawson, C.N. (1995) Effects of microbial metabolic lag in contaminant transport and biodegradation modeling. Wat Resource Research 31:553-563.

Yabusaki, S.B., Wood, B.D., and Steefel, C.I. (in prep.) Multidimensional, multicomponent subsurface transport in nonuniform velocity fields: Code verification using an advective reactive streamtube approach.

Yee, H.C. (1987) Construction of explicit and implicit symmetric TVD schemes and their applications. J Comput Phys 68:151-179.

Yeh, G.T. and Tripathi, V.S. (1989) A critical evaluation of recent developments in hydrogeochemical transport models of reactive multichemical components. Wat Resource Research 25:93-108.

Yeh, G.T. and Tripathi, V.S. (1991) A model for simulating transport of reactive multispecies components: Model development and demonstration. Wat Resource Research 27:3075-3094.

Zalesak, S.T. (1987) A preliminary comparison of modern shock-capturing schemes: Linear advection. In: Advances in Computer Methods for Partial Differential Equations, VI, (R. Vichnevetsky and R.S. Stepelman, eds.), IMACS, Rutgers University, New Brunswick, NJ:15-22.

Zheng, C. and Bennett, G.D. (1995) Applied Contaminant Transport Modeling: Theory and Practice. Van Nostrand Rheinhold, New York.

Zysset, A. and Stauffer, F. (1992) Modeling of microbial processes in groundwater infiltration systems. In Mathematical Modelling in Water Resources, (T.F. Russell, R.E. Ewing, C.A. Brebbia, W.G. Gray, and G.F. Pinder, eds.), Computational Mechanics Publications, Billerica, Mass.

Zysset, A., Stauffer, F., and Dracos, T. (1994) Modeling of chemically reactive groundwater transport. Wat Resource Research 30:2217-2228.

Chapter 3

PHYSICAL AND CHEMICAL PROPERTIES OF ROCKS AND FLUIDS FOR CHEMICAL MASS TRANSPORT CALCULATIONS

Eric H. Oelkers

CNRS/UMR 6653-Laboratoire de Géochimie
Université Paul Sabatier-CNRS-OMP
38 rue des Trente Six Ponts
31400 Toulouse, FRANCE

INTRODUCTION

As evident by the summaries presented in the various chapters in this volume, the past few years have witnessed extraordinary advances in the ability to integrate equations describing the simultaneous flow of one or more natural fluids coupled to advective/dispersive/diffusive solute transport and any number of heterogeneous or homogeneous chemical reactions. The ability to estimate the rate and extent of water-rock interaction and chemical transport is, however, also dependent on the quantitative understanding of the physical and chemical properties of the rocks and fluids participating in reactive transport properties in natural systems. Any of the physical and chemical properties required to compute reactive chemical transport in complex systems, including permeability, porosity, tortuosity, dispersivity, aqueous diffusion coefficients, thermo-dynamic equilibrium constants, and mineral dissolution/ crystallization rates are sufficiently complex functions, and their literature is of sufficient breath and scope to warrant not only a chapter in this volume, but in many cases a dedicated review volume in its own right (see, for example, *Reviews in Mineralogy* 18: Thermodynamic Modeling of Geochemical Systems, Carmichael and Eugster, 1987 or *Reviews in Mineralogy* 31: Chemical Weathering Rates of Silicate Minerals, White and Brantley, 1995). Consequently, it is not the aim of the current chapter to provide a comprehensive review of each of these properties. The aim is rather to provide, for those Earth scientists desiring to perform quantitative reactive transport calculations, a starting point in their search for those parameters required in such models, and to provide initial values for preliminary calculations. As will be emphasized below, the uncertainties in many of the required input parameters for reaction transport modeling are large, in some cases several orders of magnitude. In addition, there are often large discrepancies between those physical and chemical properties measured in the laboratory and those deduced from field measurements. It follows that the quantitative description of geochemical processes using comprehensive numerical reactive algorithms requires extensive efforts towards validating computational results using detailed field data obtained on the systems of interest.

PERMEABILITY OR HYDRAULIC CONDUCTIVITY

Fluid flow in natural porous media is generally slow and assumed to follow Darcy's law. Darcy's law (Darcy, 1856) states that the specific discharge of fluid (or Darcy velocity) $\mathbf{q}$ through a porous media is proportional to the hydraulic gradient $\nabla \mathbf{h}$ such that

0275-0279/96/0034-0003$05.00

Glossary of major symbols used in this chapter

Symbol	Definition	Dimension (Units) or value
a_i	Activity of the subscripted species	dimensionless
A	Chemical affinity defined by Equation (75)	energy/mass (cal/mol)
c_i	Concentration of the subscripted species	mass/length^3
d_e	Effective grain diameter	length (cm)
$\overline{d}$	Average grain diameter	length (cm)
D	Diffusion coefficient of the species of interest in pure solution	length^2/time (cm^2/sec)
D_i	Diffusion coefficient of the subscripted species in pure solution	length^2/time (cm^2/sec)
D_{ij}	Diffusion coefficient matrix for the subscripted species in pure solution	length^2/time (cm^2/sec)
D_i^0	Aqueous tracer diffusion coefficient of the subscripted species	length^2/time (cm^2/sec)
$\overline{D}_i$	Effective diffusion coefficient of the subscripted species in the rock	length^2/time (cm^2/sec)
D'	Coefficient of mechanical dispersion	length^2/time (cm^2/sec)
D'_L	Coefficient of longitudinal mechanical dispersion	length^2/time (cm^2/sec)
D'_T	Coefficient of transverse mechanical dispersion	length^2/time (cm^2/sec)
$\tilde{D}_L$	Coefficient of longitudinal hydrodynamic dispersion ($\tilde{D}_L = D'_L + D$)	length^2/time (cm^2/sec)
$\tilde{D}_T$	Coefficient of transverse hydrodynamic dispersion ($\tilde{D}_T = D'_T + D$)	length^2/time (cm^2/sec)
f	Formation factor	dimensionless
F	Faraday constant	9.6484×10^4 C/equiv.
g	Gravitional force	9.81 m/sec^2
h	Hydraulic head	length
J_i	Flux of the subscripted species	mass/length^2/time (mol/cm^2/sec)
k	Permeability	length^2 (darcys)
$\overline{k}$	Average permeability	length^2 (darcys)
k_b	Boltzman constant	3.2999×10^{-24}cal K^{-1}
k_+	Rate constant	mass/length^2/time (mol/cm^2/sec)
$k_{\parallel}$	Permeability parallel to strata	length^2 (darcys)
$k_{\perp}$	Permeability perpendicular to strata	length^2 (darcys)
K	Hydraulic conductivity	length/time
K°_{79}	Equilibrium constant for reaction 79	dimensionless
K°_{88a}	Equilibrium constant for reaction 88a	dimensionless
K°_{88b}	Equilibrium constant for reaction 88b	dimensionless

Symbol	Definition	Dimension (Units) or value
K_{eq}	Equilibrium constant for the overall mineral hydrolysis reaction	dimensionless
l	Average pore length	length (cm)
l_{ij}	Phenomenological transport coefficient matrix	$(mol^2/cm/cal/sec)$
l_{ij}^0	Phenomenological transport coefficient matrix at infinite dilution	$(mol^2/cm/cal/sec)$
M	Function defined by Equation (59)	dimensionless
M_{H2O}	Molecular weight of H_2O	18.015 gm/mol
M-O	Potentially reactive mineral surface site	dimensionless
M^{+Z_M}	Metal ion involved in reaction (84)	dimensionless
n	Stoichiometric coefficient	dimensionless
n_f	Formation constant (see Eqn. 29)	dimensionless
n_H	Stoichiometic number of hydrogen ions in reaction (79)	dimensionless
n_s	Number of aqueous species in the system	dimensionless
P^*	Rate controlling precursor complex	dimensionless
P_d	Dispersional Peclet number defined by Equation (73)	dimensionless
P_e	Diffusional Peclet number defined by Equation (54)	dimensionless
q	Specific discharge or Darcy velocity	length/time (cm/sec)
Q	Reaction quotient defined by Equation (76)	dimensionless
$\mathbf{r}$	Mineral dissolution rate per unit surface area	$mass/length^2/time$ $(mol/cm^2/sec)$
$\mathbf{r}_+$	Forward mineral dissolution rate per unit surface area	$mass/length^2/time$ $(mol/cm^2/sec)$
r	Grain radius	length (cm)
$\bar{r}$	Average grain radius	length (cm)
r_p	Average pore radius	length (cm)
r_s	Stokes law radius	length (cm)
R	Gas constant	1.98719 cal $mol^{-1}K^{-1}$
R_R	Reynolds number defined by Equation (3)	dimensionless
$\bar{s}$	Interfacial surface area per unit volume of porous media	$length^{-1}$
$\bar{s}_0$	Interfacial surface area per unit volume of solid	$length^{-1}$
T	Absolute temperature	K°
U	Pore flow velocity ($U = q/\phi$)	length/time
V_i	Molecular volume of subscripted solute at its boiling point	$length^3$
$\bar{X}$	Average solute front distance	length
z_i	Charge of the subscripted species	dimensionless
$\bar{z}$	Function defined by Equation (60)	$length^2/time$ (cm^2/sec)
Z_M	Charge of the metal M	(equiv./mol)

Symbol	Definition	Dimension (Units) or value
α_L	Longitudinal dispersivity ($\alpha_L = \tilde{D}_L/U$)	length
α_T	Transverse dispersivity ($\alpha_T = \tilde{D}_T/U$)	length
α_V	Vertical transverse dispersivity ($\alpha_V = \tilde{D}_V/U$)	length
δ_{ij}	Function equal to 1 if $I = j$, but equal to 0 if $I \neq j$	dimensionless
ϕ	Porosity	dimensionless
ϕ_e	Effective directional porosity (see Eqn. 28)	dimensionless
Φ	Electrical potential	energy/equivalent
γ_i	Activity coefficient of the subscripted species	length^3/mass
η	Viscosity of pure water	(centipoises)
μ_j	Chemical potential of the subscripted species	energy/mass (cal/mol)
ν	Kinematic viscosity of water or solution ($\nu = \eta/\rho$)	length^2/time (stokes)
λ_i^0	Limiting equivalent conductance of the subscripted species	(cm^2/Ω/equiv)
π	Pi	3.14159
ρ	Fluid density	mass/length^3
σ	Temkins average stoichiometric number (see Eqn. 77)	dimensionless
σ_K	Variance of the natural logarithm of hydraulic conductivity	dimensionless
τ_e	Tortuosity factor	dimensionless

Note: to prevent duplication of terms, some original symbols used in cited literature have been changed.

$$q = -K\,\nabla h = -\frac{kg\rho}{\eta}\nabla h = -\frac{kg}{\nu}\nabla h \tag{1}$$

where K stands for a proportionality constant called hydraulic conductivity, k denotes the permeability, g designates gravitational force (9.81 m/sec^2), ρ designates the fluid density, η signifies the dynamic viscosity of the fluid, and ν refers to the kinematic viscosity of the fluid ($\nu = \eta/\rho$). Hydraulic conductivity and permeability are thus related by

$$K = \frac{kg\rho}{\eta} = \frac{kg}{\nu} \tag{2}$$

Hydraulic head (h) is a function of both elevation (z) and fluid pressure (p) according to

$$h = z + \frac{p}{\rho g} \tag{2a}$$

Units of permeability are length^2, and are commonly expressed in darcys. One darcy is the permeability leading to a specific discharge of 1 cm/s for a fluid with a viscosity of 1 cp experiencing a 1 atm/cm hydraulic gradient and is equal to 0.987×10^{-8}cm^2. The hydraulic conductivity has units of velocity (length/time). For water at 20°C, a permeability of one darcy corresponds to a hydraulic conductivity of 9.613×10^4 cm/sec.

In general, permeability is independent of the flowing substance, but hydraulic conductivity depends on fluid viscosity and density. Consequently, research that is concerned with one type of near-isothermal flowing substance, such as near surface groundwater commonly uses hydraulic conductivity, research that is concerned with various fluids or non-isothermal conditions such as oil-field processes commonly uses permeability. Note that for rapidly flowing fluids in porous media, fluid flow becomes turbulent and a non-linear function of hydraulic gradient. The flow rate at which the flow begins to deviate from a Darcy's law behavior is observed to be when the Reynolds number for fluid flow, R_R, given by

$$R_R = \frac{2rq}{\nu} \tag{3}$$

where r denotes the radius of the grains in the porous media, exceeds unity (Bear, 1972).

Permeability is generally categorized based on rock type. Numerous summaries of the porosities of natural materials have presented some form of Figure 1 (Lerman, 1979; Freeze and Cherry, 1979; Dagan, 1989). This figure illustrates that the permeability of natural porous materials fall over a range of greater than 15 orders of magnitude. In addition, the permeability of any single rock type ranges over at least three to four orders of magnitude. For a given hydraulic gradient, the quantity of fluid and thus the quantity of dissolved substances passing through any system is directly proportional to its permeability. It follows that uncertainties in the permeability of natural materials is one of the greatest uncertainties in the accurate characterization of the extent and consequences of reactive flow in natural systems.

The permeability of a porous material is a measure of the ease by which a fluid can pass through it. The overall flow through a porous material is a combination of flow through the pore spaces, microcracks, and grain boundaries of the bulk material and the flow along more widely spaced fractures. The relative importance of each of these factors depends on the rock. For relatively porous materials, such as many unconsolidated materials and sedimentary rocks, flow through pore spaces dominates, and contribution to

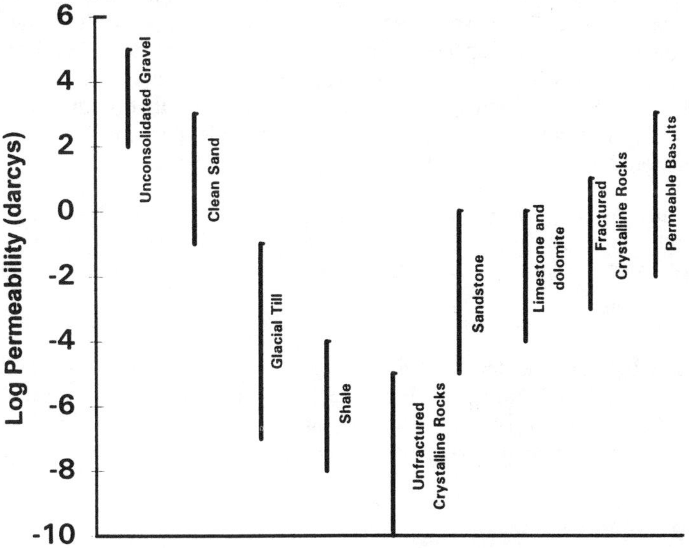

Figure 1. Schematic illustration of the permeability of various rock types (after Freeze and Cherry 1979).

permeability of cracks, fissures, and fractures tend to be negligible. In contrast, for relatively non-porous materials, such as most igneous and metamorphic rocks, fluid flow through pores is negligible. Thus the bulk of fluid flow through these materials in natural systems passes through cracks, fissures, and fractures, implying a fundamental difference in the nature of permeability in highly porous versus low porosity materials.

Permeability in sedimentary rocks

An exhaustive number of studies have generated permeabilities of sedimentary rocks in both the laboratory and in the field. The permeabilities of several sand, gravel, and clay aquifers, as measured in the field are presented together with their average porosities in Table 1. It can be seen in this table that the permeabilities of these aquifers range from 0.01 to 1500 darcys, a range of over five orders of magnitude. The bulk of the values tend to fall in the range 1 to 600 darcys. The permeabilities of a variety of sedimentary rocks obtained from laboratory measurements are listed in Table 2. The permeabilities listed in this table range from 10^{-4} to 4.4 darcys, which is lower by approximately three orders of magnitude than corresponding permeability values of the mostly unconsolidated materials listed in Table 1. The origin of these differences are (1) the effects of consolidation and cementation, (2) the existence of heterogeneities including the possible presence of highly permeable zones that increase the overall permeability in natural systems, and (3) differences in the way permeability is measured in the laboratory versus the field.

Owing to the demands of the petroleum industry, a large amount of work has been aimed at generating methods to independently predict permeability in sedimentary rocks. These studies include evaluation of the effect on sandstone permeability of bulk porosity (Carman, 1937; Berg, 1970; Bloch, 1991), grain size or grain size distribution (Krumbein and Monk, 1942; Marshall, 1958; Chilingar, 1964; Masch and Denny, 1966; Berg, 1970; Beard and Weyl, 1973; van Baaren, 1979; Inverson and Satchwell, 1989; Alyamani and Sen, 1993; Panda and Lake, 1994), surface area of the intergranular volume (Carman, 1937; Johnson et al., 1987; Sheng and Zhou, 1988; Schwartz and Banavar, 1989), formation factor (Archie, 1942; Katz and Thompson, 1986; Johnson et al., 1987; Hazlett, 1989; Avellaneda and Torquato, 1991), mineralogical composition (Howard, 1992), geometric attributes of pore space in thin section (Koplik et al., 1984; Doyen, 1988; McCreesh et al., 1988), and acoustic wave velocity (Marion et al., 1989; Iverson, 1990, Vernik and Nur, 1991). In general, these studies have found that permeability increases dramatically with increasing porosity, increasing particle size, increasing particle size homogeneity, and decreasing intergranular volume surface area.

Because porosity is more readily measured than permeability in sedimentary basins, much work has focused on generating empirical equations to relate these two physical parameters (Clarke, 1979). This relationship commonly takes the form

$$k = a_k \phi^{b_k} \tag{4}$$

where a_k and b_k designate constants and ϕ stands for the porosity. For example, Bourbie and Zinzer (1985) found that for the Fontainebleau sandstone a_k and b_k are equal to 381 darcys and 3.05 respectively, if ϕ is greater than 0.09, but 7.93×10^7 darcys and 7.73, respectively, if ϕ is less than 0.09. Note that Equation (4) is also equivalent to the equation

$$\log k = a'_k + b_k \log \phi \tag{5}$$

where $a'_k = \log a_k$, which is a form also commonly found in the literature.

Table 1: Selected values of permeabilities and effective porosities of clastic aquifers (after Gelhar et al., 1992)

Rock Type	Site	Effective Porosity (percent)	Permeability (darcys)			Reference
Heterogeneous sand and gravel	Columbus, Mississippi	35	1.04	to	104.00	Adams and Gelhar (1991)
Glaciofluviatile sands and gravel	Hanford, Washington	10	276.25			Cole (1972)
Glaciofluvial sand	Borden	38	0.01	to	1.04	Egboka et al. (1983)
Glaciofluviatile sand	Borden	33	7.49			Freyberg (1986)
Medium to corse sand	Cape Cod, Massachuetts	39	135.20			Garbedian et al. (1991)
Layered gravel and silty sand	Glatt Valley, Switzerland	25	95.68	to	64.48	Hoehn (1983)
Fluvial sands	Las Cruses, New Mexico	42	9.93			Kies (1981)
Alluvium (clay, silt, sand, and gravel)	La Junta, Colorado	20	24.96	to	436.80	Konikow and Bredehoef (1974)
Sand	Poland	24	3.22	to	15.60	Kreft et al (1974)
Sand and gravel with clay wedges	Berkeley, California	30	93.60			Lau et al. (1957)
Fine sand and glacial till	Amherst, Massachuetts	40	2.50	to	312.00	Leland and Hillel (1981)
Gravel with cobbles	Heretaunga, New Zealand	22	301.60			New Zealand Ministry of Works
Alluvium (gravel)	Heretaunga, New Zealand	22	320.67			and Development (1977)
Sandstone	Clipstone, United Kingdom	32-48	0.25	to	14.56	Oakes and Edworthy (1977)
Sand with some clay and silt	Mobile, Alabama	25	5.30	to	52.00	Papadopulos and Larson (1978)
Sand	Chalk River	38	2.08	to	20.80	Pickens and Grisak (1981)
Glacial outwash	Long Island, New York	53	78.00			Plinder (1973)
Glaciofluvial sand	Borden	33	7.49			Rajaram and Gelhar (1991)
Sand, gravel, and silt	Palo Alto Bay, California	25	52.00	to	130.00	Roberts et al. (1981)
Alluvial sediments	Barstow, California	27	21.84	to	1040.00	Robson (1974)
Clay, sand, and gravel	Lyon France	14	676.00	to	1560.00	Rousselot et al. (1978)
Glaciofluvial sand	Borden	35	4.99	to	7.90	Sykes et al. (1983)
Sand, silt, and clay	Mobile, Alabama	25	2.60	to	52.00	Sykes et al. (1983)
Gravel	Aefligen, Switzerland	17	624.00			Werner et al. (1983)
Sand and Gravel	Burdekin Delta, Australia	32	572.00			Weibenga et al. (1967)
Gravel, sand, and silt	Tucson, Arizonia	38	598.00			Wilson (1971)
Sand	Southern Maryland	35	30.16	to	90.48	Wood (1981)

Table 2. Selected reported values of laboratory-measured sedimentary rock permeabilities.

Rock Type	Porosity (percent)	Permeability (darcys)	Reference
Page sandstone			Panda and Lake (1994)
(Grain Flow)	23	3.50×10^{-0}	
(Wind Ripple)	19.5	1.60×10^{-0}	
(Interdune)	17.4	1.20×10^{-0}	
Berea Sandstone	18.4	1.0×10^{-1}	Howard (1992)
Nugget Sandstone	6.3	0.1×10^{-3}	
Portland Sandstone	20.0	0.81×10^{-3}	
Coconino Sandstone	13.9	6.1×10^{-2}	
Bradford Sandstone	14.8	2.7×10^{-3}	Wyllie and Spangler (1952)
Brea Sandstone	19.0	3.83×10^{-1}	
Woodbine Sandstone	25.6	4.4×10^{-0}	Winsauer et al. (1952)
Reppetto Sandstone	19.1	3.6×10^{-2}	
Oil Creek Sandstone	6.7	4×10^{-3}	
Fontainebleau Sandstone	5.2	4×10^{-3}	Doyen (1988)
	7.5	1.3×10^{-2}	
	9.7	5.3×10^{-2}	
	15.2	5.68×10^{-1}	
	18.0	5.91×10^{-1}	
	19.5	1.11×10^{-0}	
	22.1	7.82×10^{-1}	
Witchita Limestone	21.6	3.4×10^{-1}	Archie (1952)
Witchita Limestone	10.1	7.7×10^{-3}	
Charles Limestone	8.4	1.0×10^{-3}	Murray (1960)
Red River Dolomite	6.3	1.0×10^{-3}	Murray (1960)
Turner Valley Dolomite	27.8	2.9×10^{-1}	
Red River Dolomite	11.9	1.6×10^{-2}	

Empirical expressions relating grain size distribution to permeability have been given by numerous researchers including Hazen (1893), Krumbein and Monk (1942), Masch and Denny (1966), and Bloch (1991). According to Hazen (1893) the permeability is related to the effective grain size of the porous media (d_e) according to

$$k = \varpi d_e^2 \tag{6}$$

where $\bar{c}$ refers to an empirical constant. Harleman et al. (1963) suggested the constant in this equation is equal to 6.54×10^4. The equation of Krumbine and Monk (1942) relates permeability to the geometric mean grain diameter ($\bar{d}$), and the standard deviation of this diameter ($\tilde{\sigma}_{\bar{d}}$) and is given by

$$k = 760\bar{d}^2 \, exp\left(-1.31\tilde{\sigma}_{\bar{d}}\right) \, . \tag{7}$$

Beard and Weyl (1973) found a good correspondence between laboratory measured permeability values of artificially packed natural sand and those generated using the Krumbine and Monk (1942) equation. The equation given by Bloch (1991) can be expressed as

$$log\, k = \bar{c}_1 + \bar{c}_2\bar{d} + \bar{c}_3\tilde{s}^{-1} + \bar{c}_4\tilde{r} \tag{8}$$

where $\bar{c}_1$, $\bar{c}_2$, $\bar{c}_3$, and $\bar{c}_4$ designate constants, $\bar{d}$ refers again to the average grain diameter, $\tilde{s}$ represents the Trask sorting coefficient, and $\tilde{r}$ stands for the rigid grain content. Using this equation, Bloch (1991) reproduced the permeability of Yacheng sandstones to within ±0.75 of an order of magnitude.

Many theoretical models of permeability in porous media start from the Hagen-Poiseuille law, which gives the average laminar flow velocity in a smooth-walled tube according to (Bird et al., 1960)

$$U = -\frac{r_t^2 g}{8\nu}\nabla h \tag{9}$$

where r_t designates the radius of the tube. Because the total discharge is the product of the flow rate and the cross sectional area flow area, and for a set of parallel tubes the normalized cross sectional area of this set of tubes is equal to the porosity, one can write

$$q = -\phi\frac{r_t^2 g}{8\nu}\nabla h \tag{10}$$

As it is difficult in natural systems to directly characterize the tube radius in Equations (9) and (10), r_t in Equation (10) can be eliminated by combining an expression for the total porosity of this system given by

$$\phi = \frac{\pi r_t^2 x}{V} \tag{11}$$

with an equation for the total liquid-solid interfacial surface area per unit volume of porous media ($\bar{s}$):

$$\bar{s} = \frac{2\pi r_t x}{V} \tag{12}$$

to yield

$$\frac{\phi}{\bar{s}} = \frac{r_t}{2} \tag{13}$$

where V refers to the volume of porous media and x designates the length of the parallel tubes. Comparing the combination of Equations (10) and (13) with Equation (1) yields the following expression for permeability

$$k = \frac{\phi^3}{2\bar{s}^2} \tag{14}$$

The form of this equation is the classic Kozeny (1927) equation which is more commonly presented in the form

$$k = c_k \frac{\phi^3}{\bar{s}^2} \tag{15}$$

where c_k is referred to as the Kozeny constant. Theoretically this constant depends on the cross sectional shape of the fluid flow tubes ($c_k = 0.5$ for a circle, 0.512 for a square, and 0.597 for an equilateral triangle). Equation (15) can be extended by taking account of the fact that in a real porous medium the flow paths may be longer than the direct flow path. By defining the tortuosity, τ as the ratio of the direct flow path to the true flow path, a more general version of Equation (15) is obtained and can be written as

$$k = c_k \frac{\phi^3}{\tau\bar{s}^2} \tag{16}$$

Numerous modifications of this expression have been proposed over the years. The most popular of these was proposed by Carman who (1) assumed that $c_k/\tau = 1/5$, and (2) set $\bar{s} = \bar{s}_0(1-\phi)$, where $\bar{s}_0$ denotes the liquid-solid interfacial area per unit volume of solid yielding

$$k = \frac{\phi^3/(1-\phi)^2}{5\bar{s}_0^2} \tag{17}$$

which is commonly referred to as the Kozeny-Carman equation. For spheres having a uniform radius of r_s this specific surface area term becomes

$$\bar{s}_0 = \frac{4\pi r_s^2}{4/3\pi r_s^3} = 3/r_s \tag{18}$$

Combining Equations (17) and (18) leads to

$$k = \frac{\phi^3 r_s^2}{45(1-\phi)^2} \tag{19}$$

This equation suggests that at constant porosity, permeability is proportional to the square of the grain size, which itself is consistent with Equations (6) and (7) given above. Over the years numerous modifications have been proposed to the Kozeny-Carman equation towards improving its predictive ability (Cornell and Katz, 1953; Macmullin and Muccini, 1956; Payatakes et al., 1973; Scheidegger, 1974; Dullien, 1992; Panda and Lake, 1994).

The porosity of a given rock unit within a sedimentary basin tends to decrease with increasing depth due to compaction. In addition, the formation of clay minerals in sedimentary sandstones tends to block pore throats, further hindering fluid flow. An example of permeabilities of single rock formation with depth is depicted in Figure 2. These data, obtained from measurements performed on the Garn Formation, were reported by Ehrenburg (1990). The shallow decrease of permeability with depth from

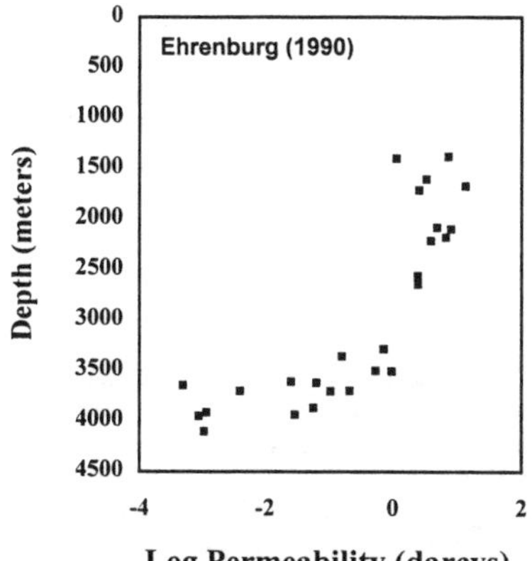

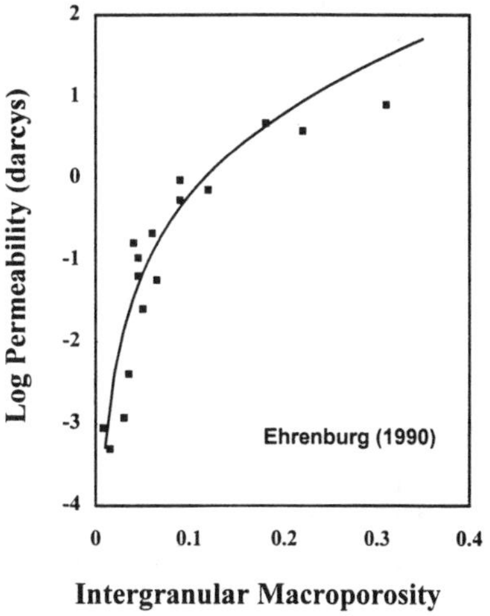

Figure 2 (left). Variation of permeability with depth in the Garn Formation. The symbols represent data reported by Ehrenburg (1990).

Figure 3 (right). Variation of Garn Formation permeability as a function of intergranular macroporosity. The symbols represent data reported by Ehrenburg (1990) and the curve represents predicted values generated using the Kozeny-Carman equation (Carman, 1937) assuming a surface area of 200/cm.

1500 to 3000 m is attributed to porosity reduction, but the dramatic porosity decrease at greater depths is attributed to clay mineral formation. Variation of Garn Formation permeabilities as a function of intergranular macroporosity are depicted in Figure 3. A systematic decrease in permeability with porosity is apparent. Ehrenburg (1990) noted a close correspondence between these data and permeabilities generated using the Kozeny-Carman equation assuming a surface area of 200 cm^{-1}. The results of this calculation are illustrated by the solid curve shown in the figure. Ehrenburg (1990) also noted that similar calculations performed using helium measured porosities yielded less satisfactory results. It should be emphasized, however, that the Kozeny-Carman equation is inconsistent with a variety of other permeability data. For example, the variation of Yangheng Field sandstone permeability with porosity is depicted in Figure 4. The dashed curve in this figure represents an attempt to fit these data using the Kozeny-Carman equation. It is apparent that the shape of this dashed curve is inconsistent with the measured permeabilities. In contrast these data appear to be more consistent with the solid line representing the fit $\log \boldsymbol{k} = -6 + 35\,\phi$, where permeability is in given in darcys.

Neuzil (1994) published a critical review of clay and shale permeabilities. It was found that permeabilities of these rocks varied over the range 10^{-11} to 10^{-3} darcys. Permeability measured both parallel and perpendicular to the bedding was considered. Laboratory measured clay and shale permeabilities increase approximately an order of magnitude for each 13% increase in porosity, consistent with

$$\log \boldsymbol{k} = -10 + 8\phi \tag{20}$$

where permeability is given in darcys. The uncertainty associated with this correlation is ±2 orders of magnitude. Analysis of field measured results indicate that the permeability of non-fractured clays and shales closely correspond to their laboratory measured counterparts, which is consistent with the work of Brace (1980) and Davis (1988). Neuzil (1994) thus suggested that permeabilities generated using a correlation such as given by

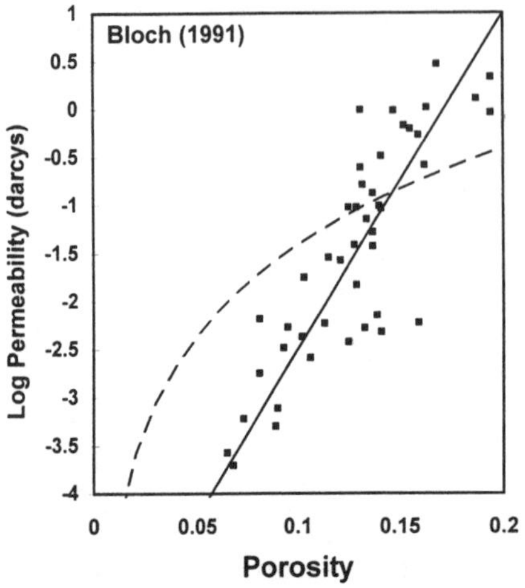

Figure 4. Variation of Yacheng Field permeability as a function of porosity. The symbols represent data reported by Bloch (1991), the dashed curve represents results of calculation performed using the Kozeny-Carman equation (Carman, 1937) assuming a surface area of 800/cm, and the solid line is consistent with log $\boldsymbol{k} = -6 + 35\ \phi$.

Equation (20) could be used to accurately represent clay and shale permeability in both laboratory and field environments. Clays and shales of basin-wide scale typically contain fractures, which can considerably increase permeability. Brace (1980) noted that the small scale permeability of well consolidated rocks, such as clays and shales, measured on samples in the laboratory can be several orders of magnitude lower than the permeability of the same rock unit, measured on a large scale by well tests on the natural system. Bethke (1989) argued that the permeability of clays and shales measured in the laboratory were orders of magnitude lower than their field scale counterparts. Bredehoeft et al. (1983) reported that the Pierre Shale has a permeability of 10^{-8} darcys at the laboratory scale, but a permeability of 10^{-4} darcys on the field scale. It follows that, natural fractured clays and shales may have permeabilities several orders of magnitude larger than those computed using Equation (20). The existence of fractures in shales can also lead to a scale dependence of permeabilities. For example, Neuzil (1993) reported a permeability generated from kilometer scale sections of the Pierre shale, which has an average porosity of 30%, to be on the order of $10^{-8.5}$ darcys, but Bredehoeft et al. (1983) generated a permeability of 10^{-4} darcys from analysis of whole basin analysis, encompassing an area of hundreds of square kilometers. In most cases the scale dependency of the permeabilities of natural systems is due to the existence of heterogeneities (see Tompson and Jackson, this volume). When one goes to a larger scale there is often one or more structures (e.g. fractures) with high permeability, absent in the small scale system, which dominate the permeability of the larger system.

Permeabilities of sedimentary rocks tend to be anisotropic, even on small scales. Several studies have shown that average sandstone permeabilities can be significantly greater parallel to the bedding than perpendicular. Piersol et al. (1940) noted from a survey of laboratory-measured sedimentary rock permeabilities, that the mean ratio of horizontal to vertical permeability is 1.5, with ~12% having ratios greater than 3. Values of laboratory measured permeabilities of several sedimentary rocks exhibiting anisotropy, listed in Table 3, are consistent with this observation. The presence of horizontal stratification in sedimentary rocks implies that the permeability of large scale systems should all be considered anisotropic, and anisotropy is commonly much greater in natural systems than in the homogeneous systems considered in most laboratory analyses. For example, the presence of clay or shale layers in a sandstone typically leads to vertical per-

Table 3. Selected values of laboratory-measured directional permeabilities of sedimentary rocks (after Davis, 1969).

Rock Type	Porosity (percent)	Horizontal Permeability (darcys)	Vertical Permeability (darcys)	Reference
Coarse Grained Arkose	10.9	5.5×10^{-4}	3.8×10^{-4}	Rima et al. (1962)
Fine Grained Arkose	14.4	1.6×10^{-3}	1.6×10^{-4}	
Medium Grained Arkose	25.6	1.1×10^{-3}	5.5×10^{-4}	
Coarse Grained Conglomerate	17.3	4.9×10^{-4}	3.8×10^{-4}	
Siltstone	9.7	1.2×10^{-4}	1.6×10^{-4}	
Cromwell Sandstone	16.6	4.1×10^{-1}	1.7×10^{-1}	Muskat (1937)
Gilcrest Sandstone	27.4	8.0×10^{-1}	5.6×10^{-1}	
Prue Sandstone	11.4	3.4×10^{-3}	4.7×10^{-4}	
Wilcox Sandstone	15.6	7.6×10^{-2}	3.6×10^{-1}	
Silty Clay	40.5	2.8×10^{-4}	1.1×10^{-5}	Johnson and Morris (1962)
Silty Clay	44.1	1.1×10^{-5}	1.1×10^{-5}	
Sand	42.9	18.2	0.6	
Sand	42.4	1.05	0.27	
Marine Sand	41.0	55.0	38.5	MacGary and Lambert (1962)
Aluvium Sand	51.3	25.3	26.4	
Aluvium Sand	46.3	13.2	16.5	
Loess	50.0	0.22	0.33	

meabilities that are orders of magnitude lower than corresponding horizontal permeabilities. This point can be illustrated by calculating the permeability of a section of strata consisting of interstratified layers, comprised of a sandstone having a permeability of 0.1 darcy, and a shale, having a permeability of 1×10^{-5} darcy. In accord with Leonards (1962) and Freeze and Cherry (1979), the effective permeability parallel and perpendicular to the stratification ($\boldsymbol{k}_{\parallel}$ and $\boldsymbol{k}_{\perp}$, respectively) can be computed from

$$\boldsymbol{k}_{\parallel} = \frac{\sum_i \boldsymbol{d}_i \boldsymbol{k}_i}{\sum_i \boldsymbol{d}_i} \qquad \text{and} \qquad \boldsymbol{k}_{\perp} = \frac{\sum_i \boldsymbol{d}_i}{\sum_i \boldsymbol{d}_i / \boldsymbol{k}_i},$$

where $\boldsymbol{d}_i$ and $\boldsymbol{k}_i$ refer to the width and the permeability of each layer. If the interstratified layer consists of equal parallel percentages of sandstone and shale, the effective permeability parallel to this strata is 0.05 darcys, but the effective permeability perpendicular to this strata is 2×10^{-5} darcys. Moreover these equations can be used to illustrate the effects of heterogeneities in natural systems on the overall permeability of a rock formation. For example, if a 1 meter wide shale having a permeability of 1×10^{-5} darcy contained just one single cm wide band of sandstone with a permeability of 0.1 darcy, the overall permeability in the direction of the sandstone band would be 1.01×10^{-3} darcy. In this case the existence of just a single cm wide sandstone leads to a two order of magnitude permeability increase of the shale.

Permeability of igneous and metamorphic rocks

In contrast to the large amount of work aimed at quantifying the permeabilities of unconsolidated and sedimentary rocks, relatively little work has been reported on igneous and metamorphic rocks. As emphasized by Norton and Knapp (1977), the fact that alteration minerals in high temperature systems tend to form along the walls of fractures and in veins demonstrates that fluid flow in these systems is controlled by fracture distribution. In addition the widespread hydrothermal alteration of deep Earth systems attests to the relatively high fracture permeability of igneous and metamorphic rocks at high pressures and temperatures. The permeability of a set of parallel fractures of length L, oriented in the direction of interest can be described using (Norton and Knapp, 1977; Phillips, 1991)

$$k = \frac{n d_f^3}{12L} \tag{21}$$

where n refers to the number of fractures per centimeter, and d_f stands for the fracture aperture. For a set of randomly oriented fractures, the permeability becomes (Dagan, 1989)

$$k = \frac{n d_f^3}{18L} \tag{22}$$

It follows that a rock having just a single fracture of 10^{-4} cm in width per square centimeter will have a permeability of ~10^{-5} darcy. This value can be compared to several of the permeabilities of igneous and metamorphic rocks obtained from laboratory measurement listed in Table 4, which are as low as 10^{-12} darcys or less. This comparison illustrates the likelihood that the permeability of most igneous and metamorphic rocks in natural systems are controlled by fracture distribution. Consequently, permeability measurements performed in the laboratory, even of large fractured samples may have little to do with the effective permeability at depth. For example, Pratt et al. (1974) reported that the permeability of the Sherman granite as measured by a field scale pump test was 10^{-3} darcys, but when a core of this granite was measured in the laboratory its permeability was found to be less than 10^{-6} darcys. The effect of fracture permeability is also responsible for the variation of permeability as a function of effective pressure, as reported by Brace et al. (1968) and Fyfe et al. (1978) (see below).

Another factor which can increase the permeability of igneous and metamorphic rocks is weathering. The combination of near surface weathering and fracturing will increase permeabilities by two to four orders of magnitude (Davis and DeWeist, 1966; Davis, 1969). In general these effects are apparent only within the first 20 m of the surface, but can extend down to as much as 100 m in tropical zones. In addition it has be proposed that mineral dissolution can lead to dramatic increases in permeability in crystalline rocks. For example, Davis (1969) considered the infiltration of a silica free rainwater into a fracture of a quartzite. By assuming ten litres of water pass through each meter of fracture length per year, and its silica concentration increases uniformly from 0

Table 4. Selected reported values of igneous and metamorphic rock permeability measured on core samples in the laboratory at 25°C.

Rock	Porositity (percent)	Permeability (darcys)	Reference
New Ukranian Granite	0.33	$0.15x10^{-8}$	Zaraisky and Balashov (1995)
Graite-Aplinite	1.09	$0.14x10^{-5}$	
Graniodiorite	0.74	$0.15x10^{-5}$	
Diorite	2.21	$1.00x10^{-10}$	
Olivine Basalt	4.50	$0.15x10^{-4}$	
Leucoctatic Granite	1.30	$1.00x10^{-6}$	
Gabbro-Dolerite	0.86	$0.48x10^{-9}$	
Andesite-dacite	6.70	$0.13x10^{-7}$	
White Marble	0.42	$0.46x10^{-6}$	
Dark Marble	0.22	$1.30 x10^{-9}$	
Zeolitized Volcanic Tuff	39	$4.0x10^{-5}$	Keller (1960)
Pumiceous Volcanic Tuff	40	$1.15x10^{-2}$	
Frable Volcanic Tuff	36	$1.40x10^{-3}$	
Welded Volcanic Tuff	14	$3.3x10^{-4}$	

to 10 ppm at a depth of 10 m, he estimated that the fracture width would increase from .01 to .048 cm over the course of 10^5 years. It follows from Equation (21) that this increase in the width of a single fracture would lead to a permeability increase from 8.3 to 920 darcys. Although this rate of fracture width increase may be high due to lack of consideration of quartz dissolution kinetics and to an optimistic choice of input parameters, this example illustrates that reactions due to weathering can substantially increase the permeability of a low porosity rock. A more recent and comprehensive calculation reported by Steefel and Lasaga (1994), estimated permeability changes due to kinetically controlled reaction rates and thermally driven convective flow in a granitic hydrothermal system. Starting from a homogeneous system, they concluded that the permeability would increase and decrease by an order of magnitude, respectively, in the heating and cooling leg of a slowly convecting system over the course of 75,000 years.

In an attempt to generate an effective method to predict the permeabilities of crystalline rocks, Brace (1977) proposed the equation

$$k = \frac{r_p^2}{k_0} f^{1.5} \tag{23}$$

where r_p designates the average pore radius, k_0 refers to a shape factor which can vary from 2 to 3, and f denotes the formation factor (see below). It offers the advantage to determine permeability from formation factors, which are relatively easy to determine from resistivity measurements. This equation was shown to predict the permeability of crystalline, as well as several sandstones and ceramics, to within a factor of 2, over a range of 9 orders of magnitude in ***k*** (Brace, 1977).

Table 5. Selected values of field-measured igneous and metamorphic rock permeabilities.

Rock	Porosity (percent)	Permeability (darcys)	Reference
Savannah River Schists and Gravels	-	3.6×10^{-2}	Webster et al. (1970)
Fractured Granite	-	1×10^{-2} to 1	Gobert (1982)
Fractured Le Cellier Granite	2-8	30 to 90	Dieulin (1980)
Sherman Granite	.002	1×10^{-3}	Pratt et al. (1974)
Fractured Gneiss Schist	-	5×10^{-3}	Marine (1966)
Unfractured Gneiss Schist	-	5×10^{-6}	

Permeability as a function of pressure and temperature in crystalline rocks

Fluid flow passes through the pores, cracks and fissures of a crystalline rock. Because pressure and temperature changes the nature of these void spaces, it also changes their permeability. In their work on the Westerly granite, Brace et al. (1968) observed that the laboratory measured permeability of this rock decreased systematically from 3.5×10^{-7} darcys to 4.2×10^{-9} darcys in response to increasing pressure from 50 to 4000 bars at 25°C. A similar decrease in permeability with increasing pressure is observed at higher temperatures. Zonov et al (1989) noted that the permeability of New Ukrainian granite decreased by approximately one order of magnitude with increase pressure from 200 to 500 bars at temperatures ranging from 400° to 600°C. The permeability of Varzob granite was found to decrease from between one and two orders of magnitude with increasing pressure from 500 to 1500 bars at all temperatures from 25° to 600°C.

The variation of crystalline rock permeabilities as a function of temperature is less clear. Zaraisky and Balashov (1994) observed that the permeability of New Ukrainian granite increased from 10^{-15} to 10^{-8} darcys, gabbro-dolerite increased from 10^{-15} to $10^{-10.5}$ darcys, and medium grained white marble increased from $10^{-11.5}$ to 10^{-8} darcys with increasing temperature from 25 to 700°C at one bar. Shmonov et al (1994) however, found three different isobaric permeability behaviors with increasing temperature to 600°C: (1) the permeability of a granite and a basalt was found to increase monotonically with increasing temperature, (2) the permeabilities of marble and serpentinite samples were found to minimize at a temperature of ~300°C, and (3) the permeabilities of amphibolite, and limestone samples were found to decrease monotonically with increasing temperature. Moreover, Zaraisky and Balashov (1994) observed a slight minima with increasing temperature at ~150°C in the isobaric permeability of Akchatu granite for all pressures between 50 and 800 bars. Vitivtova and Shmonov (1992) observed a similar minima for other granites. This behavior was attributed to the partial disruption of the pore channel system due to the expansion of minerals at low temperature, and the development of microcracks due to thermo-elastic stress at higher temperatures.

AQUEOUS DIFFUSION

Because diffusional transport through solids is relatively slow at the temperatures encounters in crustal and surficial processes (see Hoffmann et al., 1974, or Hart, 1981) aqueous diffusion is the dominant chemical transport mechanism in both non or slowly advecting systems, and for chemical transport to and from high permeability zones such as cracks and fissures. Traditionally, aqueous diffusional transport in porous rocks is computed using Fick's law which can be expressed as

$$J_i = -\overline{D}_i \nabla c_i \tag{24}$$

where J_i, $\overline{D}_i$, and c_i designate the diffusional flux, the effective diffusion coefficient of the rock, and the concentration, respectively of the ith aqueous species. Note that for diffusion coefficients in cm^2/sec, and flux in $mol/cm^2/sec$, the units of concentration in this expres-sion are mol/cm^3. Thus the units of concentration, such as molality, molarity, and mole fraction, commonly adopted in reactive transport calculations must be converted to be consistent with Equation (24).

The effective diffusion coefficient of a dissolved aqueous species is a function of both physical rock properties, and the concentrations of each species in solution. The effect of physical rock properties stem from the fact that (1) aqueous diffusion occurs only through the pore spaces in the rock, and (2) the diffusional transport path may be much longer than the straight line distance in a direction of interest. Taking account of these effects, Equation (24) can be rewritten as

$$J_i = -\frac{\phi D_i}{\tau_e^2} \nabla c_i \tag{25}$$

where D_i stands for the effective diffusion coefficient of the ith aqueous species in a pure solution, τ_e designates the effective tortuosity, and ϕ again represents the porosity of the rock. Note that $\overline{D}_i = \phi D_i / \tau_e^2$.

Tortuosity and formation factors

Strictly, the tortuosity of a system is defined as the ratio of the mean path length of a dissolved species to the straight line distance of the overall path. In practice, however, it includes all factors that alter diffusional transport in a porous material including the pore-throat constrictions, isolated pores, dead-end pore paths, and grain surface effects. Consequently, measurement of tortuosity is normally performed indirectly by either comparing the diffusional flux through a pore solution with the corresponding flux through a pure solution (Garrels et al. 1949; Berner, 1969; Schott, 1972; Norton and Knapp, 1977), or by comparing the electrical resistivity of an electrolyte solution in a porous rock with the corresponding resistivity of the pure solution (Klinkenburg, 1951). Note that special care needs to be taken to eliminate possible effects of sorption reactions when measuring indirectly effective tortuosities (Manheim, 1970). Both diffusional flux and electrical resistivity ratios yield a parameter termed formation factor, designated by the symbol f, which has been defined according to

$$f = \tau_e^2 / \phi \tag{26}$$

such that diffusional flux in a porous media is given by

$$J_i = -\frac{D_i}{f}\nabla c_i \tag{27}$$

where D_i again refers to the diffusion coefficient of the ith species in pure solution. Another term commonly adopted in the literature is the effective directional porosity, ϕ_e, which is equal to f^{-1} such that

$$J_i = -\phi_e D_i \nabla c_i \tag{28}$$

Formation factors are commonly estimated using the empirical relationship

$$f = \phi^{-n_f} \tag{29}$$

where n_f designates a formation constant. If n_f = 2 then Equation (29) is equivalent to Archie's Law (Archie, 1942). Archie (1952) reported that n_f = 2 for packed sand and n_f = 1.9 for consolidated sandstones. Berner (1980) generated a value of n_f = 1.8 from analysis of the formation factors of a variety of deep sea sediments reported by Manheim and Waterman (1974). Several studies have focused on the formation factors of igneous and metamorphic rocks. Norton and Knapp (1977) measured effective directional porosities of a variety of isotropic igneous and metamorphic rocks. Formation factors computed from these results are depicted in Figure 5 as a function of porosity on a log-log plot. The solid line through these data is consistent with n_f=2. Brace et al. (1965) observed that the formation factors of a variety of crystalline rocks were consistent with n_f = 2 over the porosity range from 0.1 to 10%. Diffusion of non-sorbing species through a variety of crystalline rocks was measured by Skagius and Neretnieks (1986). Formation factors generated by these authors are depicted in Figure 6 as a function of porosity on a log-log plot. The solid line through these data are consistent with n_f = 2. Similarly, formation factors of altered marine basalts, illustrated in Figure 7 are also consistent with n_f = 2. Although there is considerable scatter in Figures 5 to 7, it seems reasonable to adopt n_f = 2 as a first approximation for the calculation of the effect of solute path and porosity on diffusional transport in both crystalline and sedimentary rocks. It is nevertheless clear from these figures that formation factors generated from Archie's law have associated uncertainties of at least ± one order of magnitude.

As is the case for permeability, formation factors can be anisotropic due to the effect of rock fabric or grain alignment on diffusional pathways. This is most certainly the case for claystones and shales, which commonly have grains oriented parallel to deposition. Norton and Knapp (1977) found that the tortuosities of quartzites, shales and limestones can vary by as much as a factor of seven depending on sample orientation.

Diffusional transport in electrolyte solutions

The effective diffusion coefficient of a species or an chemical component in pure solution can be computed by considering the forces driving diffusional flux in a multi-component solution. The forces driving the diffusion of an aqueous species within a solvent reference frame (Anderson and Graf, 1976; Anderson, 1981; Brady 1983) stem from both chemical potential and electronic potential gradients. In accord with Miller (1966) (see also Lichtner, this volume) the diffusional flux of the ith aqueous species can be computed from

$$J_i = \sum_{j=1}^{n_s} l_{ij}\left(-\nabla\mu_j - z_j\nabla\Phi\right) \tag{30}$$

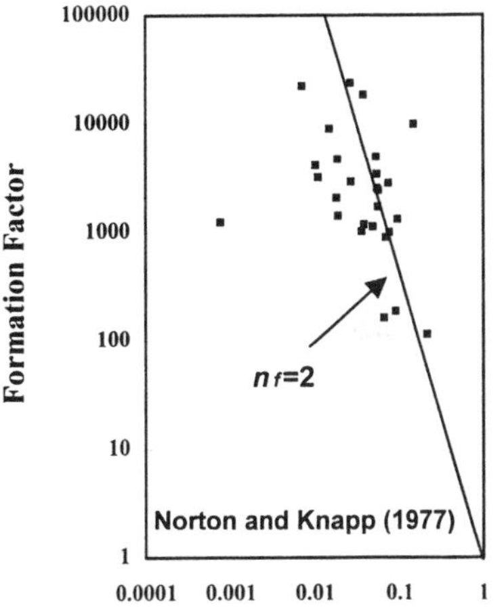

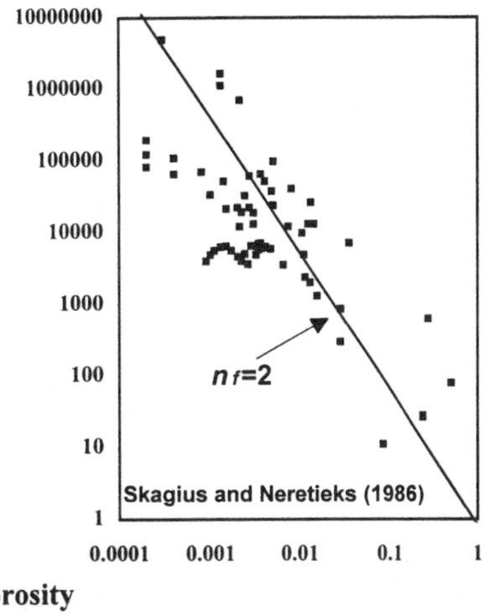

Figure 5 (left). Formation factors for igneous and metamorphic rocks as a function of porosity. The symbols represent data generated from measurements reported by Norton and Knapp (1977), and the line corresponds to $f = \phi^{-2}$.

Figure 6 (right). Formation factors for igneous and metamorphic rocks as a function of porosity. The symbols represent data generated from measurements reported by Skagius and Neretieks (1986) and the line corresponds to $f = \phi^{-2}$.

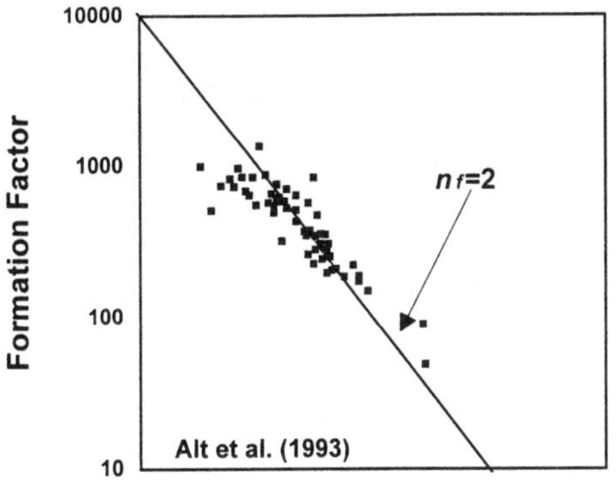

Figure 7. Formation factors for marine basalts from drill hole 896A of the deep sea drilling project (Alt et al., 1993) as a function of porosity. The symbols represent measured formation factors and the line corresponds to $f = \phi^{-2}$.

where l_{ij} stands for a phenomenological transport coefficient matrix, n_s designates the number of aqueous species in solution, $\nabla\mu_j$ represents the chemical potential gradient of the jth species, and $\nabla\Phi$ refers to the electronic potential gradient of the solution. The electronic potential gradient, which is not readily measurable in electrolyte solutions, can be eliminated from Equation (30) by taking account of the zero current constraint, which can be expressed as

$$\sum_{i=1}^{n_s} z_i J_i = 0 \tag{31}$$

Combining Equations (30) and (31) leads to (Felmy and Weare, 1991a)

$$\nabla \Phi = -\frac{\sum_{k=1}^{n_s}\sum_{l=1}^{n_s} z_k l_{kl} \nabla \mu_l}{\sum_{k=1}^{n_s}\sum_{l=1}^{n_s} z_k z_l l_{kl}} \tag{32}$$

which can be further combined with Equation (30) to yield

$$J_i = \sum_{j=1}^{n_s} l_{ij} \left(-\nabla \mu_j + z_j \frac{\sum_{k=1}^{n_s}\sum_{l=1}^{n_s} z_k l_{kl} \nabla \mu_l}{\sum_{k=1}^{n_s}\sum_{l=1}^{n_s} z_k z_l l_{kl}} \right) \tag{33}$$

In general, Equation (33) can be used together with mass action constraints to compute diffusional fluxes of all species present in electrolyte solutions. In practice, however, this is not possible for any but rather simplified systems at 25°C owing to the lack of the experimental data required for the regression of the l_{ij} parameters as a function of solution concentration. Application of this complete formalism was presented by Felmy and Weare (1991a), who regressed the experimental data of Rard and Miller (1979a,b; 1980) and Miller et al. (1980, 1984, 1986) using activity coefficient derivatives generated from the Pitzer (1979) equations to generate a model for diffusion in the system Na-K-Ca-Cl-SO_4-H_2O at 25°C.

In an attempt to generate an effective set of equations for the application of Equation (33) to other systems of geologic interest, Lasaga (1979) and Balashov (1994) argued that in most geological systems, the off-diagonal terms of the phenomenological transport coefficient matrix are negligible compared to their on-diagonal counterparts, and thus $l_{ij} \sim 0$ when $l \neq j$. In addition, for relatively dilute solutions, the on-diagonal terms can be approximated to be equal to their limiting counterparts, and thus it may be assumed that $l_{ii} \approx l_{ii}^0$ where l_{ii}^0 stands for l_{ii} in an infinitely dilute solution. The tracer phenomenological transport coefficients can be generated from corresponding tracer diffusion coefficients using (Miller, 1966; Anderson, 1981)

$$l_{ii}^0 = \frac{D_i^0 c_i}{RT} \tag{34}$$

where D_i^0 refers to the tracer diffusion coefficient of the subscripted species, R denotes the gas constant. Taking account of these assumptions and combining Equations (33) and (34) leads to

$$J_i = -\frac{D_i^0 c_i}{RT} \nabla \mu_i + \left(\frac{z_i D_i^0 c_i}{RT \sum_{k=1}^{n_s} z_k^2 D_k^0 c_k} \sum_{k=1}^{n_s} z_k D_k^0 c_k \nabla \mu_k \right) \tag{35}$$

Chemical potential gradients can be converted to concentration gradients by first taking account of $\nabla \mu_i = \frac{\partial \mu_i}{\partial c_i} \nabla c_i$, which when substituted into Equation (35) leads to

$$J_i = -\frac{D_i^0 c_i}{RT}\frac{\partial \mu_i}{\partial c_i}\nabla c_i + \left(\frac{z_i D_i^0 c_i}{RT \sum_{k=1}^{n_s} z_k^2 D_k^0 c_k} \sum_{k=1}^{n_s} z_k D_k^0 c_k \frac{\partial \mu_k}{\partial c_k} \nabla c_k \right) \tag{36}$$

The derivative $\frac{\partial \mu_i}{\partial c_i}$ can be deduced from the definition of chemical potential given by

$$\mu_i = \mu_i^0 + RT\, ln\, c_i + RT\, ln\, \gamma_i \tag{37}$$

where γ_i refers to the activity coefficient of the subscripted species and μ_i^0 represents the chemical potential in the standard state yielding

$$\frac{\partial \mu_i}{\partial c_i} = \frac{RT}{c_i} + RT\frac{\partial\, ln\, \gamma_i}{\partial c_i} \tag{38}$$

Combining Equation (38) with Equation (36) results in

$$J_i = -D_i^0 \left(1 + \frac{\partial\, ln\, \gamma_i}{\partial\, ln\, c_i} \right) \nabla c_i + \left(\frac{z_i D_i^0 c_i}{\sum_{k=1}^{n_s} z_k^2 D_k^0 c_k} \sum_{k=1}^{n_s} z_k D_k^0 \left(1 + \frac{\partial\, ln\, \gamma_k}{\partial\, ln\, c_k} \right) \nabla c_k \right) \tag{39}$$

which can be combined with the traditional form of Ficks law given by

$$J_i = -\sum_{j=1}^{n_s} D_{ij} \nabla c_j \tag{40}$$

where D_{ij} designates the diffusion coefficients of the ith species in response to the jth concentration gradient in a pure electrolyte solutions, to yield

$$D_{ij} = \delta_{ij} D_i^0 \left(1 + \frac{\partial\, ln\, \gamma_i}{\partial\, ln\, c_i} \right) - \left(\frac{z_i D_i^0 c_i}{\sum_{k=1}^{n_s} z_k^2 D_k^0 c_k} z_j D_j^0 \left(1 + \frac{\partial\, ln\, \gamma_j}{\partial\, ln\, c_j} \right) \right) \tag{41}$$

where δ_{ij} stands for a function equal to 1 if $I = j$, but equal to 0 if $I \neq j$. Lasaga (1979) and Johnson (1981) argued that the activity coefficient term in Equation (41) could be neglected because the ionic strength of most pore water solutions vary only slightly with respect to distance. Felmy and Weare (1981b) noted, however, that this approximation was invalid because the term in the diffusion coefficient equation is the logarithmic derivative of the activity coefficient with respect to concentration rather than distance. Felmy and Weare(1991b) further demonstrated that the adoption of the approximation, $\left(\frac{\partial\, ln \gamma_j}{\partial\, ln c_j} \right) = 0$ as proposed by Lasaga (1979) and Johnson (1981) could lead to significant errors in computing aqueous diffusion coefficients. Both Lasaga (1979) and Felmy and Weare (1991a,b) coupled Equation (41) with mass action constraints to generate flux equations in terms of the concentrations of aqueous components. This coupling is generally performed directly by reactive transport algorithms (see for example Lichtner 1985, 1988; this volume).

Equation (41) implies that the diffusional transport of each charged species depends on the tracer diffusion coefficient, concentration, and concentration gradient of each charged species in solution. This result is a direct consequence of the need to maintain electrical neutrality throughout the system and can lead to situations where the diffusional transport of one charged species that is dominated by the concentration gradient of another charged species. This effect should be most pronounced in solutions where there are large differences between the concentration gradients of charged species. As the bulk of natural waters contain many species, many of which (for example Na^+ and Cl^- in sea water) are orders of magnitude greater in concentration than others, wide variations in concentration gradients can be present in some natural systems. Failure to account for the effect of electrical potential in these systems can lead to large errors in mass transport calculations.

Note that for the case of neutral aqueous species, where it is commonly assumed that $\gamma_i = 1$ and that these species do not effect the activity coefficients of ionic species, Equation (41) reduces to

$$D_{ij} = \delta_{ij} D_i^0 \tag{42}$$

suggesting that for neutral aqueous species, diffusion coefficients can be approximated by their limiting counterparts, and that they are independent of the concentration gradient of all dissolved species.

Estimation of aqueous tracer diffusion coefficients

It follows from the equations summarized above, that the diffusion coefficients of an aqueous species in an electrolyte solution can be estimated from corresponding values of aqueous tracer diffusion coefficients. Aqueous tracer diffusion coefficients for ionic species are readily determined from limiting equivalent conductances using the Nernst-Einstein equation given by

$$D_i^0 = \frac{\lambda_i^0 RT}{|z_i| F^2} \tag{43}$$

where F designates the Faraday constant, and λ_i^0 stands for the limiting equivalent conductance of the subscripted ion. Limiting equivalent conductances and corresponding aqueous tracer diffusion coefficients have been tabulated for a number of ions at 25°C by Harned and Owen (1958), Robinson and Stokes (1959), and Li and Gregory (1974). Robinson and Stokes (1968) reported limiting equivalent conductances of ~20 aqueous ions up to 100°C. Quist and Marshall (1965) generated limiting equivalent conductances of ~10 aqueous ions at temperatures up to 400°C. Taking account of these data, Nigrini (1970) generated correlations between limiting equivalent conductances and (1) third law entropies, (2) entropies of hydration, and (3) charge to radius ratios to enable prediction of λ_i^0 values for a wide variety of ions at 25°C, and a corresponding states correlation for their extrapolation to 300°C. Oelkers (1987) and Oelkers and Helgeson (1988) took account of the same limiting equivalent conductance data and the electrical conductances measured as a function of temperature and pressure by Franck (1956), Quist et al. (1963, 1965), Quist and Marshall (1965, 1966, 1968a,b,c,d), Ritzert and Franck (1968), Dunn and Marshall (1969), and Frantz and Marshall (1982, 1984), to develop equations and correlations to calculate limiting equivalent conductances of over one hundred aqueous ionic species at temperatures from 0° to 1000°C and pressures from 1 to 5000 bars.

Calculated values of the aqueous tracer diffusion coefficient of Na^+ reported by Oelkers and Helgeson (1988) are depicted as a function of temperature at pressures from 1 to 5000 bars in Figure 8. It can be seen in this figure that these values increase by ~2 orders of magnitude with increasing temperature from 0° to 1000°C. In contrast, they tend to decrease only slightly with increasing pressure. Aqueous tracer diffusion coefficients for Na^+ decrease by less than 40% with increasing pressure from 1000 to 5000 bars over this whole temperature range. Computed aqueous tracer diffusion coefficients for several different ions are compared as a function of temperature in Figure 9. It can be seen in this figure that the diffusion coefficients for different ions tend to exhibit similar values, these values are generally within a factor of two of one another for any given temperature. The only exceptions are the aqueous tracer diffusion coefficients for hydrogen and hydroxide ions at relatively at low temperatures. At these conditions $D^0_{H^+}$ and $D^0_{OH^-}$ are up to an order of magnitude greater than that of other ions.

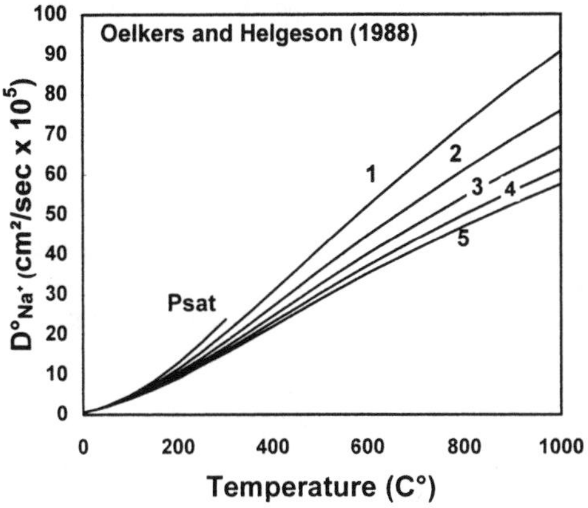

Figure 8. Aqueous tracer diffusion coefficients for Na^+ as a function of temperature at the indicated pressures in kilobars as reported by Oelkers and Helgeson (1988). Note Psat corresponds to pressures along the liquid vapor coexistence curve for H_2O.

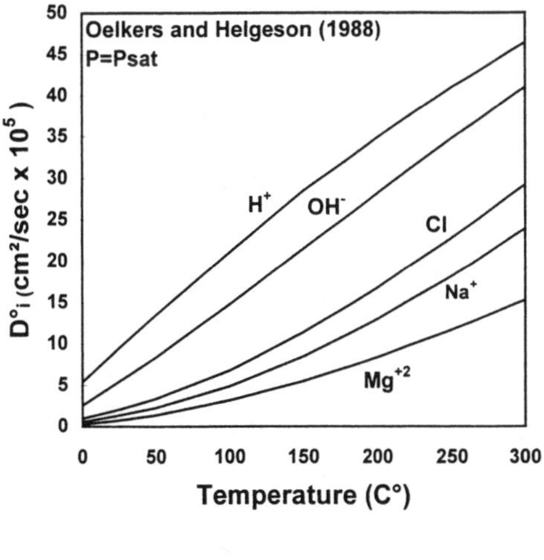

Figure 9. Aqueous tracer diffusion coefficients for Na^+, Cl^-, H^+, OH^- and Mg^{+2} as a function of temperature at Psat as reported by Oelkers and Helgeson (1988)—see the caption of Figure 8.

This phenomena is due to the effect of an H-O-H to H-O-H jump mechanism that increases the mobility of these species (Oelkers and Helgeson, 1989). Although the aqueous tracer diffusion coefficients of most charged species have similar values at any given temperature, the differences among these values may, in fact, can lead to significant effects on the diffusional transport of many charged species due to electroneutrality constraints in multicomponent electrolyte solutions (see below).

For neutral aqueous species including ion pairs, tracer diffusion coefficients cannot be generated from relatively easy to perform conductance measurements. Nevertheless, there exists an extensive array of data on the tracer diffusion coefficients of neutral dissolved gases and organic molecules in aqueous solution at temperatures up to 100°C (cf. International Critical Tables, 1929; Vivian and King, 1964; Unver and Himmelblau, 1964; Himmelblau, 1964; Witherspoon and Saraf, 1965; Wise and Houghton, 1966, 1968; Bonoli and Witherspoon, 1968; Sahores and Witherspoon, 1966; Maharajah and Walkley, 1973; Hayduk and Laudie, 1974; Tyn and Claus, 1975a,b; Hailour and Sandall, 1984; Tang and Sandall, 1985; Eastal and Woolf, 1985; Tominaga et al. 1986; Jahne et al. 1987; Versteeg and van Swaaij, 1988; Tominaga and Matsumoto, 1990). Several equations have been proposed to estimate diffusion coefficients for neutral species. The first equation developed for this purpose was the Stokes-Einstein equation, which assumes the dissolved species is a spherical particle dragged through a media of constant viscosity. The resulting equation is given by

$$D_i^0 = \frac{kT}{6\pi\eta r_s} \tag{44}$$

where k denotes the Boltzman constant, η refers to the dynamic viscosity of pure water, and r_s stands for the Stokes law radius of the dissolving particle. Also working from a hydrodynamic approach, Othmer and Thakar (1953) developed a correlation given by

$$D_i^0 = \frac{14.0\text{x}10^{-5}}{\eta V_i^{0.6}} \tag{45}$$

and Wilke and Chang (1955) developed a correlation given by

$$D_i^0 = \frac{7.4\text{x}10^{-8}(\chi M_{H2O})^{0.5}T}{\eta V_i^{0.6}} \tag{46}$$

where χ represents an association number equal to 2.6 for water, M_{H2O} signifies the molecular weight of water, and V_i corresponds to the molar volume of the subscripted solute at its boiling point. Each of these two correlations was found to reproduce the experimental data with average absolute errors of 11%. Hayduk and Laudie (1974) revised these two equations for dissolved substance in water leading to the correlation given by

$$D_i^0 = \frac{13.26\text{x}10^{-5}}{\eta^{1.4} V_i^{0.589}} \tag{47}$$

This equation was reported to reproduce aqueous diffusivities at 25°C to within a 5.8% average absolute error, and was also used to predict their variation on temperature to ~70°C. Note that for V_i in cm^3/mol, η in centipoise, and T in Kelvins, Equations (45) to (47) yield diffusion coefficients in cm^2/sec. Oelkers (1991) developed a correlation among the tracer diffusion coefficients of groups of aqueous organic polymers. These correlations were used together with a modified Arrhenius equation given by

$$D_i^0 = A_{D,i}\exp\left(-\frac{E_{A,i,T_r} + \tilde{a}(T-\theta)^{-\tilde{n}}}{RT}\right)$$

where $A_{D,i}$, E_{A,i,T_r}, and $\tilde{a}$ are temperature independent parameters, θ and $\tilde{n}$ denote solvent parameters equal to 228 K and 1.5, respectively, to generate aqueous tracer diffusion coefficients for over fifty aqueous organic compounds at temperatures from 0° to 350°C.

Relatively little data is available on aqueous ion complexes, or for numerous other neutral species essential for modeling geochemical transport in natural systems, including SiO_2^0, $Al(OH)_3^0$, $NaCl^0$, $MgSO_4^0$,... etc. Applin (1987) generated tracer diffusion coefficients of 2.2×10^{-5} cm²/sec and 1×10^{-5} cm²/sec for $Si(OH)_4^0$ and $Si_2O(OH)_6^0$, respectively at 25°C and 1 bar. Ildefonse and Gabis (1976) reported a diffusion coefficient of 2.4×10^{-4} cm²/sec for aqueous silica at 550°C and 1000 bars. Applin and Lasaga (1984) generated the tracer diffusion coefficients of 1.11×10^{-5} cm²/sec and 1.08×10^{-5} cm²/sec for $Mg(SO)_4^0$ and $Na(SO)_4^-$, respectively, at 25°C and 1 bar. To overcome the dearth of available data, Nigrini (1970) developed a model to calculate tracer diffusion coefficients for ion pairs through the extension of ideas presented by Wishaw and Stokes (1954). In this model, Nigrini (1970) assumed an ion pair to be a spherical ellipsoid consisting of two single ions designated 1 and 2, for which $D_1^0 \geq D_2^0$ and generated the equation

$$D_{12}^0 = \frac{D_1^0}{\Psi\left(1+\left(\frac{D_1^0}{D_2^0}\right)^3\right)^{1/3}} \tag{48}$$

where D_{12}^0 refers to the aqueous tracer diffusion coefficient of the ion pair, and Ψ designated a spherical factor given by

$$\Psi = \frac{\left(1-\beta^2\right)^{1/2}}{\beta^{2/3}\, ln\left(\frac{1+\left(1-\beta^2\right)^{1/2}}{\beta}\right)} \tag{49}$$

where β is given by

$$\beta = \frac{1}{1+\frac{D_1^0}{D_2^0}} \tag{50}$$

Balashov (1994) used this equation, together with corresponding aqueous tracer diffusion coefficients for ionic species reported by Oelkers and Helgeson (1988) to compute aqueous tracer diffusion coefficients for HCl^0, $NaCl^0$, KCl^0, $MgCl^+$, $MgCl_2^0$, $CaCl^+$, $CaCl_2^0$ and $NaOH^0$ at temperatures from 25° to 1000°C and pressures up to 5 kbar. It was noted that computed tracer diffusion coefficients are relatively close in value to corresponding values of H_2O liquid lattice self diffusion coefficients generated by Labotka (1991).

Uphill and downhill diffusion in electrolyte solutions

Because (1) in accord with Equation (41) the diffusion coefficient for each charged species is coupled to all other charged species, and (2) the tracer diffusion coefficient is different for each charged species, the diffusional flux of a charged species is not always down its diffusion gradient. This point can be illustrated for the case of an aqueous solution containing 1 mol/kg NaCl and 0.01 mol/kg $MgCl_2$. The diffusion coefficient matrix ($\boldsymbol{D_{ij}}$) for this solution at 25°C and 1 bar, computed using Equation (41) together with aqueous tracer diffusion coefficients taken from Oelkers and Helgeson (1988) and by assuming $(\partial \ln\gamma_i/\partial \ln c_i) = 0$ are listed in Table 6. For the purpose of illustration, ion pairing is considered to be negligible in this example. At these conditions the aqueous tracer diffusion coefficients of Na^+ and Cl^-, are 1.4×10^{-5} cm²/sec and 2.1×10^{-5} cm²/sec, respectively. Because the tracer diffusion coefficient of Cl^- is greater that that of Na^+, the more mobile anion tends to pull the slower moving Na^+ and any other cation in solution. This phenomena is expressed in the diffusion coefficient matrix by the positive $\boldsymbol{D}_{Mg^+,Cl^-}$ term, and the fact that the absolute values of $\boldsymbol{D}_{Mg^+,Cl^-}$, and $\boldsymbol{D}_{Cl^-,Mg^+}$ are significantly greater than $\boldsymbol{D}_{Mg^+,Na^+}$, and $\boldsymbol{D}_{Na^+,Mg^+}$. The fact that these cross terms are non-zero, implies that Mg^{2+} will diffuse in this solution, even in the absence of Mg^{2+} concentration gradients. This effect is illustrated in Figure 10 which depicts calculated values for the diffusional flux of Na^+, Cl^-, and Mg^{2+} in this solution assuming it has no Mg^{2+} concentration gradient, but a NaCl concentration gradient ranging from 10^{-4} to 10^{-2} mol/kg/cm. A positive flux in this figure corresponds to a flux down the NaCl concentration gradient. It can be seen in this figure that the flux of Mg^{2+} in this system is approximately two orders of magnitude lower than that of NaCl which is approximately the difference in the total concentrations of NaCl and $MgCl_2$ in solution. This result nevertheless demonstrates that diffusional transport of charged species can occur in the absence of its own concentration gradient.

Table 6. Diffusion coefficient matrix ($\boldsymbol{D_{ijj}}$) for an aqueous solution containing 1 mol/kg NaCl and 0.01 mol/kg $MgCl_2$ at 25°C and 1 bar using Equation (41) together with the aqeous tracer diffusion coefficients taken from Oelkers and Helgeson (1988). This illustrative example was computed assuming $(\partial \ln\gamma_i/\partial \ln c_i) = 0$ and aqeous complex formation is negligible. $\boldsymbol{D_{ij}}$ values in this table have units of cm²/sec.

	-	***j***	-
i	Na^+	Cl^-	Mg^{2+}
Na^+	8.46×10^{-6}	8.31×10^{-6}	-6.3×10^{-6}
Cl^-	8.33×10^{-6}	8.5×10^{-6}	9.52×10^{-6}
Mg^{2+}	-6.3×10^{-8}	9.5×10^{-8}	7.93×10^{-6}

These non-zero off diagonal diffusion matrix terms can also lead to diffusion of a species up its concentration gradient ('uphill' diffusion). This can be illustrated using this same example of an aqueous solution containing 1 mol/kg NaCl and 0.01 mol/kg $MgCl_2$ at 25°C. The flux of Mg^{2+} in this system, having a $+10^{-3}$ mol/kg/cm NaCl concentration gradient are illustrated as a function of the $MgCl_2$ concentration gradient in Figure 11. If the $MgCl_2$ concentration gradient is in the same direction as the NaCl concentration gradient it is shown to be as a positive gradient, but if the $MgCl_2$ concentration gradient is in the opposite direction as the NaCl concentration gradient it is shown to be a negative gradient in Figure 11. It can be seen in this figure that the diffusional flux of Mg^{2+} is in the direction of the NaCl concentration gradient for all $MgCl_2$ concentration gradients greater than -4×10^{-6} mol/kg/cm. Thus, Mg^{2+} diffuses up its concentration gradient if the $MgCl_2$ concentration gradient is greater than -4×10^{-6} mol/kg/cm but less than 0, thus exhibiting 'uphill diffusion'. In contrast, Mg^{2+} diffuses down its concentration gradient, thus exhibiting 'downhill' diffusion, if the $MgCl_2$ concentration gradient is less than -4×10^{-6} mol/kg/cm (that is when this gradient strongly opposes the NaCl gradient), or is greater than 0 (when $MgCl_2$ and NaCl have concentration gradients in the same direction). It follows that the effect of off-diagonal terms in the diffusion coefficient matrix can lead to significant effects in the transport of electrolytes exhibiting relatively small concentration gradients. The significance of this effect depends on the relative concentration gradients of the system; they are most significant in systems exhibiting large concentration gradients such as a waste repository of high ionic strength material surrounded by dilute groundwater, ore formation adjoining fissures, and contact metamorphic zones. In contrast, although seawater is rich in Na^+ and Cl^-, their concentration gradient is generally small in most near surface systems and thus the significance of this effect tends to be small in oceanic and diagenetic systems.

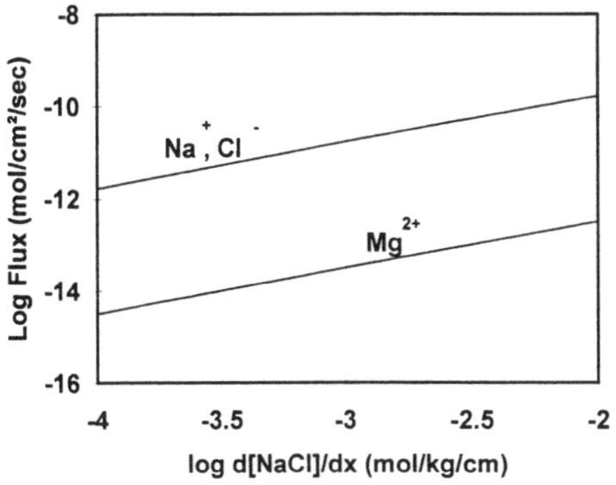

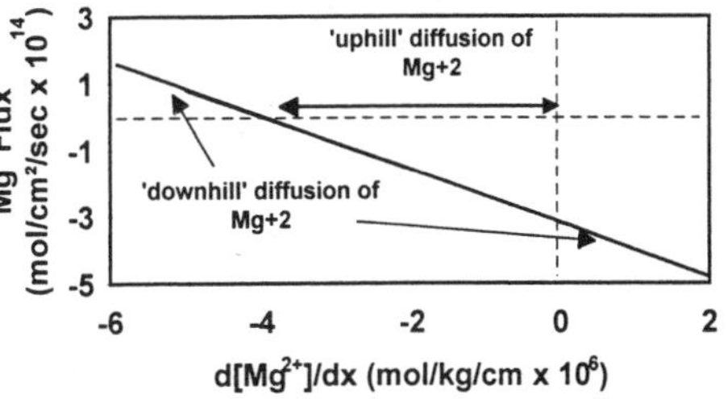

Figure 10 (left). Logarithm of Na^+, Cl^-, and Mg^{2+} flux as a function of NaCl concentration gradient in a system containing 1 mol/kg NaCl, 0.01 mol/kg $MgCl_2$ at 25°C. For this calculation it is assumed the concentration gradient of $MgCl_2 = 0$, $(\partial ln\gamma_i/\partial lnc_i) = 0$, and aqueous complex formation is negligible (see text).

Figure 11 (right). Flux of Mg^{2+} as a function of $MgCl_2$ concentration gradient in a system containing 1 mol/kg NaCl and 0.01 mol/kg $MgCl_2$ at 25°C. For this calculation it is assumed that the concentration gradient of NaCl = 1×10^{-3} mol/km/cm, $(\partial ln\gamma_i/\partial lnc_i) = 0$, and aqueous complex formation is negligible. If the $MgCl_2$ concentration gradient is in the same direction as the NaCl concentration gradient it is shown to be positive, but if the $MgCl_2$ concentration gradient is in the opposite direction as the NaCl concentration gradient it is shown to be negative. Thus if both the $MgCl_2$ concentration gradient and the Mg^{2+} flux are of the same sign this species is exhibiting 'uphill' diffusion but if they have opposite signs this species is exhibiting 'downhill' diffusion.

MECHANICAL AND HYDRODYNAMIC DISPERSION

Mechanical dispersion is a process which tends to diminish concentration gradients within fluids flowing through porous media. This process arises from the fact that (1) the velocity of solutes flowing in a porous medium varies according to flow path,

(2) friction between mineral grains and fluids leads to relatively slower flow rates along grain boundaries, and (3) at each pore space juncture there is both a mixing and a splitting of the flow streams. In general, the apparent solute flux due to the effects of mechanical dispersion is computed using a form of Fick's law given by

$$J_{D',i} = -D'\nabla c_i \tag{51}$$

where $J_{D',i}$ designates the mechanical dispersive flux, and D' denotes the coefficient of mechanical dispersion. Due to the nature of the process, the magnitude of mechanical dispersion depends on direction relative to the fluid flow, even in isotropic media. In its most general form, dispersion coefficients are treated as a tensor. In common practice, however, they are treated by considering dispersion in two directions, one in the direction of flow and the second normal to the flow. Dispersion in the flow direction is termed longitudinal dispersion and is characterized by a longitudinal mechanical dispersion coefficient represented by D'_L, and dispersion perpendicular to the flow direction is termed transverse (or lateral) dispersion and is characterized by a transverse mechanical dispersion coefficient, designated D'_T. Because molecular diffusion is governed by the same equation, and it leads to similar temporal and spatial concentration gradients as mechanical dispersion, molecular diffusion and mechanical dispersion coefficients are commonly summed together to create hydrodynamic dispersion coefficients such that

$$\tilde{D}_L \equiv D'_L + D \tag{52}$$

and

$$\tilde{D}_T \equiv D'_T + D \tag{53}$$

where $\tilde{D}_L$ and $\tilde{D}_T$ refer to the coefficients of longitudinal and transverse hydrodynamic dispersion, and D denotes the diffusion coefficient of the solute species in pure solution. The magnitude of hydrodynamic dispersion relative to molecular diffusion is commonly characterized using the diffusional Peclet number, which is a dimensionless ratio of the flow velocity versus the molecular diffusion coefficient given by

$$P_e = \frac{2\bar{r}U}{D} \tag{54}$$

where P_e refers to the diffusional Peclet number, $\bar{r}$ stands for the average particle radius of the porous material, and U refers to the pore flow rate ($U = q/\phi$, where q denotes the specific discharge).

Laboratory scale dispersion

The bulk of work performed measuring dispersivity in the laboratory has been performed on homogeneous materials such as columns of glass beads or sand. Bear (1969, 1972) notes different transport zones for these types of homogeneous materials, which are related to the diffusional Peclet number:

Zone I: This zone is defined when P_e is less than 0.4, and is the region where molecular diffusion dominates. In this zone, the amount of time required to traverse a pore space by diffusional transport is less than that required by flow.

Zone II: This zone is defined when $0.4 \geq P_e \geq 5$, and is the region where molecular diffusion is the same order of magnitude as that of mechanical dispersion, and both need to be considered.

Zones III, IV, V: When P_e is greater than 5, mechanical dispersion dominates molecular diffusion. Bear (1972) suggested that coefficients of longitudinal hydrodynamic dispersion in this zone can be approximated using an empirical expression of the form

$$\frac{\tilde{D}_L}{D} = \kappa (P_e)^{n_\kappa} \tag{55}$$

where κ and n_κ represent constants equal to 0.5 and 1.2, respectively when the diffusional Peclet number is less than ~100, but equal to 1.8 and 1.0, respectively when the diffusional Peclet number is greater than ~100.

These three zones are shown as a function of average particle diameter and fluid flow rate in Figure 12. It can be seen in this figure that for an average grain diameter of 0.1 cm, molecular diffusion dominates mechanical dispersion for all pore flow velocities of ~10 meters per year or less, but mechanical dispersion dominates at pore flow velocities of greater than ~120 meters per year. Note that pore flow velocities are greater than the discharge rate. For example, a rock with a 10% porosity will have a discharge rate equal to one tenth of its pore flow velocity.

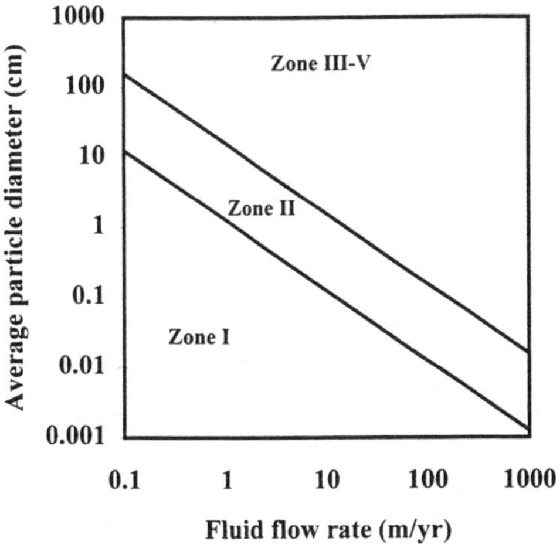

Figure 12. Diagram illustrating the dispersional transport zones described by Bear (1969) as a function of mean particle diameter and fluid flow rate in meters per million years: Zone I is where diffusion dominates over longitudinal mechanical dispersion, Zone II is where molecular diffusion and longitudinal mechanical dispersion are of the same order of magnitude, and Zones III to V are where longitudinal mechanical dispersion dominates over molecular diffusion. Calculations depicted in this figure was made assuming a diffusion coefficient of 1×10^{-5} cm^2/sec, which is characteristic for many aqueous species at 25°C. These zones will tend to migrate upward with increasing temperature owing to increasing molecular diffusion coefficients.

A large number of formulations have been developed to predict theoretically values of the longitudinal and transverse dispersion coefficients (Bear, 1969; Scheidegger, 1974). The resulting equation depends greatly on the model used to represent the porous media. Dispersion in a single tube was considered by Taylor (1953, 1954), Aris (1956) resulting in

$$\frac{\tilde{D}_L}{D} = 1 + P_e^2 / 192 \tag{56}$$

Turner (1959) and Philip (1963) considered the dispersion in a parallel plate duct obtaining

$$\frac{\tilde{D}_L}{D} = 1 + 8\left(\frac{Uh}{D}\right)^2 / 945 \tag{57}$$

where **h** designates the halfwidth of the duct. Perhaps the most comprehensive model was developed by Saffman (1959, 1960) who considered the porous media to be statistically homogeneous and isotropic, and to consist of a randomly distributed set of points connected to neighboring points by linear pores. The path of each particle was assumed to be a random walk. The probability distribution of the displacement of each particle was computed after a given time period. The equation resulting from this analysis is given by

$$\tilde{D}_L = \frac{D}{3} + \frac{3}{80}\frac{r_p^2 U^2}{D} + \frac{l^2 U^2}{4}\int_0^1 \left(3\cos^2\theta - 1\right)^2 \frac{M\coth M - 1}{zM^2}\, d(\cos\theta) \qquad (58)$$

where r_p refer to the radius of the pores, l stands for the average pore length, U again denotes the average pore velocity,

$$M = \frac{3}{2}\left(\frac{Ul\cos\theta}{z}\right) \qquad (59)$$

and

$$z = D + \frac{\left(3 r_p U \cos\theta\right)^2}{48D}. \qquad (60)$$

Taking account of this same model Saffman (1960) obtained the equation for transverse dispersion given by

$$\tilde{D}_T = \frac{D}{3} + \frac{1}{80}\frac{r_p^2 U^2}{D} + \frac{9 l^2 U^2}{8}\int_0^1 \cos\theta\left(1 - \cos^2\theta\right)\frac{M\coth M - 1}{zM^2}\, d(\cos\theta) \qquad (61)$$

Although Equations (58) to (61) contain several geometric parameters, Equation (58) has been demonstrated to yield a remarkable correspondence to measured longitudinal dispersion coefficients in uniform sand columns (see below and Fig. 13).

Dispersivity data obtained in the laboratory are commonly depicted by the ratio $\tilde{D}_L/D$ versus the diffusional Peclet number (Eqn. 54). One such example of this type of plot is illustrated in Figure 13. This figure, modified after that of Pfannkuch (1963) contains data obtained from packed columns of glass beads or sand having average radii ranging from 0.005 to 0.343 cm. No correlations are apparent with average particle radius. In contrast, $\tilde{D}_L/D$ are apparently a single function of P_e. At $P_e < 1$, this ratio is effectively constant, with a value slightly less than unity. This is a consequence of the fact that D is the molecular diffusion coefficient measured in the bulk fluid, whereas $\tilde{D}_L$

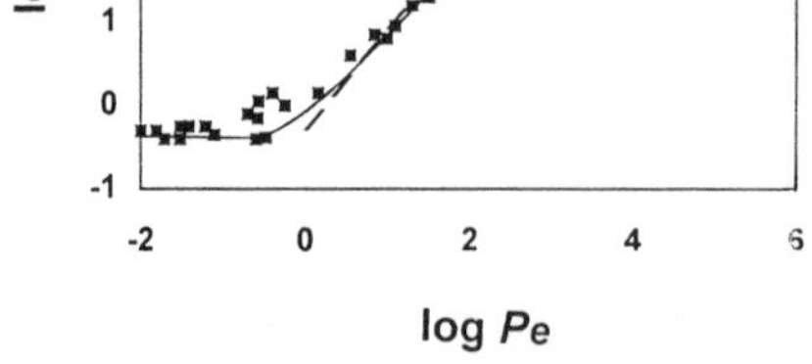

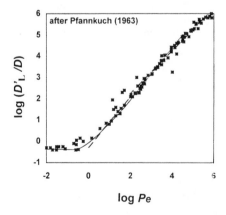

Figure 13. Variation of the ratio of the coefficient of longitudinal hydrodynamic dispersion versus the molecular diffusion coefficient plotted as a function of the diffusional Peclet number. The symbols correspond to data obtained on columns of glass beads, plastic spheres, sand, and natural porous rock summarized by Pfankuch (1963). The dashed lines in this figure corresponds to

$$\frac{D_L}{D} = 0.5 P_e^{1.2} \text{ for } P_e < 100, \text{ but } \frac{D_L}{D} = 1.8 P_e$$

for $P_e > 100$. The solid line represents the fit of the equations of Staffman (1960) (Eqns. 58 to 60) assuming $l/r_p = 5$.

at these low Peclet numbers is equal to the molecular diffusion coefficient observed in the bulk media, where diffusional transport is slower due to porosity and tortuosity (see above). At higher diffusional Peclet numbers, the ratio $\tilde{D}_L/D$ varies almost linearly with P_e. The solid lines drawn through these data are consistent with (Bear, 1972)

$$\frac{\tilde{D}_L}{D}=0.5P_e^{1.2} \text{ for } P_e \le 100, \tag{62}$$

but

$$\frac{\tilde{D}_L}{D}=1.8P_e \text{ for } P_e \ge 100. \tag{63}$$

The dashed line drawn through the data in Figure 13 correspond to values generated using Equations (58) to (60) assuming $l/r_p = 5$.

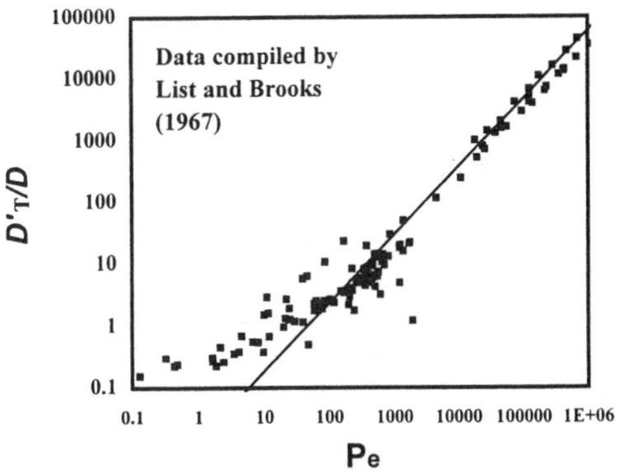

Figure 14. Variation of the ratio of the coefficient of transverse hydrodynamic dispersion versus the molecular diffusion coefficient plotted as a function of the diffusional Peclet number. The symbols correspond to data obtained on columns of glass beads, plastic spheres, sand, and natural porous rock reported by Bernard and Wilhelm (1950), Grane and Gardner (1961), Blackwell (1962), Simpson (1962), Harleman and Rumer (1963), Hoopes and Harleman (1965), and List and Brooks (1967). The solid line in this figure corresponds to

$$\frac{\tilde{D}_L}{D}=P_e^{1.15}.$$

A figure illustrating laboratory measured transverse hydrodynamic dispersion coefficients for similar homogeneous materials are depicted as a function of their diffusional Peclet number in Figure 14. The distribution of data on this figure is very similar to that exhibited by the longitudinal dispersion data shown in Figure 13. At low diffusional Peclet numbers the ratio $\tilde{D}_L/D$ tend to a constant value of slightly less than 1, but at higher P_e this ratio is apparently a linear function of the Peclet number. The curve drawn through these data points for Peclet numbers greater than ~30 is consistent with

$$\frac{\tilde{D}_T}{D}=0.015P_e^{1.10}. \tag{64}$$

A comparison of Figures 13 and 14 illustrates that transverse hydrodynamic dispersion coefficients tend to be lower than their longitudinal counterparts for all P_e>1. Because of the similar shapes of the data displayed on these two figures exhibits similar distributions, the ratio $\tilde{D}_T/\tilde{D}_L$ is often taken to be a constant in chemical mass transport calculations. By combining these two expressions for dispersion coefficients, it follows that

$$\frac{\tilde{D}_T}{\tilde{D}_L}=0.03P_e^{-0.1} \text{ when } 30 \le P_e \le 100, \tag{65}$$

but

$$\frac{\tilde{D}_T}{\tilde{D}_L}=0.008P_e^{0.1} \text{ when } P_e \ge 100. \tag{66}$$

These equations imply that the ratio $\tilde{D}_T/\tilde{D}_L$ is on the order of 0.01 to 0.03 for all Peclet numbers greater than ~30. As the Peclet number decreases below 30 the influence of aqueous diffusion becomes significant and this ratio increases reaching a maximum of 1 when P_e is less than ~ 0.5. The observation that transverse hydrodynamic dispersion coefficients are significantly lower than their longitudinal counterparts is consistent with field data (see below). The distribution of scatter in Figures 13 and 14 implies that the overall uncertainty of $\tilde{D}_T$ and $\tilde{D}_L$ values generated using Equations (62) to (64) is on the order of ±0.25 log units. It should be emphasized, however, that several studies have suggested that laboratory measured hydrodynamic dispersivities do depend somewhat on grain size. Blackwell (1962) reported the longitudinal and transverse dispersion coefficients at constant Peclet number of columns of fine sand were as much as an order of magnitude larger than those for coarse sand. Dullien (1992) suggested this difference could be the result of the differences in size distribution of the various sand fractions used in their study.

Field scale dispersion (Macrodispersion)

Dispersivity coefficients obtained from the field, which are generally measured on heterogeneous materials, may be far different from those generated in the laboratory on idealized homogeneous materials. Moreover, the existence of heterogeneities on natural systems leads to field measured dispersivity coefficients that depend on the system scale (Gelhar et al., 1979; Klotz et al., 1980; Pickens and Grisak, 1981). Gelhar et al. (1992) published a critical review of dispersivities obtained from field measurements on both porous and fractured rock formations. They noted hydrodynamic dispersivity in three directions: the direction of flow (longitudinal hydrodynamic dispersion coefficients), normal to the flow but parallel to the Earth's surface (horizontal transverse hydrodynamic dispersion coefficients), and normal to the flow but perpendicular to the surface (vertical transverse hydrodynamic dispersion coefficients). These data were reported as dispersivities symbolized by α_L, α_T, and α_V, respectively where $\alpha_L \equiv D_L / U$, $\alpha_T \equiv D_T / U$, $\alpha_V \equiv D_V / U$, and D_V stands for the value of $\tilde{D}_T$ in the vertical direction. Values α_L tabulated by Gelhar et al. (1992) are depicted in Figure 15, where it can be seen that this product systematically increases almost linearly with the scale of the measurement. The observation that longitudinal dispersivity depends on the scale of the system is contrary to the results presented above for dispersion on isotropic materials obtained in the laboratory. The field measurements were obtained at conditions where P_e < 100, where laboratory measurements suggest that

$$\alpha_L = D_L^{\prime} U = 0.5 D P_e^{1.2} U \tag{67}$$

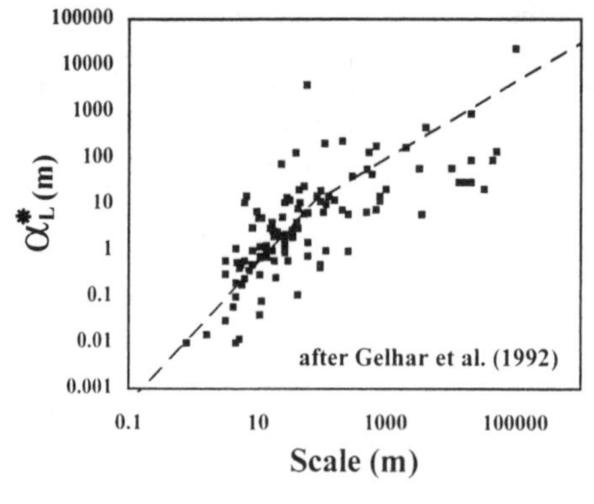

Figure 15. Variation of macrodispersivity as a function of the scale of dispersivity measurements. Symbols represent data summarized by Gelhar et al. (1992); the dashed line corresponds to the fit of these data by Neuman (1990).

and thus α_L would be a constant controlled by only the molecular diffusion coefficient, the Peclet number, and the flow rate, and thus be independent of the scale of the field area. Furthermore, as can be seen by comparing Figures 13 and 15, field-measured dispersivities can be orders of magnitude greater that laboratory-measured counterparts. For this reason, these two types of dispersivities are commonly given different names: laboratory-measured dispersivities can be referred to as microdispersivities, and field-measured dispersivities are commonly referred to as macrodispersivities. To distinguish microdispersivities from macrodispersivities in this chapter, field measured macro-dispersivities are distinguished by the addition of an asterisk to their symbols.

The problem of the origin of scale dependent filed measured dispersion coefficients was first addressed theoretically by Mercado (1967) who based his work on a stratification model. Within this model, the rock formation is assumed to consist of several uniform layers, each having a distinct permeability. Flow is confined to the individual layers and cannot cross the interlayer boundaries. Transport in each layer was assumed to be due exclusively to fluid advection; diffusional and microdispersive transport was considered negligible. Because each layer has a distinct permeability, the fluid velocity and thus the solute transport rate is different in each layer. Taking account of these assumption, Mercado (1967, 1984) generated the following equation for the effective longitudinal hydrodynamic macrodispersivity (α_L^*)

$$\alpha_L^* = \frac{1}{2}\left(\partial_{\mathbf{k}} / \bar{\mathbf{k}}\right)^2 \bar{\mathbf{x}} \tag{68}$$

where $\partial_{\mathbf{k}}$ refers to the standard deviation of the permeability, $\bar{\mathbf{k}}$ represents the average permeability, and $\bar{\mathbf{x}}$ denotes the average solute front distance. Equation (68) suggests that the longitudinal hydrodynamic macrodispersivity will vary linearly with system scale, which is consistent with the data illustrated in Figure 15. This observation is consistent with the concept that the origin of the variation of dispersivity with system scale stems from the existence of heterogeneities in natural systems. More recent theoretical efforts have relaxed many of the assumption made by Mercado (1967, 1984) resulting in similar equations (Gelhar et al., 1979; Gelhar and Axness, 1983; Dagan, 1984; Gelhar, 1987; Neuman et al., 1987; Naff et al., 1988; Neuman and Zhang, 1990, Zhang and Numan, 1990). For example, Neuman et al. (1987) derived the following equation for the effective longitudinal dispersivity of an isotropic media

$$\alpha_L^* = \frac{3}{8}\sigma_{\mathbf{K}}{}^2 \bar{\mathbf{x}} \tag{69}$$

where $\sigma_{\mathbf{K}}$ designates the variance of the natural logarithm of hydraulic conductivity. Corresponding equations generated for anisotropic media yield similar expressions, each exhibiting linear or semi-linear dependence of the system scale. Direct application of these theoretically generated macrodispersivity equations requires independent knowledge of the spatial and directional variation of permeability, which is not generally known. Consequently, Neuman (1990) generated an empirical fit of the field data illustrated in Figure 15 given by

$$\alpha_L^* = 0.0169\bar{\mathbf{x}}^{1.53} \quad \text{for} \quad \bar{\mathbf{x}} \le 100 \text{ m} \tag{70}$$

and

$$\alpha_L^* = 0.32\bar{\mathbf{x}}^{0.83} \quad \text{for} \quad \bar{\mathbf{x}} \ge 100 \text{ m} \tag{71}$$

where $\bar{\mathbf{x}}$ again designates the length scale of the system, and α_L is given in meters. Neuman (1990) suggested that these can be considered universal equations for calculating longitudinal dispersivity in natural systems. The 95% confidence limits of this fit is reported to be ±0.4 log units or less for length scales to 100 m but as high as 1.5 orders of magnitude for larger length scales. Although the universality of this equation was questioned by Gelhar et al. (1992), who suggested that the distribution of data in Figure 15 results from a family of dispersivity/scale curves corresponding to different degrees of formation heterogeneity, it does provides a first approximation for computing longitudinal dispersivities in systems for which there is no available data.

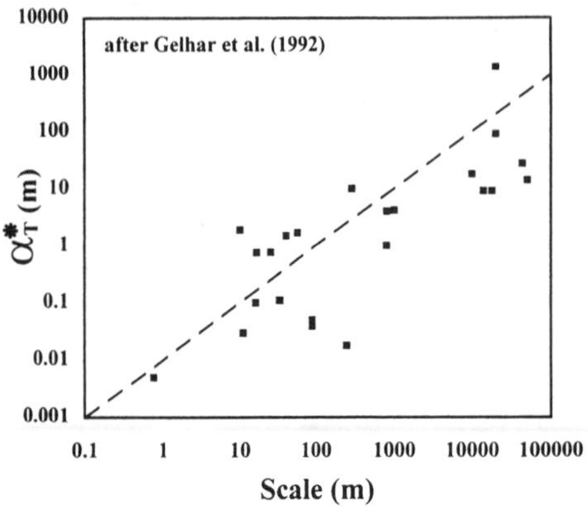

Figure 16. Variation of the horizontal tangential macrodispersivity as a function of the scale of the dispersivity measurement. The symbols represent data summarized by Gelhar et al. (1992), but the line drawn through the data are consistent with $\alpha_T^* = 0.01\bar{x}$.

Values of horizontal tangential macrodispersivities, α_T, tabulated by Gelhar et al. (1992) are illustrated as a function of the scale of measurement in Figure 16. These data exhibit a similar pattern as the α_L data depicted in Figure 15. Similar to the theoretical development presented above for longitudinal hydrodynamic macrodispersivities, taking account various assumptions for the nature of the porous media Gelhar and Axness (1983) and Neuman et al. (1987) generated a variety of semi-linear relationships between system scale and transverse hydrodynamic macrodispersivity. For example, Neuman et al. (1987) derived the following equation for the transverse hydrodynamic macrodispersivity (α_T^*) in a isotropic media

$$\alpha_T^* = \frac{\sigma_K^2 \bar{x}}{15}\left(1+\frac{4\tilde{D}_T}{\tilde{D}_L}\right)\frac{1}{P_d} \tag{72}$$

where P_d signifies a dispersive Peclet number given by

$$P_d = \frac{2\bar{T}U}{\tilde{D}_L} \tag{73}$$

Again such theoretical equations are difficult to apply to natural systems as they require a detailed understanding of spatial and directional heterogeneities of the natural material. The line drawn through these data are consistent with

$$\alpha_T^* = 0.01\bar{x}$$

with an average uncertainty of ±0.4 log units. In the absence of more detailed information, this empirical equation affords a simple first approximation to the effective horizontal hydrodynamic dispersivity of natural rock formations.

The ratio of the horizontal tangential macrodispersivities versus their corresponding longitudinal macrodispersivities are depicted as a function of length scale of measurement in Figure 17. No clear trend is apparent in these data. Although there is considerable scatter in these data, it seems reasonable, in the absence of measured values in a given system to adopt the average of these values, α^*_T/α^*_L = 0.33, for mass transport calculations. It is interesting to note that this ratio is within the range defined from laboratory dispersion measurements (see above).

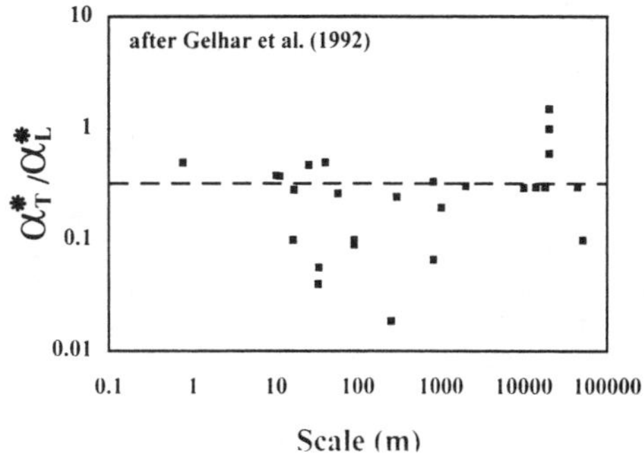

Figure 17. Variation of the ratio of the horizontal tangential macrodispersivity to the longitudinal macrodispersivity as a function of the scale of the dispersivity measurement. The symbols represent data generated using summarized by Gelhar et al. (1992), but the curve corresponds to the average value of this ratio α^*_T/α^*_L = 0.33.

Gelhar et al. (1992) also noted that measured α^*_T tend to be larger than corresponding α^*_V values. It should, however, be emphasized that the only available α^*_V data is for sedimentary systems which will tend to have grains oriented parallel to the Earth's surface. Grain orientation results in fundamental differences in the pore void paths in the vertical versus the horizontal direction. The ratio α^*_V/α^*_L are depicted as a function of length scale of measurement in Figure 18. As was observed for the case of the ratio α^*_T/α^*_L illustrated in Figure 17, no clear trend is apparent in these data. Although there is considerable scatter in these data, it seems reasonable, in the absence of measured values for a given system to adopt the average of these values α^*_V/α^*_L = 0.04 for mass transport calculations in sedimentary environments. Note, however, as evident by the considerable scatter apparent in Figure 18, large uncertainties accompany the use of this average value in chemical mass transport calculations. It seems reasonable to expect, however, that in crystalline rock there would be little or no difference between transverse macrodispersivity measured in the horizontal versus the vertical direction. In this case, such differences would be controlled by preferential grain or fracture orientations, which would not necessarily be orientated in the horizontal or the vertical plane.

The use and application of micro- versus macro-dispersivity in reactive transport calculations remains problematical. If it were possible to perform such calculations using

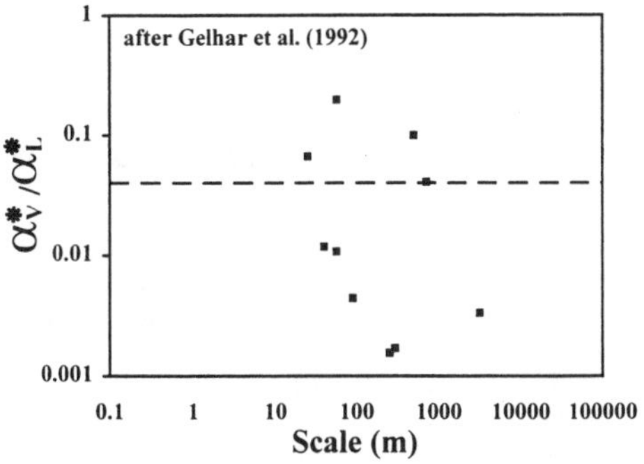

Figure 18. Variation of the ratio of the vertical tangential macrodispersivity to longitudinal macrodispersivity as a function of the scale of the dispersivity measurement. Symbols represent data generated using summarized by Gelhar et al. (1992), but the curve corresponds to the average value of this ratio, α^*_V/α^*_L = 0.04.

a spatial grid in which each representative elemental volume (REV) contained only homogeneous materials, microdispersivities could be used to describe the dispersion phenomena in each REV, and the interactions among the various REVs in the system would lead to an accurate computed macrodispersivity (see Schwartz, 1977). This is likely to be impossible for most natural systems, which are not sufficiently characterized to identify spatial and directional heterogeneities. Consequently, a macrodispersivity characteristic of the REV size may produce more accurate results in reactive transport calculations.

RATES OF MINERAL/WATER INTERACTION

The successful application of reaction transport algorithms to calculating the chemical evolution of natural systems requires accurate methods to compute the rates of mineral/fluid surface reactions. Towards this goal, a substantial quantity of work has been aimed at characterizing mineral dissolution rates over the past two decades. Summaries of this recent work, together with some recent work on precipitation rates, are provided in White and Brantley (1995) and Brady (1996). Although, in theory, it is possible to directly use laboratory measured dissolution/crystallization kinetics to compute the rates of mineral-fluid reactions in natural systems, there are may be significant discrepancies between rates measured in the laboratory and those measured in the field, even for relatively simple minerals (White, 1995). The main sources of these discrepancies are likely due to (1) the effects of chemical affinity and solution composition on the dissolution rates, (2) possible differences between the surface areas of minerals in the natural environment versus the laboratory, and (3) biological interaction. Moreover, there are significant discrepancies among laboratory measured rates themselves. One such example is illustrated in Figure 19a, which depicts the specific dissolution rates of quartz at acidic to neutral conditions generated from equations proposed in the literature over the past sixteen years as a function of temperature from 0 to 300°C. The different curves in the figure correspond to rates at 'far from equilibrium conditions' computed using expressions generated from distinct sets of experimental data. Far from equilibrium conditions are commonly defined by

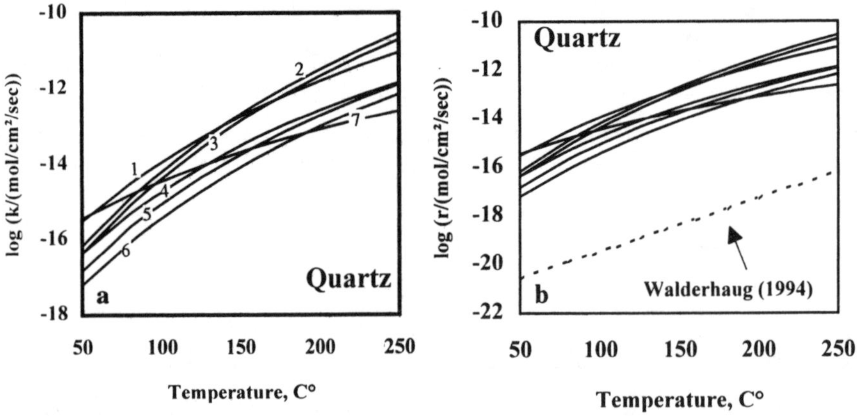

Figure 19. Quartz dissolution rates at near to neutral pH as a function of temperature. (a) Curves computed for steady state dissolution at far from equilibrium conditions by adopting equations and parameters reported by (1) Dove and Crerar (1990), (2) Knauss and Wolery (1989), (3) Tester et al. (1994), (4) Rimstidt and Barnes (1980), (5) Murphy and Helgeson (1989), (6) Gislason et al. (1996), and (7) Brady and Walther (1990). (b) Solid curves correspond to the same calculations as in (a), but the dotted curve illustrates the variation of quartz precipitation rates obtained from the interpretation of Jurassic North Sea sandstones as a function of temperature reported by Walderhaug (1994).

$$A >> RT, \tag{74}$$

where R designates the gas constant, T refers to absolute temperature, and A refers to the chemical affinity of the dissolving mineral, which is a measure of the distance from equilibrium a fluid is from the mineral of interest, can be defined by

$$A = -RT\ ln(\frac{Q}{K_{eq}}) \tag{75}$$

where K_{eq} refers to the equilibrium constant for the overall dissolution reaction and Q designates the reaction quotient defined by

$$Q \equiv \prod_{i=1}^{n_s} a_i^{\tilde{n}_i} \tag{76}$$

where a_i designates the activity of the subscripted ion, and $\tilde{n}_i$ represents the stoichiometric number of the ith aqueous species in the overall hydrolysis reaction, and n_s denotes the number of species in the system. It is commonly assumed that rates are independent of the aqueous activities of the elements contained in the mineral, and thus chemical affinity, at far from equilibrium conditions, but such may not be the case for many multi-oxide silicates (see below).

It can be seen in Figure 19a that proposed quartz rate constant equations yield dissolution rates that vary by as much as three orders of magnitude from one another at high and low temperature. Some of this discrepancy can be resolved if one explicitly considers the pH dependence of these rates or if one fits globally all of the rate data available for quartz dissolution (Dove, 1995; Cadoré, 1995). It should be emphasized, however, that far less experimental data are available for the bulk of the other major rock forming minerals, many of which have far more complex dissolution rate behaviors than that of quartz. Consequently, the scatter indicated by the curves in Figure 19a, which is on the order of two to three orders of magnitude, should be indicative of the uncertainties associated with other laboratory measured 'far from equilibrium' dissolution/precipitation rates.

The magnitude of the uncertainties associated just with the laboratory measured dissolution/precipitation rates can lead to large uncertainties in the computed quantity of chemical mass transfer in natural systems. On such example can be seen in Figure 20, which illustrates the porosity of Jurassic North Sea sandstones as a function of depth computed using some of the quartz dissolution rates illustrated in Figure 19a. Computed quartz porosities depicted in this figure were computed using the schematic model outlined by Oelkers et al. (1996), which consists of mica promoted quartz dissolution at stylolite boundaries, coupled to diffusional transport, and kinetically controlled quartz precipitation. Quartz dissolution rates at the near to equilibrium conditions found in sedimentary basins were computed from the 'far from equilibrium' rates using Equation (77) (see below). It can be seen in Figure 20 that the computed porosity for these petroleum reservoir rocks at a depth of 4 km vary from 5 to nearly 20% depending on the chosen dissolution rate expression. Although the absolute value of the results shown in Figure 20 are dependent on the boundary conditions of the model calculations, and consequently cannot be used to distinguish the relative quality of the dissolution rates given in Figure 19a, they illustrate the uncertainties associated with adopting dissolution rates reported in the literature for the quantification of chemical mass transport in natural systems. For this particular example, a sandstone porosity of less than ~10% is considered to be non-economic, but a sandstone porosity of greater than ~10% may produce substantial quantities of crude petroleum. Thus the uncertainty level exhibited by the curves in Figure 19a results in an unacceptably high uncertainty in the reactive transport calculations in this system, which might only be improved using direct field evidence.

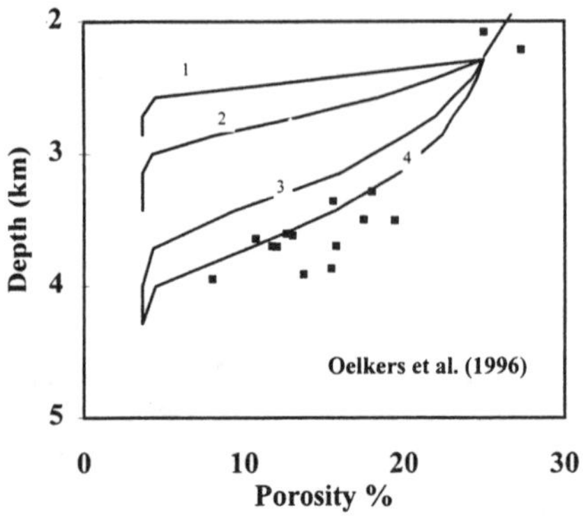

Figure 20. Computed porosities of Jurassic North Sea sandstones as a function of depth (after Oelkers et al., 1997) generated by adopting quartz dissolution/precipitation rates generated using equations and parameters reported by (1) Dove and Crerar (1990), (2) Tester et al. (1994) (3) Murphy and Helgeson (1989), and (4) Gislason et al. (1996). The symbols represent measured sandstone porosities reported by Ehrenburg (1990).

Experimentally measured quartz dissolution rates are compared with field measured quartz precipitation rates, reported by Walderhaug (1994), in Figure 19b. These 'field measured' rates were obtained from petrographic analysis of North Sea reservoir sandstones by adopting the assumptions (1) precipitation occurred continuously over the last 80 million years, (2) all quartz grains were continuously bathed in reactive solution, and (3) the reactive surface area is equal to its geometrically estimated counterpart. It can be seen in this figure that these 'field measured' precipitation rates are several orders of magnitude lower than the experimentally measured dissolution rates. Direct comparison of these two rates would be possible using equations given below if the silica concentration of the reactive fluid were known.

In general, the dissolution or precipitation of a mineral is a multistep process consisting of the transport of reactants through a fluid phase to and from the surface of the mineral, coupled to reactions at and/or near the mineral surface releasing or consuming aqueous solute species (see Murphy et al., 1989; Alkattan et al., 1996). For the case of numerous non-silicate minerals, such as halite or calcite at neutral to acidic conditions, the rate of surface release of solutes from the mineral surface is rapid, and their overall dissolution or precipitation rates is controlled by the transport of material to or from this surface from the bulk pore fluid through a surface boundary layer. Equations describing chemical transport through boundary layers are given by Lichtner (this volume). In contrast, the dissolution/precipitation rates of most silicate minerals at ambient conditions are sufficiently slow so that their overall rates are controlled by the release or incorporation of solute species at or near the mineral surfaces. The rates of such reactions are proportional to the surface areas of the reactive minerals.

Reactive surface area

A major challenge in reactive transport calculations is the accurate characterization of reactive mineral surface areas in natural systems. Two major factors are responsible for uncertainties of reactive surface area in natural systems are (1) the effect of surface roughness, and (2) channeling of reactive fluid flow. Surface roughness (λ) is defined as the ratio of the true reactive surface area to the equivalent geometric surface area of a hypothetical smooth surface encompassing the actual surface (Helgeson et al., 1984). Generally the true reactive surface area is measured by microscopic techniques such as

BET gas sorption. Most recently reported laboratory measured dissolution/crystallization rates are normalized to surface areas obtained using such microscopic techniques. There are significant differences between the surface roughness of (1) freshly ground minerals, (2) mineral samples used in laboratory dissolution experiments, and (3) naturally weathered materials. Anbeek (1992) found that freshly ground silicate mineral surfaces had surface roughness values ranging from 2.5 to 11. Blum (1994) reported surface roughness of 9±6 for crushed feldspar samples. White and Peterson (1990) found a average roughness of 7 for unweathered clays, oxides, carbonates, and silicates. Surface roughness of mineral samples measured in the laboratory may be even lower owing to the use of a variety of surface cleaning techniques. For example, the surface roughness of crushed, cleaned feldspar samples used for dissolution experiments by Oelkers et al. (1994) Gautier et al. (1994) and Oelkers and Schott (1995) was less than 4. In contrast, naturally weathered samples exhibit substantially larger surface roughness. Anbeek (1992) reported surface roughness values of 130 to 2600 for naturally weathered silicate from glacial deposits; White and Peterson (1990) found surface roughness of naturally weathered sand sized particles ranged from 50 to 200 and surface roughness of Merced soils ranged from 100 to 1000.

Uncertainties in reactive surface area due to possible effects of fluid channeling are likely far greater than those associated with surface roughness. This can be illustrated using a simple example. The surface area of spherical grains with a diameter d is given by (Anbeek, 1992; White, 1995): $\mathbf{s} = (6/d)\lambda$, where $\mathbf{s}$ stands for the true reactive surface area. The surface area of a meter cube of spherical sand grains with a diameter of 0.2 mm, and a representative surface roughness of 10 is 300,000 m^3. In contrast, the surface area of a single smooth fracture parallel to one of the sides of the cube is 2 m^3. Thus if all the surfaces of this cubic meter of rock were bathed in reactive fluid, the system will have a reactive surface area of 300,000/m but if all of the fluid in the cubic meter of rock is channeled in a single fracture, and no mass transport between the bulk rock and the fracture fluid occurred, the reactive surface area would be 2/m, more than five orders of magnitude lower. Most systems likely lie between these two extremes. The fact that both permeabilities and dispersivities obtained from field measurements tend to be scale dependent (see above) implies that fluid channeling is common in natural systems. The degree to which the reactive surface area of a particular system tends to be consistent with the concept that most of the reactive fluid is channeled and thus isolated from the bulk of the potential reactive surfaces or pervasive and thus in contact with most grains can be deduced from petrographic analysis.

Variation of mineral hydrolysis rates with chemical affinity

The variation of surface controlled mineral dissolution and precipitation reactions can be described by considering transition state theory. In accord with transition state theory as applied to minerals, the overall rate of a mineral dissolution reaction per unit surface area (**r**) can be described using (Aagaard and Helgeson, 1977, 1982; Lasaga, 1981; Helgeson et al., 1984)

$$\mathbf{r} = \mathbf{r}_+ (1 - \exp(-A / \sigma RT)) \tag{77}$$

where $\mathbf{r}_+$ designates the forward dissolution rate per unit surface area, and σ stands for Temkin's average stoichiometric number equal to the ratio of the rate of destruction of the activated or precursor complex relative to the overall dissolution rate. The term in the parentheses in Equation (77) takes account of the effects of back reaction as equilibrium is approached, and assures that $\mathbf{r} = 0$ at equilibrium. Equation (77) can equivalently describe dissolution and precipitation reactions. In the present chapter, $\mathbf{r}_+$ is taken to be negative for dissolution, and the chemical affinity is computed using the equilibrium constant, K_{eq} for the overall dissolution reaction. Taking account these conventions, chemical affinity is positive for undersaturation, zero for equilibrium, and negative for supersaturation. It follows from Equation (77) that **r** is then negative for dissolution, zero at equilibrium, and positive for precipitation. The variation of $\mathbf{r}/\mathbf{r}_+$ as a function of chemical affinity consistent with Equation (77) is depicted in Figure 21. It can be seen in this figure that this equation predicts that $\mathbf{r}/\mathbf{r}_+$ values minimize with increasing chemical affinity, reaching a minimum equal to $-\mathbf{r}_+$ when $A/\sigma RT > 2$. At these 'far from equilibrium' conditions the dissolution rate is independent of chemical affinity, although they may depend on aqueous solute composition through their effect on $\mathbf{r}_+$. For cases where $\mathbf{r}_+$ is independent of solute composition this region is referred to as the 'dissolution plateau'. In contrast precipitation rates generated using Equation (77) increase continuously with an increasing absolute value of A. The curve in Figure 21 illustrates that dissolution/precipitation rates measured in the laboratory at 'far from equilibrium' conditions can be far greater than at the near to equilibrium conditions characteristic of most natural systems.

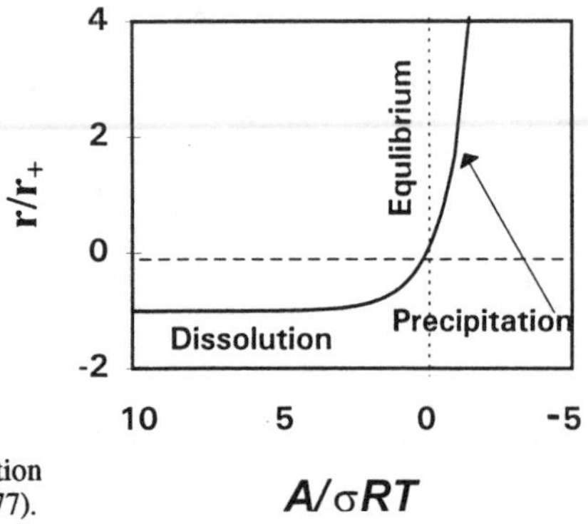

Figure 21. Variation of mineral dissolution/precipitation rates as a function of $A/\sigma RT$ computed using Equation (77).

Although Lasaga (1995) argues that the form of Equation (77) is not valid because of a 'fatal error' made in the original derivation of this equation by Boudart (1976), Lasaga and coworkers have used an identical mathematical form to fit the dissolution and crystallization rates of kaolinite (Nagy et al., 1990, 1991), and a similar equation to describe the variation of gibbsite dissolution rates on chemical affinity (Nagy and Lasaga, 1992, 1993). The justification of these latter applications was based on empirical grounds.

Within the context of transition state theory, the forward reaction rate, $\mathbf{r}_+$, is equal to the product of two factors, the concentration of a rate controlling surface complex, sometimes termed 'precursor complex' ($P^\bullet$) and the rate of destruction of this precursor to form reaction products (Wieland et al., 1988; Stumm and Wieland, 1990). This concept is consistent with

$$\mathbf{r}_+ = k_{P^\bullet}\{P^\bullet\} \tag{78}$$

where $k_{P^\bullet}$ refers to a rate constant consistent with the $P^\bullet$ precursor complex and $\{P^\bullet\}$ stands for its concentration. The applicability of Equations (75) to (78) to mineral hydrolysis is greatly facilitated by knowledge of the precursor complex identity and formation reactions, as such information can be used to relate $\{P^\bullet\}$ in Equation (78) to measurable quantities such as aqueous solute concentrations.

The identity and formation reactions of rate controlling precursor complexes apparently depend on the number and relative strength of the bonds within each mineral structure (Schott and Oelkers, 1995). The dissolution of an single (hydr)oxide and some multi oxide silicates (e.g. anorthite, Oelkers and Schott, 1995b) requires the breaking of only one type of cation-oxygen bond. It follows that the rate controlling precursor complex for such solids can be formed by adsorption reactions involving H^+, OH^-, or any other species that tend to weaken this single bond type (Zindler et al., 1986, Furrer and Stumm, 1986). An example of one such reaction can be expressed as

$$n_H H^+ + M\text{-}O = P_H^\bullet \tag{79}$$

where M-O represents a potentially reactive surface site, n_H represents a stoichiometric coefficient, and $P_H^\bullet$ a precursor complex. Taking account of the law of mass action for this adsorption reaction and the fact that there are a limited number of total possible adsorption sites on the (hydr)oxide surface, an equation describing the variation of rates can be expressed as (cf. Oelkers et al., 1994)

$$r = k_+ \left(\frac{a_{H^+}^{n_H}}{1 + K_{79}^\bullet a_{H^+}^{n_H}} \right) \left(1 - exp(-A/\sigma RT)\right) \tag{80}$$

where $K^\bullet_{79}$ designates the equilibrium constant for the subscripted reaction, k_+ denotes a rate constant($k_+ = k_{P^\bullet} K^\bullet_{79}$), and a_i again stands for the aqueous activity of the subscripted species. Note that Equation (80) has two limits; when the surface is saturated with precursor complexes ($K_{79}^\bullet a_{H^+}^{n_H} >> 1$) Equation (80) reduces to

$$r = k_{P^\bullet} \left(1 - exp(-A/\sigma RT)\right) \tag{81a}$$

but when the surfaces contain relatively few precursor complexes ($K_{79}^\bullet a_{H^+}^{n_H} << 1$), Equation (80) reduces to

$$r = k_+ a_{H^+}^{n_H} \left(1 - exp(-A/\sigma RT)\right) \tag{81b}$$

In addition, Equation (81b) itself reduces to

$$r = k_+ a_{H^+}^{n_H} \tag{82}$$

at 'far from equilibrium' conditions. It follows from Equations (80) to (82) that the variation of dissolution rates at constant pH for minerals having rate controlling precursor complexes formed by Reaction (79), will be a fairly simple function of chemical affinity, equivalent to that illustrated in Figure 21. The degree to which these equation can describe the dissolution rates of several minerals can be seen in Figure 22, where constant pH dissolution rates of quartz, anorthite, hematite, and halite are depicted as a function of chemical affinity. The lines drawn through the data points, consistent with Equations (80) and (81a,b) illustrate a close correspondence between the experimental data and the behavior predicted using transition state theory assuming their rate controlling precursor complex is formed by the simple sorption of H^+, OH^-, and/or H_2O.

In contrast, the dissolution mechanism of some multi-oxide minerals require the breaking of several different types of cation-oxide bonds. As some of these cation-oxygen bonds are easier to break than others, the key to hydrolysis is the destabilization and ultimate destruction of the strongest bond that is essential to a viable mineral structure. In these cases, the rate controlling precursor complex can involve both (1) exchange reactions that break the more reactive cation-oxygen bonds that are not essential to the structure, thus better exposing the bonds essential to the structure to hydrolysis, and (2) adsorption reactions that weaken these bonds. For the case of aluminosilicate hydrolysis, the key to dissolution is the destruction of the framework containing both silicon-oxygen and aluminum-oxygen bonds. Work on albite (Oelkers and Schott, 1992; Oelkers et al., 1994), K-feldspar (Gautier et al., 1994), kaolinite (Devidal, 1992, 1994, 1996), kyanite (Oelkers and Schott, 1994, 1996a), and analcime (Murphy et al., 1996) suggest that for these minerals the destruction of tetrehedrally coordinated Al-O bonds are relatively rapid in comparison to the destruction of tetrehedrally coordinated Si-O bonds. This destruction of Al-O bonds leads the formation of a silica-rich precursor complex via the exchange of aluminum for aqueous hydrogen ions according to the reaction

$$3\,\boldsymbol{n}\,\mathrm{H}^{+} + \mathrm{M\text{-}O} = \boldsymbol{P^{\bullet}} + \boldsymbol{n}\,\mathrm{Al}^{+3} \tag{83}$$

where M-O again represents a potentially reactive surface site and $\boldsymbol{n}$ again represents a stoichiometric coefficient. In the general case where the rate controlling precursor complex is created by a metal for aqueous hydrogen exchange reaction, this equation becomes

$$Z_{\mathrm{M}}\,\boldsymbol{n}\,\mathrm{H}^{+} + \mathrm{M\text{-}O} = \boldsymbol{P^{\bullet}} + \boldsymbol{n}\,\mathrm{M}^{+Z_{\mathrm{M}}} \tag{84}$$

where Z_M designates the charge on the metal ion. Taking account of the law of mass action for this exchange reaction and the fact that there are a limited number of total possible exchange sites on the aluminosilicate surface, an equation describing the dissolution rates of the minerals that follow this mechanism can be expressed as (Oelkers et al., 1994)

$$\mathrm{r} = \boldsymbol{k}_{+} \frac{\left(\dfrac{\boldsymbol{a}^{Z_{\mathrm{M}}}_{\mathrm{H}^{+}}}{\boldsymbol{a}_{\mathrm{M}^{+Z_{\mathrm{M}}}}}\right)^{\boldsymbol{n}}}{1 + \boldsymbol{K}^{\bullet}_{84}\left(\dfrac{\boldsymbol{a}^{Z_{\mathrm{M}}}_{\mathrm{H}^{+}}}{\boldsymbol{a}_{\mathrm{M}^{+Z_{\mathrm{M}}}}}\right)^{\boldsymbol{n}}} \left(1 - \boldsymbol{exp}\left(-\boldsymbol{A}/\sigma \boldsymbol{RT}\right)\right) \tag{85}$$

where $\boldsymbol{K}^{\bullet}_{84}$ designates the equilibrium constant for Reaction (84) and $\boldsymbol{k}_{+}$ denotes a rate constant ($\boldsymbol{k}_{+} = \boldsymbol{k}_{\boldsymbol{P}^{\bullet}}\,\boldsymbol{K}^{\bullet}_{84}$). Note Equation (85) reduces to Equation (81a) in the limit when the surface is saturated with precursor complexes, but reduces to

$$\mathrm{r} = \boldsymbol{k}_{+}\left(\frac{\boldsymbol{a}^{Z_{\mathrm{M}}}_{\mathrm{H}^{+}}}{\boldsymbol{a}_{\mathrm{M}^{+Z_{\mathrm{M}}}}}\right)^{\boldsymbol{n}} \left(1 - \boldsymbol{exp}\left(-\boldsymbol{A}/\sigma \boldsymbol{RT}\right)\right) \tag{86}$$

when the surface contains relatively few precursor complexes

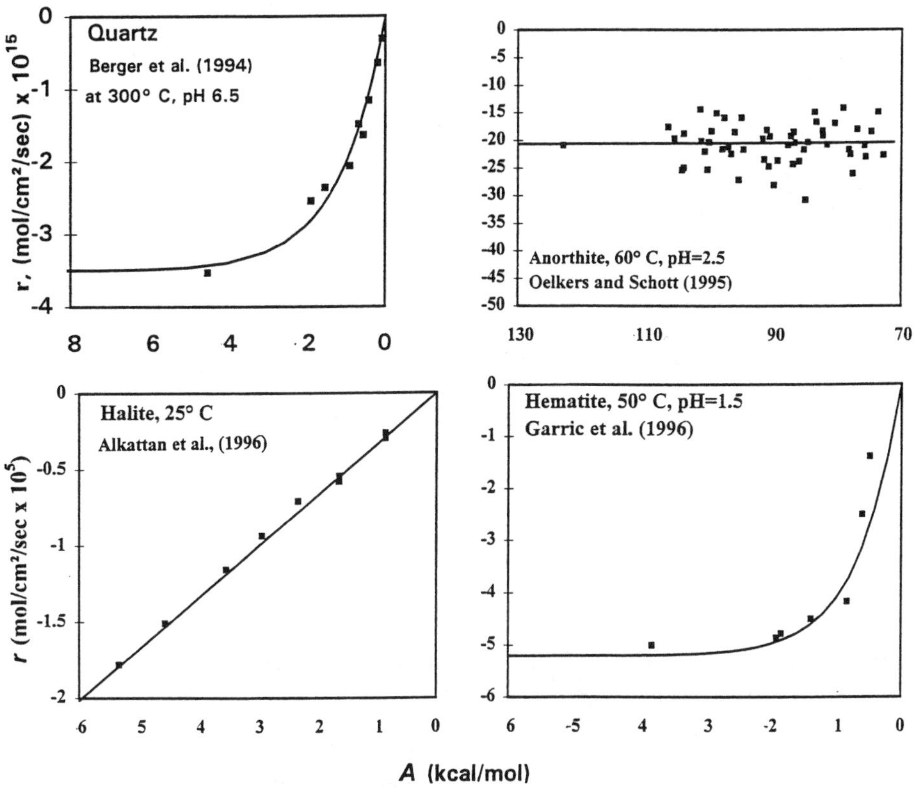

Figure 22. Steady state dissolution rates of quartz, anorthite, halite qnd hemqtite, as a function of chemical affinity at the temperatures and pH indicated in the diagrams. For the case of anorthite, constant pH rates were generated, assuming the rates depend on $a_{H^+}{}^{1.5}$. The symbols correspond to experimentally measured dissolution rates taken from the references listed in the figure, but the curves were generated using Equation (81). The apparent linear dependence of halite dissolution rates on affinity stems from the close to equilibrium conditions at which the experiments were performed.

$$\left(K^{\bullet}_{84} \left(\frac{a^{Z_M}_{H^+}}{a_{M^{+Z_M}}} \right)^{n} << 1 \right)$$

Furthermore, at far from equilibrium conditions Equation (86) becomes

$$r = k_{+} \left(\frac{a^{Z_M}_{H^+}}{a_{M^{+Z_M}}} \right)^{n} \qquad (87)$$

The identity of M and the value of n in Equations (85) to (87) depends on the stoichiometry of the exchange reaction (Reaction 84). A list of M and n for a variety of minerals is listed in Table 7.

Table 7. Parameters to be used together with Equations (85) to (87) describing the variation of mineral dissolution rates as a function of solution composition and chemical affinity.

Mineral	M	*n*	**Reference**
Quartz	*	-	Berger et al. (1994)
Albite	Al	0.333	Oelkers et al. (1994)
K-feldspar	Al	0.333	Gautier et al. (1994)
Anorthite	*	-	Oelkers and Schott (1995b)
Muscovite	Al	0.333	Oelkers et al. (1994) (deduced from variation of rates on pH)
Kaolinite	Al	1	Oelkers et al. (1994) (Acid pH)
Kaolinite	Al	2	Devidal et al. (1996) (Basic pH)
Analcime	Al	2	Murphy et al. (1996)
Kyanite	Al	0.5	Oelkers and Schott (1996)
Enstatite	Mg	0.25	Oelkers and Schott (1996)
Wollastonite	Ca	0.25	Deduced from variation of rates on pH (see Fig. 26).
Monazite	*	-	Oelkers et al. (1995)
Halite	*	-	Alkattan et al. (1996)
Calcite	*	-	Sjöberg and Rickard (1984)
Hematite	*	-	Garric et al. (1996)

* An asterisk indicates rate controlling precursor complexes are not formed by reactions containing aqueous metals of found in the mineral and thus the dissolution rates of these minerals are independent of the aqueous concentration of constituent metals at far from equilibrium conditions.

The variation of the steady state dissolution rates of several multi-oxide minerals at constant pH as a function of chemical affinity are depicted in Figure 23. A characteristic shape is apparent. At extremely far from equilibrium conditions in the limit where the mineral surface is completely saturated with precursor complexes

$$\left(K^{\bullet}_{84} \left(\frac{a_{\mathrm{H}^+}^{Z_M}}{a_{\mathrm{M}^{+Z_M}}} \right)^{n} >> 1 \right)$$

the rates are consistent with Equation (81a) and are independent of chemical affinity and the aqueous activity of M^{+Z_M}. This region of dissolution rate independence with both chemical affinity and $a_{M^{+Z_M}}$, corresponding to a 'dissolution

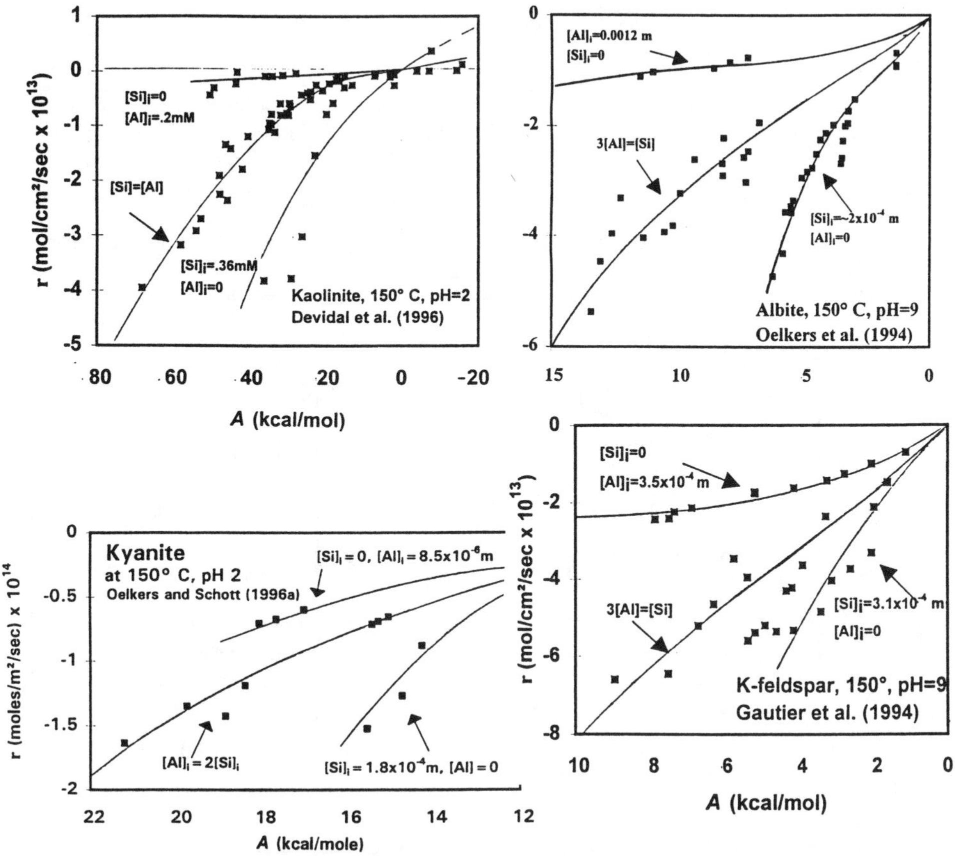

Figure 23. Steady state dissolution rates of kaolinite, albite, kyanite and K-feldspar as a function of chemical affinity at the constant temperature and pH listed in the diagrams. The symbols correspond to experimentally measured dissolution rates taken from the reference listed in the figure, but the curves were generated using Equation (85).

plateau' may not be apparent for all minerals and for all pH values. Of the various minerals depicted in Figure 23, this region is most apparent for kaolinite. At closer to equilibrium conditions $\left(K^{\bullet}_{84} \left(\frac{a_{H^+}^{Z_M}}{a_{M^{+Z_M}}} \right)^{n} \approx 1 \right)$, and these constant pH rates are proportional to $a_{M^{+Z_M}}$. As a result at these conditions, the rates in Figure 23 'appear' to depend on chemical affinity. Finally, at close to equilibrium conditions $A < 2\sigma RT$ the effect of saturation state dominates the rate behavior as a function of chemical affinity. The apparent variation of the rates illustrated in Figure 23 with chemical affinity stems from the effect of those aqueous metal ions involved in the precursor forming reaction (Reaction 84). This effect is manifested in Equations (85) to (87) by the term

$\left(\frac{a^{Z_M}_{H^+}}{a_{M^{+Z_M}}}\right)^n$. The effect of the term $\left(1-\boldsymbol{exp}\left(-\boldsymbol{A}/\sigma \boldsymbol{RT}\right)\right)$ is negligible at chemical affinities greater than ~2 kcal/mol. This observation is confirmed by the distribution of data points depicted in Figure 24, which illustrates the logarithm of constant pH, 'far from equilibrium' dissolution rates ($\boldsymbol{A}$ > 2 kcal per mol) as a function of the corresponding logarithm of $a_{M^{+Z_M}}$. The linear distribution of data in these figures confirms the fact that dissolution rates at these conditions depend only on $a_{M^{+Z_M}}$, and are not effected by $\boldsymbol{A}$. Because $a_{M^{+Z_M}}$ can appear in both the law of mass action for the formation of the

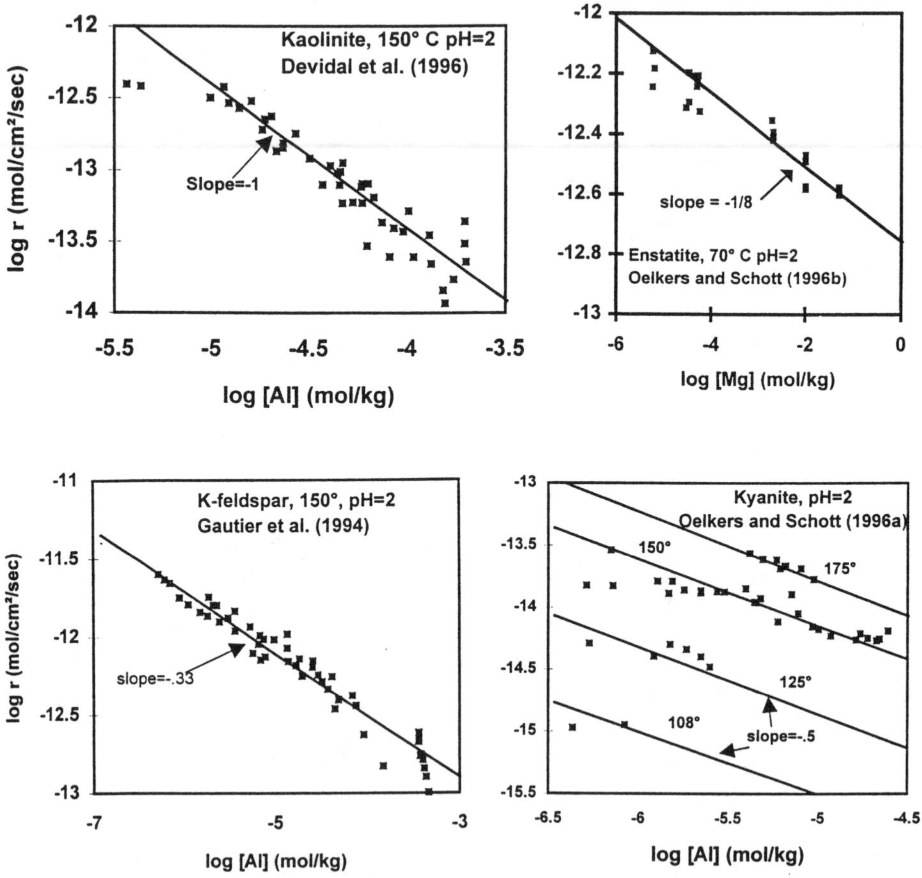

Figure 24. Logarithms of steady state far from equilibrium dissolution rates of kaolinite, enstatite, kyanite, and K-feldspar as a function of the logarithm of the total concentration of the indicated species. The symbols correspond to experimentally measured dissolution rates at the constant temperatures and pHs listed in the diagrams, and taken from the reference listed in the figure. The linear curves in these diagrams is consistent with Equation (87) and the parameters given in Table 7.

rate-controlling precursor complex (for example Al^{+3} in the case of albite) and the equation defining the chemical affinity, it can be difficult to distinguish between the two factors when attempting to interpret experimental data. Such difficulties have lead several to propose distinct rate controlling steps at different chemical affinities (Burch et al., 1993; Cama et al., 1994). It should be emphasized, however, that the origin of the chemical affinity term that appears in Equations (77), (80), (80), (85), and (86) is the effect of the reverse reaction on the overall dissolution rate. This effect is expected to be negligible except at relatively near to equilibrium conditions $\boldsymbol{A/\sigma RT}$<2.

Variation of dissolution rates as a function of pH

More attention has been focused on the variation of mineral dissolution rates on pH than any other possible solution compositional parameter (cf. Chou and Wollast, 1984, 1985; Knauss and Wolery, 1986 1988 1989, Carroll-Webb and Walther, 1988, Brady and Walter, 1989, 1992). The variation of far from equilibrium mineral dissolution rates with pH depends on the reaction mechanism. For minerals whose dissolution is controlled by precursor complexes formed by simple adsorption reactions, the variation of rates with pH is controlled by the sorption reactions to make the rate controlling precursor complexes. For the case where a minerals dissolution is controlled by rate controlling precursor complexes formed at various pH by the absorption of H^+, H_2O, or OH^- in accord with Reaction (79),

$$n_W\, H_2O + M\text{-}O = P^{\bullet}_W \tag{88a}$$

and

$$n_{OH}\, OH^- + M\text{-}O = P^{\bullet}_{OH} \tag{88b}$$

where n_W, and n_{OH} designate stoichiometric constants, and $P^{\bullet}_W$ and P_{OH} represent precursor complexes, the far from equilibrium dissolution rate as a function of pH could be expressed as

$$r_+ = k_H \frac{a_{H^+}^{n_H}}{1+K^{\bullet}_{79} a_{H^+}^{n_H}} + k_W \frac{a_{H_2O}^{n_W}}{1+K^{\bullet}_{88a} a_{H_2O}^{n_W}} + k_H \frac{a_{OH^-}^{n_{OH}}}{1+K^{\bullet}_{88b} a_{OH^-}^{n_{OH}}} \tag{89}$$

where k_H, k_W, and k_{OH} refer to a rate constant for the destruction of each of the three precursor complexes and $K^{\bullet}_{88a}$ and $K^{\bullet}_{88b}$ stand for the equilibrium constant for reactions 88a and 88b, respectively. Note that when the mineral surfaces contain relatively few precursor complexes, it assumed that the activity of water is equal to unity, and the activities of H^+ and OH^- are approximated by their concentrations this equation reduces to an empirical equation of the form (see Brantley and Chen, 1995)

$$r = k_H [H^+]^{n_H} + k_W + k_{OH} [OH^-]^{n_{OH}} \tag{90}$$

where $[H^+]$ and $[OH^-]$ refer to the concentration of the indicated aqueous species. This empirical equation can readily describe the common U-shape of dissolution rates as a function of pH. Several examples are shown in Figure 25, where a close correspondence can be seen between the measured dissolution rates and the lines corresponding to a fit of the data. It should be emphasized, however, that the general applicability of Equation (90) is limited in reactive transport calculations due to the fact that is not currently possible to predict the values of k_H, k_W, k_{OH}, n_W, and n_{OH} in the absence of experimental data.

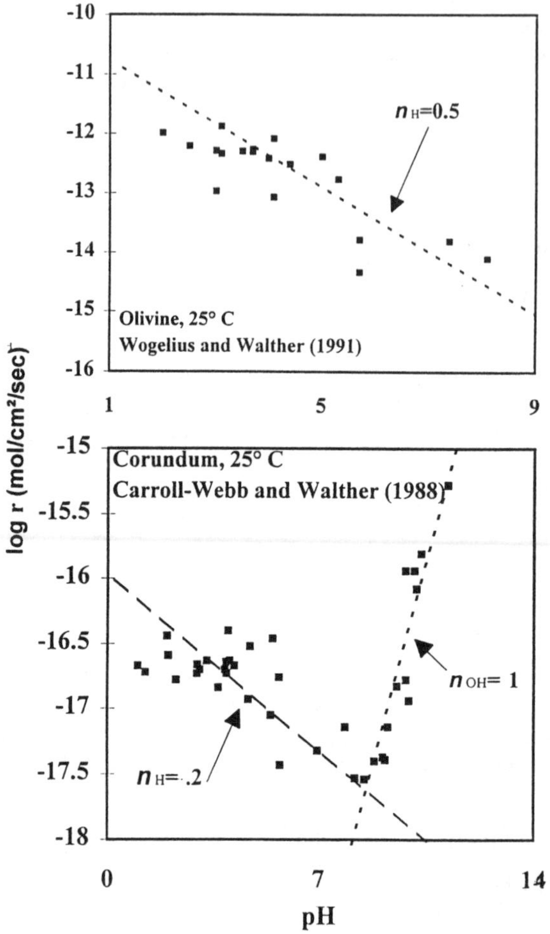

Figure 25. Steady state far from equilibrium dissolution rates of olivine and corundum as a function of pH at 25°C. The symbols correspond to experimentally measured dissolution rates taken from the reference listed in the figure, but the curves were generated (using Eqn. 90).

The variation of far from equilibrium dissolution rates for those multi-oxide minerals, whose dissolution involves the formation of rate controlling precursor complexes by the exchange of a metal for an aqueous hydrogen can be directly computed from Equation (87) by first calculating the aqueous activity ratio $\left(\frac{a_{H^+}^{Z_M}}{a_{M^{+Z_M}}}\right)$. Several comparisons between experimentally measured steady state dissolution rates as a function of pH and those generated using Equation (87) are depicted in Figure 26. A close correspondence can be seen between the computed curves and the experimental data. It should be emphasized that for these minerals, the characteristic U-shaped dependence of rates with pH is a result of the formation of various aqueous metal-hydroxide complexes in solution. It is for this reason that these dissolution rates as a function of pH exhibit a similar behavior to the solubilities of their corresponding metal-hydroxide minerals. In addition, a close correspondence can be seen between the steady state dissolution rates of albite as a function of pH at temperatures ranging from 25° to 300°C. This latter observations suggests that Equation (87) can be used to predict the pH variation of far from equilibrium dissolution rates of those multi-oxide minerals whose rates are controlled by precursor complexes formed by Reaction (84) over wide ranges of temperature.

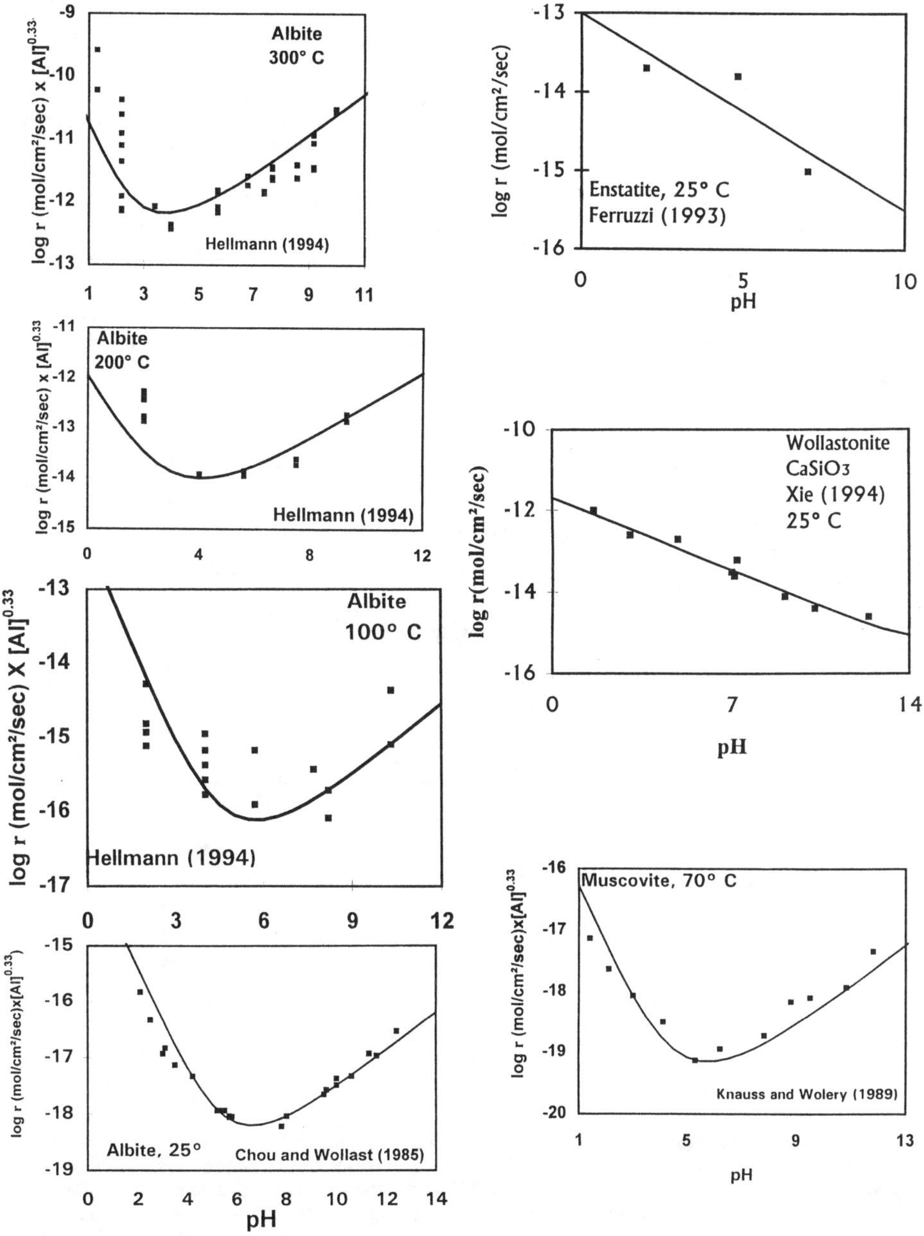

Figure 26. Steady state far from equilibrium dissolution rates of albite, enstatite, muscovite, and wollastonite as a function of pH at constant temperature. The symbols correspond to experimentally measured dissolution rates taken from the reference listed in the figure, but the curves were generated using Equation (87), parameters given in Table 7, and species distribution calculations computed using EQ3 (Wolery, 1983)—see caption of Figure 27.

Variation of dissolution rates in the presence of organic acids

The validity and utility of the expressions presented above can be demonstrated by comparing dissolution rates calculated using Equations (86) to (88) with corresponding experimental results obtained in the presence of organic acids. In addition, the effects of organic acids on dissolution reaction may be representative of effects of biogenetic activity on rates. Figure 27 compares experimentally measured plagioclase dissolution rates as a function of pH, acetate and oxalate concentration reported by Welch and Ullman (1993) with corresponding values calculated with Equation (87) together with activity ratios (a^3_{H+}/a_{Al+3}) generated from the EQ3 computer code (Wolery, 1983). All these experiments were performed at far from equilibrium conditions, so that these constant temperature rates should only depend on the rate constant, k_+, and the (a^3_{H+}/a_{Al+3}) ratio of the fluid phase. This activity ratio was computed from distribution of species calculations assuming the presence of

H^+, Na^+, Ca^{+2}, $SiO_2{}^0$, Cl^-, $NO_3{}^-$, $HNO_3{}^0$, OH^-, Al^{+3}, $Al(OH)^{+2}$, $Al(OH)_2{}^+$, $Al(OH)_3{}^0$, $Al(OH)_4{}^-$, $Al(CH_3COO)^{+2}$, $Al(CH_3COO)_2{}^+$, $H(CH_3COO)^0$, CH_3COO^-, $Al(C_2O_4)^+$, $Al(C_2O_4)_3{}^{-3}$, $C_2O_4{}^{-2}$, $H(C_2O_4)^-$, $H_2(C_2O_4)^0$, and H_2O,

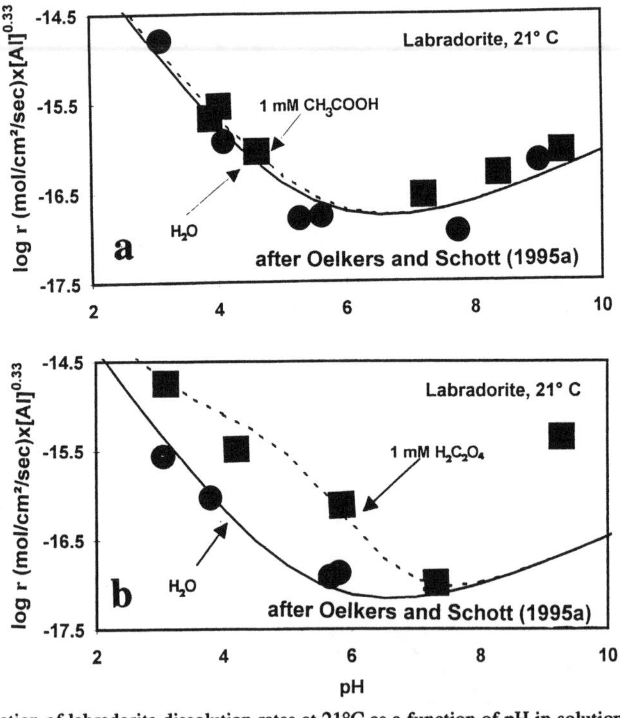

Figure 27. Variation of labradorite dissolution rates at 21°C as a function of pH in solutions free of organic acids, and (a) acetic acid and (b) oxalic acid. The y-axis in this figure is taken to be the logarithm of the product of the steady state dissolution rate in mol/cm2/sec and the total aqueous aluminum concentration in mol/kg taken to the 1/3 power (after Oelkers and Schott, 1995a). This product allows these data to be accurately presented in this two dimensional representation (see Oelkers *et al.* 1994). The symbols represent experimentally determined steady state dissolution rates reported by Welch and Ullman (1993) in organic free (circles) and organic rich (squares) solutions. Both the solid lines, that correspond to dissolution rates in organic free solutions and the dashed curves, that correspond to solutions containing organic ligands, were computed using Equation (87), together with species distribution calculations to compute (a^3_{H+}/a_{Al+3}), and the assumption that $A >> RT$ for all solutions—see text.

in solution using thermodynamic parameters available in the literature. It can be seen in Figure 27 that, with the exception of a single datum obtained in oxalate-rich solutions at pH = 9.3, there is a close correspondence between the experimental data and the computed results. The poor correspondence between the calculation and this one datum may be due to (1) the presence of a mixed ligand aluminum oxalate-hydroxide complex that is not considered in the speciation model, (2) a change in dissolution mechanism at these conditions, or (3) experimental uncertainty.

The close correspondence between the curves and symbols in Figure 27 is particularly noteworthy as the shape and position of both of the curves in Figures 27a and 27b, respectively, were computed using only one adjustable parameter, the value of k_+ in Equation (87). The shape and position of these curves is controlled totally by the degree of formation of aluminum and hydrogen bearing aqueous complexes. It follows from the results shown in Figure 27 that the reason why acetate ion has a relatively small effect on plagioclase dissolution rates is because aluminum acetate complexes do not dominate the distribution of aqueous aluminum. In contrast, the enhanced effect of oxalate on plagioclase dissolution rates appears to be directly attributable to the dramatically stronger formation of aqueous aluminum oxalate complexes. In addition, the results shown in Figure 27 suggest that the effect of oxalate on aluminosilicate dissolution rates should greatly diminish at high pH. Taking account of these results and those reported in the references listed in Table 7, it seems reasonable to presume that the variation of the dissolution rates of a large variety of minerals with chemical affinity and solution composition can be predicted accurately with transition state based equations such as Equations (80), (81), (84) and (85) together with the results of independently generated aqueous species distribution calculations.

CONCLUSIONS

The data and equations summarized above are meant to serve as a starting point for quantifying the permeabilities, dispersivities, diffusion coefficients, and mineral dissolution/precipitation rate expressions required for quantitative reactive transport calculations in natural systems. As emphasized above, each of these properties are associated with large uncertainties. Not only are the uncertainties associated with the laboratory measurement of each of these properties is great, but for the case of each of these properties other than aqueous diffusion, there are large differences between those values measured in the field and those measured in the laboratory. The lone exception, aqueous diffusional transport, stems from the fact that far less effort has been made towards quantifying directly this phenomena from natural systems. Thus it is currently not known how well (or poorly) laboratory generated equations and parameters describe diffusional transport in natural systems.

Thus, in the best case situation, even by adopting the best constrained independent input parameters, one must accept at least one order of magnitude uncertainties in each parameter, including the permeability, diffusivity, dispersivity, and mineral dissolution/precipitation rates. It follows that results of reactive transport calculations performed using such independently constrained parameters will also likely have uncertainties of several orders of magnitude, at best. Whether or not this level of uncertainty is acceptable likely depends on the system and problem addressed. Much uncertainty can be overcome by obtaining system and/or site specific physical and chemical parameters (e.g. directly measuring the permeability of the rock formation of interest). For many practical situations, however, direct measurement of system properties will be impractical. For these latter cases, reactive transport calculations may be best used to test the feasibility of certain possible reactive transport scenarios, rather than to gather quantitative data.

It is clear from the summary presented above that there is still much to be learned about the physical and chemical properties of natural reactive transport systems. Recent advances in the ability to accurately integrate the coupled reactive transport equations, and the availability of comprehensive reactive transport algorithms allow for the improved interpretation of natural systems. Such improved interpretations can be used to better constrain permeabilities, dispersivities, reactive surface areas, and mineral reaction rates in natural systems, and to design 'field experiments' for their direct measurement. For example, a tracer test performed using a non-reactive tracer can be used together with reactive transport algorithms to generate in situ permeablities and dispersivities for a system of interest. Variation of these properties with scale yields information on the degree of heterogeneity, and thus the degree of fluid channeling in the system. A subsequent tracer test performed using a reactive tracer can then be performed to determine in situ reaction rates. It seem likely that the interpretation of both natural phenomena and 'field experiments' with currently available reactive transport algorithms will lead to dramatic advances in our understanding of the physical and chemical properties of natural systems in the coming years.

ACKNOWLEDGMENTS

This manuscript is contribution No. 60 of the Dynamique et Bilan de la Terre (DBT)/Fluides dans la Croûte program of the CNRS. First I am indebted to Stacey Callahan for cheerful encouragement during the long months required to complete this chapter, and to Vala Ragnarsdottir and Bernie Wood, who generously hosted me during two visits to use the Bristol University library. I thank Carl I. Steefel, Peter C. Lichtner, and Jacques Schott for providing reviews of this chapter, and Jacques Schott, Jean-Louis Dandurand, Dimitri A. Sverjensky, William M. Murphy, Per-Arne Bjørkum, Olav Walderhaug, Stephano Salvi, Robert Gout, Marwan Alkattan, Christophe Monnin, Jean-Luc Devidal, Gilles Berger, Gleb Pokrovski, Igor Diakonov, and Jean-Marie Gautier for helpful discussions during the course of this study. Support from Centre National de la Recherche Scientifique is gratefully acknowledged.

REFERENCES

Aagaard P, Helgeson HC (1977) Thermodynamic and kinetic constraints on the dissolution of feldspars. Geol Soc Am Abstr with Program 9:873

Aagaard P, Helgeson HC (1982) Thermodynamic and kinetic constraints on reaction rates among minerals and aqueous solutions: I. Theoretical considerations. Am J Sci 282:237-285

Alkattan M, Oelkers EH, Dandurand JL, Schott J (1996) The dissolution kinetics of halite. I. the effect of temperature, saturation state and the presence of trace cations. Submitted to Chem Geol

Alt JC et al. (1993) Initial Reports: Costa Rica Rift. Proc Ocean Drilling Progr 148:1-352

Alyamani MS, Sen Z (1993) Determination of hydraulic conductivity from complete grain size distribution curves. Groundwater 31:551-555

Anderson DE, Graf DL (1976) Multicomponent electrolyte diffusion. Ann Rev Earth Planet Sci 4:95-121

Anderson DE (1981) Diffusion in electrolyte mixtures. Rev Mineral 8:211-260

Anbeek C (1992) Surface roughness of minerals and implications for dissolution studies. Geochim Cosmochim Acta 56:1461-1469

Applin KR (1987) The diffusion of dissolved silica in dilute aqueous solution. Geochim Cosmochim Acta 51:2147-2151

Applin KR, Lasaga AC (1984) The determination of SO_4^{-2}, $NaSO^{4-}$ and $MgSO_4^0$ tracer diffusion coefficients and their application to diagenetic flux calculations. Geochim Cosmochim Acta 48:2151-2162

Archie G (1942) The electrical resistivity log as an aid in determining some reservoir characteristics. Trans A I M E 146:54-62

Archie G (1952) Introduction to petrophysics of reservoir rocks. Am Assoc Petrol Geol Bull 34:943-961

Aris R (1956) On the dispersion of a solute in a fluid flowing through a tube. Proc Roy Soc London A235:67

Avellaneda M, Torquato S (1991) Rigorous link between fluid permeability and electrical conductivity, and relaxation times for transport in porous media. Phys Fluids 3:2529-2540

Balashov VN (1994) Diffusion of electrolytes in hydrothermal systems: free solution and porous media. In: Fluids in the Crust: Equilibrium and Transport properties. Chapman Hall, London, p 215-251

Bear J (1969) Hydrodynamic dispersion. In: Flow Through Porous Media, DeWiest RJM (ed) Academic Press, New York, 109-200

Bear J (1972) Dynamics of Fluids in Porous Media. Elsevier, New York, 764 p

Beard D, Weyl P (1973) Influences of texture on porosity of unconsolidated sand. Am Assoc Petrol. Geol Bull 349-369

Berg RR (1970) Method of determining permeability from reservoir rock properties. Gulf Coast Assoc Geol Soc Trans 20:303-317

Berger G, Cadoré E, Schott J, Dove P (1994) Dissolution rate of quartz in Pb and Na electrolyte solutions. Effect of the nature of surface complexes and reaction. Geochim Cosmochim Acta 58:541-551

Bernard R, Wilhelm R (1950) Turbulant diffusion in fixed beds of packed soil. Chem Eng Prog 46:233-244

Berner RA (1969) Migration of ion and sulfur within aerobic sediments during early diagenesis. Am J Sci 267:19-42

Berner RA (1980) Early Diagenesis: A Theoretical Approach. Princeton Univ Press, Princeton, NJ, 241 p

Bethke CM (1989) Modeling subsurface flow in sedimentary basins. Geol Rundsch 78:129-154.

Bird RB, Stewart WE, Lightfoot EN (1960) Transport Phenomena. J Wiley & Sons, New York, 708 p

Blackwell RJ (1962) Laboratory studies of macroscopic dispersion phenomena. Soc Petrol Eng J 225:1-8

Bloch S (1991) Empirical prediction of porosity and permeability from reservoir rock properties. Gulf Coast Assoc Geol Soc Trans 20:303-317

Blum AE (1994) Feldspars in weathering. In: Parsons I (ed) Feldspars and Their Reactions. NATO ASI C421:595-629, Kluwer, Dordrecht, The Netherlands

Blum AE, Stillings LL (1995) Feldspar dissolution kinetics. Rev Mineral 31:291-351

Bonoli L, Witherspoon PA (1968) Diffusion of aromatic and cycloparaffin hydrocarbons in water from 2 to 60°C. J Phys Chem 72:2532-2534

Bourbie T, Zinszner B (1985) Hydraulic and acoustic properties as a function of porosity in Fontainebleau sandstone. J Geophys Res 90:11524-11532

Brace WF (1977) Permeability from resistivity and pore shape. J Geophys Res 95:7072-7090

Brace WF (1980) Permeability of crystalline and argillaceous rocks. Int'l J Rock Mech Min Sci 17:241-425

Brace WF, Orange AS, Madden TR (1965) The effect of pressure on the electrical resistivity of water saturated rocks. J Geophys Res 70:5669-5678

Brace WF, Walsh JG, Frangos WT (1968) Permeability of granite under high pressure. J Geophys Res 73:2225-2236

Brady JB (1983) Intergranular diffusion in metamorphic rocks. Am J Sci 283A:181-200

Brady PV (1996) Physics and Chemistry of Mineral Surfaces. CRC Press, Boca Raton, FL, 368 p

Brady PV, Walther JV (1988) Controls on silicate dissolution in neutral and basic pH at 25°C. Geochim Cosmochim Acta 53:2823-2830

Brady PV, Walther JV (1990) Kinetics of quartz dissolution at low temperatures. Chem Geol 82:283-297

Brady PV, Walther JV (1992) Surface chemistry and silicate dissolution at elevated temperatures. Am J Sci 292:639-658

Brantley SL, Chen Y (1995) Chemical weathering rates of pyroxenes and amphiboles. Rev Mineral 31:119-172

Bredehoeft JD, Neuzil CE, Milly PCD (1983) Regional flow in the Dakota Aquifer: A study of the role of confining layers US Geol Surv Water Supply Paper 2237, 45 p

Burch TE, Nagy KL, Lasaga AC (1993) Free energy dependence of albite dissolution kinetics at 80°C, pH 8.8. Chem Geol 105:137-162

Cama J, Ganor J, Lasaga AC (1994) The kinetics of smectite dissolution. Mineral Mag 58A:140-141.

Carman PC (1937) Fluid flow through granular beds. Trans Inst Chem Eng 15:150-166

Carroll-Webb SA, Walther HV (1988) A surface complexation model for the pH dependance of corundum and kaolinite dissolution rates. Geochim Cosmochim Acta 52:2609-2623

Carmichael ISE, Eugster HP (1987) Thermodynamic Modeling of Geological Materials: Minerals, Fluids and Melts. Rev Mineral 17, 499 p

Chilingar G (1964) Relationship between porosity, permeability, and grain size distribution of sands and sandstones. In: Deltaic and Shallow Marine Deposits. van Staaten L (ed) Elsevier, New York, p 71-75

Chou L, Wollast R (1984) Study of the weathering of albite at room temperature and pressure with a fluidized bed reactor. Geochim Cosmochim Acta 48:2205-2217

Chou L, Wollast R (1985) Steady state kinetics and dissolution mechanisms of albite. Am J Sci 285:963-993

Clarke RH (1979) Reservoir properties of conglomerates and conglomeratic sandstones. Am Assoc Petrol Geol Bull 63: 799-803

Cole JA (1972) Some interpretations of dispersion measurements in aquifiers. In: Groundwater Pollution in Europe. Cole JA (ed) Water Resource Assoc, Redding, England, p 86-95

Cornell D, Katz DL (1953) Flow of gases through consolidated porous media. Ind Eng Chem 45:2145-2154

Dagan G (1984) Solute transport in heterogeneous porous media. J Fluid Mech 145:151-177

Dagan G (1989) Flow and transport in porous formations. Springer-Verlag, Berlin

Darcy H (1856) Les fontaines publique de la ville de Dijon. Dalmont, Paris, 647 p

Davis SN (1969) Porosity and permeability of natural materials. In: Flow Through Porous Media, DeWiest RJM (ed) Academic Press, New York, p 53-87

Davis SN (1988) Sandstones and shales. In The geology of North America, v. O-2 Hydrology. Back W, Rosenhein JR, Seaber PR (eds) p 323-332, Geol Soc Am, Boulder CO

Davis SN, DeWeist RJM (1966) Hydrology. J Wiley & Sons, New York

Devidal J-L, Dandurand J-L, Schott J (1992) Dissolution and precipitation kinetics of kaolinite as a function of chemical affinity (T = 150° C, pH = 2 and 7.8) In: Water Rock Interaction. Kharaka YK, Maest AS (eds) A A Balkema, Rotterdam, 1:93-96

Devidal J-L (1994) Solubilité et cinétique de dissolution/précipitation de la kaolinite en milieu hydrothermal. Approche expérimentale et modélisation. PhD Dissertation, Univ Paul Sabatier, Toulouse, France

Devidal J-L, Dandurand J-L, Schott J (1996) An experimental study of the dissolution and precipitation kinetics of kaolinite as a function of chemical affinity. Submitted to Geochim Cosmochim Acta

Dieulin A (1980) Propagation de pollution dans un aquifere alluvial, L'effet de parcours. PhD dissertation, Univ Pierre et Marie Curie, Paris

Dove PM, Crerar DA (1990) Kinetics of quartz dissolution in electrolyte solutions using a hydrothermal mixed flow reactor. Geochim Cosmochim Acta 54:955-970

Dove PM (1995) Kinetic and thermodynamic controls on silica reactitivy in weathering environments. Rev Mineral 31:235-290

Doyen P (1988) Permeability, conductivity, and pore geometry of sandstones. J Geophys Res 93:7729-7740

Dullien FAL (1992) Porous Media, Fluid Transport, and Pore Structure, 2nd edn. Academic Press, New York, 396 p

Dunn LA, Marshall WL (1969) Electrical conductances of aqueous sodium iodide and the comparitive thermodynamic properties of aqueous sodium halide solution to 800°C and 4000 bar. J Phys Chem 73:723-728

Easteal AJ, Woof LA (1985) Pressure and temperature dependance of tracer diffusion coefficients of methanol, ethanol, acetonitrile, and formamide in water. J Phys Chem 89:1066-1069

Egboka BCE, Cherry JA, Farvolden RN, Frind EO (1983) Migration of contaminants in groundwater at a landfill: A case study, 3, Tritium as an indicator of dispersion and recharge. J Hydrol 63:51-80

Ehrenburg SN (1990) Relationship between diagenesis and reservoir quality in sandstones of the Garn formation, Haltenbaken, mid-Norwegian continental shelf. Am Assoc Petrol Geol Bull 74:1538-1558

Femley AR, Weare JH (1991a) Calculation of multicomponent ionic diffusion from zero to high concentration: I. The system Na-K-Ca-Cl-SO_4-H_2O at 25°C. Geochim Cosmochim Acta 55:113-131

Femley AR, Weare JH (1991b) Calculation of multicomponent ionic diffusion from zero to high concentration: II. Inclusion of associated ion species. Geochim Cosmochim Acta 55:133-144

Ferruzzi GG (1993) The character and rates of dissolution of pyroxenes and pyroxenoids. MS Thesis, Univ California, Davis, CA

Franck EU (1956) Hochverdichter Wasserdampf III. Ioendissoziation von KCl, KOH, und H_2O in uberkritischem wasser. Zeit Phys Chem 8:192-206

Frantz JD, Marshall WM (1982) Electrical conductances and ionization constants of calcium chloride and magnesium chloride in aqueous solutions at temperatures to 600°C and pressures to 4000 bars. Am J Sci 282:1666-1693

Frantz JD, Marshall WM (1984) Electrical conductances and ionization constants of salts, acids and bases in supercritical aqueous fluids. I. hydrochloric acid from 100° to 700°C and at pressures to 4000 bars. Am J Sci 284:651-667

Freeze RA, Cherry JA (1979) Groundwater. Prentice-Hall, Englewood Cliffs, NJ, 604 p

Freyberg DL (1986) A natural gradient experiment on solute transport in a sand aquifer, 2, Spatial moments and the advection and dispersion of non-reactive tracers. Water Resources Res 22:2031-2046

Furrer G, Stumm W (1986) The coordination chemistry of weathering: Dissolution kinetics of Al_2O_3 and BeO Geochim Cosmochim Acta 50:1847-1860

Fyfe WS, Price NJ, Thompson AB (1978) Fluids in the Earth's Crust. Elsevier, Amsterdam

Garabedian SP, LeBlanc DR, Gelhar LW, Celia MA (1991) Large scale gradient tracer test in sand and gravel, Cape Cod, Massachuetts, 2, Analysis of tracer moments for a non reactive tracer. Water Resources Res 27:911-924

Garrels RM, Dreyer, RM, Howland AL (1949) Diffusion of ions through intergranular spaces in water saturated rocks. Geol Soc Am Bull 60:1809-1828

Gautier J-M, Oelkers EH, Schott J (1994) Experimental study of K-feldspar dissolution rates as a function of chemical affinity at 150°C and pH 9. Geochim Cosmochim Acta 58:4549-4560

Gelhar LW (1987) Stochastic analysis of solute transport in saturate and unsaturated media. NATO ASI Ser E128:657-700

Gelhar LW, Axness CL (1983) Three dimensional stochastic analysis of macrodispersion in aquifers. Water Resources Res 19:161-180

Gelhar LW Gutjahr AL, Naff, RL (1979) Stochastic analysis of microdispersion in aquifers. Water Resources Res 15:1387-1397

Gelhar LW, Welty C, Rehfeldt KT (1992) A critical review of data on field scale dispersion in aqueifers. Water Resources Res 28:1955-1974

Gisalson SR, Heaney PJ, Oelkers EH, Schott J (1996) Dissolution rate and solubuility of quartz and quartz/moghanite mixtures (chalchedony). Submitted to Geochim Cosmochim Acta

Goblet P (1982) Interpretation d'experiences de tracage en milleu granitique (site B) Rept LHM/RD/82/11, Centre Info Ecole Natl Sup des Mines de Paris, Fontainebleau, France

Grane FE, Gardner GHF (1961) Measurements of traverse dispersion in granular media. J Chem Eng Data 6:283-287

Haimour N, Sandall OC (1984) Molecular diffusivity of hydrogen sulfide in water. J Chem Eng Data 29:20-22

Harleman DRF, Rumer R (1963) Longitudinal and lateral dispersion in an isotropic porous media. J Fluid Mech 16:1-12

Harned HS, Owen BB (1958) The Physical Chemistry of Electrolyte Solutions. Reinhold, New York, 645 p

Hart SR (1981) Diffusion compensation in natural silicates Geochim Cosmochim Acta 45:279-291

Hayduck W, Laudie H (1974) Prediction of diffusion coefficients for non-electrolytes in dilute aqueous solutions. A I Ch E J 20:611-615

Hazen A (1893) Some physical properties of sands and gravels. Massachusetts State Board of Health 24th Annual Report

Hazlett W (1989) Correlation of reservoir rock properties in porous media. PhD dissertation, Texas A&M University, College Station, TX

Helgeson HC, Murphy WM, Aagaard P (1984) Thermodynamic and kinetic constraints on reaction rates among minerals and aqueous solutions: II. Rate constants, effective surface area, and the hydrolysis of feldspar. Geochim Cosmochim Acta 48:405-2432

Hellmann R (1994) The albite-water system: Part I. The kinetics of dissolution as a function of pH at 100, 200 and 300°C. Geochim Cosmochim Acta 58:95-611

Himmelblau DM (1964) Diffusion of dissolved gases in liquids. Chem Rev 56:527-550

Hoehn (1983) Geological interpretation of local scale tracer observations in a river ground water infiltration system. Draft Report, Swiss Fed Inst Reactor Res (EIR) Wüenlingen, Switzerland

Hoffman AW, Giletti BJ, Yoder, HS, Yund RA (1974) Geochemical transport and kinetics Carnegie Inst Washington Publ 634, Washington DC

Hoopes JA, Harleman DRF (1965) Waste water recharge and dispersion in porous media. Mass Inst Tech Hydraulics Lab Rept 75:55-60

Howard JJ (1992) Influence of authigenic-clay minerals on permeability. In: Origin, Diagenesis, and Petrophysics of Clay Minerals in Sandstones. S E P M Spec Pub 47:257-264

Ildefonse JP, Gabis U (1976) Experimental study of silica diffusion during metasomatic reactions inthe presence of water at 550°C and 1000 bars. Geochim Cosmochim Acta 40:297-303

International Critical Tables (1929) McGraw-Hill Book Co, New York

Inverson W (1990) Permeability estimation from sonic versus porosity gradients. Soc Explor Geophys Int'l Meeting, Extended Abstr 1:106-109

Inverson W, Satchwell R (1989) Permeability: An elusive goal of production geophysics. Soc Explor Geophys Int'l Meeting, Extended Abstracts 1:576-578

Jahne B, Heinz G, Dietrich W (1987) Measurement of diffusion coefficients of sparingly soluable gases in water. J Geophs Res 90:10767-10776

Johnson AI, Morris DA (1962) Physical and hydrologic properties of water-bearing deposits from core holes in the Los Banos-Kettleman City area, California. US Geol Surv Open-File Rept, Denver, CO

Johnson D, Koplik J, Dashen R (1987) Theory of dynamic permeability and tortuousity in fluid saturated fluid media. J Fluid Mech 176: 379-402

Johnson KS (1981) The calcualtion of ion pair diffusion coefficients: A comment. Mar Chem 10:195-208

Katz A, Thompson A (1986) Quantitative prediction of permeability in porous rocks. Phys Rev B 34:8179-8181

Keller GV (1960) Physical properties of tuffs of the Oak springs formation, Nevada. US Geol Surv Prof Paper 400-B

Kies B (1981) Solute transport in unsaturated field soil and in groundwater. PhD Dissertation New Mexico State Univ, Las Cruces, NM

Klinkenberg LJ (1951) Analogy between diffusion and electrical conductivity in porous rocks. Geol Soc Am Bull 62:559-567

Klotz D, Seiler KP, Moser H, Neumaier F (1980) Dispersivity and velocity relationships from laboratory and field experiements. J Hydrology 45:169-184

Knauss KG, Wolery TJ (1986) Dependence of albite dissolution kinetics on pH and time at 25°C and 70°C. Geochim Cosmochim Acta 50:2481-2497

Knauss KG, Wolery TJ (1988) The dissolution kinetics of quartz as a function of pH and time at 70°C. Geochim Cosmochim Acta 52:43-53

Knauss KG, Wolery TJ (1989) Muscovite dissolution kinetics as a function of pH and time at 70°C. Geochim Cosmochim Acta 53:1493-1502

Konikow LF, Bredehoft JD (1974) Modelling flow and chemical quality changes in an irregated stream aquifer system. Water Resources Res 10:546-562

Koplik J, Lin C, Vermette M. (1984) Conductivity and permeability from microgeometry. J Appl Phys 56:3127-3131

Kozeny J (1927) Über kapillare Leitung des Wassers im Boden. Sitzungber Oesterr Akad Weiss Math Naturwiss Kl Abt/ 2a. 136:271-307

Kreft A, Lenda A, Turek B, Zuber A, Czauderna K (1974) Determination of effective porosities by the two well pulse method. Isot Tech Groundwater Hydrol Proc Symp 2:295-312

Krumbein W, Monk G (1942) Permeability as a function of the size of unconsolidated sands. A I M E Petrol Trans 151:153-163

Labotka TC (1991) Chemical and physical properties of liquids. Rev Mineral 26:43-104

Lasaga AC (1979) The treatment of multi-component diffusion and ion pairs in diagenetic fluxes. Am J Sci 279:324-346

Lasaga AC (1981) Transition state theory. Rev Mineral 8:135-169

Lasaga AC (1995) Fundamental approaches in describing mineral dissolution and precipitation rates. Rev Mineral 31:23-86

Lau LK, Kaufman WJ, Todd DK (1957) Studies of dispersion in a radial flow system, Canal Seepage Research:Dispersion Phenomena in Flow through Porous Media, Progress. Rept I E R Ser 93 Dept Eng and Scholl Pub Health, Univ California, Berkeley, CA

Leland DF, Hillel D (1981) Scale effects on measurement of dispersivity in a shallow unconfined aquifer, paper presented at Chapman Conference on Spatial Veriability in Hydrologic Modelling AGU, Fort Collins, CO

Leonards GA (1962) Engineering properties of soils. In: Foundation Engineering. Leonards GA (ed) McGraw-Hill, New York, p 66-240

Lerman A (1979) Geochemical Processes in Water and Sediment Environments. John Wiley & Sons, New York, 481 p

Li YH, Gregory S (1974) Diffusion of ions in seawater and in deep-sea sediments. Geochim Cosmochim Acta 38:703-714

Lichtner PC (1985) Continuum model for simultaneous chemical reactions and mass transport in hydrothermal systems. Geochim Cosmochim Acta 49:779-800

Lichtner PC (1988) The quasi stationary state approximation to coupled mass transport and fluid-rock interaction in porous media. Geochim Cosmochim Acta 52:143-165

List EJ, Brooks NH (1967) Lateral dispersion in saturated porous media. J Geophys Res 72:2531-2541

MacGary LM, Lambert TW (1962) Reconnaissance of ground water resources of the Jackson Purchase region, Kentucky, US Geol Surv Hydrologic Atlas 13

Macmullin RB, Muccini GA (1956) Characteristics of porous beds and structures. A I Ch E J. 2:393-405

Maharajh DM, Walkley J. (1973) The temperature dependence of the diffusion coefficients of Ar, CO_2, CH_4, CH_3Cl, CH_3Br, and CH_3Cl_2F in water. Can J Chem 51:944-952

Manheim FT (1970) The diffusion of ions in unconsolidated sediments. Earth Planet Sci Lett 9:307-309

Manheim FT, Waterman LS (1974) Diffusimitry (diffusion constant estimation) on sediment cores by resistivity probe. Initial Reports of the Deep Sea Drilling Project, 22:662-670

Marshall, TJ (1958) A relation between permeability and the size distribution of pores. J Soil Sci 9:1-8

Marine IW (1966) Hydraulic correlation of fracture zones in buried crystalline rocks at the Savana river plant, near Aiken, South Carolina. US Geol Surv Prof Paper 50-D:223-227

Marion D, Nur A, Alabert F (1989) Modelling the relationships between sonic velocity, porosity, permeability, and shaliness in sand, shale, and shaly sand. SPWLA Logging Symp Proc 13:1-12

Masch FD, Denny KJ (1966) Grain size distribution and its effect on the permeability of unconsolidated sands. Water Resources Res 2:665-677

McCreesh C, Etris E, Brumfeild D, Ehrlich R (1988) Relating thin sections to permeability, mercury porosimetry, formation factor, and tortuousity. Am Assoc Petrol Geol Bull 72: 221-222

Mercado A (1967) The spreading pattern of injected water in a permeability stratified aquifer. Symposium of Haifa: Artifical recharge and Management of aquifers. IASH-AISH Pub 72:23

Mercado A (1984) A note on micro and macrodispersion. Groundwater 22:790-791

Miller DG (1966) Application of irreversible thermodynamics to electrolyte solutions. I determination of ionic transport coefficients l_{ij} for isothermal vector transport processes in binary systems. J Phys Chem 70:2693-2659

Miller DG, Rard JA, Eppstein LB, Robinson RA (1980) Mutual diffusion coefficients, electrical conductances, osmotic coefficients, and ionic transport coefficients lij for aqueous $CuSO_4$ at 25°C. J Soln Chem 9:467-496

Miller DG, Rard JA, Eppstein LB, Albright JG (1984) Mutual diffusion coefficients, and ionic transport coefficients l_{ij} of $MgCl_2$-H_2O at 25°C. J Phys Chem 88:5739-5748

Miller DG, Ting AW, Rard JA, Eppstein LB (1986) Ternary diffusion coefficients of the brine systems NaCl(0.5M)-Na_2SO_4(0.5M)-H_2O and NaCl(0.489)-$MgCl_2$-H_2O (seawater composition) at 25°C. Geochim Cosmochim Acta 50:2397-2403

Murphy WM, Helgeson HC (1987) Thermodynamic and kinetic constraints on reaction rates among minerals and aqueous solutions. III. Activated complexes and pH-dependence of the rates of feld-spar, pyroxene, wollastonite, and olivine hydrolysis. Geochim Cosmochim Acta 51:3137-3153

Murphy WM, Oelkers EH, Lichtner PC (1989) Surface reaction versus transport control of mineral dissolutionand growth rates in geochemical processes. Chem. Geol. 78:357-380

Murphy WM, Pabalan RT, Prikryl JD, Goulet CJ (1996) Reaction kinetics and thermodynamics of dissolution and growth of analcime and clinoptilolite, Am J Sci 296:(in press)

Murray RC (1960) Origin of porosity in carbonate rocks. J Sed Pet 30:58-84

Muskat M (1937) The flow of homogeneous fluids through porous media. McGraw-Hill, New York.

Naff RL, Yeh T-CJ, Kemblowski MW (1988) A note on the recent natural gradient tracer test at the Borden site. Water Resources Res 24:2099-2103

Nagy KL, Blum AE, Lasaga AC (1992) Dissolution and precipitation kinetics of kaolinite at 80°C and pH 3: The dependence on solution saturation state. Am J Sci 291:649-686

Nagy KL, Lasaga AC (1992) Dissolution and precipitation kinetics of gibbsite at 80°C and pH 3: The dependence on solution saturation state. Geochim Cosmochim Acta 56:3093-3111

Nagy KL, Lasaga AC (1993) Simultaneous precipation kinetics of kaolinite and gibbsite at 80°C and pH 3. precipitation kinetics of gibbsite at 80° and pH 3: The dependence on solution saturation state. Geochim Cosmochim Acta 57:4329-4335

Nagy KL, Steefel CI, Blum AE, Lasaga AC (1992) Dissolution and precipitation kinetics of kaolinite: Initial results at 80 °C with application to porosity evolution in a sandstone. In: Meshri ID, Ortoleva PJ (eds) Prediction of Reservoir Quality through Chemical Modeling. Am Assoc Petrol Geol Memoir 49:85-101

Neuman SP (1990) Universial scaling of hydraulic conductivies and dispersivities in geological media. Water Resources Res 26:1749-1758

Neuman SP, Winter CL, Newman CM (1987) Stocastic theory of field-scale Fickian dispersion in anisotropic porous media. Water Resources Res 23:453-466

Neuman SP, Zhang Y-K (1990) A quasi-linear theory of non-Fickian and Ficikian subsurface dispersion. I. Theoretical analysis and application to isotropic media. Water Resources Res 26:887-902

Neuzil CE (1993) Low fluid pressure within the Pierre Shale: A transient response to erosion. Wat. Res.Res. 29:2007-2020

Neuzil CE (1994) How permeable are clays and shales? Wat. Res.Res. 30:145-150

New Zealand Ministry of Works and Development (1977) Movement of contaminents into and through the Heretaunge Plains aquifer, report, Wellington, New Zealand

Nigrini A (1970) Diffusion in rock alteration systems. I. Prediction of limiting equivalent ionic conductances at elevated temperatures. Am J Sci 269:65-85

Norton D, Knapp R (1977) Transport phenomena in hydrothermal systems: The nature of porosity. Am J Sci 277:913-936

Oakes DB, Edworthy DJ (1977) Field measurement of dispersion coefficients in the United Kingdom. In: Groundwater Quality, Measurement, Prediction, and Protection. Water Resources Centre, Redding, UK, p 327-340

Oelkers EH (1987) Calculation of the transport and thermodynamic properties of aqueous species at elevated pressures and temperatures. PhD Dissertation Univ California, Berkeley, 221 p

Oelkers EH (1991) Calculation of diffusion coefficients for aqueous organic species at temperatures from 0 to 300°C. Geochim Cosmochim Acta 55:3515-3529

Oelkers EH, Helgeson HC (1988) Calculation of the thermodynamic and transport properties of aqueous species at high temperature and pressure: Aqueous tracer diffusion coefficients of ions to 1000°C and 5 kb. Geochim Cosmochim Acta 52:63-85

Oelkers EH, Helgeson HC (1989) Calculation of the transport properties of aqueous species at pressures to 5 KB and temperatures to 1000°C. J Soln Chem 18:601-640

Oelkers EH, Schott J (1992) The dissolution rate of albite as a function of chemical affinity and the stoichiometry of activated complexes in aluminosilicate dissolution reactions. Geol Soc Am Abstr with Program 24:A-207

Oelkers EH, Schott J, Devidal J-L (1994) The effect of aluminum, pH, and chemical affinity on the rates of aluminosilicate dissolution reactions. Geochim Cosmochim Acta 58:2011-2024

Oelkers EH, Schott J (1994) Experimental study of kyanite dissolution rates as a function of Al and Si concentration. Mineral Mag 58A:659-660

Oelkers EH, Helgeson HC, Shock EL, Sverjensky DA, Johnson JW, Pokrovskii VA (1995) Calculation of the Gibbs free energies of minerals, gases and aqueous species to 1000°C and 5 KB. Journal of Chemical and Physical Reference Data 24:1401-1560

Oelkers EH, Schott J (1995a) The dependence of silicate dissolution rates on their structure and composition. Water Rock Interaction. Oleg C, Kharaka YK (eds) A A Balkema, Rotterdam, p 153-156

Oelkers EH, Schott J (1995b) Experimental study of anorthite dissolution and the relative mechanism of feldspar hydrolysis. Geochim Cosmochim Acta 59:5039-5053

Oelkers, EH, Bjørkum PA, Murphy WM (1996) A petrographic and computational investigation of quartz cemention and porosity reduction in North Sea sandstones Am J Sci 296:420-452

Oelkers EH, Schott J (1996a) Experimental study of kyanite dissolution as a funciton of chemical affinity and solution composition. Submitted to Geochim Cosmochim Acta

Oelkers EH, Schott J (1996b) Experimental study of enstatite as a funciton of chemical affinity and solution composition (in preparation)

Othmer DF, Thakar MS (1953) Correlating diffusion coefficients in liquids. Ind J Chem Eng 46:598-608

Panda MN, Lake LW (1994) Estimation of single phase permeability from parameters of particle size distribution. Am Assoc Petrol Geol Bull 78:1028-1039

Papadopulos SS, Larson SP (1978) Aquifer storage of heated water. II. Numerical simulation of field results. Groundwater 16:242-248

Paytakes AC, Chi T, Turian R (1973) A new model for granular porous media. AIChE. J. 19:58-76.

Pfankuch HO (1963) Rev Inst Franc Pet 18:215

Philip JR (1963) Aust J Phys 16:287

Phillips OM (1991) Flow and reactions in premeable rocks. Cambridge Universisty Press, Cambridge, UK

Pickens JF, Grisak GE (1981) Scale dependent dispersion in a stratified granular aquifer. Water Resources Res 17:1191-1211

Piersol RJ, Workman LE, Watson MC (1940) Porosity, total liquid saturation, and permeability of Illinois oil sands. Illinois Geol Surv Rep Invest 67

Pitzer KS (1979) Activity coefficients in electrolyte solutions. CRC Press, Boca Raton, FL

Plinder GF (1973) A galerkin-finite element simulation of groundwater contamination on Long Island. Water Resources Res 9:1657-1669

Pratt HR, Black AD, Brace, WF, Norton DL (1974) In situ joint permeability of a granite. EOS Am Geophys Union Trans 55:433

Quist AS, Franck EU, Jolley HR, Marshall WL (1963) Electric conductances of aqueous solutionsat high temperature and pressure. I. The conductances of potassium sulfate-water solutions from 25 to 800°C and at pressures up to 4000 bars. J Phys Chem 67:2453-2458

Quist AS, Marshall WL (1965) Assignment of limiting equivalent conductances for single ions to 400°C. J Phys Chem 69:2984-2987

Quist AS, Marshall WL (1966) Electrical conductances of aqueous solutions at high temperatures and pressures. III. The conductances of potassium bisulfate solutions from 0 to 700°C and at pressures to 4000 bars. J Phys Chem 70:3714-3725

Quist AS, Marshall WL (1968a) Electrical conductances of aqueous sodium chloride solutions from 0 to 800°C and at pressures to 4000 bars. J Phys Chem 72:684-703

Quist AS, Marshall WL (1968b) Electrical conductances of aqueous hydrogen bromide solutions from 0 to 800°C and at pressures to 4000 bars. J Phys Chem 72:1545-1552

Quist AS, Marshall WL (1968c) Electrical conductances of aqueous sodium bromide solutions from 0 to 800°C and at pressures to 4000 bars. J Phys Chem 72:2100-2105

Quist AS, Marshall WL (1968d) Ionization equilibria in ammonia-water solutions to 700°C and to 4000 bars of pressure. J Phys Chem 72:3122-3128

Quist AS, Marshall WL, Jolley HR (1965) Electrical conductances of aqueous solutions at high temperatures and pressures. II. The conductances and ionization constants of sulfuric acid-water solutions from 0 to 800°C and at pressures to 4000 bars. J Phys Chem 69:2726-2735

Rajaram H, Gelhar LW (1991) Three-dimensional spatial moments analysis of the Borden tracer test. Water Resources Res 27:1239-1251

Rard JA, Miller DG (1979a) The mutual diffusion coefficients of $NaCl-H_2O$ and $CaCl_2-H_2O$ at 25°C from Raleigh interferometry. J Soln Chem 8:701-716

Rard JA, Miller DG (1979b) The mutual diffusion coefficients of $Na_2SO_4-H_2O$ and $MgSO_4-H_2O$ at 25°C from Raleigh interferometry. J Soln Chem 8:755-766

Rard JA, Miller DG (1980) The mutual diffusion coefficients of $BaCl_2-H_2O$ and $KCl-H_2O$ at 25°C from Raleigh interferometry. J. Chem. Eng. Data 25:211-215

Rima DR, Meisler H, Longwill S (1962) Geology and hydrology of the Stockton Formation in southeastern Pennsylvania. Pennsylvania Topographic Geol Surv, Ground Water Rept W-14

Rimstidt JD, Barnes HL (1980) The kinetics of silica-water interactions. Geochim Cosmochim Acta 44:1683-1700

Ritzert G, Franck EU (1968) Elekrisch Leifahigkeit wassriger Losungen bei hohen Temperaturen und Drucken, I. KCl, $BaCl_2$, $Ba(OH)_2$, und $MgSO_4$ bis 750°C und 6 Kbar. Ber Bunsen Physik Chem 72:798-807

Roberts PV, Reinhard M, Hopkins GD, Summers RS (1981) Advection-dispersion-sorption models for simulating the transport of organic contaminents. Paper presented at Int'l Conf Groundwater Quality Research, Rice University, Houston, TX

Robinson RA, Stokes R (1968) Electrolyte Solutions. Butterworths, London

Robson SG (1974) Application of digital profile modelling techniques to groundwater solute transport at Barstow, California. US Geol Surv Water Resources Invest 46-73

Rousselot D, Sauty JP, Gaillard B (1978) Etude hydrogéologique de la zone industreille de Blyes-Saint Vulbas, rapport préliminaire no. 5: Caracteristiques hydrodynamique du système aquifère. Bur Res Geol Min, Rept Jal 77/33, Orleans, France

Saffman PG (1959) A theory of dispersion in a porous medium. J Fluid Mech 6:321-349

Saffman PG (1960) Dispersion due to molecular diffusion and macroscopic mixing in flow through a network of capillaties. J Fluid Mech 7:194-208

Sahores JJ, Witherspoon PA (1966) Diffusion of light hydrocarbons in water from 2°C to 80°C. In: Advances in Organic Geochemistry. Int'l Ser Monogr Earth Sci 219-230

Scheidegger AE (1974) The Physics of Flow Through Porous Media, 3rd edn. University of Toronto Press, Toronto, 313 p

Schott J (1972) Sur la mesure des coefficients de diffusion ordinaire des sels ionisés dans les lits poreux. C R Acad Sci Paris 275:795-798

Schott J, Oelkers EH (1995) Dissolution and crystallization rates of silicate minerals as a function of chemical affinity. Pure Appl Chem 67:603-610

Schwartz FW (1977) Macrodispersion in porous media: The controlling factors. Water Resources Res 13:743-754

Schwartz L, Banavar J (1989) Transport properties of disordered continuous systems. Phys Rev B39:11965-11970

Sheng P, Zhou M-Y (1988) Dynamic permeability in porous media. Phys Rev Lett 61:1591-1594

Shmonov VM, Vitovtova VM, Zarubina IV (1994) Permeability of rocks at elevated temperatures and pressures. In: Fluids in the Crust: Equilibrium and Transport Properties. Chapman Hall, London, 285-313

Simpson ES (1962) Tranverse dispersion in liquid flow through porous media. US Geol Surv Prof Paper 441-C, 30 p

Sjöberg EL Rickard DT (1987) Temperature dependance of calcite dissolution kinetics between 1 and 62°C at pH 2.7 to 8.4 in aqueous solutions. Geochim Cosmochim Acta 48:485-493

Skagius K, and Neretneiks I (1986) Porosities and Diffusivities of some nonsorbing species in crystalline rocks. Water Resources Res 22:389-398

Steefel CI, Lasaga AC (1994) A coupled model for transport of multiple chemical species and kinetic precipitation/dissolution reactions with applications to reactive flow in single phase hydrothermal systems. Am J Sci 294:529-592

Sykes JF, Pahwa DS, Ward DS, Lantz DS (1983) The validation of SWENT, a geosphere transport model, in Scientific Computing. Stapleman R (ed) IMAES/North-Holland, Amsterdam, p 351-361

Tang A, Sandall OC (1985) Diffusion coefficient of chlorine in water at 25-60°C. J Chem Eng Data 30:189-191

Taylor GI (1953) Proc. Roy Soc. London Ser A 219:186

Taylor GI (1954) Proc Phys Soc London 67:857

Tester JW, Worley WG, Robinson BA, Grigsby CO, Feerer JL (1994) Correlating quartz dissolution rates in pure waterfrom 25 to 625°C. Geochim Cosmochim Acta 58:2407-2420

Tominaga T, Matsumoto S (1990) Limiting interdiffusion coefficients of some hydroxylic compounds in water from 265 to 433 K. J Chem Eng Data 35:45-47

Tominaga T, Matsumoto S, Ishii T (1986) Limiting interdiffusion coefficients of some aeromatic hydrocarbon compounds in water from 265 to 433 K. J Phys Chem 90:139-143

Turner GA (1959) Chem Eng Sci 10:14

Tyn MT, Calus WF (1975a) Diffusion coefficients in dilute binary liquid mixtures. J Chem Eng Data 20:106-109

Tyn MT, Calus WF (1975b) Temperature and concentration dependence of mutual diffusion coefficients of some binary liquid systems. J Chem Eng Data 20:310-316

Unver AA, Himmelblau DM (1964) Diffusion coefficients of CO_2, C_2H_4, C_3H_6, and C_4H_8 in water from 6° to 65°C. J Chem Eng Data 9:428-431

Van Baaren J (1979) Quick-look permeability estimates using sidewall samples and porosity logs. SPWLA 6th European Symp Trans, 1-10

Vernik L, Nur A (1991) Lithology prediction and storage/transport properties evaluation in clastic sedimentary rocks using sizemic velocities Am Assoc Petrol Geol Bull 75:384

Versteeg GF, van Swaaij WPM (1988) Solubility and diffusivity of acid gases (CO_2, N_2O) in aqueous alkanolamine solutions. J Chem Eng Data 33:29-34

Vitovtova VM, Shmonov VM (1982) Permeability of rocks at pressures to 2000 kg/cm^{-2} and temperatures to 600°C Doklady Akad Nauk SSR 266:1244-1248 (in Russian)

Vivian JE, King CJ (1964) Diffusivities of slightly soluable gases in water. A I Ch E J 10:220-221

Walderhaug O (1994) Precipitation rates for quartz cement in sandstones determined by fluid inclusion microthermometry and temperature history modelling. J Sed Res A64:324-333

Webster DS, Procter JF, Marine JW (1970) Two well tracer test in fractured crystalline rock. US Geol Surv Water Supply Paper 1544-I

Werner et al. (1983) Nutzung von Grundwasser für Warmepumpen Versickerrungstest Aegfigen Versuch 2 1982/1983, Water Energy Management Agency of Bern, Switzerland

White AF (1995) Chemical Weathering rates of silicate mienrals in soils. Rev Mineral 31:407-462

White AF, Peterson ML (1990) Role of reactive surface area characterization in geochemical models. Chemical modelling of aqueous systems II, In: Melchior DC, Bassett RL (eds) Am Chem Soc Symp Ser 416:416-475

White AF, Brantley SL (1995) Chemical weathering rates of silicate minerals. Rev Mineral 31:1-583

Wiebenga WA et al. (1967) Radioisotopes as groundwater tracers. J Geophys Res 72:4081-4091.

Wieland E, Werhli B, Stumm W (1988) The coordination chemistry of weathering: III A potential generalization of dissolution rates of minerals. Geochim Cosmochim Acta 52:1969-1981

Wilke CR, Chang P. (1955) Correlation of diffusion coefficients in dilute solutions. A I Ch E J 1:264-273

Wilson LG (1971) Investigationon the subsurface disposal of waste effulents at inland sites. Res Develop Progress Rept 650, US Dept Interior, Washington, DC

Winsauer WO, Shearin HH, Masson PH, Williams M (1952) Resistivity of brine saturated sands in relation to pre geometry. Am Assoc Petrol Geol Bull 36:253-277

Winshaw BF, Stikes RH (1954) The diffusion coefficients and conductances of some concentrated electrolyte solutoins at 25°C. J Am Chem Soc 76:2065-2071

Wise DL, Houghton G (1966) Diffusion coefficients ten slightly soluble gases in water at 10-60°C. Chem Eng Sci 21:999-1010

Wise DL, Houghton G (1968) Diffusion coefficients of neon, krypton, xenon, carbon monoxide and nitric oxide in water at 10-60°C. Chem Eng Sci 23:1211-1216

Witherspoon PA, Saraf DN (1965) Diffusion of methane, ethane, propane, and n-butane in water from 25 to 43°. J Phys Chem 69:3752-3755

Wogelius RA (1991) Olivine dissolution at 25°C: Effects of pH, CO_2, and organic acids. Geochim Cosmochim Acta 55:943-954

Wolery TJ (1983) EQ3NR, A computer program for geochemical aqueous speciation-solubility calculations: Users guide and documentation. UCRL-53414., Lawrence Livermore National Laboratory, Livermore, CA

Wood W (1981) A geochemical method of determining dispersivity in regional groundwater systems. J Hydrol 54:209-224

Wyllie MRJ, Spangler MB (1952) Application of electrical resistivity measurements to the problem of fluid flow in porous media. Am Assoc Petrol Geol Bull 36:359-403

Xie Z (1994) Surface properties of silicates, their solubility and dissolution kinetics. PhD Dissertation, Northwestern Univ, Evanston, IL

Zaraisky GP, Balashov VN (1994) Thermal decompaction of rocks. In: Fluids in theCrust: Equilibrium and Transport Properties. Chapman Hall, London, p 253-284

Zhang Y-K, Neuman SP (1990) A quasi linear theory of non-Fickian and Fickian subsurface dispersion. 2. Application to anisotropic media and the Bordon site. Water Resources Res 26:903-913

Zindler B, Furrier G, Stumm W (1986) The coordination chemistry of weathering. II. Dissolution of Fe(III) oxides. Geochim Cosmochim Acta 50:1861-1870

Zonov SV, Zaraisky GP, Balashov VN (1989) The effect of thermal decompaction on permeability of granites, whith lithostatic pressure being slightly in excess of fluid pressure. Dokl Akad Nauk SSR 307:191-195 (in Russian)

Chapter 4

MULTICOMPONENT ION EXCHANGE AND CHROMATOGRAPHY IN NATURAL SYSTEMS

C. A. J. Appelo

Faculty of Earth Sciences, Free University
De Boelelaan 1085
1081 HV Amsterdam, The Netherlands

INTRODUCTION

Ritchie (1966) noted that it was a geologic setting that first evoked the need for a chromatographic explanation. He was referring to experiments by soil chemists who in the middle of the 19th century measured the exchange of NH_4^+ versus Ca^{2+} in soil samples. Ritchie then continued to lament that geologists are reluctant to accept the concepts of chromatography, and he invited chromatographers to apply their expertise to geological problems. Multicomponent ion exchange was indeed not given much attention in geochemistry, and soil chemistry textbooks that explain how to calculate multicomponent exchange equilibria are still rare.

There is, on the other hand, not one soil scientist who will deny the importance of ion exchange in regulating soil solution compositions. It is not reluctance to apply chromatographic theory in soil science, but rather the feeling that processes are more complicated than can be solved by the relatively simple solutions of chromatographers. Hydrogeochemists face even greater problems than soil scientists, as they have few sampling points, lack analyses of the solid material, and have not much knowledge of the flowpath and residence time of their water sample in the subsoil. The lack of material information has probably stimulated even more the search for concepts which can explain the observed response of the subsoil to water quality changes during artificial recharge. The foremost chemical process in these transient systems is multicomponent chromatography.

Multicomponent ion exchange is the basis of chromatography. It involves the competition of *all* the ions in pore water for the soil exchanger(s). Because natural exchangers show different selectivity for different cations, the ratio of sorbed over solute concentration is variable for individual cations. This implies that retardation (or transport velocity) is different for different species, and that transport will induce separation of solutes according to individual transport velocities. By definition chromatography uses this behavior in a laboratory column to separate components from a complex mixture.

The pioneering studies on multicomponent chromatography of Sillén (1951), of Klein, Tondeur and Vermeulen at the seawater conversion laboratory (Klein et al., 1967), and of Helfferich and Klein (1970), form the basis for a quantitative description. The theory was rapidly investigated for application to enhanced oil recovery (Pope et al., 1978; DeZabala et al., 1982). Numerical modeling of subsoil transport was almost immediately combined with multicomponent exchange formulations (Rubin and James, 1973; Valocchi et al., 1981a, b; and many others more recently, cf. Yeh and Tripathi, 1989, and Mangold and Tsang, 1991). Hydrochemists have at an early stage recognized that $NaHCO_3$ waters in coastal areas are connected with cation exchange (Foster, 1950; Back, 1966; Chapelle and Knobel, 1983). It may be that the analytical tools of the chromatographers were somewhat neglected as numerical models were applied very directly to subsoil transport problems.

0275-0279/96/0034-0004$05.00

However, the chromatographic concepts are also appearing in the hydrogeological literature (Schweich and Sardin, 1981; Charbeneau, 1981; 1988; Appelo et al., 1993, 1994b; Schweich et al., 1993).

The purpose of this chapter is to show how decisive the role of multicomponent cation exchange in hydrochemistry can be and to illustrate chromatographic concepts with practical examples. Applications so far have been mainly related to man-induced effects in aquifers such as artificial recharge or injection tests. However, examples of natural water qualities that are the result of chromatography are now also available (Beekman, 1991; Manzano et al., 1992; Stuyfzand, 1993; Appelo, 1994a; Walraevens and Cardenal, 1994; Hansen and Postma, 1995). The intricacies of nature require calculations with numerical models to be able to *match* what is observed. On the other hand, having chromatographic concepts and simple tools helps us to gain a fundamental understanding of the processes involved. It may also help us to recognize the broader applicability of this field in geochemistry.

The chapter begins with the basic equations to calculate multicomponent equilibria and shows that calculating these equilibria is meaningful. Much of the chromatographic theory is related to front type and front development, and this is treated for one component. The concepts are illustrated with column experiments from the literature or compared with numerical model results. Then multicomponent chromatography is considered, in the sense of having a system of linear equations that can be solved by the method of characteristics, or similar, recently developed mathematical tools. Examples from the hydrochemical literature demonstrate applicability.

EXCHANGE EQUILIBRIA AND CALCULATIONS

Soil chemical laboratories express cation exchange capacity (*CEC*) of soils in meq/100g (dry) sample. To obtain meq/l pore water, the *CEC* is multiplied with 10 times the bulk density ($10 \times \rho_b$, g/cm^3) to give meq/dm^3 sediment, and then divided by the water filled porosity (ε, -). A *CEC* of 1 meq/100g, which is representative for a fine sand, amounts to 60 meq/l pore water with $\rho_b = 1.8$ g/cm^3 and $\varepsilon = 0.3$. The pool of exchange-

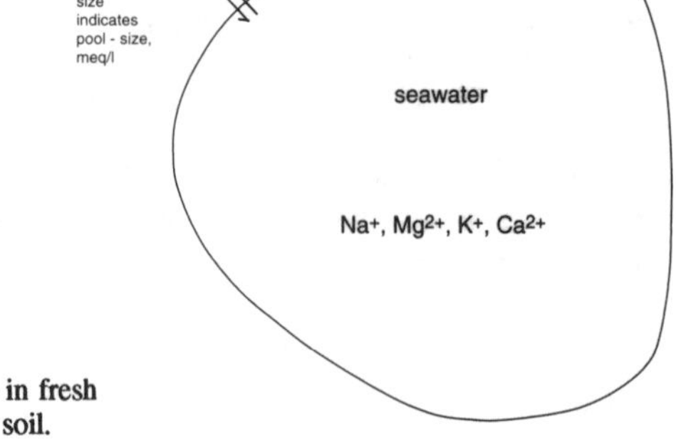

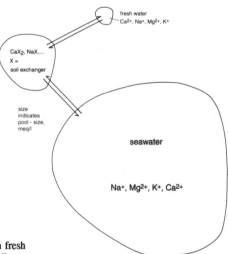

Figure 1. Comparison of solute concentrations in fresh water and seawater, and exchangeable on a sandy soil.

able cations is therefore quite high compared to concentrations in fresh water. Figure 1 shows amounts of exchangeable cations and total cation concentrations in fresh water and in seawater, and illustrates the relative proportions.

Also indicated in Figure 1 is the multicomponent character of ion exchange. The cations are in mutual competition for the exchanger sites, and the ratio of sorbed over solute concentration (the distribution coefficient K_d, -) increases with increasing selectivity. The distribution coefficient is a variable for the individual solutes, which moreover depends on the solution composition. Thus, for seawater in a fine sand, $K_d \approx 0.06$ for Na^+, while $K_d \approx 600$ for protons. In fresh water with 3 mmol/l Ca^{2+} and 1 mmol/l Na^+, $K_d \approx 0.3$ for Na^+, and at pH = 7.0, $K_d \approx 90{,}000$ for protons. The distribution coefficients were calculated from multicomponent exchanger compositions, in equilibrium with the aqueous solution. These exchanger compositions are often calculated with Newton iteration from aqueous concentrations, but analytical solution is possible by suitable rearrangement of the equations as will be shown in the following section.

Exchange equations

Equilibrium among solute and exchangeable cations is calculated with the law of mass action. For example, for Na^+ and K^+ the reaction is

$$Na^+ + K\text{-}X \leftrightarrow Na\text{-}X + K^+ \tag{1}$$

The law of mass action relates the activities as

$$K_{Na\backslash K} = \frac{[Na\text{-}X]\,[K^+]}{[K\text{-}X]\,[Na^+]} \tag{2}$$

The brackets indicate activities. For the solute ions these are a fraction of the standard state of 1 mol/kg H_2O (thus, numerically almost equal to the concentration in mol/l). For the exchangeable cations the equivalent fraction is used (Gaines and Thomas, 1953). Activity coefficients and complexation in solution and in the exchanger are neglected when in this chapter approximate solutions are calculated; they are included in the results from numerical models that are presented for comparison. The equivalent fraction $\beta_i = [i\text{-}X_{z_i}]$ of an exchangeable cation is calculated from measured concentrations and *CEC*. For example, when *CEC* = 1.2 meq/100g, and Na-X = 0.3 meq/100g, $\beta_{Na} = 0.25$. Because activity coefficients equal to 1 are assumed, $[Na\text{-}X] = \beta_{Na} = 0.25$. We also have the sum

$$\Sigma\,\beta = 1 \tag{3}$$

In general for two cations i^{z_i+} and j^{z_j+} the exchange equation is:

$$1/z_i\, i^{z_i+} + 1/z_j\, j\text{-}X_{z_j} \leftrightarrow 1/z_i\, i\text{-}X_{z_i} + 1/z_j\, j^{z_j+} \tag{4}$$

with

$$K_{i\backslash j} = \frac{\beta_i^{1/z_i}\,[j^{z_j+}]^{1/z_j}}{\beta_j^{1/z_j}\,[i^{z_i+}]^{1/z_i}} \tag{5}$$

This equation, together with (3), allows all β's to be determined in a multicomponent solution when the aqueous concentrations are known. To this end, β_j of ion j^{z_j+} is expressed as a function of β_i;

$$\beta_j = \frac{[j^{z_j+}]}{K_{i\backslash j}^{z_j}\,[i^{z_i+}]^{z_j/z_i}}\;\beta_i^{z_j/z_i} \tag{6}$$

Thus, β_k ,.... of ions k^{z_k+},.... are all expressed as function of β_i . All β's are entered in the sum

$$\beta_i + \beta_j + \beta_k + ... = 1 \tag{3a}$$

to give an equation with only β_i as the unknown. It is convenient to use a monovalent ion (Na^+) as the reference cation i^{z_i+}, because that gives a quadratic equation for the system with Na^+, K^+, Mg^{2+} and Ca^{2+} (mono- and divalent ions in general). Equation (3a) can be solved to give one positive root for β_{Na}. Subsequently the other β's are obtained from (6).

The other option—calculating the solution from known exchanger composition—follows in the same way, but now the sum of the cations in solution is the equation to solve:

$$\Sigma\, z_i c_i = N \tag{7}$$

where c_i is the concentration of i (mol/l), and N is the normality of the solution (eq/l). Again, it is advantageous to express all concentrations as function of a monovalent ion (Na^+) as it leads immediately to a quadratic (or cubic when trivalent ions are present) equation with one unknown. Example calculations can be found in Appelo and Postma (1993) and Appelo (1994b).

A particular effect of exchange equilibrium is observed if salinity changes. When exchangeable cations form a large pool that buffers the solution composition, the ratios of solute concentrations are fixed as

$$\frac{[j^{z_j+}]^{1/z_j}}{[i^{z_i+}]^{1/z_i}} = K_{i\backslash j}\frac{\beta_j^{1/z_j}}{\beta_i^{1/z_i}} \tag{5a}$$

A 10-fold change of the Na^+ concentration is accompanied by a 100-fold change of the Ca^{2+} concentration and a 1000-fold change of the Al^{3+} concentration. Dilution gives a relative increase of monovalent cations, salinity increase leads to a relative increase of polyvalent cations.

Values for the exchange coefficients $K_{Na\backslash i}$ are listed in Table 1. A range is indicated because they depend somewhat on the soil mineral and the solution composition (Bruggenwert and Kamphorst, 1982). This is implicit to the fact that different exchange sites exist on the minerals and that activity coefficients for the exchangeable cations are assumed to be 1.

The coefficient for other exchange reactions between ions i^{z_i+} and j^{z_j+} can be calculated conveniently from the values in Table 1 with

$$K_{i\backslash j} = K_{i\backslash Na} \cdot K_{Na\backslash j} \tag{8}$$

Also, it is obvious that

$$K_{i\backslash Na} = \frac{1}{K_{Na\backslash i}} \tag{9}$$

In geochemical computer models that use ion association, the exchange reaction can be calculated by splitting the exchange reaction into two association reactions of the cations with X^- (Parkhurst, 1995). A reference reaction reaction has to be specified, and all the other coefficients then follow. For example, when the reference reaction is for Na-X:

Table 1. Values for exchange coefficients with respect to Na^+ (Gaines-Thomas-convention, i.e. equivalent fraction is used for exchangeable cations). From Appelo and Postma, 1993.

Equation: $Na^+ + 1/z_i\ i\text{-}X_{z_i} \leftrightarrow Na\text{-}X + 1/z_i\ i^{z_i+}$

$$K_{Na\backslash i} = \frac{[Na\text{-}X]\,[i^{z_i+}]^{1/z_i}}{[i\text{-}X_{z_i}]^{1/z_i}\,[Na^+]} = \frac{\beta_{Na}\,[i^{z_i+}]^{1/z_i}}{\beta_i^{1/z_i}\,[Na^+]}$$

Ion i^+	$K_{Na\backslash i}$	Ion i^{2+}	$K_{Na\backslash i}$	Ion i^{3+}	$K_{Na\backslash i}$
Li^+	1.2 (0.95-1.2)	Mg^{2+}	0.50 (0.4-0.6)	Al^{3+}	0.6 (0.5-0.9)
K^+	0.20 (0.05-0.25)	Ca^{2+}	0.40 (0.3-0.6)	Fe^{3+}	?
NH_4^+	0.25 (0.2 -0.3)	Sr^{2+}	0.35 (0.3-0.6)		
Rb^+	0.10	Ba^{2+}	0.35 (0.2-0.5)		
Cs^+	0.08	Mn^{2+}	0.55		
		Fe^{2+}	0.6		
		Co^{2+}	0.6		
		Ni^{2+}	0.5		
		Cu^{2+}	0.5		
		Zn^{2+}	0.4 (0.3-0.6)		
		Cd^{2+}	0.4 (0.3-0.6)		
		Pb^{2+}	0.3		

$$Na^+ + X^- \leftrightarrow Na\text{-}X\,; \quad \log K = 0.0 \tag{10}$$

then for the other cations the reaction is

$$1/z_i\, i^{z_i+} + X^- \leftrightarrow 1/z_i\, i\text{-}X_{z_i}\,; \quad \log K = \log K_{i\backslash Na} = -\log K_{Na\backslash i} \tag{11}$$

A problem may arise when the concentration of X^- is a variable, e.g. when X^- is made proportional to a solid that may dissolve *or* precipitate, or because a mineral has a variable charge that depends on solution composition (cf. Lichtner's chapter). Charge in the aqueous solution appears no longer to be conserved because the solution loses cations if X^- increases and gains them if X^- decreases. Charge of the aqueous solution is used in geochemical models to calculate the concentration of a component (often pH), and a computation that neglects these effects may go wrong completely. However, charge must be conserved for the combination of mobile and immobile entities of the model system. An increase of immobile charge X^-, can only be due to an equivalent loss of negative charge from solution. Because the increased X^- is compensated by positive counterions that are also derived from solution, the charge transfer is balanced. A similar reasoning holds when X^- decreases. The total charge of either the mobile or the immobile part, therefore, remains equal.

Parkhurst (1995) has solved the problem of variable charge and charge transfer among solid and solution in PHREEQC by calculating explicitly the double layer counter- and co-ions that compensate the variable charge. The double layer charge is assigned to the immobile mineral to form a unit which remains electrically neutral. Similarly, when charge is associated to a specific mineral that precipitates or dissolves, an exchange model can be constructed in which only neutral molecules can be adsorbed (Appelo et al., 1996).

Determination of exchangeable cations

Much of the acceptance of chromatographic theory for soils and aquifers rests on experimental proof that exchangeable cations correspond to the calculated equilibria. Exchangeable cations are routinely analyzed by displacement with a solution that is free of the cations of interest (Page et al., 1982). The difficulty lies in separating the exchanged cations from the contribution by the soil solution and from the reactions that occur when an extractant is applied to the sediment sample.

The contribution of pore water can be high for Na^+, which often has a low distribution coefficient. Pore water contributions are sometimes reduced by displacing the pore solution with alcohol, where it is assumed that the exchangeable cations are not affected. (Note that rinsing with water is strictly taboo, since a dilution of the equilibrium solution displaces the lower charged ions from the exchange complex, cf. Eqn. 5). The influence can also be reduced by centrifuging the pore water.

Bothersome side reactions are dissolution of salts (notably gypsum) and carbonates, and oxidation of minerals (pyrite) in case of a reduced sediment. The latter reaction can be prevented by working under anaerobic conditions. Dissolution of minerals is corrected by analyzing SO_4^{2-} and HCO_3^- in the extract, and subtracting the anion concentrations from Ca^{2+} and, in case dolomite is present, from Mg^{2+}. Obviously, the displacing cation must not precipitate with this anion, which precludes use of the divalent cations Sr^{2+} and Ba^{2+}.

Monovalent ions are therefore preferred as displacers. A high concentration of at least 1M is necessary for displacement of divalent and trivalent ions, which puts a claim on purity. Ammonium and the heavier alkaline ions increase proton exchange and carbonate dissolution. It makes the HCO_3^- correction on Ca^{2+} more difficult because part of alkalinity is compensated by protons. Lithium is not commercially available in pure enough form to analyze Na^+, and it very readily peptizes clay and organic matter. Na^+ is therefore the obvious choice for displacing exchangeable cations. Exchangeable Na-X can be displaced with NH_4Cl. The technique using 1M NaCl and 1M NH_4Cl for analyzing exchangeable cations was developed by Zuur (1938), cited by Van der Molen (1958). It has been applied by Van der Molen (1958), and in studies related to poldering in The Netherlands (Hofstee, 1971). [Poldering is the reclamation of low-lying land from sea or river, practiced in the Netherlands.] We have found that it is the only rapid (batch) method that gives exchange coefficients for a large variation of aqueous solutions and aquifer sediments that are consistent with those in Table 1 (cf. Appelo et al., 1996).

For example, Van der Molen (1958) has determined exchangeable cations in two soils equilibrated with mixtures of seawater and distilled water. His results for two soils are compared in Figure 2 with model calculations using the log K_{iX} for the association reaction given in Table 2. To obtain a good fit of observed fractions over the full range from fresh water to seawater, the selectivity for Na^+ has to decrease by 0.5 log-units when going from fresh water to seawater. The selectivity coefficient for fresh water yields $\beta_{Na} = 0.6$ for seawater, whereas experimentally $\beta_{Na} \approx 0.4$.

As another example, exchangeable cations in young soils that were poldered from the sea are plotted as a function of seawater fraction in Figure 3 (Groen, 1991). Only Cl^- was analyzed in the soil solution, and the other ions were estimated assuming that the soil solution was a mixture of seawater and 3 mM $Ca(HCO_3)_2$ water. The seawater fraction was based on Cl^- concentration. Thick lines are model exchange compositions using the coefficients from Table 2. Thin, dotted lines give compositions when the coefficients from

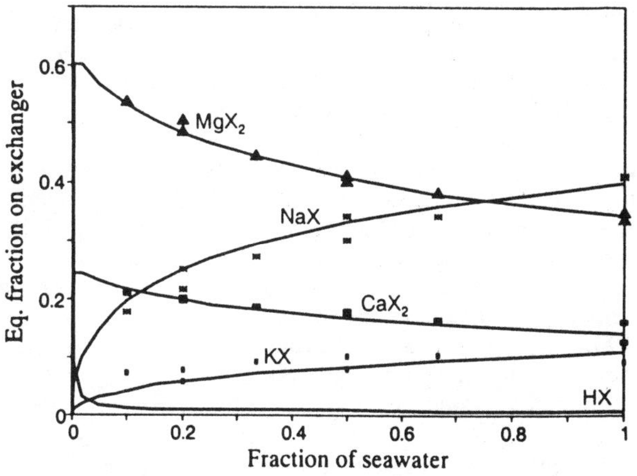

Figure 2. Fractions of exchangeable cations on two soils in mixtures of seawater and distilled water. Analyzed fractions indicated by single points; lines are modeled concentrations with exchange coefficients from Table 2. [Used by permission of the editor of *Water Resources Res.*, from Appelo (1994)].

Table 2. Association coefficients for the reaction $1/z_i\ i^{z_i+} + X^- \leftrightarrow 1/z_i\ i\text{-}X_{z_i}$ in Van der Molen's exchange experiment with fresh water and seawater (Appelo, 1994a).

	Na^+	K^+	Mg^{2+}	Ca^{2+}
log K_{IX}	$-0.5+0.5(1-\beta_{Na})$	0.902	0.307	0.465

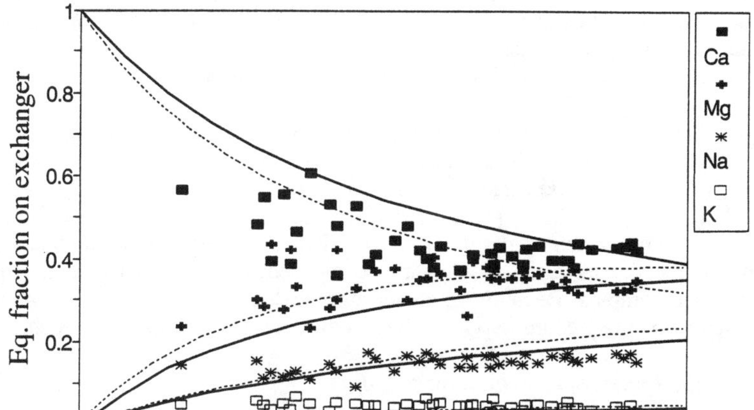

Figure 3. Fractions of exchangeable cations in polder soils as function of Cl^- concentration (seawater fraction) in the soil solution. Analyzed fractions from Groen (1991). Model lines: thick lines with variable selectivity for Na^+ (exchange coefficients from Table 2), dotted lines with 'standard' exchange coefficients (from Table 1).

Table 1 are used. The selectivity for K^+ has been increased by decreasing $K_{Na\backslash K}$ to 0.1, as it fits better for the illitic Dutch soils. There is clearly a large spread among the determined values, and the lines only indicate the trends in these values. On the other hand, doing a calculation with an estimated soil solution composition which is not far wrong, shows that the basic theory of multicomponent exchange calculations is correct.

One may speculate why the multicomponent exchange equilibria work so well, while it should not be doubted that precise measurements on individual exchangers necessitate use of activity coefficients (Elprince and Babcock, 1975; Bichkova and Soldatov, 1985; Franklin and Townsend, 1988; Sposito, 1994). It certainly can not be explained with double layer theory. The apparent selectivity, calculated from excess Na^+/Ca^{2+} in the double layer, is about equal to the values in Table 1 when the solution has seawater normality (Bolt, 1967). However, a dilution has not much effect on the ratio of mono- and divalent ions in the double layer. The apparent selectivity (log $K_{Na\backslash Ca}$) therefore increases markedly upon dilution; this in contrast to observations as in Figures 2 and 3.

CHROMATOGRAPHIC PATTERNS

With functional equilibrium relations among solute and sorbed cations, the effect of water quality changes must be considered. This is studied under the heading of ion-chromatography in chemistry and chemical engineering (DeVault, 1943; Sillén, 1951; Glückauf, 1955; Tondeur, 1969; Helfferich and Klein, 1970; Vermeulen et al., 1984; Rhee et al., 1989). The theory has been applied and verified in numerous laboratory experiments with a variety of soils and sediments (Pope et al., 1978; DeZabala et al., 1982; Schweich et al., 1983; Rainwater et al., 1987; Beekman and Appelo, 1990; Appelo et al., 1990; Bond and Phillips, 1990; Griffioen et al., 1992; Bürgisser et al., 1993; Cerník et al., 1994; Scheidegger et al., 1994). A chromatographic origin has been recognized for water quality changes during well injections (Valocchi et al., 1981a,b; Charbeneau, 1988; Bjerg et al., 1993). Also with large scale freshening of saline aquifers, a sequence of water qualities develops along a flow path that can be related to multicomponent cation exchange (Appelo and Willemsen, 1987; Beekman, 1991; Stuyfzand, 1993; Appelo, 1994a; Walraevens and Cardenal, 1994). It is clear that the theory can be applied in hydrogeology, and that it deserves attention as an important basic process for regulating water quality in aquifers.

The following parts of this chapter illustrate the development of chromatographic patterns, and show how to calculate them in a simple way and compare these also with numerical models. First, the basic equations for single solute transport are presented and illustrated. The shape of the sorption or exchange isotherm is of fundamental importance for solute transport and determines the front type along a flowline and during passage along an observation well. This also holds for multicomponent transport, but here the equations rapidly become complex. Recourse to numerical models is then appropriate, also because more complex boundary conditions and other chemical processes can be included.

Single solute transport, broadening fronts

The advection-dispersion-reaction equation in one dimension gives the change in concentration with time for a solute species:

$$\frac{\partial c}{\partial t} = -v\frac{\partial c}{\partial x} + D\frac{\partial^2 c}{\partial x^2} - \frac{\partial q}{\partial t} \tag{12}$$

where c is concentration in solution (mol/l), t is time (s), v is pore water flow velocity (m/s), x is distance (m), D is the dispersion coefficient (m^2/s), and q is the sorbed concentration, expressed as a pore water concentration (mol/l). In the case of ion exchange, $q_i = \beta_i \cdot CEC \cdot 10\rho_b/\varepsilon$ as was noted before.

Analytical solutions can be obtained for varying q/c, but the dispersion coefficient D must be assumed zero. This is what we shall pursue here, because we already concluded that with multicomponent exchange $K_d = q/c$ for the individual elements is variable.

With $D = 0$ and the slope of the isotherm dq/dc, Equation (12) becomes:

$$\left(1 + \frac{\mathrm{d}q}{\mathrm{d}c}\right)\frac{\partial c}{\partial t} + v\frac{\partial c}{\partial x} = 0 \tag{13}$$

A constant concentration means dc = 0, and therefore:

$$\mathrm{d}c = \left(\frac{\partial c}{\partial t}\right)\mathrm{d}t + \left(\frac{\partial c}{\partial x}\right)\mathrm{d}x = 0 \tag{14}$$

or

$$\left(\frac{\partial x}{\partial t}\right)_c = -\left(\frac{\partial c}{\partial t}\right)\left(\frac{\partial x}{\partial c}\right) \tag{15}$$

On combining with Equation (13), v_c, or the velocity of a constant concentration c, is obtained as

$$v_c = \frac{v_{\mathrm{H_2O}}}{1 + \dfrac{\mathrm{d}q}{\mathrm{d}c}} \tag{16}$$

for clarity the pore water flow velocity is indicated as $v_{\mathrm{H_2O}}$. Equation (16) is similar to the retardation equation in that it shows that the velocity of a solute is retarded due to sorption. However, the velocity is now variable for different concentrations and depends on the slope dq/dc.

The distance x traveled by a retarded solute c_i can be simply calculated and compared with that traveled by a conservative solute c_a (which has dq_a/dc_a = 0). In both cases we have $x = v\,t$, and therefore

$$x_{c_i} = \frac{x_{c_a}}{1 + \dfrac{\mathrm{d}q_i}{\mathrm{d}c_i}} \tag{17}$$

Figure 4 illustrates front development for a conservative solute a, and solutes i and j whichhave $q_i = c_i$ (R = 2) and $q_j = \sqrt{c_j}$ ($R = 1 + 1/2\sqrt{c}$). The initial concentration of all solutes is 1 at all x, and water enters at x = 0 with $c_i = c_j = c_a = 0$. Substance a has been flushed till 70 m, the location of the advective front. The position of 2 concentrations of each of the three substances is given in Table 3. Flushing of i is retarded, the concentration drops to zero at x = 70/(1 + 1) = 35 m. Flushing of j starts earlier than of i, the concentration begins to decrease at x = 70/(1 + 0.5) = 46.7 m. Smaller concentrations travel slower as the slope dq/dc increases in accordance with Equation (17), and these concentrations have moved less. Chemicals a and i show *indifferent* fronts, meaning that all concentrations have identical velocity. Chemical j shows a *broadening* or *diffuse* front upon elution, also called a *wave* or a *simple wave.*

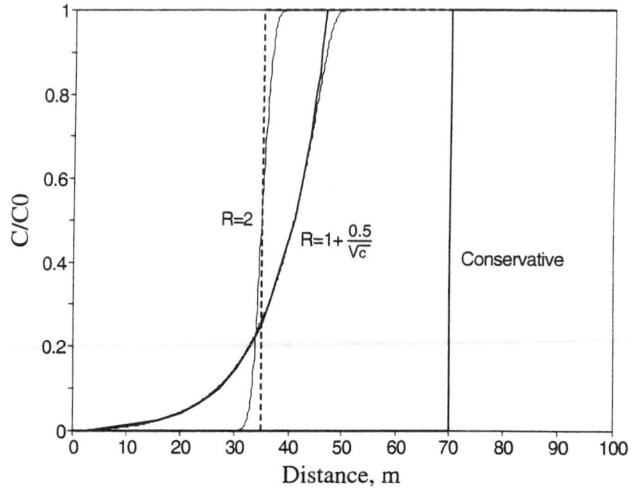

Figure 4. Flushing of a conservative chemical (q_a = 0 for all c_a), a linearly sorbed chemical with R = 2 ($q_i = c_i$), and a chemical with variable R ($q_j = \sqrt{c_j}$). The initial condition is c = 1 for all x, and c = 0 at x = 0, t > 0. Thin lines are from numerical model with Δx = 0.1 m.

Table 3. Example calculations for the position of two concentrations of three substances with different isotherms during flushing along a flowline.

c	$q_a = 0$ (dq/dc=0)	$q_i = c_i$ (dq/dc=1.0)	$q_j = c_j^{1/2}$ (dq/dc = $0.5c^{-1/2}$)
1.0	70	35	46.7
0.5	70	35	41.0

Results of a finite difference model are given for i and j with flux boundary conditions at x = 0 and dc/dx = 0 at x = ∞ (cf. example 9.14 in Appelo and Postma, 1993). The concentration of j starts to decrease in a singular point (the derivative dc/dx is not defined). It can be seen that the numerical model shows dispersion around this point where c_j starts to fall. The rest of the c_j profile is correctly followed.

Equation (17) also gives a fairly good approximation in case dispersion is present, at least when its effect is overruled by chemical dispersion (Bolt, 1982). Figure 5 shows the same case as Figure 4, with additional dispersivity α = 3.3 m. The numerical model is now more correct as it uses the numerical dispersion to mimick part of the physical dispersion (cf. Appelo and Postma, 1993). The analytical solution of Lindstrom et al. (1967) is plotted for R = 2, but it is indiscernible from the numerical model in Figure 5. The approximation for $R = 2$ by Equation (17) is incorrect as it assumes that dispersion increases linearly with x (dispersion increases with $\sqrt{x}$).

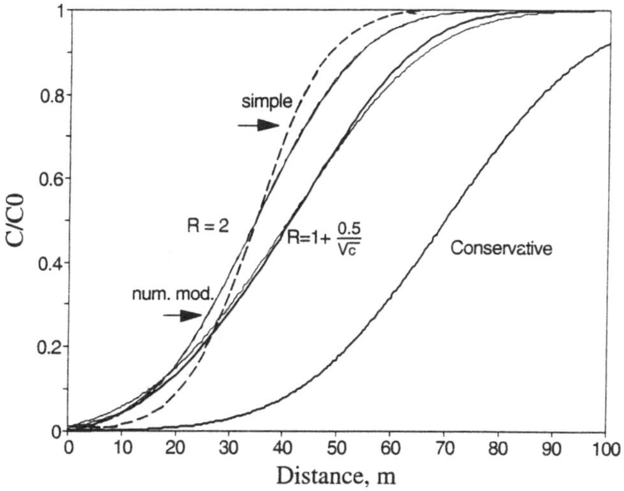

Figure 5. Flushing of two chemicals as in Figure 4, but with dispersion. Dispersivity $\alpha = 3.3$ m. Thin lines are from numerical model ($\Delta x = 0.5$ m) and from the Lindstrom et al. (1967) analytical solution for $R = 2$. Note that the simple approximation with Equation (17) for $R = 2$ is incorrect.

Sharp fronts

When chemical j (with $q_j = \sqrt{c_j}$) enters a pristine aquifer, then again, small concentrations advance more slowly than high concentrations. This would mean that the higher concentrations of j would arrive earlier at some point than the small concentrations. This is clearly impossible. Rather, the high concentrations push up the smaller concentrations, and all concentrations arrive simultaneously. Such arrivals are evocatively denoted as *self-sharpening* fronts, or *shocks*. They form the basis for *displacement* chromatography.

The front position can be calculated from the integral mass balance:

$$\Delta c \cdot (x_{H_2O} - x_j) = \Delta q \cdot x_j \qquad (18)$$

where x_{H_2O} is the position of the conservative front, and x_j the position of the sharp front for j. Δc and Δq are the changes in respectively the solute and sorbed concentrations over the front. With $x = v\,t$ the velocity of the front is obtained as in Equation (16):

$$v_c = \frac{v_{H_2O}}{1 + \dfrac{\Delta q}{\Delta c}} \qquad (19)$$

Figure 6 illustrates the fronts for chemical a and two concentrations of j along the 100 m flowline. The calculations for the front positions are given in Table 4. Note how the front for $c_j = 0.2$ ($R = 3.2$) is retarded more than for $c_j = 1.0$ ($R = 2$). Once more results of the numerical model are shown for comparison, but now the chemical term helps in counteracting numerical dispersion. The numerical model therefore almost coincides with the sharp front approximation.

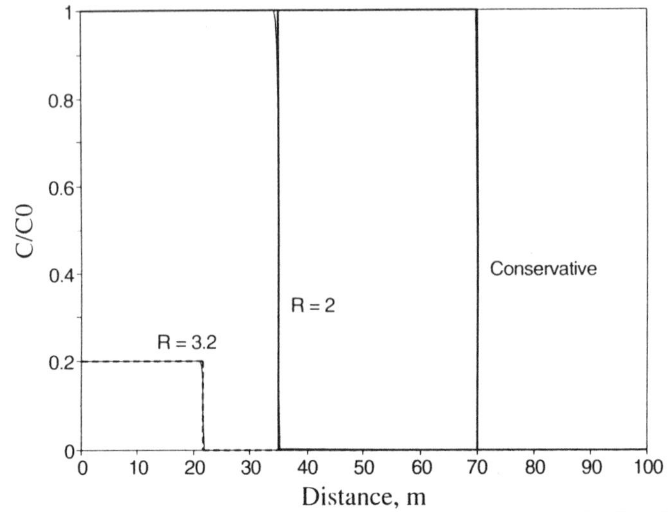

Figure 6. Transport of the chemical j from Figure 4 ($q_j = \sqrt{c_j}$). in a clean flowtube at two concentrations. Note the different retardations that follow from different $\Delta q / \Delta c$ at the two concentrations. Thin lines from numerical model ($\Delta x = 0.5$ m) almost coincide with the sharp front solution.

Table 4. Sharp front positions for two concentrations of substance j which has ($q_j = \sqrt{c_j}$).

c	x_{H_2O}	q	$\Delta q/\Delta c$	R	x_j	dq/dc
0.2	70	0.44	2.24	3.24	47.9	1.18
1.0	70	1.0	1.0	2	61.5	0.5

Two-cation exchange

With exchange, q is expressed as fraction β of *CEC*, as noted before. For a system with two cations, only one equation [(16) or (19)] suffices because the other cation is automatically obtained as the complement. Bond and Phillips (1990) have measured concentration profiles with Na^+ or K^+ displacing Ca^{2+} in an unsaturated soil column during unsteady flow infiltration. The exchange isotherms for the two ions are given in Figure 7.

The exchange isotherm for Na^+ is concave at all concentrations, hence small concentrations have higher velocity than high concentrations. For Na^+ entering the soil column, a broadening front results. For K^+ the exchange isotherm at 0.2N shows a selectivity reversal that Mansell et al. (1993) have modeled using an additional exponent for the solute concentration ratio. The isotherm has a convex part at low K^+ concentrations, and a concave part at high concentrations. Small concentrations will therefore give a sharp front. However, because with increasing K^+ concentrations the slope of the isotherm increases again, the high concentrations are retarded more, thus leading to a broadening front for the highest concentrations. The turnover point lies where the isotherm slope is equal to the slope at $K^+ = 0$.

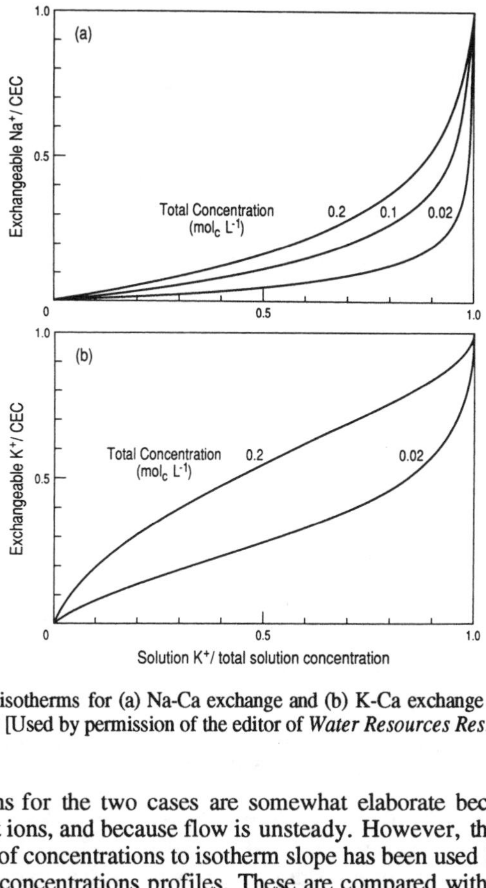

Figure 7. Exchange isotherms for (a) Na-Ca exchange and (b) K-Ca exchange at different normality *N* (indicated as $mol_c\ L^{-1}$) [Used by permission of the editor of *Water Resources Res.*, from Bond and Phillips (1990)].

The calculations for the two cases are somewhat elaborate because the exchange is among heterovalent ions, and because flow is unsteady. However, the above procedure of relating retardation of concentrations to isotherm slope has been used by Bond and Phillips (1990) to calculate concentrations profiles. These are compared with experimental data in Figure 8. Very clear is the fact that the Na^+ profile is broadening, while the K^+ front is sharp over a large domain and broadening for the highest concentrations only.

Column elution curves

It is equally simple to calculate elution curves for a column. The relative arrival time t/t_0 is calculated at fixed $x = L$, the end of column, with Equation (16):

$$t/t_0 = 1 + \mathrm{d}q/\mathrm{d}c \tag{20}$$

where t_0 is the time used by a conservative substance to pass through the column. The ratio t/t_0 is known as the *throughput ratio* (Vermeulen et al., 1984). It can also be expressed as relative volumes V/V_0, where V_0 is the pore volume of the column. If Equation (20) is condensed to just the slope of the isotherm, it indicates the pore volumes that are needed *after* the conservative front has passed. It has been termed the ψ condition (Sillén, 1951), or the *flushing factor* V^* (Appelo and Postma, 1993):

$$V^* = V/V_0 - 1 = \mathrm{d}q/\mathrm{d}c \tag{21}$$

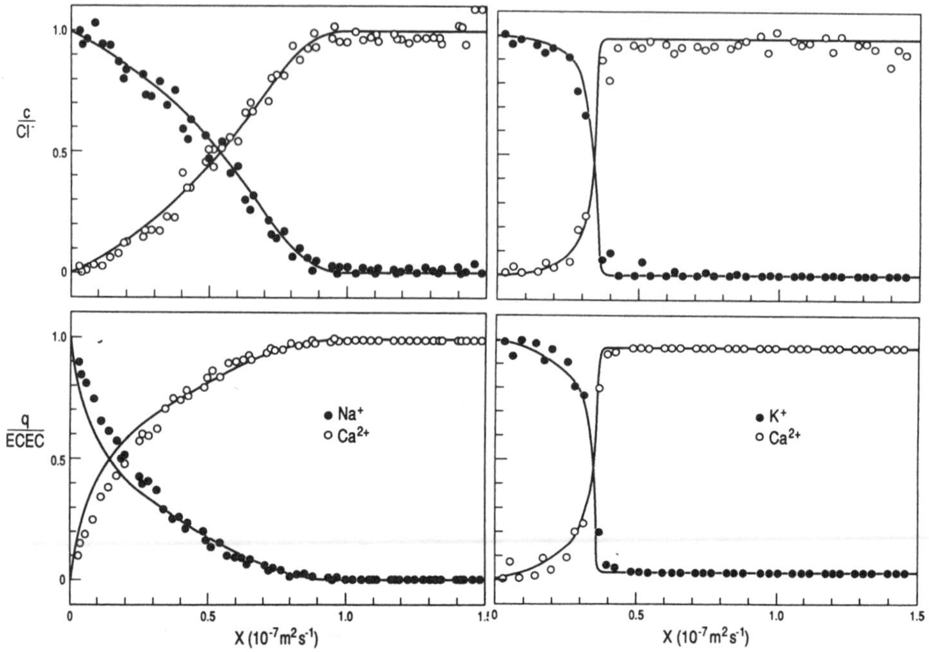

Figure 8. Comparison of measured (symbols) and modeled (lines) distributions of solution and sorbed cation concentrations following the entry of 0.2 mol/l solution of NaCl (left) and KCl (right). The X axis is a normalized distance that includes the varying flux in the unsaturated column (the columns are between 0.2 and 0.35 cm in length). The cation concentrations are normalized to total anion concentration and *CEC*. [Used by permission of the editor *Water Resources Res.*, from Bond and Phillips (1990)].

The equation is for broadening fronts, when the slope of the isotherm is smaller at the initial condition of the column than at the end. If the slope of the isotherm *decreases* towards the final condition, the final condition pushes the earlier compositions forward, and a sharp front results. The integral mass balance applies again, and a *sharp front flushing factor* is defined as:

$$V^s = \Delta q / \Delta c \tag{22}$$

Column elution curves are qualitatively pictured in Figure 9.

Sorption isotherms from elution curves

The sorption isotherm can be deduced from a measured elution curve by integration of Equation (21). For sharp fronts obviously only one point of the isotherm is obtained. For broadening fronts the whole isotherm over the range from start to final concentration can be derived (Glückauf, 1945; Sillén, 1950) For a broadening elution curve the change of q from c_1 to c_2 is:

$$q_2 - q_1 = \int_{q_1}^{q_2} dq = \int_{c_1}^{c_2} V^* \, dc \tag{23}$$

Thus from a starting condition with known q_1, q_2 can be obtained by integrating the elution curve. The procedure is illustrated in Figure 10. Note that the integration is along the vertical axis from c_1 to c_2. Also note that the integration gives a negative number when performed from c_1 (high) to c_2 (low), which agrees to the negative sum $q_2 - q_1$.

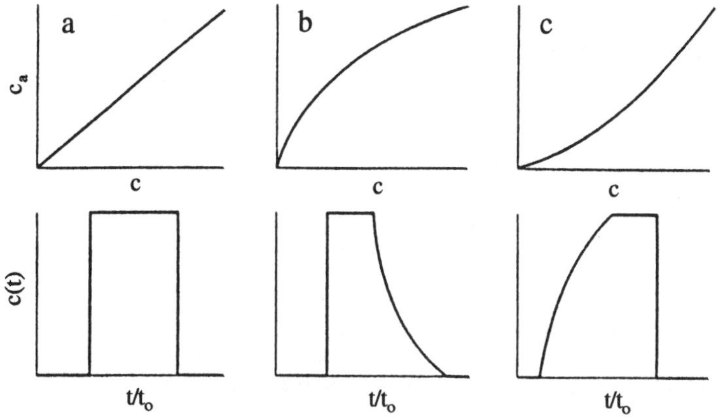

Figure 9. Schematic representation of the response of a column for different sorption isotherms (top). The column breakthrough of a step concentration change without dispersion effects is shown (below) for different isotherms: (a) linear, (b) convex, and (c) concave. [Used by permission of the editor of *Environmental Science and Technology*, from Bürgisser et al. (1993)].

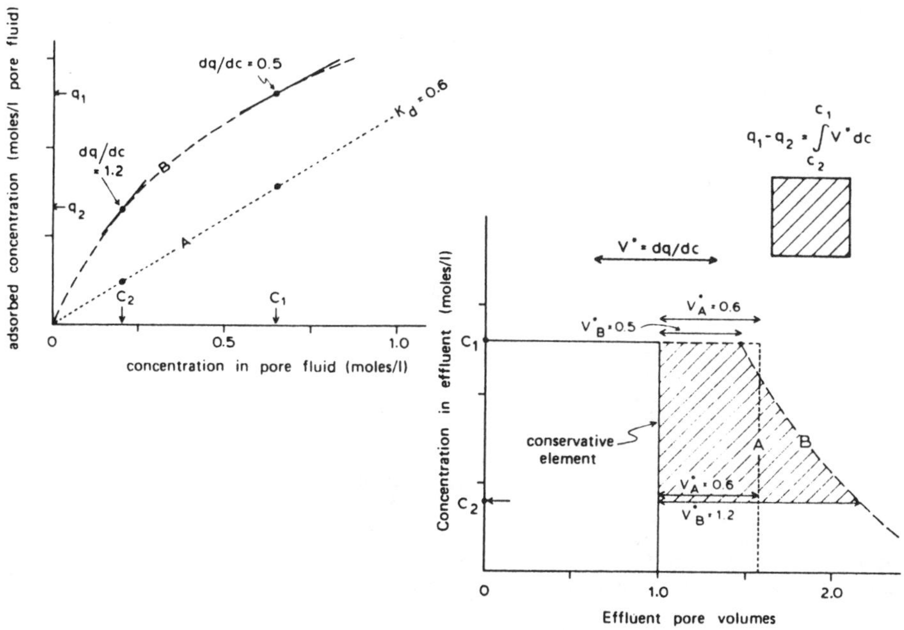

Figure 10. A broadening elution curve can be used to obtain the sorption isotherm, as illustrated for substance B. *Left:* isotherms for A (linear) and B (convex). *Right:* Elution curves for A and B, and integration area to obtain points of the curved isotherm of B. [Used by permission of Balkema, from Appelo and Postma (1993)].

The method was advocated as soon as the character of chromatographic behavior was fully realized (DeVault, 1943; Glückauf, 1945; Sillén, 1950). Several investigators have described the good agreement between sorption isotherms from elution curves and single points obtained from batch experiments (Glückauf, 1949; Duncan and Lister, 1949; Ekedahl et al., 1950; Faucher et al., 1952, 1954; Merriam et al., 1952). However, it was also found that the chromatographic method could be badly in error, and in "complete disagreement with the reliable data from the equilibrium method" (Merriam and Thomas, 1956). Reference to the method in the chemical engineering literature has disappeared since. Clearly, it is not the mathematical description of the process but improper experimentation (too high flow velocity) or incorrect interpretation (neglect of dispersion) that may result in bad data.

In recent years the method has been re-advocated for obtaining sorption or exchange isotherms for natural materials (Schweich et al., 1983; Kool et al., 1989; Griffioen et al., 1992; Bürgisser et al., 1993, 1994; Scheidegger et al., 1994). The principal attraction is that it allows to measure an entire, possibly non-linear isotherm in a one time procedure that takes not much more time than a batch experiment needs for a single point of the isotherm. A column experiment also permits to use a high solid-solution ratio that may be necessary to measure sorption when K_d is very low. Bürgisser et al. (1993) have measured a sorption isotherm of Cd^{2+} on silica sand by the chromatographic method, as shown in Figures 11 and 12. The column experiment has been accomplished at two different flow velocities to show that kinetic effects are absent. The sorption isotherm that was calculated from the elution curve is in very good agreement with results from batch experiments (Fig. 11). With a density of the sand $\rho = 2.31$ g/cm^3 and porosity $\varepsilon = 0.45$, the K_d for Cd^{2+} decreases from approximately 5 to only 0.85 at respectively 1.0 μM and 100 μM Cd^{2+}. Also dq/dc decreases from 4.2 to 0.4 between these two concentrations.

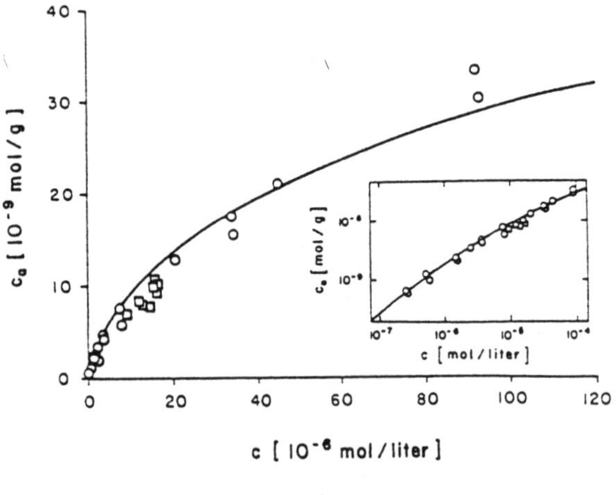

Figure 11. Adsorption isotherm of Cd^{2+} on sand (125 to 250 μm) in 10 mM $NaNO_3$. Symbols from batch experiments, solid line from column elution shown in Figure 12. [Used by permission of the editor of *Environmental Science and Technology,* from Bürgisser et al. (1993)].

Bürgisser et al. analyzed their experiment as a Cd^{2+} sorption process for which the isotherm needs to be found by experimentation. It is illustrative to interpret it as an exchange process of Cd^{2+} for Na^+ from the background electrolyte. The exchange reaction is:

$$Na^+ + 1/2\ Cd\text{-}X_2 \leftrightarrow Na\text{-}X + 1/2\ Cd^{2+} \quad (24)$$

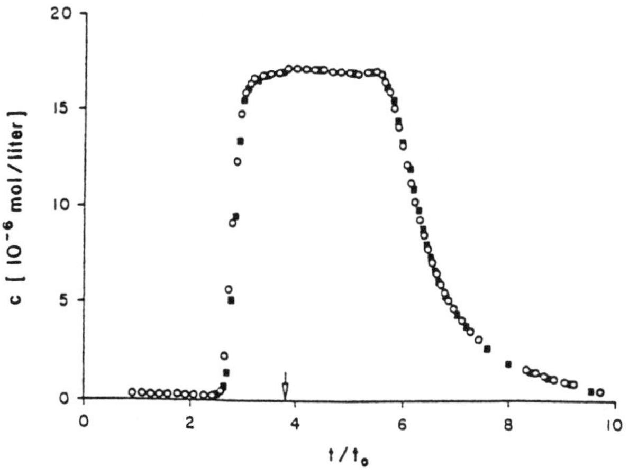

Figure 12. Breakthrough curves of Cd^{2+} in 10 mM $NaNO_3$ from a column packed with sand. Step input of 17.5 μM Cd^{2+} for $0 < t/t_0 < 3.8$. Two symbols are for two different flow velocities. [Used by permission of the editor of *Environmental Science and Technology* from Bürgisser et al. (1993)].

With $K_{Na\backslash Cd} = 0.4$ from Table 1, observed exchangeable Cd^{2+} at 17.5 μM Cd^{2+} is obtained with *CEC* = 0.056 meq/kg sand. The calculated isotherm is shown in Figure 13. It has an initial steep slope that flattens rapidly when concentrations increase above 17.5 μM. Observed exchangeable Cd^{2+} (from batch experiments and from a column experiment with 175 μM Cd^{2+}) is higher. The higher concentrations can be fitted by increasing the *CEC* to 0.146 meq/l, and increasing $K_{Na\backslash Cd}$ to 0.9 (note the very small exchange capacity of this almost pure cristobalite). This isotherm has a smaller slope at low concentrations that shows more gradual increase towards the higher isotherm points. The small initial slope gives rapid elution of the low concentrations, as is illustrated in Figure 14. Comparison with the observed elution curve suggests that two sorption sites exist on the sand, one with

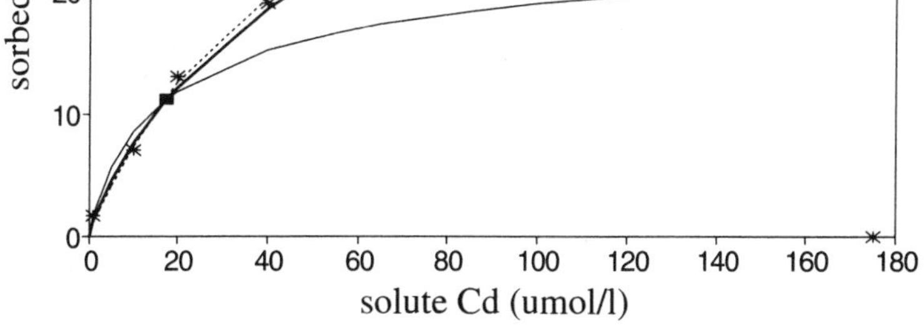

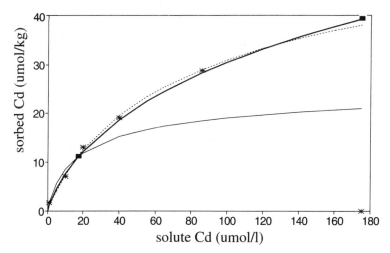

Figure 13. Cd^{2+} sorption on cristobalite sand interpreted as an ion-exchange process against Na^+ from 10mM $NaNO_3$ solution. Filled squares: data from 2 column experiments with sharp fronts; other symbols: batch experiments (Bürgisser et al., 1993). Thin line: $K_{Na\backslash Cd} = 0.4$, *CEC* = 0.056 meq/kg; dotted line: $K_{Na\backslash Cd} = 0.9$, *CEC* = 0.146 meq/kg; thick line: combined isotherm with two selectivity sites (see text).

a standard K that dominates elution of low concentrations < 5 μmol/l, and one with reduced selectivity for Cd^{2+} that determines elution of higher concentrations. The combined isotherm with 0.028 meq/kg standard sites ($K_{Na\backslash Cd}$ = 0.4) and 0.165 meq/kg low sorption sites ($K_{Na\backslash Cd}$ = 1.55) is also shown in Figure 13, and the corresponding elution curve in Figure 14.

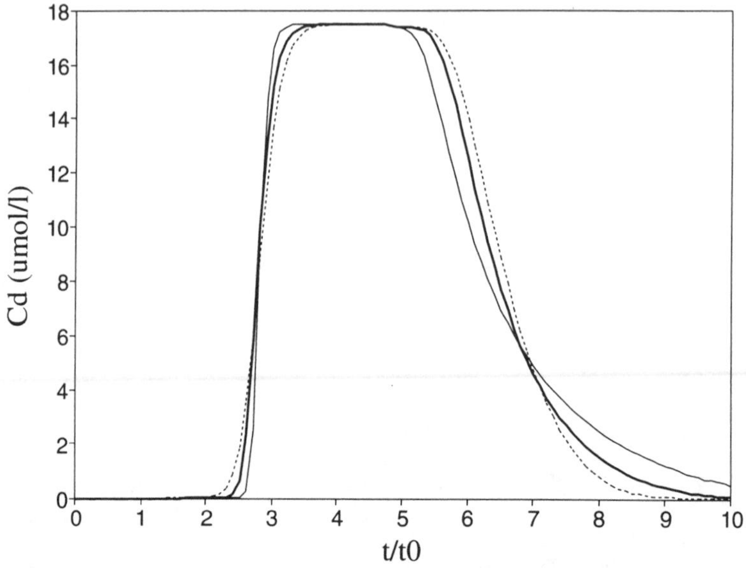

Figure 14. Calculated breakthrough curves for the column experiment of Bürgisser et al. (1993), shown in Figure 12. Three different exchange isotherms have been used, plotted in Figure 13.

The example is of interest as the chromatographic method complements the information from batch experiments. A sound interpretation requires that dispersion is very small (vL/D > 100, as in the example of Bürgisser et al.), or that the elution curve is modeled numerically with a least squares optimization of the adjustable parameters (Kool et al., 1989; Van Veldhuizen et al., 1995).

MULTICOMPONENT CHROMATOGRAPHY

With more than two exchanging cations, the isotherm slope dq/dc for each q_i becomes a variable for all c_j. Still, the mass balance given by Equation (21) for broadening fronts, or by Equation (22) for sharp fronts must hold for each ion. In other words, the flushing factors must be equal for all ions so that in a given solution composition all concentrations have the same velocity. In that case is:

$$V^{*} = \frac{dq_i}{dc_i} = \frac{dq_j}{dc_j} = ... = \frac{dq_n}{dc_n} \tag{25}$$

or for a sharp front:

$$V^{s} = \frac{\Delta q_i}{\Delta c_i} = \frac{\Delta q_j}{\Delta c_j} = ... = \frac{\Delta q_n}{\Delta c_n} \tag{26}$$

If all ions have to have equal velocity only restricted compositions are possible. When equal velocity exists for all ions according to Equation (25) or (26) the composition is said to be *coherent* (Helfferich and Klein, 1970). The problem is to find these compositions and their succesion when a column in a certain initial condition, is changed by injection of a solution with another composition.

Klein et al. (1967) noted that $V^* + 1$ in Equation (25) is identical to pore volumes eluted. At the corresponding point of the elution curve only dq/dc for the multicomponent solution has to be solved to find c. They have shown that taking the derivative of the log transform of the exchange equilibrium facilitates the problem. Equation (26) can also be solved if only shock fronts occur (Appelo et al., 1993). In both cases the number of roots for V^* or V^s is equal to n-1, where n is the total number of exchanging cations. In case of broadening fronts, these roots are succesively followed from small V^* to large V^*, stepping off when an intermediate composition is reached, and stepping on again when $V^* + 1$ equal the eluted pore volumes. For a system with homovalent cations, the complete sequence can be deduced once and for all by using the H transformation (Helfferich, 1967; Helfferich and Klein, 1970).

Self-similar solution

It is also possible to view the whole system as a Riemann problem and solve it by using the method of characteristics (Tondeur, 1969; Rhee et al., 1970, 1989) or by using the self-similar solution developed by Lax (1957, 1973) and Smoller (1983) as has been done by Charbeneau (1988) and by Thoolen and Hemker (1991). We follow here the treatment by Charbeneau (1988) as it allows for a particularly elegant and quick solution of the system of chromatographic equations.

For a multicomponent system, in which $q_i = f(c_{i,j,...})$, a set of mass balance equations must be solved. Thus, if n species compete for the exchanger sites, (13) becomes:

$$\begin{aligned}
&\frac{\partial c_1}{\partial t} + \frac{\partial q_1}{\partial c_1}\frac{\partial c_1}{\partial t} + \frac{\partial q_1}{\partial c_2}\frac{\partial c_2}{\partial t} + \;.....\; + \frac{\partial q_1}{\partial c_{n-1}}\frac{\partial c_{n-1}}{\partial t} + v\frac{\partial c_1}{\partial x} = 0 \\
&\qquad\qquad\vdots \qquad\qquad\qquad\qquad \vdots \\
&\frac{\partial c_{n-1}}{\partial t} + \frac{\partial q_{n-1}}{\partial c_1}\frac{\partial c_1}{\partial t} + \frac{\partial q_{n-1}}{\partial c_2}\frac{\partial c_2}{\partial t} + \;.....\; + \frac{\partial q_{n-1}}{\partial c_{n-1}}\frac{\partial c_{n-1}}{\partial t} + v\frac{\partial c_{n-1}}{\partial x} = 0
\end{aligned} \tag{27}$$

We define $y = x/v$ (the water flow velocity v is steady), and write (27) in matrix form as:

$$(\boldsymbol{I} + \boldsymbol{Q})\frac{\partial c}{\partial t} + \boldsymbol{I}\frac{\partial c}{\partial y} = 0 \tag{28}$$

where $\boldsymbol{I}$ is the identity matrix, $\boldsymbol{Q}$ the matrix given by the differential coefficients $(\partial q_i/\partial c_j)_{i,j=1,...,n-1}$, and $\partial c/\partial t$ and $\partial c/\partial y$ are column vectors. The matrix $(\boldsymbol{I} + \boldsymbol{Q})$ has eigenvalues λ which are found from setting the determinant

$$|\, \boldsymbol{I} + \boldsymbol{Q} - \lambda \boldsymbol{I} \,| = 0 \tag{29}$$

It can be shown that all roots are real and distinct for the multicomponent exchange equations. The roots λ are equal to the flushing factors $V^* + 1$, and thus represent retardation factors. They may be ordered from $\lambda_1 < \lambda_2 < ... < \lambda_{n-1}$.

The Equations (27) form a quasi linear system of partial differential equations of first order. The initial and boundary conditions which make its solution a Riemann problem are:

$$\begin{aligned} y > 0, t = 0: c = c^{init}, q = q^{init} = f(c^{init}) \\ y = 0, t > 0: c = c^{inj}, q = q^{inj} = f(c^{inj}) \end{aligned} \tag{30}$$

Now a self-similar variable $\theta = \theta(t/y)$ is defined such that $c = c(\theta)$. It can be used to transform (28) into

$$\left[(\boldsymbol{I} + \boldsymbol{Q})\frac{\partial \theta}{\partial t} + \boldsymbol{I}\frac{\partial \theta}{\partial y} \right] \frac{\mathrm{d}c}{\mathrm{d}\theta} = 0 \tag{31}$$

This equation can have only a non-trivial solution if the determinant of the coefficient matrix is zero. The eigenvalues defined by Equation (29) are identical to

$$\lambda = -\frac{\partial \theta / \partial y}{\partial \theta / \partial t} \tag{32}$$

For each eigenvalue, say λ_k, the vector $\mathrm{d}c/\mathrm{d}\theta$ is a right eigenvector of $(\boldsymbol{I} + \boldsymbol{Q})$:

$$\frac{\mathrm{d}c}{\mathrm{d}\theta} = r^k(c) \tag{33}$$

This equation is the essence of the solution of the Riemann problem. Since there are n-1 eigenvalues for $(\boldsymbol{I} + \boldsymbol{Q})$, there are also n-1 eigenvectors and n-1 intermediate compositions. Each eigenvector represents a part of the compositions that have to be followed from c^{init} to c^{inj} when going in the upstream direction along x. When going from c^{init} in the upstream direction (x decreases), first the eigenvector that belongs to λ_1 is integrated. Along this path, $x_{H_2O}/x = \lambda_1$. When λ_1 of the first intermediate composition is reached, that composition is maintained until $x_{H_2O}/x = \lambda_2$. Then the second eigenvector is taken until λ_2 of the second intermediate composition is reached, etc. The intermediate compositions are called *plateaux*, and the interconnections via the eigenvectors are termed *paths* (Helfferich and Klein, 1970).

The above sequence is valid for broadening waves along the eigenvectors, i.e. λ_k of the kth eigenvector increases upstream. For shock fronts the eigenvalue λ_k that belongs to the k+1th composition is smaller than λ_k of the kth composition. In that case the Rankine-Hugoniot set of Equation (26) must be used.

The general situation for a stepover on the plateau composition or on the following eigenvector is shown in Figure 15. The curved lines are for eigenvectors along which λ increases, and which show gradual composition changes. The straight lines are for shocks, where the composition abruptly changes from the one at the downstream end to the plateau point, or from this plateau point to the upstream composition.

In principle, the slope of the isotherm of any binary couple of cations that participates in the chromatographic sequence can be inspected to see whether a shock or a broadening front will occur. If dq/dc increases in the upstream direction, a wave develops. If it decreases, a shock occurs. Rule based sequences can be given (Vermeulen et al., 1984). For example, the selectivity of natural exchangers for Na^+, K^+, Mg^{2+} and Ca^{2+} increases in the given order (Table 1). When, going upstream from x_{H_2O}, a decrease of an ion is accompanied by increase of a less selected ion, a wave develops. When it is accompanied by increase in concentration of a more selected ion, a shock occurs. In terms of eigenvalues (or flushing factors + 1) of the system the following holds for the kth transition (Charbeneau, 1988; cf. Figure 15, c^p is the plateau composition at the intersection point).

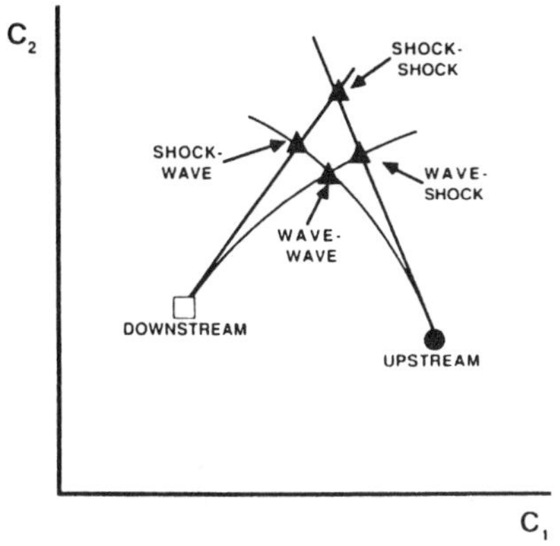

Figure 15. Possible locations of intersection points (plateaux) of two eigenvectors [Used by permission of the editor of *Water Resources Research,* from Charbeneau (1988)].

wave-wave:

$$\lambda_{k+1}(c^u) \geq \lambda_{k+1}(c^p) > \lambda_k(c^p) \geq \lambda_k(c^d)$$

wave-shock:

$$\lambda_{k+1}(c^u) \geq \lambda_{k+1}(c^p) > \lambda_k(c^p) < \lambda_k(c^d)$$

shock-wave: (34)

$$\lambda_{k+1}(c^u) < \lambda_{k+1}(c^p) > \lambda_k(c^p) \geq \lambda_k(c^d)$$

shock-shock:

$$\lambda_{k+1}(c^u) < \lambda_{k+1}(c^p) > \lambda_k(c^p) < \lambda_k(c^d)$$

[The sequences will be illustrated below.]

For a system with all cations of the same charge (homovalent exchange) all the paths can be made straight by suitable mapping (Helfferich and Klein, 1970; Rhee et al., 1970, 1989). It is obtained setting the parameters θ and κ_i:

$$\theta = 1 + \sum_{i=1}^{n} c_i K_{i\backslash n} \tag{35}$$

and

$$\kappa_i = c_i (K_{i\backslash n} - 1) \tag{36}$$

so that

$$\beta_j = \frac{c_j}{\sum_{i=1}^{n} c_i K_{i\backslash j}} = \frac{\kappa_j K_{j\backslash n}}{\theta (K_{j\backslash n} - 1)} \tag{37}$$

This makes β_j a simple function of θ. The further development requires rather elaborate calculations, cf. Rhee et al. (1989).

For a heterovalent system, which will be most often encountered in nature, the paths are curved, and Equation (33) needs to be integrated. This can be done as shown by

Charbeneau for a system with three cations, and therefore with two eigenvectors. The integral is:

$$c^{k+1} = c^{k} + \int_{\theta_k}^{\theta_{k+1}} r^{k}\, d\theta \tag{38}$$

It can be approximated, using the trapezoidal rule:

$$c^{k+1} = c^{k} + \frac{1}{2}[r^{k}(c^{k}) + r^{k}(c^{k+1})]\Delta\theta_k \tag{39}$$

where $\Delta\theta_k = \theta_{k+1} - \theta_k$. With the notation $r^i(k)$ representing the average i^{th} eigenvector along the kth composition path, (38) can be written

$$c^{k+1} = c^{k} + \Delta\theta_k r^i(k) \tag{40}$$

Thus, $r^i(k)$ is the ith eigenvector, averaged in the kth characteristic direction. There is one Equation (40) for each composition path. They are connected by the requirement that

$$c^{inj} - c^{init} = \sum_{j=1}^{n-1} \Delta\theta_j r^j(j) \tag{41}$$

Charbeneau (1988) has solved Equation (40) and (41) for two 3-cation systems, the laboratory experiment of Rainwater et al. (1987) and the field injection of Valocchi et al. (1981a, b). The latter will be discussed fully in the following parts. In principle, the sharp front Equations (26) and the differential Equation (33) for a broadening wave can be used for systems with more than 3 cations. This approach is now being implemented in a computer program (Van Veldhuizen and Hendriks, pers. comm.). The great advantage is that an exact solution is obtained of the chemical process during transport, without ghost peaks that may arise due to numerical oscillations, or spreading due to artificial dispersion that sometimes hamper the more common numerical solutions. Also, this program indicates exactly whether a shock or a wave occurs.

FIELD EXAMPLES OF ION CHROMATOGRAPHY

The case of Valocchi et al. (1981)

Valocchi et al. (1981a,b) injected fresh water in a brackish water aquifer shown in outline in Figure 16. The *CEC* of the recharged silty sand aquifer was 750 meq/l. The native brackish water had high Na^+ and Mg^{2+} concentrations, while the recharge water composition was dominated by Na^+ and Ca^{2+} but at about 10 times lower concentrations. Table 5 gives the water compositions, the exchangeable cations associated with these water compositions, and the difference in the exchangeable cation concentrations.

The injection of fresh water is accompanied by a loss of Na^+ and Mg^{2+} from the exchanger, and an increase of $Ca\text{-}X_2$. It is of interest to note that this loss of Na-X takes place although Na^+ as percentage of the three cations is higher in the injected water than in the native water. This is entirely due to the lower salinity of the injected solution. In equilibrium with a given exchanger composition, the solute ratios $[Na^+]/[Mg^{2+}]^{\cdot}$ and $[Na^+]/[Ca^{2+}]^{\cdot}$ remain the same, independent of the salinity of the solution as follows from Equation (5). When Na^+ is diluted 6.5 times from 86.5 to 13.3 mmol/l, Mg^{2+} and Ca^{2+} are diluted $(6.5)^2 = 42$ times, cf. Table 5.

Table 5. Mass balance calculations for exchanging ions in the injection experiment by Valocchi et al. (1981a): observation well S23. Concentrations in mmol/l.

Ion	Water			Sediment		
	Initial	Diluted	Injected	Initial	Injected	**DIF**
Na^+	86.5	13.3	9.4	160	56	+104
Mg^{2+}	18.2	0.43	0.5	142	41	+101
Ca^{2+}	11.1	0.26	2.13	153	306	-153

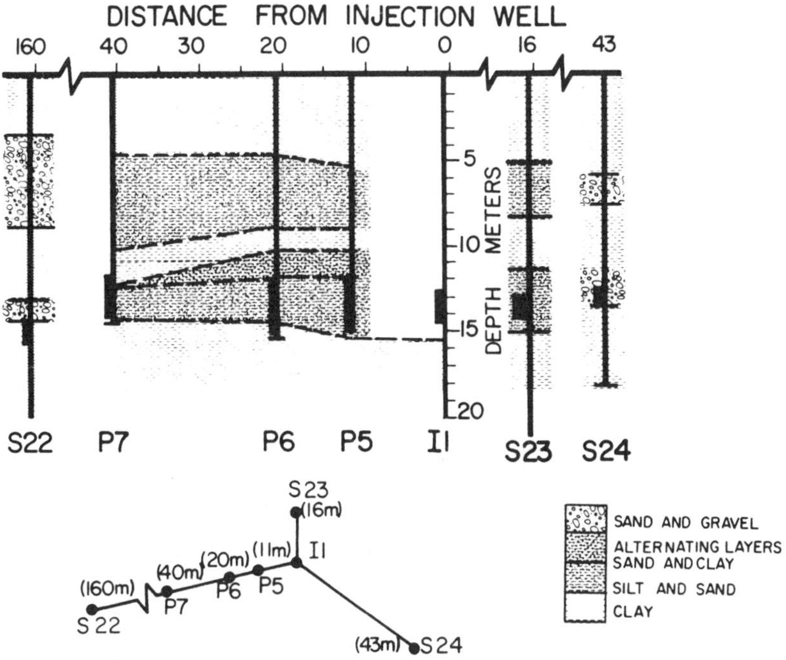

Figure 16. Outline of aquifer used for injecting fresh water by Valocchi et al. (1981a). [Used by permission of the editor of *Water Resources Research.*]

In this three-cation case, two eigenvectors must be followed from the initial to the injected composition, and one intermediate composition should be found between the (diluted) initial and injected compositions. Because it involves a displacement of Na^+ and Mg^{2+} by Ca^{2+}, and because Ca^{2+} is selected above the other two cations, the slope of the isotherm of any of the two binary pairs decreases in the upstream direction. Therefore, shock fronts are expected.

Figure 17 shows the trace of the Mg^{2+} and Ca^{2+} concentrations observed by Valocchi et al. (1981a) in well S23 at 16 m from the injection well. The full line is the computer simulation by Valocchi et al., the dotted is line a chromatographic pattern calculated for sharp fronts by Appelo et al. (1993). Concentrations drop when one pore volume (295 m^3 from injection well to S23) has been injected (the value of 295 m^3 was found by numerical modeling of the Cl^- breakthrough curve; Appelo et al. used 260 m^3). The Ca^{2+} and Mg^{2+} concentrations remain at the (diluted) initial equilibrium concentrations until the first shock arrives with V^s = 25.3 (after 25.3 × 295 + 295 = 7758 m^3). Then the Na^+ concentration drops to almost the final concentrations, and the surplus of Mg^{2+} is flushed from the exchanger until the final concentration arrives with V^s = 113.2 (after 113.2 × 295 + 295 = 33689 m^3).

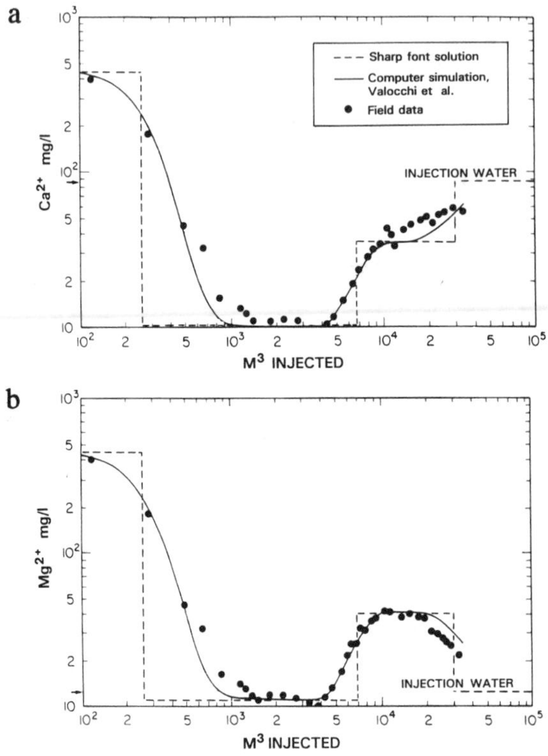

Figure 17. Concentrations of Ca^{2+} and Mg^{2+} in water from observation well S23 during injection of fresh water (Valocchi et al., 1981). The results of the sharp front approximation are shown hatched [Used by permission of the editor of the *Journal of Hydrology,* from Appelo et al. (1993)].

The condition for a shock front is given by Equation (34). Eigenvalues (flushing factors) are smaller for the upstream composition (which arrives later at the observation point). Figure 18 shows the eigenvalues for the three compositions of the Valocchi case and the transitions at the shock fronts. The first shock is pushed by $(V^*_1)_2 < (V^*_1)_1$. The second shock similar by V^*_2. Since also $V_2^s > V_1^s$, the criteria for shock fronts are fully met.

Side reactions in the Valocchi case. Valocchi et al. did not refer to minerals in the aquifer, but calcite is ubiquitous and can be expected to be present. The decrease of Ca^{2+} concentrations during the chromatographic sequence will then be compensated by calcite dissolution:

$$H^+ + CaCO_3 \rightarrow Ca^{2+} + 2\ HCO_3^- \qquad (42)$$

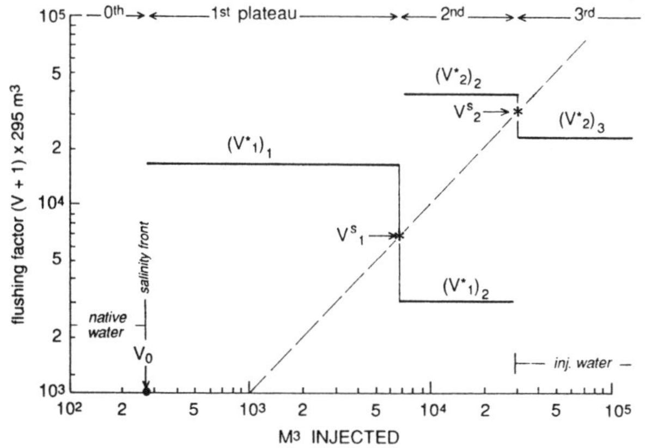

Figure 18. Flushing factors V^* and V^S for compositions of the Valocchi case. Hatched line indicates where flushing factor $(V + 1)$ is equal to injected volume. Note that $V^* + 1 = \lambda$. [Used by permission of the editor of the *Journal of Hydrology*, from Appelo et al. (1993)].

The increased availability of Ca^{2+} will accelerate the chromatographic sequence. This also follows from the smaller change $V^s = \Delta q_{Ca}/\Delta c_{Ca}$, since

$$\Delta q_{Ca} = \Delta Ca\text{-}X_2 + \Delta CaCO_3 \qquad (43)$$

where $\Delta CaCO_3$ is the change in calcite concentration in the aquifer sediment (expressed as mol/l pore water). It is possible to calculate fronts for a combination of exchange and dissolution and precipitation (Harmsen and Bolt, 1982; Bryant et al., 1987; cf. also Lefèvre et al., 1993). Here, we illustrate the effect with a numerical simulation of the complete reaction, assuming that calcite is not exhausted

In principle the loss of Ca^{2+} from solution is sufficient to drive the dissolution reaction. However, the amount that dissolves is small, because pH rapidly increases to 10. This is not observed in aquifers, because of proton buffering. Reaction (42) indicates that some form of proton buffering will be present, that will keep pH at about 7 to 8. Proton buffering may be due to desorption of protons from oxides and organic matter, or perhaps due to desorption of complexes, in combination with reaction of oxides (Griffioen, 1993). For the model calculation done here, it was assumed that a constant CO_2 pressure of 0.01 atm was the source of protons. Figure 19 shows model results obtained with the hydrogeochemical transport model PHREEQM (Appelo and Postma, 1993) when the exchange reaction and the calcite equilibrium reaction are both included.

About 0.94 mmol/l calcite dissolves over the interval of the chromatographic sequence which lasts 170 pore volumes. This amounts to $170 \times 0.94 \times 295 = 47.14$ kmol calcite in total. Figure 19 also shows the modeled Ca^{2+} concentration for the case without calcite dissolution. The main difference lies in the higher Ca^{2+} concentration in the initial dilution plateau that is reached when calcite equilibrium is included, and also in that the complete reaction finishes more rapidly. However, the overall pattern is not much different, which points to the dominance of the cation exchange reaction. One may speculate that dissolution of calcite has been an actual reaction in the Valocchi experiment. In that case the exchange capacity will be probably somewhat higher ($CEC \approx 1000$ meq/l) than noted by Valocchi et al. (1981a).

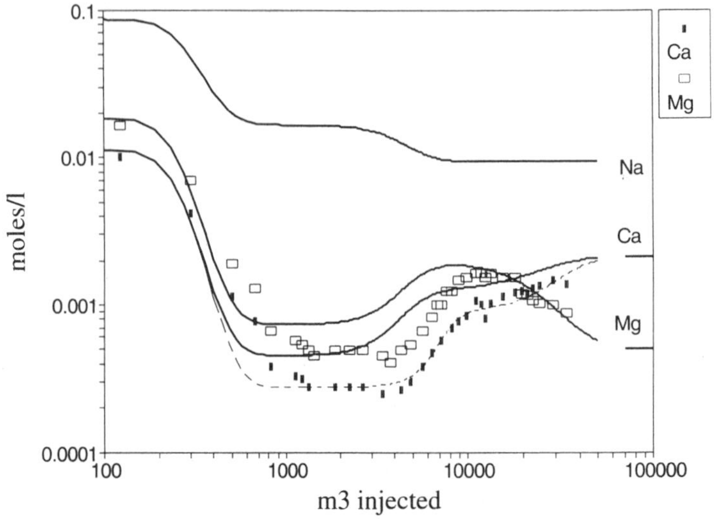

Figure 19. Modeling the Valocchi et al. injection with the hydrogeochemical transport model PHREEQM, ion exchange and calcite dissolution, P_{CO_2} = 0.01 atm. Thin dotted line is Ca^{2+} concentration in a model without calcite equilibrium.

Inverting water compositions. It is interesting to invert the water compositions in the Valocchi case and inject brackish water into a fresh aquifer. The initial composition is given in Table 6. Now Na^+ and Mg^{2+} increase in the upstream direction, and Na-X and Mg-X_2 must increase at the cost of Ca-X_2. Figure 20 shows the solute concentrations for the ideal chromatographic case without dispersion.

Table 6. Inverted compositions for the injection experiment by Valocchi et al. (1981a): Brackish water into fresh water aquifer. Concentrations in mmol/l.

Ion	Water Initial	Diluted	Injected
Na^+	9.4	41.6	86.5
Mg^{2+}	0.5	9.83	18.2
Ca^{2+}	2.13	41.9	11.1

There are two eigenvectors and one plateau composition. Because the less selected ions increase upstream, the paths are waves. The first one starts when the first eigenvalue of the first composition is equal to the number of pore volumes that have been injected $[(V^*_1)_1 + 1 = 2.56]$. It goes on until the first plateau composition is reached $[(V^*_1)_2 + 1 = 3.46]$. This composition is maintained until the second eigenvalue is equal to the injected pore volumes. Then the second eigenvector is followed up to the injected composition. The development of the eigenvalues for this system is given in Figure 21. Note that the overall response is much quicker than in the real case discussed earlier, because higher concentrations in the injected solution allow a larger Δc.

Effects of salinity pulses

The response to groundwater salinity changes is clearest with a dilution because Δc's are limited by N, and the flushing factors are accordingly larger. It can nevertheless be observed in the passage of the salt pulse when the exchangeable cations continue to dominate over solute cations (Bjerg et al., 1993; Ceazan et al., 1989). Ceazan et al. (1989) injected NH_4Br in an aquifer and analyzed the passage of the pulse in an observation well 1.5 m downstream. The arrival of Br^- was accompanied by an increase of N from 0.5 to

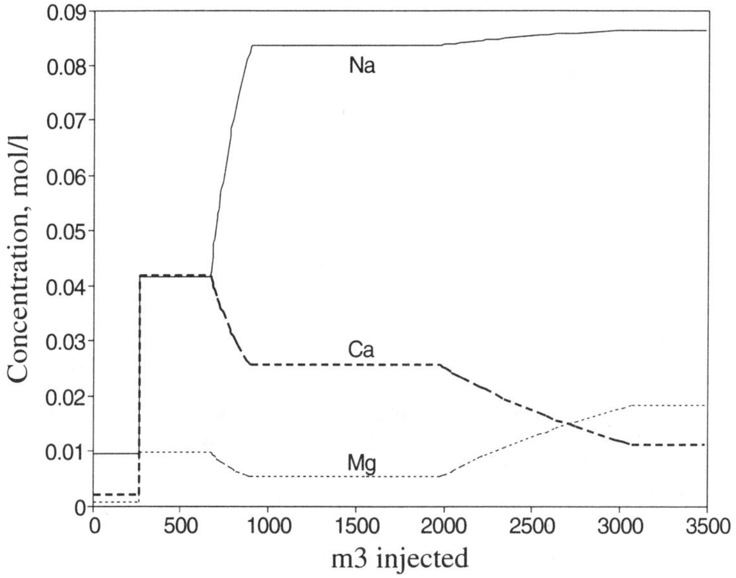

Figure 20. Concentrations in the Valocchi case when brackish water is injected in the fresh aquifer. Based on calculations with MIE (Van Veldhuizen and Hendriks, pers. comm.).

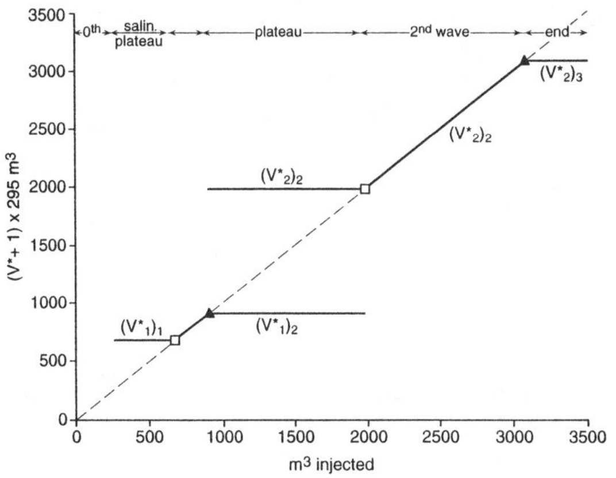

Figure 21. Flushing factors ($V^* + 1$) for the inverted Valocchi case. Eigenvector path starts at □ and ends at ▲.

1.8 meq/l, as shown in Figure 22. The ratio of the divalent cations over Na^+ changed from 0.33 in the native water to a maximum of 0.78 at the peak of the Br^- concentration, strictly conforming to Equations (5) and (7) (Fig. 23).

Salinity pulses are naturally occurring in soils during sequences of dry and wet years. A particular example has been noted by Hansen and Postma (1995) in a 4 to 5 m thick unsaturated zone near the coast of Denmark. Acidification has caused dissolution of Al^{3+} from minerals in the sand, and Al^{3+} concentrations as high as 0.8 mmol/l were observed. When in a dry year the Cl^- concentration increases, equilibrium with the exchange complex requires that Al^{3+} increases by n^3 when Na^+ increases by n. Because the Al^{3+} concentration in the soil solution is in equilibrium with gibbsite, the increase leads to precipitation of

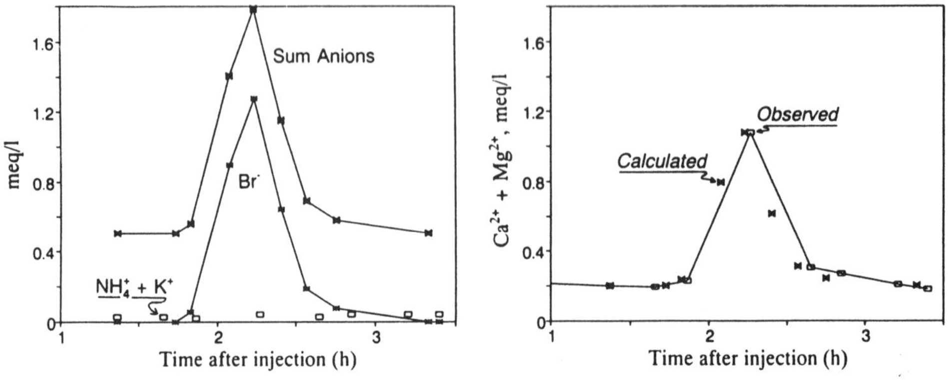

Figure 22 (left). Concentrations of ions in the NH4Br injection experiment of Ceazan et al. (1989).
Figure 23 (right). The observed (□) and calculated (✱) concentrations of Ca^{2+} and Mg^{2+} during the passage of the Br^- peak in the injection experiment of Ceazan et al. (1989). [Figures 22 and 23 used by permission of the editor of *Ground Water,* from Appelo (1994b)].

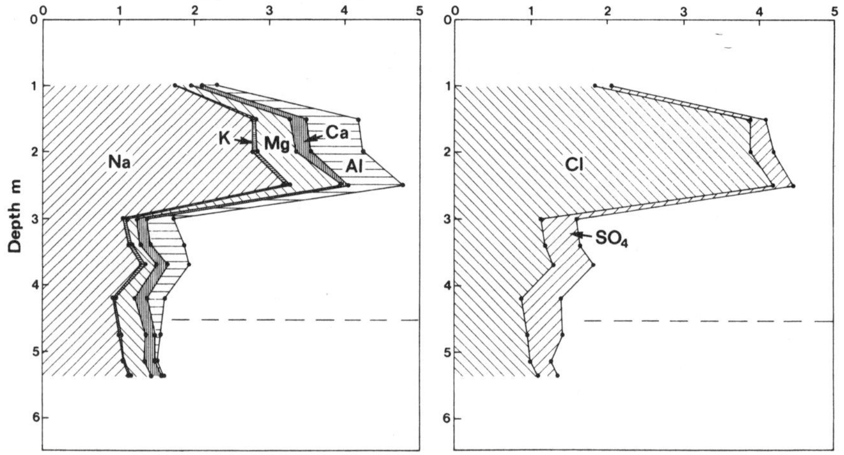

Figure 24. Solute concentrations with depth in an acidified profile at Klosterhede. The hatched line gives the groundwater table. [Used by permission of the editor of *Water Resources Research,* from Hansen and Postma (1995)].

$Al(OH)_3$. Consequently, OH^- is removed from solution, and the salt pulse is at the same time an acid pulse.

Figure 24 shows an observed profile, note that Al in the soil solution is highest at the highest Cl^- concentration. Figure 25 shows calculations of a salt pulse through the profile, with cation exchange and gibbsite equilibrium. Essentially the reaction mechanism is as noted above. The solute concentrations of the salt pulse rapidly equilibrate with the exchange complex in the upper part of the profile. Na^+ is sorbed, and Mg^{2+} and particularly Al^{3+} are desorbed. The equilibrated composition then travels indifferently through the soil. Figure 25 shows the composition when the salt pulse has arrived at -4 m. Note that the exchangeable cations do not change over the salt pulse but that the relative proportions of the solute ions differ quite markedly.

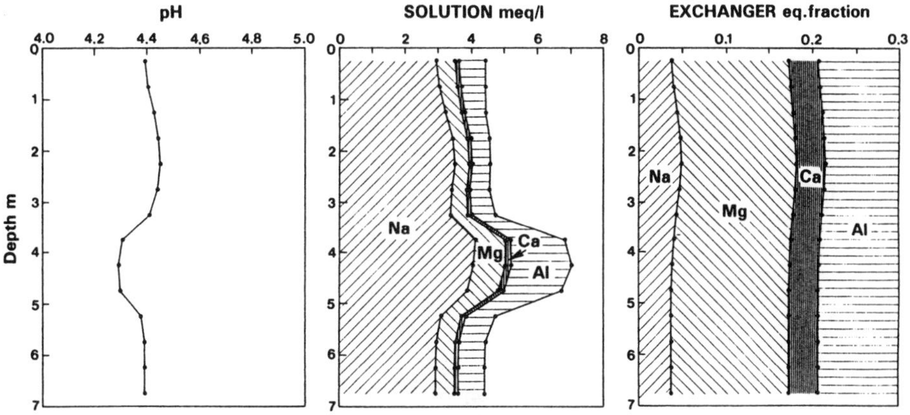

Figure 25. Modeling the movement of a salt pulse through an acidified profile. The pulse had doubled NaCl at zero depth. [Used by permission of the editor of *Water Resources Research,* from Hansen and Postma (1995)].

Freshening of saline aquifers

$NaHCO_3$ waters are found in many Tertiary aquifers along the coast of western Europe (Walraevens and Cardenal, 1994) and eastern U.S.A. (Foster, 1950; Back, 1966; Chapelle and Knobel, 1983). The water quality has been related to cation exchange of Na^+ for Ca^{2+} from fresh $Ca(HCO_3)_2$ water. The freshening of the (initially) saline aquifers was induced by the drop of sea level during the Pleistocene. The aquifers also contain $Mg(HCO_3)_2$ waters, which have been related to transformation of Mg-calcite into a purer carbonate (Chapelle and Knobel, 1983). An attractive alternative to the carbonate reactions is to assume a chromatographic sequence similar to the Valocchi case (Appelo, 1994a). A soil exchanger in equilibrium with seawater has also increased $Mg\text{-}X_2$ (cf. Fig. 2 and Fig. 3). Freshening wiii lead to a sequence in which Na^+ is removed first and then Mg^{2+}.

The sequence is present in an exemplary form in the Aquia aquifer (Maryland, U.S.A.). The aquifer is shown in outline in Figure 26. Observed water qualities along an averaged flowpath even show a K^+ peak between the high Na^+ and Mg^{2+} concentrations (Fig. 27). The pattern could be modeled as a cation exchange process, as shown by the model lines in Figure 27.

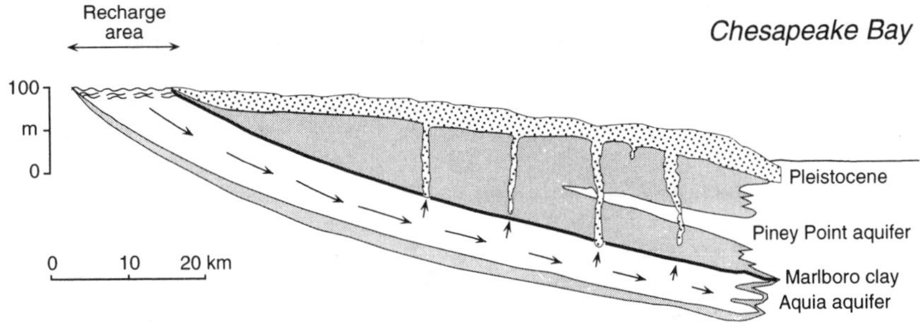

Figure 26. Cross section of the Aquia aquifer, Maryland, USA. [Used by permission of the editor of *Water Resources Research,* from Appelo (1994b)].

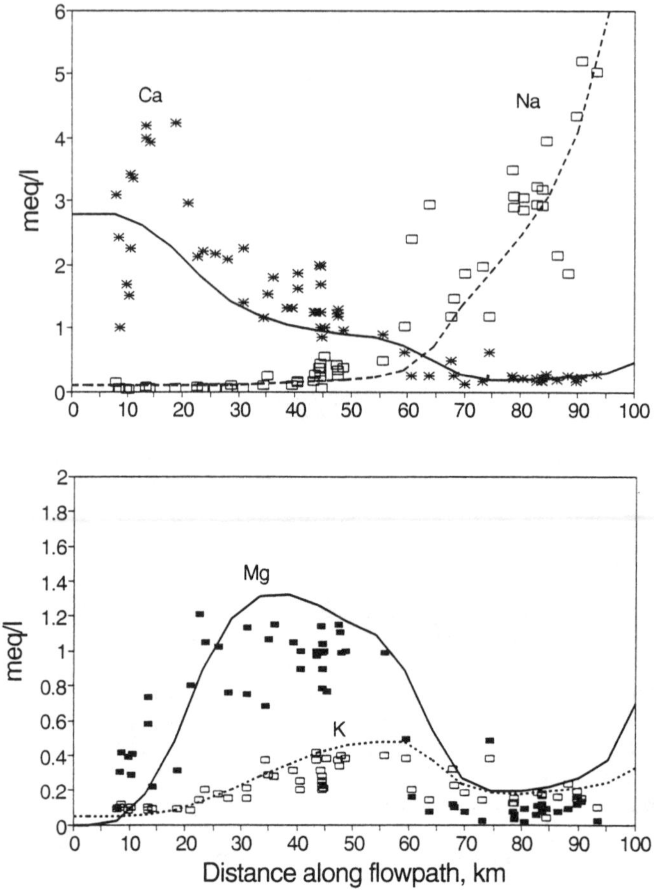

Figure 27. Along an average flowpath in the Aquia aquifer, a sequence of Na^+, K^+, Mg^{2+} and Ca^{2+} concentration peaks are observed that can be modeled as a chromatographic sequence. Data points from Chapelle and Knobel (1983).

The Aquia aquifer is probably the most ideal natural analogue of a laboratory column. Its thickness is less than 100 m over a length of 90 km (Fig. 26), which makes it similar to a hair-thin chromatographic column of, say 1 dm length. Dispersion is therefore relatively reduced and the chromatographic transitions remain visible. The sharpness of the pattern is also preserved because during freshening all the transitions are shocks. Dispersion that is inevitable at the scale of an aquifer, is counteracted by the cation exchange reaction. The pattern is apparently stable over the 100 ka that were calculated to be necessary for its establishment (Appelo, 1994a).

Diffusion profiles tend to show only reduced chromatographic separations in space, because diffusion of the exchange forcing cations is about as rapid as the exchanged cations (Appelo and Willemsen, 1987). However, the salinity effect is very clear in a fresh water diffusion profile observed by Manzano et al. (1992). Manzano et al. (1992) extracted pore water from a silty Holocene aquitard in the Llobregat delta in Spain. The aquitard contains saline water and is freshening from below by diffusion of fresh water from the underlying aquifer. The measured cation concentrations are compared with concentrations calculated for conservative mixture of fresh and saline water in Figure 28.

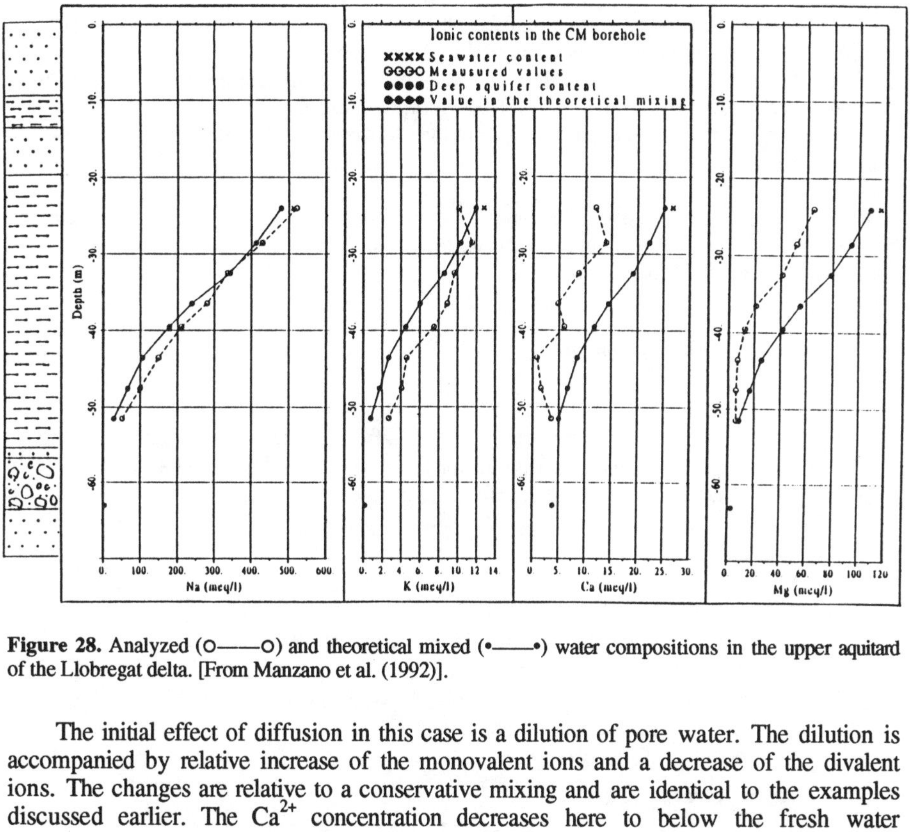

Figure 28. Analyzed (o——o) and theoretical mixed (•——•) water compositions in the upper aquitard of the Llobregat delta. [From Manzano et al. (1992)].

The initial effect of diffusion in this case is a dilution of pore water. The dilution is accompanied by relative increase of the monovalent ions and a decrease of the divalent ions. The changes are relative to a conservative mixing and are identical to the examples discussed earlier. The Ca^{2+} concentration decreases here to below the fresh water concentration, similar to the Valocchi case. Manzano et al. have modeled the concentration profiles and observed that varying $K_{i\backslash j}$ for the exchange reaction had only minor influence on the calculated solute profiles. This can now be understood fully, because in the model calculations the solute ratio $i^+/\sqrt{j^{2+}}$ is constant by the exchange equilibrium with $K_{j\backslash i}\beta_i/\sqrt{\beta_j}$ (Eqn. 5). The latter quotient is balanced by the (constant) initial saline water composition, in which the ratios of the β's compensate for changes in $K_{j\backslash i}$. This will change the available mass of i-X_{z_i}, but these are only noted with transport and in the onset of the next composition path that is followed in the chromatographic pattern.

SUMMARY

We presented an overview of chromatographic effects on natural water qualities. We started with multicomponent ion exchange and have shown that simple equilibrium calculations are valid for natural waters and (correctly) measured exchangeable cations. Chromatography involves the separation of ions due to different transport velocities for different concentrations. The fundamental equation for calculating transport velocity is mass balance. Most simply, it can be stated in the form $dq = V\,dc$. It contends that a change in sorbed concentration must go with a change in solution concentration over a fixed solution volume (or time). The change in solution concentration can be infinitesimal as in the equation, and it will lead to a wave; the solution volume is given by the change of slope of the isotherm dq/dc. It can also be finite, and the solution volume is given by $\Delta q/\Delta c$; this will give a shock-like change in concentrations. For multicomponent exchange and transport, the dq/dc term must be equal for all solutes that participate in the solid-solution equilibria. This relationship can be exploited to deduce chromatographic waves and shocks, the separations of ions, and the velocity of compositions and composition transitions in general. Advances in the mathematical description of these systems have been presented. They were found to be applicable to various hydrochemical field situations with transient water qualities.

In many natural settings the effects of a salinity pulse are the most obvious. These can be simply calculated. The ratio of heterovalent ions in solution after a change in total concentration (eq/l) are found from:

$$\left(\frac{c_i^{1/z_i}}{c_j^{1/z_j}}\right) = \text{constant} \tag{44}$$

When the salinity pulse develops into a block, a full chromatographic sequence will be observed. These sequences were illustrated for the three cation injection cases of Valocchi et al. (1981a,b) and for a freshening aquifer.

Symbols

Greek letters

β_i	Equivalent fraction of *CEC* occupied by cation *i* (-)
ε	(Water filled) porosity (-)
θ	Self-similar variable, $\theta = \theta(c, (vt/x))$ (mol/l)
λ	Eigenvalue of the matrix $(\boldsymbol{I} + \boldsymbol{Q})$ (-)
ρ_b	Bulk density (g/cm^3)

Latin letters

c_i	Solute concentration of *i* (mol/l)
c	Vector of solute concentrations
CEC	Cation exchange capacity (meq/100g soil, or mol$_c$/l pore water)
$\boldsymbol{I}$	Identity matrix
N	Total cations of the solution (eq/l)
$\boldsymbol{Q}$	Coefficient matrix of multicomponent exchange $(\partial q_i/\partial c_j)_{i,j=1,\ldots,n-1}$
q_i	Sorbed or exchangeable concentration of *i* (mol/l pore water)
r^k	k^{th} eigenvector
V^*	Flushing factor (-)
V^s	Sharp front flushing factor (-)
v	Pore water flow velocity (m/s)
z_i	Charge on ion i^{z_i+}

ACKNOWLEDGMENTS

Peter Lichtner and Rien van Veldhuizen are gratefully acknowledged for critical reading of the manuscript. The EEC funded Palaeaux project provided financial support.

REFERENCES

Appelo CAJ, Willemsen A (1987) Geochemical calculations and observations on salt water intrusions. I. J Hydrol 94:313-330

Appelo CAJ, Willemsen A, Beekman HE, Griffioen J. (1990) Geochemical calculations and observations on salt water intrusions. II. J Hydrol 120:225-250

Appelo CAJ, Hendriks JA, Van Veldhuizen M (1993) Flushing factors and a sharp front solution for solute transport with multicomponent ion exchange. J Hydrol 146:89-113

Appelo CAJ, Postma D (1993) Geochemistry, Groundwater and Pollution. Balkema, Rotterdam, 536 p

Appelo CAJ (1994a) Cation and proton exchange, pH variations, and carbonate reactions in a freshening aquifer. Water Resources Res 30:2793-2805

Appelo CAJ (1994b) Some calculations on multicomponent transport with cation exchange in aquifers. Ground Water 32:968-975

Appelo CAJ, Verweij E, Schäfer H (1996) A hydrogeochemical transport model for an oxidation experiment with pyrite/calcite/exchangers containing sand. Geochim Cosmochim Acta (submitted)

Back W (1966) Hydrochemical facies and groundwater flow patterns in northern part of Atlantic coastal plain. US Geol Surv Prof Paper 498-A, 42 p

Beekman HE (1991) Ion chromatography of fresh- and seawater intrusion. PhD dissertation, Free Univ, Amsterdam, 198 p

Beekman HE, Appelo CAJ (1990) Ion chromatography of fresh- and salt-water displacement: laboratory experiments and multicomponent transport modeling. J Contam Hydrol 7:21-37

Bichkova VA, Soldatov VS (1985) A method for predicting ion exchange equilibria in ternary ion exchange systems. React Polymers 3:207-215

Bjerg PL, Ammentorp HC, Christensen TH (1993) Model simulations of a field experiment on cation exchange-affected multicomponent solute transport in a sandy aquifer. J Contam Hydrol 12:291-311

Bolt GH (1967) Cation exchange equations used in soil science - a review. Neth J Agric Sci 15:81-103

Bolt GH (ed) (1982) Soil Chemistry. B. Physico-Chemical Models. Elsevier, Amsterdam, 527 p

Bond WJ, Phillips IR (1990) Approximate solutions for cation transport during unsteady, unsaturated soil water flow. Water Resources Res 26:2195-2205

Bruggenwert MGM, Kamphorst A (1982) A survey of experimental information on cation exchange in soil systems. In Bolt GH (ed) Soil Chemistry, B. Physico-Chemical Models. 141-203. Elsevier, Amsterdam

Bryant SL, Schechter RS, Lake LW (1986) Interactions of precipitation/dissolution waves and ion exchange in flow through permeable media. Am Inst Chem Eng J 32:751-764

Bürgisser C, Cerník M, Borkovec M, Sticher H (1993) Determination of nonlinear adsorption isotherms from column experiments: an alternative to batch studies. Environ Sci Technol 27:943-948

Bürgisser C, Scheidegger AM, Borkovec M, Sticher H (1994) Chromatographic charge density determination of materials with low surface area. Langmuir 10:855-860

Ceazan ML, Thurman EM, Smith RL (1989) Retardation of ammonium and potassium transport through a contaminated sand and gravel aquifer: the role of cation exchange. Environ Sci Technol 23:1402-1408

Cerník M, Barmettler K, Grolimund D, Rohr W, Borkovec M, Sticher H (1994) Cation transport in natural porous media on laboratory scale: multicomponent effects. J Contam Hydrol 16:319-337

Chapelle FH, Knobel LL (1983) Aqueous geochemistry and exchangeable cation composition of glauconite in the Aquia aquifer, Maryland. Ground Water 21:343-352

Charbeneau RJ (1981) Groundwater contaminant transport with adsorption and ion exchange chemistry: method of characteristics for the case without dispersion. Water Resources Res 17:705-713.

Charbeneau RJ (1988) Multicomponent exchange and subsurface solute transport: characteristics, coherence, and the Riemann problem. Water Resources Res 24:57-64.

DeVault D (1943) The theory of chromatography. J Am Chem Soc 65:532-540.

DeZabala EF, Vislocky JM, Rubin E, Radke CJ (1982) A chemical theory for linear alkaline flooding. Soc Petrol Eng J 245-258

Duncan JF, Lister BAJ (1949) Ion exchange studies. I. The sodium-hydrogen system. J Chem Soc (London) 3285-3296

Ekedahl E, Högfeldt E, Sillén LG (1950) Kinetics and equilibria of ion exchange. Nature 166:723-724

Elprince AM, Babcock KL (1975) Prediction of ion-exchange equilibria in aqueous systems with more than two counter-ions. Soil Sci 120:332-338.

Faucher JA, Southworth RW, Thomas HC (1952) Adsorption studies on clay minerals. I Chromatography on clays. J Chem Phys 20:157-160

Faucher JA, Thomas HC (1954) Adsorption studies on clay minerals. IV The system montmorillonite-cesium-potassium. J Chem Phys 22:258-261

Foster MD (1950) The origin of high sodium bicarbonate waters in the Atlantic and Gulf coastal plains. Geochim Cosmochim Acta 1:33-48

Franklin KR, Townsend RP (1988) Multicomponent ion exchange in zeolites. Pt 3 Equilibrium properties of the sodium/potassium/cadmium—zeolite X system. J Chem Soc Faraday Trans 1 84:687-702

Gaines GL, Thomas HC (1953) Adsorption studies on clay minerals. II. A formulation of the thermodynamics of exchange adsorption. J Chem Phys 21:714-718

Glückauf J (1945) Adsorption isotherms from chromatographic measurements. Nature 156:748

Glückauf J (1949) Theory of chromatography. VI Precision measurements of adsorption and exchange isotherms from column-elution data. J Chem Soc (London) 3280-3285

Glückauf J (1955) Principles of operation of ion-exchange columns. in: Ion exchange and its applications. Soc Chem Ind (London) 34-46

Griffioen J (1993) Multicomponent cation exchange including alkalinization/acidification following flow through sandy sediment. Water Resources Res 29:3005-3019

Griffioen J, Appelo CAJ, Van Veldhuizen M (1992) Practice of chromatography: deriving isotherms from elution curves. Soil Sci Soc Am J 56:1429-1437

Groen KP (1991) History of Salt Water Investigations in The Netherlands (in Dutch). Flevobericht 321, Min van Waterstaat, Lelystad

Hansen BK, Postma D (1995) Acidification, buffering, and salt effects in the unsaturated zone of a sandy aquifer, Klosterhede, Denmark. Water Resources Res 31:2795-2809

Harmsen K, Bolt GH (1982) Movement of ions in soil. I. Ion exchange and precipitation. Geoderma 28:85-101

Harmsen K, Bolt GH (1982) Movement of ions in soil. II. Ion exchange and dissolution. Geoderma 28:103-116

Helfferich F (1967) Multicomponent ion exchange in fixed beds. Ind Eng Chem Fund 6:362-364

Helfferich F, Klein G (1970) Multicomponent Chromatography. M Dekker, New York,

Hofstee J (1971) Methods of Analysis (in Dutch). Rijksdienst IJsselmeerpolders, Kampen, The Netherlands

Klein G, Tondeur D, Vermeulen T (1967) Multicomponent ion exchange in fixed beds. Ind Eng Chem 6:339-351

Kool JB, Parker JC, Zelazny LW (1989) On the estimation of cation exchange parameters from column displacement experiments. Soil Sci Soc Am J 53:1347-1355

Lax P (1957) Hyperbolic systems of conservation laws II. Comm Pure Appl Math 10:537-566

Lax P (1973) Hyperbolic Systems of Conservation Laws and the Mathematical Theory of Shock Waves. S I A M, Philadelphia, PA, 48 p

Lindstrom FT, Haque R, Freed VH, Boersma L (1967) Theory on the movement of some herbicides in soils: linear diffusion and convection of chemicals in soils. J Environm Sci Tech 1:561-565

Lefèvre F, Sardin M, Schweich D (1993) Migration of strontium in clayey and calcareous sandy soil: precipitation and ion exchange. J Contam Hydrol 13:215-229

Mansell RS, Bloom SA, Bond WJ (1993) A tool for evaluating a need for variable selectivities in cation transport in soil. Water Resources Res 29:1855-1858

Mangold DC, Tsang C-F (1991) A summary of subsurface hydrological and hydrochemical models. Rev Geophys 29:51-79

Manzano M, Custodio E, Carrera J (1992) Fresh and salt water in the Llobregat delta aquitard. Proc 12th Saltwater Intrusion Mtg, Barcelona. CIMNE, Barcelona, 207-238

Merriam CN, Southworth RW, Thomas HC (1952) Ion exchange mechanism and isotherms from deep bed performance. J Chem Phys 20:1842-1846

Merriam CN, Thomas HC (1956) Adsorption studies on clay minerals. VI. Alkali ions on attapulgite. J Chem Phys 24:993-995

Page AL, Miller RH, Keeney DR (1982) Methods of soil analysis. Pt 2. Chemical and Microbiological Properties, 2nd edition. Soil Sci Soc Am, Madison, Wisconsin, 1159 p

Parkhurst DL (1995) User's guide to PHREEQC - A computer program for speciation, reaction-path, advective transport, and inverse geochemical calculations. U S Geol Surv Water Resources Inv 95-4227, 143 p

Pope GA, Lake LW, Helfferich FG (1978) Cation exchange in chemical flooding: Part 1 - Basic theory without dispersion. Soc Petrol Eng J 418-433

Rainwater KA, Wise WR, Charbeneau RJ (1987) Parameter estimation through groundwater tracer tests. Water Resources Res 23:1901-1910

Rhee H-K, Aris R, Amundson NR (1970) On the theory of multicomponent chromatography. Phil Trans Roy Soc (London) 267A:419-455

Rhee H-K, Aris R, Amundson NR (1989) First order partial differential equations. Vol. II. Prentice Hall, Englewood Cliffs, NJ, 548 p

Ritchie AS (1966) Chromatography as a natural process in geology. In Giddings JC, Keller RA (eds) Adv Chromatogr 3:119-134. M Dekker, New York

Rubin J, James RV (1973) Dispersion-affected transport of reacting solutes in saturated porous media: Galerkin method applied to equilibrium-controlled exchange in unidirectional steady water flow. Water Resources Res 9:1332-1356

Scheidegger A, Bürgisser CS, Borkovec M, Sticher H, Meeussen H, Van Riemsdijk W (1994) Convective transport of acids and bases in porous media. Water Resources Res 30:2937-2944

Schweich D, Sardin M (1981) Adsorption, partition, ion exchange and chemical reaction in batch reactors or in columns—a review. J Hydrol 50:1-33

Schweich D, Sardin M, Gaudet J-P (1983) Measurement of a cation exchange isotherm from elution curves obtained in a soil column: preliminary results. Soil Sci Soc Am J 47:32-37

Schweich D, Sardin M, Jauzein M (1993) Properties of concentration waves in presence of nonlinear sorption, precipitation/dissolution, and homogeneous reactions. 1. Fundamentals. Water Resources Res 29:723-733

Sillén LG (1950) Theory of sorption columns. Nature 166:722-723

Sillén LG (1951) On filtration through a sorbent layer. IV. Arkiv Kemi 2:477-498

Smoller J (1983) Shock waves and reaction-diffusion equations. Springer-Verlag, New York

Sposito G (1994) Chemical equilibria and kinetics in soils. Oxford Univ Press, New York, 268 p

Stuyfzand PJ (1993) Hydrochemistry and hydrology of the coastal dune area of the Western Netherlands. PhD dissertation, Free Univ, Amsterdam, 366 p

Thoolen PMC, Hemker PW (1991) Approximation methods for n-component solute transport and ion-exchange. Rept 91-9, Dept Mathematics, Univ Amsterdam, 26 p

Tondeur D (1969) Théorie des colonnes d'échange d'ions. PhD dissertation, Nancy, 166 p

Valocchi A.J, Street RL, Roberts PV (1981a) Transport of ion-exchanging solutes in groundwater: chromatographic theory and field simulation. Water Resources Res 17:1517-1527

Valocchi AJ, Roberts PV, Parks GA, Street RL (1981b) Simulation of the transport of ion-exchanging solutes using laboratory-determined chemical parameter values. Ground Water 19:600-607

Van der Molen WH (1958) The Exchangeable Cations in Soils Flooded with Seawater. Staatsdrukkerij, Den Haag, 167 p

Van Veldhuizen M, Appelo CAJ, Griffioen J (1995) The (least-squares) quotient algorithm as a rapid tool for obtaining sorption isotherms from column elution curves. Water Resources Res 31:849-857

Van Veldhuizen M, Hendriks J (in prep) MIE, a computer program for analytical solution of multicomponent, heterovalent chromatography. Dept Mathematics, Free Univ, Amsterdam

Vermeulen Th, LeVan MD, Hiester NK, Klein G (1984) Adsorption and ion exchange. In: Perry's Chemical Engineer's Handbook, 6th edn, McGraw-Hill, New York, Ch 16.

Walraevens K, Cardenal J (1994) Aquifer recharge and exchangeable cations in a Tertiary clay layer (Bartonian clay, Flanders-Belgium). Mineral Mag 58A:955-956.

Yeh GT, Tripathi VS (1989) A critical evaluation of recent developents of hydrogeochemical transport models of reactive multichemical components. Water Resources Res 25:93-108

Zuur AJ (1938) In: Trans 2nd Comm and Alkali-Subcomm, Int'l Congress Soil Science, Helsinki B:66-67

Chapter 5

SOLUTE TRANSPORT MODELING UNDER VARIABLY SATURATED WATER FLOW CONDITIONS

D.L. Suarez and J. Šimůnek

U.S. Salinity Laboratory, ARS
U.S. Department of Agriculture,
450 W Big Springs Road
Riverside, California 92507 U.S.A

INTRODUCTION

A proper representation of important physical and chemical processes and soil properties in the rootzone is critical to modeling the transport of major solute species in the vadose zone. A proper modeling effort is also needed for prediction of the solution chemistry of groundwater systems, since these are usually recharged via the vadose zone. Hydrological models for water flow, both saturated and variably saturated, and models describing chemical processes in earth materials have been mostly developed independently. This independent development may have been due primarily to the lack of sufficient computational capability in the past, but is also due in part to the requirements of distinct scientific disciplines. To date most approaches consist of chemical models with simple description of water flow, or water flow models with simplified descriptions of chemical processes. The solute transport models developed in the hydrological sciences have mostly considered only one solute, and assumed only simplified chemical processes, most often with linearized expressions which can be incorporated directly into the transport equation. Chemical modeling has been predominantly based on assumptions that the system is at thermodynamic equilibrium. Realistic modeling of the vadose zone requires representation not only of physical and chemical processes but also of such biological processes as microbial respiration and plant water uptake.

In this paper we will briefly describe variably saturated water flow, including the modeling concepts of chemical effects on hydraulic properties and root water uptake and growth. We will also briefly discuss the traditional approaches for modeling reactive transport of single solute species including the concepts of both physical and chemical nonequilibrium. Our major emphasis is on modeling of CO_2 transport and production, and in discussing multicomponent solute transport. We also include discussion of processes required for application of multicomponent models to field systems, and describe some of the features of the UNSATCHEM model which meets at least some of these requirements. Example simulations are given to demonstrate the importance that root water uptake, CO_2 dynamics and kinetics of calcite precipitation have on prediction of solute concentration and distribution in the unsaturated zone.

UNSATURATED WATER FLOW

Governing equation

The Richards' equation is widely used to describe one dimensional water movement in partially saturated porous media. Application of the model to near-surface environments requires use of a source-sink term. The following relation is of general

0275-0279/96/0034-0005$05.00

applicability:

$$\frac{\partial\theta}{\partial t} = \frac{\partial}{\partial z}[K(\frac{\partial h}{\partial z} + 1)] - S \tag{1}$$

where h is the water pressure head [L], θ is the water content [L^3L^{-3}] (positive upwards), K is the hydraulic conductivity [LT^{-1}], t is time [T], z is the spatial coordinate [L] and S is the sink/source term [T^{-1}], which is used here to represent water uptake rate by plant roots. This equation assumes a rigid porous media, that the air phase does not affect the liquid flow process, and neglects the effects of thermal or solution density gradients. A discussion of the assumptions and limitations of this approach is presented in Nielsen et al. (1986).

Hydraulic characteristics

The hydraulic conductivity under unsaturated conditions is dependent on the water content, which is in turn related to the pressure head. A widely used representation of the unsaturated soil hydraulic properties is the set of closed-form equations formulated by van Genuchten (1980), based on the capillary model of Mualem (1976). Soil water retention and hydraulic conductivity functions are given by

$$\theta(h) = \theta_r + \frac{\theta_s - \theta_r}{(1 + |\alpha h|^n)^m} \tag{2}$$

and

$$K(h) = K_s K_r = K_s S_e^{1/2} \left[1 - (1 - S_e^{1/m})^m\right]^2 \tag{3}$$

respectively, where

$$m = 1 - 1/n \qquad n > 1 \tag{4}$$

$$S_e = \frac{\theta - \theta_r}{\theta_s - \theta_r} \tag{5}$$

and where θ_r and θ_s denote residual and saturated water contents [L^3L^{-3}], respectively, K_s is the saturated conductivity [LT^{-1}], K_r is the relative hydraulic conductivity [-], S_e is relative saturation [-] and m [-], n [-], and α [L^{-1}] are the empirical parameters of the hydraulic characteristics. The hydraulic functions are determined by a set of 5 parameters, θ_r, θ_s, α, n, and K_s. Use of the model requires optimizing the parameters from experimental water retention, pressure head, and saturated conductivity data. This parameter optimization can be performed using the RETC code (van Genuchten et al., 1991).

Alternative models for the prediction of the relative hydraulic conductivity of unsaturated soils include the earlier, widely used Brooks and Cory (1964) formulation. The relative saturation, S_e, in that case is given by

$$S_e = \begin{cases} 1 & \alpha h \leq 1 \\ (\alpha h)^{-\beta} & \alpha h > 1 \end{cases} \tag{6}$$

and the relative hydraulic conductivity by

$$K(h) = K_s S_e^{3+2/\beta} \tag{7}$$

While it is easier to obtain the Brooks and Cory parameters, the van Genuchten model is considered to provide a better match to the experimental data (Stankovich and Lockington, 1995). Other water retention models include those described by Hutson and Cass (1987) and Russo (1988), among others. These hydraulic functions are used as input to the Richards' equation for prediction of water movement and water content.

Chemical effects on hydraulic conductivity

Among the implicit assumptions when using the above equations is that the hydraulic properties are not affected by the composition of the solution phase. That this assumption is not valid is evidenced in the numerous studies that have documented the effects of solution composition on saturated hydraulic conductivity. Elevated levels of exchangeable sodium result in swelling of smectitic clays. Dispersion of clay, migration and subsequent blocking of pores results from low electrolyte and presence of exchangeable sodium. This process is readily observed in the natural development of clay pan layers in soils. The process has also been related to the distribution of divalent cations, with Mg^{2+} being more susceptible to dispersion than Ca^{2+}. In addition, it has been determined that elevated levels of pH adversely impact saturated hydraulic conductivity (Suarez et al., 1984).

Suarez and Šimůnek (1996) considered that the equations developed above could be optimal representations and that the chemical effects on hydraulic properties could be represented by the use of a reduction function, r, given by

$$r = r_1 r_2 \tag{8}$$

where r_1 is the reduction due to the adverse effects of low salinity and high exchangeable sodium fractions on the clay and r_2 is the adverse effect of pH. The r_1 term is given by McNeal (1968) as

$$r_1 = 1 - \frac{cx^n}{1 + cx^n} \tag{9}$$

where c and n are empirical factors, and x is defined by

$$x = f_m 3.6 * 10^{-4} ESP^* \ d^* \tag{10}$$

where f_m is the mass fraction of montmorillonite in the soil, d^* is an adjusted interlayer spacing and ESP^* is an adjusted exchangeable sodium percentage (percentage of the total negative exchange charge of the soil that is neutralized by Na^+). The term d^* is defined by

$$\begin{aligned} d^* &= 0 && C_0 > 300 \text{ mmol}_c\text{L}^{-1} \\ d^* &= 356.4\,(C_0)^{-0.5} + 1.2 && C_0 \leq 300 \text{ mmol}_c\text{L}^{-1} \end{aligned} \tag{11}$$

and the term ESP^* is given by

$$ESP^* = ESP_{\text{soil}} - (1.24 + 11.63\log C_0) \tag{12}$$

The reduction factor r_2, for the adverse effect of pH on hydraulic conductivity, was calculated from the experimental data of Suarez et al. (1984) after correcting for the adverse effects of low salinity and high exchangeable sodium using the r_1 values:

$$\begin{aligned} r_2 &= 1 && \textit{for } pH < 6.83 \\ r_2 &= 3.46 - 0.36\,pH && \textit{for } pH \in \langle 6.83, 9.3 \rangle \\ r_2 &= 0.1 && \textit{for } pH > 9.3 \end{aligned} \tag{13}$$

In view of the differences among soils, these specific corrections may not be generalized predictors of the soil hydraulic conductivity but they illustrate the changes in K that may affect infiltration and solute movement under various chemical conditions. However, it is not yet certain that these parameters need to be characterized for each soil, as do the hydraulic characteristics in Equations (2)-(5).

The relations described by Equations (9)-(13) were based on data for soils of mixed mineralogy where the processes of both dispersion and swelling were likely important. Russo and Bresler (1977) modeled the effect of electrolyte concentration and Na/Ca ratio on the unsaturated hydraulic conductivity. Their model uses the concept that clay (mostly smectites) swelling results in a decrease in pore diameter and subsequent reduction in hydraulic conductivity. The number of clay platelets of Ca saturated smectite was given as

$$N_i^{\text{Ca}} = 4 + \log_{10}[\alpha(h_i/\rho g)] \tag{14}$$

where α is an empirical parameter (= 1.0), h_i is the pressure head (which can be calculated by the hydraulic functions given earlier), ρ is the density of water and g is the gravitational constant. The number of platelets in a mixed Na-Ca system at the same pressure head is then given by

$$\begin{aligned} N_i &= N_i^{\text{Ca}} && \text{for } 0 \leq ESP < ESP' \\ N_i &= N_i^{\text{Ca}} - \Delta N_i && \text{for } ESP' \leq ESP \leq ESP'' \\ N_i &= N_i^{\text{Na}} && \text{for } ESP > ESP'' \end{aligned} \tag{15}$$

where ESP′ is the critical value above which the number of platelets in a particle decreases and ESP′′ is the value at which there is only a single platelet in a particle (Russo and Bresler, 1977). Using the data of Shainberg and Otoh (1968) and assuming that N_i=1.0 above ESP=35, Equation (15) can be replaced by

$$N_i = N_i^{Ca} - 0.09ESP \quad \text{for} \quad ESP \leq 35 \tag{16}$$

The volume of water retained by the clay in each unit of soil can be determined from

$$u_i^{Ca} = (S/2N_i^{Ca})[2b_i + b_0(N_i^{Ca} - 1)]\,\rho_b \tag{17}$$

for Ca saturated soil. When ESP > ESP´´, then

$$u_i = (S/2N_i^{Ca})\,[2b_i + b_0(N_i - 1) - 2b_i 0.09ESP]\,\rho_b \tag{18}$$

where S is now the specific surface area of the soil, b_0 is the thickness of the water film between two clay platelets in a particle (= 9 nm), b_i is the distance between two adjacent particles, and ρ_b is the bulk density of the soil. The value for b_i can be calculated from diffuse layer theory (Bresler, 1972), as a function of exchange composition and total salt concentration.

The pore space available for water flow, H_i, is given by

$$H_i = [\theta_i - (u_i - u_i^{Ca})]\,v_i \tag{19}$$

where v_i is the volumetric water content of the Ca saturated soil. The effective porosity ϵ_i is given by

$$\epsilon_i = \epsilon_t - (H_{i-1}/V) \qquad i = 1,2.....n; \quad H_0 = 0 \tag{20}$$

The hydraulic conductivity at a given water content is then calculated using the equation of Marshall (1958)

$$K(\epsilon_i) = \rho \frac{g}{8\eta}\,\epsilon_i^2\, l^{-2} \sum_{i=1}^{n} (2i-1)a_i^2\,, \quad a_i > a_{i+1} \tag{21}$$

where η is the viscosity of water , l=n-i+1, and a_i is the mean equivalent radius of the pores. The term a_i is given by

$$a_i = (\epsilon_i v_i / \pi m_i)^{0.5} \tag{22}$$

where

$$m_i = \epsilon_0 v_i / \pi a_{i0}^2 \tag{23}$$

and a_{i0}=2γ/P_i and γ is the surface tension of the soil solution.

Both the Suarez and Šimůnek (1996) and the Russo and Bresler (1977) formulations are based on experiments in which the adverse conditions were imposed after measurements under ideal chemical conditions. It is not likely that the chemical and physical processes which cause the reduction in hydraulic conductivity are reversible; thus the models may not be suited for prediction of improvements in hydraulic properties as a result of improvement in the chemical environment (such as reclamation of soils high in exchangeable Na soil by addition of gypsum).

Chemical processes of dissolution and precipitation can also affect hydraulic properties by changes in porosity. Carnahan (1990) proposed equations to account for changes in porosity resulting from dissolution and precipitation under saturated flow. It is not clear how these changes in porosity relate to changes in variably saturated water flow, which is more dependent on the changes in pore size distribution.

Within the past few years flow models have been developed to provide two (and three)-dimensional solutions to the Richards' equation, which can be represented by

$$\frac{\partial \theta}{\partial t} = \frac{\partial}{\partial x_i}[K(K_{ij}^A \frac{\partial h}{\partial x_j} + K_{ij}^A)] - S \tag{24}$$

where x_i, x_j are the spatial coordinates, K_{ij}^A are components of a dimensionless aniosotropy tensor K^A, K is the unsaturated hydraulic conductivity function [LT^{-1}], and S is again the sink/source term.

ROOT WATER UPTAKE AND ROOT GROWTH

The process of evaporation and plant transpiration exerts a major influence on the solution composition and associated water and solute distributions in the unsaturated zone, particularly in arid and semiarid environments. In these instances neither water flow, solute transport, nor solution composition can be modeled without consideration of the evapotranspiration process. As a result, where plant water uptake is important, only agriculturally oriented models which take into account the influence of root water uptake are suitable for modeling water flow and solute transport.

Various models have been developed for description of water extraction by plant roots. Among the models in widespread use is one initially proposed by Feddes (1978). A detailed review of different expressions used to represent root water uptake can be found in Molz (1981). In the following expression, Šimůnek and Suarez (1993a) modified the Feddes et al. (1978) relation to include the adverse effect of osmotic stress on water uptake

$$S(h,h_\phi) = \alpha_s(h)\alpha_\phi(h_\phi)S_p \tag{25}$$

where S_p is the potential water uptake rate [$L^3L^{-3}T^{-1}$] in the root zone, $\alpha_\phi(h_\phi)$ is the osmotic stress response function [-], and h_ϕ is the osmotic head [L]. The water stress response function, $\alpha_s(h)$, is a prescribed dimensionless function of the soil water pressure head ($0 \le \alpha_s \le 1$) described by van Genuchten (1987) as

$$\alpha_s(h) = \frac{1}{1 + (\frac{h}{h_{50}})^b} \tag{26}$$

where h_{50} [L] and b [-] are empirical constants. The parameter h_{50} represents the pressure head at which the water extraction rate is reduced by 50%. Note that this formulation of the water stress response function, $\alpha_s(h)$, in contrast to the earlier expression of Feddes et al. (1978, not shown), does not consider a reduction in transpiration at water contents near saturation. The decrease in water uptake that is sometimes observed at saturation is related to oxygen stress and is more properly treated using predictions of the gas phase composition (for models that include CO_2 production and transport).

The potential water uptake rate in the root zone is expressed as the product of the potential transpiration rate, T_p [LT^{-1}], and the normalized water uptake distribution function, $\beta(z)$ [L^{-1}], which describes the spatial variation of the potential water uptake rate, S_p, over the root zone, as follows

$$S_p = \beta(z)\, T_p \tag{27}$$

There are many ways to express the function $\beta(z)$, including equations that are constant with depth, linear (Feddes et al., 1978), or exponential with a maximum at the soil surface (Raats, 1974):

$$\beta(z) = a e^{-a(L-z)} \tag{28}$$

where L is the z-coordinate of the soil surface [L] and a is an empirical constant [L^{-1}]. Alternatively, van Genuchten (1987) suggested the following depth-dependent root distribution function $\beta(z)$:

$$\begin{aligned} \beta(z) &= \frac{5}{3L_r} && L - 0.2L_r \le z \le L \\ \beta(z) &= \frac{25}{12L_r}\left(1 - \frac{L-z}{L_r}\right) && L - L_r < z < L - 0.2L_r \\ \beta(z) &= 0 && z \le L - L_r \end{aligned} \tag{29}$$

where L_r is the root depth [L]. The actual transpiration rate, T_a, is obtained by integrating the root water uptake rate over the root zone as follows

$$T_a = \int_{L-L_r}^{L} S(h, h_\phi, z)\, dz = T_p \int_{L-L_r}^{L} \alpha_s(h)\, \alpha_\phi(h_\phi)\, \beta(z)\, dz \tag{30}$$

Root growth

The root depth, L_r, can be either constant or variable during the simulation. For annual vegetation a growth model is required to simulate the change in rooting depth with time. In UNSATCHEM (Šimůnek and Suarez, 1994; Suarez and Šimůnek, 1996) the root depth is the product of the maximum rooting depth, L_m [L], and the root growth coefficient, $f_r(t)$ [-]:

$$L_r(t) = L_m f_r(t) \tag{31}$$

To calculate the root growth coefficient, $f_r(t)$, Šimůnek and Suarez (1993b) combined the Verhulst-Pearl logistic growth function with the growth degree day (*GDD*) or heat unit concept (Gilmore and Rogers, 1958). The logistic growth function is usually used to describe the biological growth at constant temperature, whereas the heat unit model is utilized for determining the time between planting and maturity of the plant. The heat unit model cannot be used directly to predict biomass during the growth stage since it would predict a linear growth with time at constant temperature. Combining the heat unit concept with the classical logistic growth function incorporates both time and temperature dependence on growth.

For the growth degree day function a suitable option is a modified version of the relation developed by Logan and Boyland (1983), who assumed that this function is fully defined by the temperature, T [K], which in turn can be expressed by a sine function to approximate the behavior of temperature during the day, and by the three temperature limits, T_1, T_2, and T_3 [K]. When the actual temperature is below the base value T_1, plants register little or no net growth. The plant growth is at a maximum level at temperature T_2, which remains unchanged for some interval up to a maximum temperature T_3, above which increased temperature has an adverse effect on growth. Based on this information, Šimůnek and Suarez (1993b) proposed the following dimensionless growth function

$$g(t) = \begin{cases} g(t) = 0 & t \le t_p; t \ge t_h \\ \frac{1}{\mathrm{T_{Bas}}}\left[\int \delta(\mathrm{T}-\mathrm{T}_1)dt - \int \delta(\mathrm{T}-\mathrm{T}_2)dt - \int \delta(\mathrm{T}-\mathrm{T}_3)dt\right] & t \epsilon (t_p, t_m) \\ g(t) = 1 & t \epsilon (t_m, t_h) \end{cases} \tag{32}$$

where T_{Bas} represents the heat units [KT] necessary for the plant to mature and the roots to reach the maximum rooting depth, t_p, t_m, and t_h represent time of planting, time at which the maximum rooting depth is reached and time of harvesting, respectively; and parameter δ [-] introduces into the heat unit concept the reduction in optimal growth due to the water and osmotic stress. The expression inside the parenthesis of Equation (32) reaches the value T_{Bas} at time t_m when roots reach the maximum rooting depth. The individual integrals in Equation (32) are evaluated only when the particular arguments are positive. Parameter δ [-] is defined as the ratio of the actual to potential transpiration rate:

$$\delta = \frac{T_a}{T_p} \tag{33}$$

Biomass or root development during the growth stage can also be expressed by the Verhulst-Pearl logistic growth function

$$f_r(t) = \frac{L_0}{L_0 + (L_m - L_0)e^{-rt}} \tag{34}$$

where L_0 is the initial value of the rooting depth at the beginning of the growth period [L] and r is the growth rate [T^{-1}].

Both growth functions (32) and (34) can be used directly to model root growth. However, to avoid the drawbacks of both concepts, as discussed above, Šimůnek and Suarez (1993b) combined these equations by substituting the growth function calculated from the heat unit concept (32) for the time factor in the logistic growth function (34):

$$t = t_m\, g(t) \tag{35}$$

where t_m is the time when *GDD* reaches the required value for the specific plant species (T_{Bas}). This value is not known a priori; only the product rt_m must be known and that can be selected, for example, so that $f_r(t)$ equals 0.99 for $g(t) = 1$.

The water uptake expressions in Equation (30) result in a decrease in water uptake by plant roots at each node where water stress exceeds a critical value. Other models, such as LEACHM (Wagenet and Hutson, 1987; and Hutson and Wagenet, 1992) and SOWATCH (Dudley and Hanks, 1991) utilize the following equation for root water extraction

$$S = [h_r + z(1 + R_c) - h - h_\Phi]\, \frac{RDF\, K(\theta)}{\Delta x\, \Delta z} \tag{36}$$

where h_r is the water head within the root, R_c is a root resistance term, h is the matric head in the soil, h_Φ is the osmotic head, RDF is the fraction of root mass at depth z, and Δx and Δz are the dimensions of the element from which water is extracted. This model assumes a stable root profile and a Darcy flux control of water movement to the plant roots. Subject to this constraint, a water deficit from a given node is compensated by water use from any other node. Cardon and Lety (1992) determined that extraction functions of the form expressed in Equation (36) are not realistic as they result in very rapid shifts between optimal and zero transpiration, in contrast to experimental data which indicates slower changes. Equation (36) also fails to predict significant decreases in transpiration under high osmotic pressure. Equations of the form described by Equation (30) provided more realistic simulations. The model described by Equation (36) assumes a mature, stable, rootzone, while Equation (30) incorporates an expression for root growth. Neither Equation (36) nor Equation (30) consider the redistribution of roots (and consequently a more realistic representation of water uptake) as a result of stress at particular depths. Such a detailed model for plant root distribution and plant water uptake may not always be essential for a chemical transport model but estimation of the evapotranspiration is essential.

HEAT TRANSPORT

Prediction of temperature in the unsaturated zone is required for prediction of water movement and water content. Plant growth, extraction of water by plant roots and evaporation of water at the soil surface are all highly dependent on temperature. In addition, soil temperature is required for calculation of the temperature dependence of the chemical, kinetic and equilibrium reactions as well as for prediction of CO_2 production and pH. Among the available heat transport models are those of Milley et al. (1980) and Nassar et al. (1992), the latter on considering heat, water and nonreactive solute transport. Šimůnek and Suarez (1994) included a heat transport routine in UNSATCHEM which is used for prediction of the factors discussed above.

CONCENTRATION/PRODUCTION/TRANSPORT OF CARBON DIOXIDE

Vadose zone models typically either consider a closed system with constant inorganic carbon, as is also commonly considered for ground water systems, or assume an open system at fixed CO_2. The first assumption is clearly not desirable as large amounts of

CO_2 are produced by plant decomposition as well as plant root respiration. Models such as LEACHM consider an open system with specification of the CO_2 concentration as an input variable. Specification of a fixed CO_2 is a marked improvement over the closed system assumption but still does not consider the spatial and temporal fluctuations. These changes are due to both changes in production of CO_2, as well as changes in the transport of CO_2, which is mostly related to changes in the air-filled porosity of the soil, but can also be related to the flow of water. In the shallow vadose zone, the quantity of CO_2 added or removed by mineral dissolution/precipitation reactions is usually relatively small compared to the production and flux values and can be neglected. Below the CO_2 production zone this process may constitute the major control on CO_2 gas concentration.

Carbon dioxide production

Šimůnek and Suarez (1993b) described a general model for CO_2 production and transport. They considered CO_2 production as the sum of the production rate by soil microorganisms, γ_s [$L^3L^{-3}T^{-1}$], and the production rate by plant roots, γ_p [$L^3L^{-3}T^{-1}$]

$$P = \gamma_s + \gamma_p = \gamma_{s0} \prod_i f_{si} + \gamma_{p0} \prod_i f_{pi} \tag{37}$$

where the subscript s refers to soil microorganisms and the subscript p refers to plant roots, Πf_i is the product of reduction coefficients dependent on depth, temperature, pressure head (the soil water content), CO_2 concentration, osmotic head and time. The parameters γ_{s0} and γ_{p0} represent, respectively, the optimal CO_2 production by the soil microorganisms and plant roots for the whole soil profile at 20°C under optimal water, solute and soil CO_2 concentration conditions [$L^3L^{-2}T^{-1}$]. The individual reduction functions are given in Šimůnek and Suarez (1993a); a discussion of selection of the values for optimal production as well as coefficients for the reduction functions is given in Suarez and Šimůnek (1993).

Carbon dioxide transport

Gas transport in the unsaturated zone includes three general transport mechanisms (Massmann and Farrier, 1992): Knudsen diffusion, multicomponent molecular diffusion and viscous flow. Thorstenson and Pollock (1989) presented equations to describe these transport mechanisms in a multicomponent gas mixture. They also presented the Stefan-Maxwell approximation of these equations, where Knudsen diffusion and viscous flow are neglected. The original equations, as well as the Stefan-Maxwell approximation, are fully coupled and generally highly nonlinear. However, Massmann and Farrier (1992) showed that gas fluxes in the unsaturated zone can satisfactorily be simulated using the single-component transport equation, neglecting Knudsen diffusion, as long as the gas permeability of the medium is greater than about 10^{-10} cm^2. They also showed that overestimation of the gas fluxes using the single component advection diffusion equation becomes quite large for permeabilities of 10^{-12} to 10^{-13} cm^2. Use of the transport equation based on Fick's law to represent diffusive fluxes seems to be justified since air permeabilities smaller than 10^{-12} cm^2 occur only for very fine grained materials or for soils close to saturation. Also, Freijer and Leffelaar (1996) showed that CO_2 concentrations and fluxes can be described by Fick's law to within 5% accuracy.

The one-dimensional carbon dioxide transport model presented by Šimůnek and Suarez (1993a) assumed that CO_2 transport in the unsaturated zone occurs in both the liquid and gas phases. Furthermore, the CO_2 concentration in the soil is governed by two transport mechanisms (Patwardhan et al., 1988), convective transport in the aqueous phase and diffusive transport in both gas and aqueous phases, and by CO_2 production and/or removal. Thus one-dimensional CO_2 transport is described by the following equation:

$$\frac{\partial c_T}{\partial t} = -\frac{\partial}{\partial z}(J_{da} + J_{dw} + J_{ca} + J_{cw}) - Sc_w + P \tag{38}$$

where J_{da} is the CO_2 flux caused by diffusion in the gas phase [LT^{-1}], J_{dw} the CO_2 flux caused by dispersion in the dissolved phase [LT^{-1}], J_{ca} the CO_2 flux caused by convection in the gas phase [LT^{-1}], and J_{cw} the CO_2 flux caused by convection in the dissolved phase [LT^{-1}]. The term c_T is the total volumetric concentration of CO_2 [L^3L^{-3}] and P is the CO_2 production/sink term [$L^3L^{-3}T^{-1}$]. The term Sc_w represents the dissolved CO_2 removed from the soil by root water uptake. This term represents the (reasonable) assumption that when plants take up water the dissolved CO_2 is also removed from the soil-water system.

The individual terms in Equation (38) can be defined as (Patwardhan et al., 1988)

$$\begin{aligned} J_{da} &= -\theta_a D_a \frac{\partial c_a}{\partial z} \\ J_{dw} &= -\theta_w D_w \frac{\partial c_w}{\partial z} \\ J_{ca} &= -q_a c_a \\ J_{cw} &= -q_w c_w \end{aligned} \tag{39}$$

where c_w and c_a are the volumetric concentrations of CO_2 in the dissolved phase and gas phase [L^3L^{-3}], respectively, D_a is the effective soil matrix diffusion coefficient of CO_2 in the gas phase [L^2T^{-1}], D_w is the effective soil matrix dispersion coefficient of CO_2 in the dissolved phase [L^2T^{-1}], q_a is the soil air flux [LT^{-1}], q_w is the soil water flux [LT^{-1}] and θ_a is the volumetric air content [L^3L^{-3}].

The total CO_2 concentration, c_T [L^3L^{-3}], is defined as the sum of CO_2 in the gas and dissolved phases

$$c_T = c_a\theta_a + c_w\theta_w \tag{40}$$

After substituting Equations (40) and (39) into (38) we obtain

$$\frac{\partial(c_a\theta_a + c_w\theta_w)}{\partial t} = \frac{\partial}{\partial z}\theta_a D_a \frac{\partial c_a}{\partial z} + \frac{\partial}{\partial z}\theta_w D_w \frac{\partial c_w}{\partial z} - \frac{\partial}{\partial z}q_a c_a - \frac{\partial}{\partial z}q_w c_w - Sc_w + P \tag{41}$$

The total aqueous phase CO_2, c_w, is defined as the sum of CO_2(aq) and H_2CO_3, and is related to the CO_2 concentration in the gas phase by (Stumm and Morgan, 1981)

$$c_w = K_H R T c_a \tag{42}$$

where K_H is the Henry's Law constant [$MT^2M^{-1}L^{-2}$], R is the universal gas constant (8.314 $kgm^2s^{-2}K^{-1}mol^{-1}$) [$ML^2T^{-2}K^{-1}M^{-1}$] and T is the absolute temperature [K]. The value of K_H as a function of temperature is taken from Harned and Davis (1943). Aqueous carbon also exists in the form of HCO_3^-, CO_3^{2-} and other complexed species, such as $CaCO_3^o$, and these species should be included in the definition of c_w. Determination of these species cannot be made without use of a complete chemical speciation program. Substituting Equation (42) into (41) gives

$$\frac{\partial R_f c_a}{\partial t} = \frac{\partial}{\partial z} D_E \frac{\partial c_a}{\partial z} - \frac{\partial}{\partial z} q_E c_a - S^* c_a + P \tag{43}$$

where R_f is the CO_2 retardation factor [-], D_E is the effective dispersion coefficient for CO_2 in the soil matrix [L^2T^{-1}], q_E is the effective velocity of CO_2 [LT^{-1}], S^* is the CO_2 uptake rate [T^{-1}] associated with root water uptake and θ_a is the volumetric air content [L^3L^{-3}]. These parameters are defined as

$$\begin{aligned} R_f &= \theta_a + K_H R T \theta_w \\ D_E &= \theta_a D_a + K_H R T \theta_w D_w \\ q_E &= q_a + K_H R T q_w \\ \theta_a &= p - \theta_w \\ S^* &= S K_H R T \end{aligned} \tag{44}$$

Equation (43) is a nonlinear partial differential equation in which all parameters, except for c_a and q_a, are either known or can be obtained from solutions of the water flow equation. The nonlinearity of Equation (43) is caused by the term P which is dependent on the CO_2 concentration, c_a. Since the model does not consider coupled water and air movement, the flux of air, q_a, is unknown and thus additional assumptions are required. One possibility is to assume that the advection of CO_2 in response to the total pressure gradient is not important compared to CO_2 diffusion, and therefore to assume a stagnant gas phase and consider only diffusive transport within the gas phase (q_a=0). Another possibility is to consider that because of the much lower viscosity of air in comparison to water, significant gas flow can be caused by a relatively small pressure gradient. Thus, only rarely will the gas phase not be at atmospheric pressure throughout the unsaturated zone. Also, under most conditions, the compressibility of the air can be neglected. Then, with the assumption that the air flux is zero at the lower soil boundary and that water volume changes in the soil profile caused by the water flow must be immediately matched by corresponding changes in the gas volume, Šimůnek and Suarez (1993) obtained

$$q_a(z) = q_w(0) - q_w(z) + \int_{L-L_r}^{z} S(z)dz \tag{45}$$

This latter assumption seems to be reasonable, since when water leaves the soil system due to evaporation and root water uptake, air enters the soil at the surface and, vice versa, when water enters the soil during precipitation and irrigation events, soil air is escaping. Only in the case of saturation (typically at the soil surface) does the condition arise that air can not escape and is compressed under the wetting front.

Two-dimensional CO_2 transport (Šimůnek and Suarez, 1994) is described by the following mass balance equation

$$\frac{\partial(c_a\theta_a + c_w\theta)}{\partial t} = \frac{\partial}{\partial x_i}(\theta_a D_{ij}^a \frac{\partial c_a}{\partial x_j}) + \frac{\partial}{\partial x_i}(\theta D_{ij}^w \frac{\partial c_w}{\partial x_j}) - \frac{\partial}{\partial x_i}(q_i c_w) - Sc_w + P \quad (46)$$

where D_{ij}^a is the effective soil matrix diffusion coefficient tensor of CO_2 in the gas phase $[L^2T^{-1}]$, D_{ij}^w is the effective soil matrix dispersion coefficient tensor of CO_2 in the dissolved phase $[L^2T^{-1}]$, and q_i is the soil water flux $[LT^{-1}]$.

REACTIVE SINGLE COMPONENT SOLUTE TRANSPORT

Local equilibrium models

These models consider individual species and their reactions with the solid phase. The assumption is that the species of interest and the solid are unaffected by the concentration of other solutes and that the chemical processes influencing the species are instantaneous. The complex processes of adsorption and cation exchange are usually accounted for by a linear (Huyakorn et al., 1991) or nonlinear Freundlich isotherms (Yeh and Huff, 1985; Šimůnek and van Genuchten, 1994), where all reactions between solid and liquid phases are lumped into the distribution coefficient K_D (Liu and Narasimhan, 1989) and possibly into the nonlinear exponent. This approach requires experimental determination of the K_D for each material and for each set of chemical conditions, such as competing ions, pH, ionic strength, complexing ions etc. That these chemical effects are not "second-order corrections" is readily demonstrated by the findings of Means et al. (1978) who determined that the unpredicted and extensive migration of ^{60}Co from waste disposal trenches was due to the failure to consider EDTA complexation of what would otherwise be highly sorbed species. Despite these severe limitations the K_D approach has the appeal of simplicity, computational efficiency and avoids the requirements of needing complete chemical descriptions. Applicability and limitations of this approach have been presented, respectively, by Valocchi (1984) and Reardon (1981) among others.

Nonequilibrium models

In many instances the equilibrium approach has not been successful. Nonequilibrium expressions have been developed for precipitation, biodegradation, volatilization and radioactive decay. These processes have been simulated by simple first- or zero-order rate constants. Successful modeling is often demonstrated by excellent fit between breakthrough curves from laboratory columns and model simulations using parameters generated from the same experiments. Such models have been developed for nitrogen transformations, pesticide degradation and radioactive decay. Aside from radioactive decay the above mentioned processes cannot be represented by such simple relations without the need for different parameters for each set of conditions. While suitable for the laboratory experiments used to generate the parameters, such approaches will not be generally suited for natural systems. In addition to the single solute models, several models exist which simulate multiple solutes involved in sequential first-order decay reactions in one dimensional (Wagenet and Hutson, 1987) as well as in two dimensional models (Gureghian, 1981; Šimůnek and van Genuchten, 1994).

Various models are available to treat multisite sorption. For example Selim et al. (1976) and van Genuchten and Wagenet (1989) consider two-site models where one site is assumed to be at local equilibrium with the solution, while sorption on the other site is described by a kinetic expression. For a linear sorption process, the two-site transport model is given by (van Genuchten and Wagenet, 1989)

$$\frac{\partial}{\partial t}(\theta + f\rho\, k_D)c = \frac{\partial}{\partial z}(\theta D \frac{\partial c}{\partial z} - qc) - \alpha\rho[(1-f)k_D c - s_k] - \theta\,\mu_l c - f\rho k_D \mu_{s,e} c \tag{47}$$

$$\frac{\partial s_k}{\partial t} = \alpha[(1-f)k_D c - s_k] - \mu_{s,k} s_k \tag{48}$$

where μ_l and μ_s are first-order degradation constants for the chemical in the solution and sorbed phases, respectively; f is the fraction of adsorption sites assumed to be at equilibrium, and the subscripts e and k refer to equilibrium and kinetic sorption sites. The adsorption model has been used successfully in several laboratory column miscible displacement studies (Rao et al., 1979; Parker and Jardine, 1986).

A related model is the two-region transport model, which deals with physical nonequilibrium conditions. This approach assumes that there are two distinct solution regions, one immobile and the other mobile. Each region is in equilibrium with a specified fraction of the solid phase, and solute exchange between the regions is treated as a first-order process. The governing equations are (van Genuchten and Wagenet, 1989)

$$\frac{\partial}{\partial t}(\theta_m + f\rho k_D)c_m = \frac{\partial}{\partial z}(\theta_m D_m \frac{\partial c_m}{\partial z} - qc_m) - \alpha(c_m - c_{im}) - (\theta\,\mu_{l,m} + f\rho\, k_D \mu_{s,m})c_m \tag{49}$$

$$[\theta_{im} + (1-f)\rho k_D]\frac{\partial c_{im}}{\partial t} = \alpha(c_m - c_{im}) - [\theta_{im}\mu_{l,im} + (1-f)\rho k_D \mu_{s,im}]c_{im} \tag{50}$$

where the subscript m refers to the mobile solution region, im to the immobile solution region, l to the liquid phase and s to the sorbed phase. The term f represents the fraction of sorption sites that equilibrate with the mobile liquid phase, and α is a first-order mass transfer coefficient governing the rate of solute exchange between the mobile and immobile solution regions. The two-region physical nonequilibrium model has been successfully applied to laboratory-scale transport experiments involving tritiated water, chloride, and organic chemicals, among others (Gaudet et al., 1977; van Genuchten et al., 1977; Nkedi-Kizza et al., 1983; Gaber et al., 1995).

Since the two-site and two-region nonequilibrium models have the same mathematical structure, it is not possible to distinguish between the processes of chemical and physical nonequilibrium, based on the fit to these models. A determination as to which of the processes is applicable requires additional experimentation, such as the use of nonadsorbing tracers (van Genuchten and Šimůnek, 1996).

The two-region nonequilibrium physical model mentioned above assumes that water flow occurs only in the macropore region. While this model has been successfully used for saturated flow conditions, it clearly is less realistic for variably saturated conditions, especially for drier situations when the macropores are empty and flow only occurs through the finer pores. Among the models which describe flow in two regions are those of Wang (1991), Zimmerman et al. (1993), and Gerke and van Genuchten (1993).

The dual-porosity model developed by Gerke and van Genuchten (1993) assumes that the Richards equation for transient water flow and the convection-dispersion equation for solute transport can be applied to each of the two pore systems as follows

$$C_f \frac{\partial h_f}{\partial t} = \frac{\partial}{\partial z}\left(K_f \frac{\partial h_f}{\partial z} - K_f\right) - \frac{\Gamma_w}{w_f} \tag{51}$$

$$C_m \frac{\partial h_m}{\partial t} = \frac{\partial}{\partial z}\left(K_m \frac{\partial h_m}{\partial z} - K_m\right) + \frac{\Gamma_w}{1 - w_f} \tag{52}$$

and

$$\frac{\partial}{\partial t}(\theta_f R_f c_f) = \frac{\partial}{\partial z}\left(\theta_f D_f \frac{\partial c_f}{\partial z} - q_f c_f\right) - \frac{\Gamma_s}{w_f} \tag{53}$$

$$\frac{\partial}{\partial t}(\theta_m R_m c_m) = \frac{\partial}{\partial z}\left(\theta_m D_m \frac{\partial c_m}{\partial z} - q_m c_m\right) + \frac{\Gamma_s}{1 - w_f} \tag{54}$$

where the subscripts f and m refer to the macropore and matrix pore systems, respectively; Γ_w and Γ_s describe the rate of exchange of water and solute, respectively, between the two regions, w_f is the volume of the macropore region relative to that of the total soil pore system. Similarly as for the first-order mobile-immobile transport models, water and solute mass transfer between the two pore systems is described with first-order rate equations:

$$\Gamma_w = \alpha_w (h_f - h_m) \tag{55}$$

$$\Gamma_s = \alpha_s (c_f - c_m) + \begin{cases} \Gamma_w w_f \theta_f c_f / \theta & \Gamma_w \geq 0 \\ \Gamma_w (1 - w_f) \theta_m c_m / \theta & \Gamma_w < 0 \end{cases} \tag{56}$$

in which α_w and α_s are first-order mass transfer coefficients for water and solute, respectively. The first term on the right-hand side of Equation (56) specifies the diffusion contribution to Γ_s, while the second term gives the convective contribution. The above variably-saturated dual-porosity transport model reduces to the first-order model for conditions of steady-state flow in the macropore region and no flow in the matrix pore system ($q_m = \Gamma_w = 0$). The simultaneous solution of the coupled Richards equations and first-order nonlinear transfer term for approximation of fluid exchange results in a complex representation. Practical application of the model requires further development of accurate and efficient numerical solutions (Gerke and van Genuchten, 1993). Subsequently Tseng et al. (1995) developed a numerically stable and relatively efficient partitioned procedure for solution of the dual-porosity problem.

COUPLED WATER FLOW AND MULTICOMPONENT MODELS

Much of the research on multicomponent solute transport is focussed on the saturated zone. Most of the developed models were based on assumptions of one-dimensional steady-state saturated water flow with a fixed water flow velocity, temperature and pH (Valocchi et al., 1981; Jennings et al., 1982; Walsh et al., 1984; Cederberg, 1985; Kirkner et al., 1985; Förster, 1986; Bryant et al., 1986; Förster and Gerke, 1988; Kirkner and Reeves, 1988; among others). Under saturated conditions, changes in water velocity, temperature and solution composition, including pH, are relatively gradual and thus these assumptions may be acceptable for many cases. In contrast the assumptions are not reasonable for unsaturated zone modeling, particularly for simulations of conditions near the soil surface.

In the past 10 to 15 years the increasing speed of computer processing has enabled development of numerical schemes and models which couple water flow and solute transport with chemical models. Among recent reviews of hydrogeochemical transport models of reactive multichemical components are those given by Kirkner and Reeves (1988), Yeh and Tripathi (1989), Rubin (1990) and Mangold and Chin-Fu Tsang (1991). Kirkner and Reeves (1988) presented an analysis of several approximation methods for solving the governing equations for multicomponent mass transport with chemical reactions. They discuss how the choice of formulation and solution algorithm should reflect the type of chemical problem encountered. Yeh and Tripathi (1989) provided a critical review of many computational methods that have been presented in the hydrologic literature for solving multicomponent, equilibrium-controlled transport.

Equilibrium models

The majority of coupled chemistry-water flow models predict concentrations based on thermodynamic equilibrium reactions. Among these extensive chemical equilibrium models are WATEQ (Truesdell and Jones, 1974), which computes solution equilibria, and MINTEQ (Allison et al., 1990), GEOCHEM (Sposito and Mattigod, 1977) and PHREEQE (Parkhurst et al., 1980) which compute solution equilibria and have the option to predict solution-solid equilibria as well. These models have been incorporated into combined transport chemical equilibria models, such as the saturated flow models described by Narasimhan et al. (1986), Liu and Narasimhan (1989), and Griffioen (1993). These coupled models are of the type classified by Yeh and Tripathi (1989) as solving sets of linear partial differential equations and nonlinear algebraic equations using a sequential iteration approach (SIA). Yeh and Tripathi (1989) consider that only those models which employ the SIA approach and which use either "total analytical concentrations of aqueous components, or the concentrations of all species, or the concentrations of all component and precipitated species as the primary dependent variables, can encompass the full complement of geochemical processes." Models which simultaneously solve a set of mixed differential and algebraic equations (DAE) or directly substitute the chemical equations into the hydrological transport equation (DSA) are regarded as requiring excessive CPU memory and time to be used for other than simple chemical descriptions. A more detailed comparison of these solution methods is presented in Steefel and MacQuarrie (this issue).

Within the past few years several models were published that can be applied to problems that include multicomponent solute transport and two-dimensional variably saturated water flow (Liu and Narasimhan, 1989; Yeh and Tripathi; 1991, Šimůnek and Suarez, 1994). Narasimhan et al. (1986) and Liu and Narasimhan (1989) developed the model DYNAMIX which can be used in conjunction with an integral finite difference program for fluid flow in variably saturated porous media. Yeh and Tripathi (1991) presented the development and a demonstration of a two-dimensional finite element hydrogeochemical transport model, HYDROGEOCHEM, for simulating transport of reactive multispecies chemicals. The HYDROGEOCHEM model consists of a matrix solver in which the user adds the desired equations and components to obtain an equilibrium solution. Generalized relations are presented by Yeh and Tripathi (1991) for use in describing kinetic processes, but were not included in the model. The model is not well suited for vadose zone simulations as it considers fixed inorganic carbon and does not allow for open system CO_2 control where pH and CO_2 can vary. Except for Šimůnek and Suarez (1994), the presented examples in all of these papers are only for steady state water movement.

The governing equation for one-dimensional convective-dispersive chemical transport under transient flow conditions in partially saturated porous media is taken as (Suarez and Šimůnek, 1996)

$$\frac{\partial \theta c_{T_i}}{\partial t} + \rho \frac{\partial \bar{c}_{T_i}}{\partial t} + \rho \frac{\partial \hat{c}_{T_i}}{\partial t} = \frac{\partial}{\partial z}[\theta D \frac{\partial c_{T_i}}{\partial z} - q c_{T_i}] \qquad i = 1, n_s \tag{57}$$

where c_{Ti} is the total dissolved concentration of the aqueous component i [ML^{-3}], $\bar{c}_{Ti}$ is the total adsorbed or exchangeable concentration of the aqueous component i [MM^{-1}], $\hat{c}_{Ti}$ is the non-adsorbed solid phase concentration of aqueous component i [MM^{-1}], ρ is the bulk density of the soil [ML^{-3}], D is the dispersion coefficient [L^2T^{-1}], q is the volumetric flux [LT^{-1}], and n_s is the number of aqueous components. The second and third terms on the left side of Equation (55) are zero for components that do not undergo ion exchange, adsorption or precipitation/dissolution. The coefficient D is the sum of the diffusion and dispersion components

$$D = \tau D_m + \lambda \frac{|q|}{\theta} \tag{58}$$

where τ is the tortuosity factor [-], D_m is the coefficient of molecular diffusion [L^2T^{-1}], and λ is the dispersivity [L]. This representation is a simplified treatment of the diffusion process. A more detailed description of diffusion requires calculation of the diffusion rates of individual species, with consideration of Coulombic interactions which maintain electroneutrality, requiring coupling of individual ion fluxes to the concentration gradients of all individual species (Lasaga, 1979). If ion pairs are utilized in the chemical model, this additional coupling of diffusion rates should also be made (Lasaga, 1979). However, in porous media, errors generated by uncertainty in determination of the tortuosity factor and velocity vectors are likely more significant for determination of solute transport than errors associated with a simplified treatment of diffusion.

Realistic modeling of the chemistry in the unsaturated zone requires consideration of various factors which are usually not considered. Among these are changes in hydraulic properties of the soil as related to the solution chemistry (discussed above), water uptake by plant roots and the spatial distribution of water uptake, temperature effects on water uptake, gas phase composition, equilibrium or reaction constants, and prediction of the dynamic changes in CO_2 concentration with time and space, all of which affect water and solute movement as well as the chemical processes for the solutes of interest.

Generalized models

The model HYDROGEOCHEM (Yeh and Tripathi, 1991) is one of the most extensive in terms of its potential to handle a wide range of problems. The model contains generalized routines for treatment of chemical equilibrium, including aqueous complexation, adsorption reactions, ion exchange, precipitation-dissolution, redox reactions, and acid-base reactions, all considering chemical equilibria. The model calculates activity coefficients using the Davies equation. Specification of the thermodynamic constants requires that users first correct the constants for the particular temperature and pressure of the case to be run. Acid-base reactions and prediction of pH is possible by calculation of the proton balance. Oxidation-reduction processes can also be treated with this model. Oxidation-reduction reactions are treated by defining electron activity as a master variable and making this entity an aqueous component subject to transport. Such calculations allow for equilibrating redox within a closed system but will not be suitable for most environments where biological processes, such as microbial degradation of organic matter are dominant.

Adsorption equilibria in HYDROGEOCHEM are obtained from the law of mass action

$$y_i = \alpha_i^y \left(\prod_{k=1}^{N_a} c_k^{a_{ik}^y} \right) \left(\prod_{k=1}^{N_s} s_k^{b_{ik}^y} \right) \qquad i = 1,2,\ldots,M_y \tag{59}$$

in which

$$\alpha_i^y = K_i^y \left[\prod_{k=1}^{N_a} (\gamma_k^a)^{a_{ik}^y} \right] \left[\prod_{k=1}^{N_s} (\gamma_k^s)^{b_{ik}^y} \right] (\gamma_i^y)^{-1} \qquad i = 1,2,\ldots,M_y \tag{60}$$

where α_i^y is the stability constant of the i th adsorbed species, K_i^y is the equilibrium constant of the i th adsorbed species, γ_i^y is the activity coefficient of the i th adsorbed species, γ_k^s is the activity coefficient of the k th adsorbent component species, γ_k^a is the activity coefficient of the k th aqueous species, M_y is the number of adsorbed species, N_a is the number of aqueous components, N_s is the number of adsorbent components, a_{ik}^y is the stoichiometric coefficient of the k th aqueous component in the i th adsorbed species, s_k is the concentration of the k th adsorbent component species, and b_{ik}^y is the stoichiometric coefficient of the k th adsorbent component in the i th adsorbed species. Alternatively HYDROGEOCHEM has the capacity to treat adsorption using the triple layer model (Davis et al., 1978; Davis and Leckie, 1978, 1980).

Generalized kinetic expressions for reactions are described by Yeh and Tripathi, (1989), (although not included in the HYDROGEOCHEM code) where for formation of a complex

$$\theta \frac{\partial x_i}{\partial t} = L(x_i) + \theta r_i^x \qquad i = 1,2,\ldots,K_x \tag{61}$$

where K_x is the number of complexed species of kinetic reactions, and r_i^x can be given by

$$r_i^x = -k_i^{bx} x_i + k_i^{fx} \prod_{k=l}^{N_a} c_k^{a_{ik}^x} \quad i = 1,2,\ldots,K_x \tag{62}$$

where k_i^{bx} is the modified back reaction rate constant for the i th reaction and k_i^{fx} is the modified forward reaction constant for the i th reaction. For sorption reactions

$$\theta \frac{\partial y_i}{\partial_t} + \frac{\partial \theta}{\partial t} y_i = \theta r_i^y \qquad i = 1,3,\ldots, K_y \tag{63}$$

with the reaction rate given by

$$r_i^y = -k_i^{by} y_i + k_i^{fy} \prod_{k=l}^{N_a} c_k^{a_{ik}^y} \prod_{k=l}^{N_s} S_k^{b_{ik}^y} \qquad i = 1,2,..,K_y \tag{64}$$

Precipitation reactions are represented by

$$\theta \frac{\partial p_i}{\partial t} + \frac{\partial \theta}{\partial t} p_i = \theta r_i^p \qquad i = 1,2,\ldots,K_p \tag{65}$$

where the reaction rate is given by

$$r_i^p = -k_i^{bp} + k_i^{fp} \prod_{k=l}^{N_a} C_k^{a_{ik}^p} \qquad i = 1,2,..,K_p \tag{66}$$

in which K_p is the number of precipitated species of kinetic reactions, k_i^{bp}, k_i^{fp} are the modified back reaction and forward reaction rate constants for the i th precipitation reaction. This generalized formulation provides great flexibility but requires a user to provide the needed input for specific vadose zone problems. This includes deciding which reactions are pertinent, and which are to be treated via equilibrium expressions.

Prediction of pH in this model is made under the assumptions of either a closed system with fixed total carbon, or possibly adding an expression for the solubility of CO_2 in water (open system). As discussed earlier neither of these is optimal, as there is important production of CO_2 in the soil, as well as losses due to diffusion and transport with the water phase. In most instances biological processes are the driving force for redox changes, and the electron balance must be expanded to include these processes. Reduction processes in the subsurface generally occur as a result of oxygen consumption and microbial production of reduced species. These equilibrium models consider only the redistribution of electrons among inorganic species, or equilibration of the system to a fixed *p*E.

Models with specified chemistry

A number of models have been developed which are limited in the number of aqueous species and solids considered, but which nonetheless include the effects of biological processes. These models simulate specific processes and provide the user with the required reaction constants. Robbins et al. (1980a,b) developed chemical precipitation-dissolution and cation exchange subroutines using equilibrium chemistry and coupled them with a one-dimensional water movement-salt transport-plant growth model. They tested their model by comparing its results with experimental data obtained from a lysimeter study.

Robbins equilibrium chemistry model was also the basis for the numerical code LEACHMS of Wagenet and Hutson (1987). The LEACHMS and SOWATCH (Dudley and Hanks, 1991) models utilize a simplified chemical routine to predict cation exchange and major ion chemistry with calcite and gypsum solid phase control. Suarez and Dudley (1996) compared the output from LEACHMS, SOWATCH and UNSATCHEM (Suarez and Šimůnek, 1996) for a simulation with application of high alkalinity, high sodium water. All models gave somewhat different results but the LEACHMS model failed to predict high alkalinity and high exchangeable sodium. This was attributed to failure to accurately converge on the proper solution to the calcite equilibrium problem, due to use of a quadratic expression with the CO_3^{2-} term rather than a third-order solution using HCO_3^-.

Russo (1986) combined the salinity model of Robbins et al. (1980a) with the transport model of Bresler (1973) to theoretically investigate the leaching of a gypsiferous-sodic soil under different soil conditions and water qualities. With the exception of the UNSATCHEM (Suarez and Šimůnek, 1996) and UNSATCHEM-2D (Šimůnek and Suarez, 1994) models, most of these models call the equilibrium chemistry routines only once at each time step without iterating between transport and chemical modules. In many cases, as was shown by Yeh and Tripathi (1991) and Šimůnek and Suarez (1994), this simplification produces noticeable numerical error. This error is particularly noted in the distribution with depth of the precipitated or dissolved solid phase. The UNSATCHEM model (Suarez and Šimůnek, 1994) also differs from other open system unsaturated zone models which consider plant water uptake in that it considers kinetic as well as equilibrium processes, includes chemical effects on hydraulic conductivity, predicts CO_2 and pH, predicts heat transport and soil temperature and corrects equilibrium or reaction expressions for temperature effects, calculates activities using the Pitzer equations at high ionic strength, and includes a larger set of possible reactions, including B adsorption, and cation exhange with organic matter as well as clay.

In comparison to the generalized coupled equilibrium-transport models, UNSATCHEM provides a limited set of potential solid phase controls. Distinction is also made between dissolution and precipitation processes, for which kinetic expressions are utilized if natural systems are not generally at equilibrium. Precipitation processes are evaluated with consideration only for phases that control solution composition under earth surface conditions, using kinetic expressions if supersaturation persists. Published data for natural systems have indicated that kinetics and poorly crystallized phases often control solution composition. For example, studies of major ion compositions in and below the rootzone of calcareous arid zone soils have indicated that calcite equilibrium is not a reasonable assumption for predicting water composition (Suarez, 1977; Suarez et al., 1992) and that a kinetic expression yields values closer to field measurements (Suarez, 1985).

As discussed earlier, most models assume either a fixed pH or a fixed CO_2, assumptions which are questionable for soils, which often exhibit rapid fluctuations in both of these variables (Suarez and Šimůnek, 1993). Although input of pH may be suitable for ex post facto simulation of data where pH has been measured, it is generally not suitable for predictive purposes. In addition there are almost no vadose zone measurements for solution pH, as this requires specialized sampling to avoid CO_2 degassing, precipitation of carbonates and oxyhydroxides and pH shifts during solution extraction, as well as during storage.

The processes of evaporation and plant transpiration serve to concentrate the salts in the remaining soil water. When these processes are combined with irrigation in arid regions, saline conditions can result, especially during transient conditions when the water content is low. These chemical conditions require that ion activities be calculated with expressions suitable for use in brines rather than the standard formulations for dilute solutions. The interaction of evapotranspiration, changing soil gas composition, ion exchange and soil-water reactions requires that we consider the potential to precipitate or dissolve various minerals. These solids and major ions (consisting mainly of Ca, Mg, Na, Cl and SO_4 and secondarily of K, alkalinity and NO_3) can accumulate in certain parts of the soil profile in such amounts that water consumption, and thus further concentration increases resulting from transpiration can be seriously reduced.

Soil temperature, which can change annually from about -10 up to +50 °C, can significantly change the thermodynamic equilibrium constants and reaction rates and therefore influence even the selection of the method used for predicting soil solution chemistry, ranging from equilibrium models to models based on kinetic expressions.

UNSATCHEM CHEMICAL MODEL

The UNSATCHEM model (Suarez and Šimůnek, 1996) includes equilibrium chemistry for aqueous species and either equilibrium or kinetic expressions for the solid phase controls. All equilibrium constants are calculated from available temperature dependent expressions. Soil temperature is calculated based on a heat flow submodel, with input of air temperature. Eight major aqueous components, consisting of Ca, Mg, Na, K, SO_4, Cl, alkalinity and B are defined, along with SO_4, CO_3, and HCO_3 complexes. Alkalinity is defined as

$$\text{Alkalinity} = [HCO_3^-] + 2[CO_3^{2-}] + 2[CaCO_3^0] + [CaHCO_3^+] + 2[MgCO_3^0] + [MgHCO_3^+] + 2[NaCO_3^-] + [NaHCO_3^0] + [H_2BO_3^-] - [H^+] + [OH^-] \tag{67}$$

where brackets represent concentrations. The reactions in the CO_2-H_2O system and complexation reactions for major ions have been described in numerous publications; thus further discussion is not needed.

Calcite precipitation

The equilibrium condition of a solution with calcite in the presence of CO_2 can be described by the expression

$$(Ca^{2+})(HCO_3^-)^2 = \frac{K_{SP}^C K_{CO_2} K_{a_1}}{K_{a_2}} P_{CO_2}(H_2O) = K_{SP}^C K_T \tag{68}$$

where parenthesis denote activities, and K_{CO2} is the Henry's law constant for the solubility of CO_2 in water, K_{a1} and K_{a2} are the first and second dissociation constants of carbonic acid in water, and $K_{SP}{}^C$ is the solubility product for calcite. To obtain equilibrium, i.e., when the ion activity product (IAP) is equal to the solubility product K_{SP}, a quantity x of Ca^{2+} and HCO_3^- must be added or removed from the solution to satisfy the equilibrium condition. The quantity x is obtained by solving the following third-order equation

$$[\mathrm{Ca}^{2+} + x]\,[\mathrm{HCO}_3^- + 2x]^2 = \frac{K_{SP}^C K_T}{\gamma_{\mathrm{Ca}^{2+}}\,\gamma_{\mathrm{HCO}_3^-}^2} \tag{69}$$

Several options are available if a kinetic condition is specified. The calcite kinetic models are almost all based on the assumption that the reaction rate is dependent on the surface area of the calcite. Of all the models, the deterministic dissolution/precipitation model of Plummer et al. (1978) is the most comprehensive, as it is based on dissolution studies over a wide range in pH (2-7.0) and CO_2 pressure (35Pa to 100kPa). The reaction rate of calcite dissolution in the absence of DOC is thus calculated with the rate equation of Plummer et al. (1978)

$$R^C = k_1\,(\mathrm{H}^+) + k_2(\mathrm{H_2CO_3}^*) + k_3(\mathrm{H_2O}) - k_4\frac{K_{a_2}}{K_{SP}^C}(\mathrm{Ca}^{2+})(\mathrm{HCO}_3^-) \tag{70}$$

where

$$k_4 = k_1' + \frac{1}{(\mathrm{H}_S^+)}\left[k_2(\mathrm{H_2CO_3^*}) + k_3(\mathrm{H_2O})\right] \tag{71}$$

and k_1, k_2, k_3, and k_1' are temperature dependent rate constants. The dissolution-precipitation rate R^C is expressed in mmol of calcite per cm^2 of surface area per s. The term (H_S^+) is the H^+ activity at the calcite surface. It is assumed that $(H_S^+) = (H^+)$ of the solution at calcite saturation when P_{CO2} at the surface equals P_{CO2} in the bulk solution.

For conditions where $pH > 8$ and $P_{CO2} < 0.01$ atm, the above Equation (70) of Plummer et al. (1978) underestimates the precipitation rate (Suarez, 1983; Inskeep and Bloom, 1985). Under these conditions it seems preferable to use the rate expression of Inskeep and Bloom (1985)

$$R^C = -k_f\,[(\mathrm{Ca}^{2+})(\mathrm{CO}_3^{2-}) - K_{SP}^C] \tag{72}$$

where k_f = 118.2 mol^{-1}m^{-2}s^{-1}, with an apparent Arrhenius activation energy of 48.1 kJ mol^{-1} for the rate constant.

The above relationships for calcite crystal growth, and all available calcite precipitation models, are based on a clean surface in the absence of "surface poisons". The inhibiting effect of dissolved organic matter (DOC) on calcite precipitation is well established and related to surface adsorption of DOC. Inskeep and Bloom (1986) reported on the effect of water-soluble organic carbon on calcite crystal growth. From their data we developed the following relationship (Suarez and Šimůnek, 1996)

$$r = \exp(-a_1 x - a_2 x^2 - a_3 x^{0.5}) \tag{73}$$

which represents the reduction of the precipitation or dissolution rate due to dissolved organic carbon, where r is the reduction constant [-], x is the DOC [μmol L^{-1}] and a_1, a_2, and a_3 are regression coefficients (0.005104, 0.000426, 0.069111, respectively). This relation provided an excellent fit to the data, with an R^2 value of 0.997. **The reduction** constant, r, is then multiplied by the R^C values calculated with either Equations (70) and (71), or Equation (72), to obtain the predicted rate constant in the presence of the specified DOC concentration.

These and other calcite rate models all consider reaction rates to be proportional to surface area. For simulation of calcite dissolution in natural systems, these models may be suitable, after adjustment for the poisoning of the surface, as discussed above. However, these rate models may not be suitable for predicting calcite precipitation rates, as the concentrations of DOC in near-surface natural environments are usually comparable to levels found by Inskeep and Bloom (1986) to completely inhibit calcite crystal growth. Furthermore, these precipitation rate experiments were usually done at calcite supersaturation levels where heterogeneous nucleation rather than crystal growth was dominant. As a result calcite precipitation rates in natural environments are not related to existing calcite surface areas, and these equations serve only as empirical fitting equations.

Recently Lebron and Suarez (1996) developed a precipitation rate model which considers the effects of dissolved organic carbon both on crystal growth and heterogeneous nucleation. The combined rate expression is given by

$$R_T = R_{CG} + R_{HN} \tag{74}$$

where R_T is the total precipitation rate, expressed in mmol $L^{-1}s^{-1}$, R_{CG} is the precipitation rate related to crystal growth, and R_{HN} is the precipitation rate due to heterogeneous nucleation. The R_{CG} term is given by

$$R_{CG} = s k_{CG}\left[(Ca^{2+})(CO_3^{2-}) - K_{SP}\right]\left[-0.14 - 0.11\log[DOC]\right] \tag{75}$$

where s is the calcite surface area, k_{CG} is the precipitation rate constant due to crystal growth, and DOC is the dissolved organic carbon in mmol L^{-1}. The R_{HN} term is given by

$$R_{HN} = k_{HN} f(SA)(\log\Omega - 2.5)(3.37x10^{-4}DOC^{-1.14}) \tag{76}$$

where k_{HN} is the precipitation rate constant due to heterogeneous nucleation, $f(SA)$ is a function of the surface area of the particles (e.g. clay) upon which heterogeneous nucleation occurs (= 1.0 if no solid phase is present), Ω is the calcite saturation value, and 2.5 is the Ω value above which heterogeneous nucleation can occur.
This equation leads to calcite precipitation rates which are independent of the calcite surface area, consistent with the experimental data of Lebron and Suarez (1996). The presence of calcite (varying surface area) does not affect the calcite precipitation rate when DOC is $\geq$ 0.10 mM. We consider Equations (75) and (76) to be the most realistic for precipitation in rootzone environments while Equations (70) and (72) are most suitable for dissolution. In the presence of DOC these equations should be used in combination with Equation (73), or similar expressions which can be developed to correct for the effects of DOC on dissolution.

Precipitation of gypsum

Precipitation / dissolution of gypsum can be described by

$$[\mathrm{Ca}^{2+}][\mathrm{SO}_4^{2-}] = \frac{IAP}{\gamma_{\mathrm{Ca}^{2+}}\gamma_{\mathrm{SO}_4^{2-}}(\mathrm{H_2O})^2} \tag{77}$$

To obtain equilibrium, i.e., when the IAP is equal to the solubility product K_{SP}^G, a quantity of gypsum, x, must be added or removed from the Ca^{2+} and SO_4^{2-} concentrations in solution, obtained by solving the quadratic equation.

Magnesium precipitation

The UNSATCHEM model considers that Mg precipitation can occur as a carbonate (either nesquehonite or hydromagnesite), or as a silicate (sepiolite). Since this is a predictive model, it considers only phases that either precipitate under earth surface conditions or occur frequently and are reactive under earth surface conditions (these need not necessarily be the thermodynamically most stable). With this consideration magnesite can be neglected, as it apparently does not form at earth surface temperatures, is relatively rare, and its dissolution rate is exceedingly small, such that its solubility has not yet been satisfactorily determined from dissolution studies at or near 25°C. Similarly, dolomite precipitation is not considered, since true dolomite appears to very rarely form in soil environments. If dolomite is present in the soil, the model uses the kinetic formulation of Busenberg and Plummer (1982) to represent the dissolution process. The dissolution rate of dolomite is very slow, especially when the solution IAP values approach within 2 to 3 orders of magnitude of the solubility product.

If nesquehonite or hydromagnesite saturation is reached, the model will precipitate the predicted Mg carbonate. The Mg carbonate precipitated, combined with calcite precipitation, will likely represent the mixed Ca Mg precipitate called protodolomite. However, the resulting solution composition is much different than that produced by simply forcing equilibrium with respect to dolomite, as the model forms this mixed precipitate (calcite + magnesium carbonate) under conditions of approximately three orders of supersaturation with respect to dolomite. This result is consistent with the high levels of dolomite supersaturation maintained in high Mg waters (Suarez, unpublished data). Precipitation (or dissolution, if present in the soil) of sepiolite is also considered by the model. Sepiolite will readily precipitate into a solid with a K_{SP}^S greater than that of well crystallized sepiolite. Formation of this mineral requires high pH, high Mg concentrations and low CO_2 partial pressure.

Precipitation of nesquehonite and hydromagnesite

At 25°C and at CO_2 partial pressures above $10^{-3.27}$ kPa, nesquehonite $MgCO_3 \cdot 3H_2O$ is stable relative to hydromagnesite. The precipitation (if saturation is achieved) or dissolution of nesquehonite (if specified as a solid phase) in the presence of CO_2 can be described by

$$\mathrm{MgCO_3 \cdot 3H_2O + CO_2(g) \rightleftarrows Mg^{2+} + 2HCO_3^- + 2H_2O} \tag{78}$$

with the solubility product K_{SP}^N defined by

$$K_{SP}^N = (\mathrm{Mg}^{2+})(\mathrm{CO}_3^{2-})(\mathrm{H_2O})^3 \tag{79}$$

Substituting the equation for Henry's law for solubility of CO_2 in water, and the equations for the dissociation of carbonic acid in water into the solubility product, we obtain:

$$(Mg^{2+})(HCO_3^-)^2 = \frac{K_{SP}^N K_{CO_2} K_{a_1}}{K_{a_2}} \frac{P_{CO_2}}{(H_2O)^2} = \frac{K_{SP}^N K_T}{(H_2O)^3} \tag{80}$$

The equation is solved for equilibrium in a manner similar to that used for calcite, with a third order equation.

The precipitation or dissolution of hydromagnesite in the presence of CO_2 can be described by

$$Mg_5(CO_3)_4(OH)_2 \cdot 4H_2O + 6CO_2(g) \rightleftarrows 5Mg^{2+} + 10HCO_3^- \tag{81}$$

Similarly the equilibrium condition is defined by

$$(Mg^{2+})^5(HCO_3^-)^{10} = \frac{K_{SP}^H K_{CO_2}^6 K_{a_1}^6 P_{CO_2}^6}{K_{a_2}^4 K_w^2} = K_{SP}^H K_T \tag{82}$$

Again the equilibrium condition is solved as described for calcite and nesquehonite.

Precipitation of sepiolite

The precipitation or dissolution of sepiolite in the presence of CO_2 can be described by

$$Mg_2Si_3O_{7.5}(OH) \cdot 3H_2O + 4.5H_2O + 4CO_2(g) \rightleftarrows 2Mg^{2+} + 3H_4SiO_4^0 + 4HCO_3^- \tag{83}$$

with the solubility product $K_{SP}{}^S$ defined by

$$K_{SP}^S = \frac{(Mg^{2+})^2(H_4SiO_4^0)^3(OH^-)^4}{(H_2O)^{4.5}} \tag{84}$$

In this instance we utilize the precipitated sepiolite solubility value given by Wollast et al. (1968) rather than the well crystallized equilibrium value. Freshly precipitated sepiolite has been prepared in the laboratory at IAP values of 10^{-35} comparable to the $K_{SP}{}^S$ listed by Wollast et al. (1968), thus we consider that a kinetic expression for precipitation is not essential for prediction of unsaturated zone solution composition. The equilibrium condition is expressed as

$$(Mg^{2+})^2(HCO_3^-)^4 = \frac{K_{SP}^S K_{CO_2}^4 K_{a_1}^4 P_{CO_2}^4 (H_2O)^{4.5}}{K_w^4 (H_4SiO_4^0)^3} = K_{SP}^S K_T^+ \tag{85}$$

Relatively little information exists on the controls on Si concentrations in soil waters, especially in arid zones. In soil systems Si concentrations are not fixed by quartz solubility but rather by dissolution and precipitation of aluminosilicates and Si adsorption onto oxides and aluminosilicates. As a result of these reactions Si concentrations in soil solution follow a U shaped curve with pH, similar to Al oxide solubility with a Si minimum around pH 8.5 (Suarez 1977b).

There are two options in UNSATCHEM to predict Si concentrations in solution. In arid land soils it is assumed that Si in solution is a simple function of pH, fitted to data from 8 arid land soils reacted at various pHs for two weeks by Suarez (1977b), as follows

$$\Sigma SiO_2 = 0.001(6.34 - 1.43\,pH + 0.0819\,pH^2) \tag{86}$$

where SiO_2 is the sum of all silica species expressed in mol L^{-1}. This relationship likely provides only a rough estimate of Si concentrations, but we consider it acceptable because it is used only to restrain Mg concentrations at high levels of evapotranspiration, when Mg concentrations become very high at low CO_2 and elevated pH.

An additional option is to consider the Si concentration to be controlled by inputs from mineral weathering and concentrated only by processes of evapotranspiration. In this case UNSATCHEM utilizes kinetic expressions for the weathering of selected silicate minerals. Options are available for two kinetic models.

Silicate weathering

Several different rate expressions have been used for feldspar dissolution, the most successful of which are variations of the Furrer and Stumm (1986) model

$$\begin{aligned} R_t &= R_H + R_L \\ R_H &= k_H (C_H^s)^n \\ R_L &= k_L\, C_L^s \end{aligned} \tag{87}$$

where R_t, R_H and R_L are the proton and ligand promoted rates, k_L and k_H are the rate constants, C_H^S and C_L^S are, respectively, the surface concentrations of protons and ligands and n is the order of the reaction.

The Furrer and Stumm (1986) model developed for oxides, is not able to simulate the silicate dissolution rates in the pH range 3 to 8 (Amrhein and Suarez, 1988). Amrhein and Suarez (1988) modified this equation by adding a rate term proportional to the uncharged surface silanol groups and substituting the term Γ, which represents the sum of the proton and hydroxyl sites for the term C_H^S in Equation (87). Combining the surface proton and hydroxyl sites into one expression was justified by the essentially equal effects of the two surface groups on the dissolution rates of plagioclase feldspars. This has been determined experimentally in the work of Chou and Wollast (1985) for albite and Amrhein and Suarez (1988) for anorthite, representing the end members in the plagioclase feldspar series. However, other studies indicate that this equality cannot be generalized, thus we use the modified rate equation

$$R_t = a[\Gamma]^{4.0} + a'[\Gamma']^{4.0} + b\,[SOH] + c\,[S\text{-}L] \tag{88}$$

where a, a', b, and c are the rate coefficients for the proton, hydroxyl, neutral, and surface-ligand sites and Γ, Γ', *SOH*, and *S-L* are the surface concentrations of proton, hydroxyl, neutral and surface ligand sites. Detailed rate data for use with Equation (88) are not available for all silicates, however in the pH range of 5 to 9 this equation can often be simplified to

$$R_t = b\,[\mathrm{SOH}] + c\,[S\text{-}L] \tag{89}$$

when R_t is expressed in mmol m^{-2} s^{-1}, b and c are equal to 2.09 x 10^{-8} (s^{-1}) and 4.73 × 10^{-6} (s^{-1}), respectively, for anorthite, and SOH, the total number of surface sites is taken as 0.12 mmol m^{-2} (Amrhein and Suarez, 1988).

Using the rate model expressed in Equation (89), and assuming the restricted pH range of 5 to 9, UNSATCHEM includes dissolution kinetics for anorthite, labradorite, albite, potassium feldspar, biotite and hornblende, based on specific surface areas of the minerals, and the assumption that the released Al is retained by formation of kaolinite. The selected rate constants, where available, are based on long-term weathering studies in experiments which minimized the effects of grinding and pretreatment artifacts.

An alternative approach to the Furrer and Stumm (1986) model is the rate model of Sverdrup and Warfvinge (1988)

$$R = k_{H^+} \frac{[\mathrm{H}^+]^n}{[\mathrm{M}]^x[\mathrm{Al}^{3+}]^y} + \frac{k\mathrm{H_2O}}{[\mathrm{Al}^{3+}]^u} + k_{CO_2} \cdot P_{CO_2} + k_{\mathrm{org}}\,[org]^{0.5} \tag{90}$$

where k is the reaction rate for the different processes, and n,x,y,u, and m are the reaction orders, all determined experimentally. Equation (90) is utilized with the reaction rates given by Sverdrup and Warfvinge (1988).

The kinetic models based on Furrer and Stumm (1986) as well as the Sverdrup and Warfvinge (1988) model do not consider a back reaction expression. This omission is acceptable only if the solution is very far from equilibrium with these phases. At earth surface temperatures all the above considered silicates are very unstable with respect to clays and oxides, and the solution phase remains very undersaturated with respect to these minerals. At elevated temperatures back reaction expressions must be added, as these mineral become the stable phases.

Cation exchange

Cation exchange is generally the dominant *chemical* process for the major cations in solution in the unsaturated zone. Cation exchange is usually treated with a Gapon-type expression of the form (White and Zelazny, 1986)

$$K_{ij} = \frac{\bar{c}_i^{y+}}{\bar{c}_j^{x+}} \frac{(c_j^{x+})^{1/x}}{(c_i^{y+})^{1/y}} \tag{91}$$

where y and x are the respective valences of species i, and j, and the overscored concentrations are those of the exchanger phase (concentration expressed in mol$_c$ mass^{-1}). It is assumed that the cation exchange capacity c_T is constant, and that for non-acid soils

$$\bar{c}_T = \overline{\mathrm{Ca}}^{2+} + \overline{\mathrm{Mg}}^{2+} + \overline{\mathrm{Na}}^{+} + \overline{\mathrm{K}}^{+} \tag{92}$$

Existing chemical models require either input of a soil specific selectivity value or use a generalized value for the selectivity coefficient. We note that the experimentally determined selectivity values are not constant, nor is the cation exchange capacity which varies as a function of pH, due to variable charge materials such as organic matter. It has been observed that soils have an increased preference for Ca^{2+} over Na^+, and Ca^{2+} over Mg^+, at low levels of exchanger phase Ca^{2+}. Suarez and Wood (1993) developed a mixing model which is able to approximate the nonconstant values of the soil selectivity coefficient by taking into account the organic matter content of the soil and using the published constant selectivity values for clay and organic matter. Calcium preference decreases as the organic matter exchanger sites (which have higher Ca preference than clays) become Ca saturated. UNSATCHEM uses this approach by solving two sets of equations for cation exchange (clay and organic matter).

Anion adsorption

Prediction of the mobility of minor element anions such as B, As and Se generally requires consideration of adsorption processes. Among the modeling approaches utilized are retardation factors (as discussed earlier), and Langmuir and Freundlich isotherms. These models generally require the input of adsorption isotherm values specific to a given pH. Such an approach is of limited value when used in transport models which need to consider a range of pH values, since anion adsorption is highly dependent on pH. Keren et al.(1981) developed a modified Langmuir expression to account for the effect of pH on B adsorption, given as

$$Q_{BT} = T\left\{1 + \frac{PR}{F(Q_T - Q_{BT})}\left[1 + K_{\mathrm{OH}}(\mathrm{OH}^-)\right]\right\}^{-1} \tag{93}$$

where Q_{BT} is the total adsorbed B, Q_T is the total B (adsorbed + in solution), T is the total adsorption capacity for B, $P=K_{\mathrm{H}}\,(\mathrm{OH}^-)10^{14}+1$, R is the solution/clay ratio, expressed in L/g clay, $F=K_{\mathrm{HB}}+K_{\mathrm{B}}(P\text{-}1)$, and K_{HB}, K_{B}, K_{OH}, are constants relating to the binding energy for the ions $B(OH)_3$, $B(OH)_4^-$, and OH^-, respectively. The term K_H is the equilibrium constant of the reaction:

$$\mathrm{B(OH)_3 + H_2O \rightleftarrows B(OH)_4^- + H^+} \tag{94}$$

This model has been incorporated into the B transport model described by Shani et al. (1992). Anion adsorption is also treated in the one-dimensional saturated water flow model TRANQL (Cederberg et al., 1985) using the constant capacitance model, in HYDROGEOCHEM using the triple layer model, and in the variably saturated flow model UNSATCHEM using the constant capacitance model.

The constant capacitance model is also able to represent the effect of pH on adsorption affinity, and in contrast to the empirical Langmuir model, is intended to provide chemical representation of the system. The model contains the following assumptions: adsorption is a ligand exchange mechanism, all surface complexes are inner sphere, no surface complexes are formed with other salts in solution. The relation between the surface charge and the surface potential is given by

$$\sigma = \frac{CSa}{F}\Psi \tag{95}$$

where C is the capacitance density(F m^{-2}), S is the specific surface area (m^2 g^{-1}), a is the suspension density (g L^{-1}), F is the Faraday constant (coulombs mol$_c^{-1}$), Ψ is potential (volts) and σ is expressed in mol$_c$L^{-1}. The intrinsic conditional equilibrium constants corresponding to B adsorption are (Goldberg and Glaubig, 1985)

$$K_+ = \frac{[SOH_2^+]}{[SOH][H^+]} \exp[F\Psi/RT] \tag{96}$$

$$K_b = \frac{[SH_2BO_3]}{[SOH][H_3BO_3]} \tag{97}$$

$$K_- = \frac{[SO^-][H^+]}{[SOH]} \exp[-F\Psi/RT] \tag{98}$$

The mass balance equations for the surface functional groups and B are given by

$$[SOH]_T = [SOH] + [SOH_2^+] + [SO^-] + [SH_2BO_3] \tag{100}$$

$$B_T = H_3BO_3 + SH_2BO_3 \tag{99}$$

and the charge balance equation for the surface is defined by

$$\sigma = [SOH_2^+] - [SO^-] \tag{101}$$

In the absence of soil specific data the model utilizes average soil constants of 9.3, -10.6, and 5.5, for the values of log K_+, log K_-, and log K_B, respectively (Goldberg, 1993). We utilize the value of 1.06 Fm^{-2} for the capacitance (C). Using the data of Goldberg and Glaubig (1986), the following expression relates the experimentally determined adsorption site density to soil surface area

$$[SOH]_T = 2.53 \cdot 10^{-7} + 4.61 \cdot 10^{-9} \cdot S \tag{102}$$

where S is expressed in $m^2\ g^{-1}$ and adsorption density in mol g^{-1}. The system is defined by input of a soil surface area, specifying initial conditions, B concentration in the input water and calculating suspension density a from water content and bulk density.

EXAMPLE SIMULATIONS USING *UNSATCHEM*

In this section two sets of simulations are presented. In the first example we evaluate the impact of different CO_2 assumptions and different calcite models (equilibrium versus kinetic) on solution chemistry. In the second example we evaluate the impact of the chemical composition on hydraulic parameters and the resultant water flow and chemical composition in the profile.

The effect of model assumptions regarding carbonate chemistry and CO_2 which influence vadose zone chemistry can be demonstrated by an example involving various simulations. In this example we consider the recharge of water having the following composition (Welton-Mohawk drainage well #616; reported in Suarez, 1977), expressed in $mmol_cL^{-1}$: Ca= 5.05; Mg, 6.34; Na, 17.2; K, 0.234; Cl, 111.3; alkalinity, 7.18; SO_4,

15.0; and NO_3, 0.8. This water is applied to a field at the rate of 15 cm/day for 0.5 days, once a week. The infiltration is into a loam soil with a K_s of 24.4 cm/day. The potential evapotranspiration is constant at 1.0 cm/day and the soil temperature is taken as 20°C. An exponential root distribution is used to describe root water uptake. The simulations were made for 365 days.

Figure 1 shows the Cl concentration distribution with time. The increase with depth is due to plant root extraction, concentrating the salts by a factor of 5. In arid zone regions evapotranspiration exerts a dominant influence on solution chemistry. The profiles in Figure 1 indicate that it takes approximately one year to stabilize the Cl concentration at the 100 cm depth. The concentration distribution in the shallow depths continue to undergo a cyclical change related to the infiltration events (data not shown). Figures 2a and 2b show the Ca and alkalinity concentrations with depth and time assuming atmospheric CO_2 throughout the profile and calcite equilibrium. This simulation corresponds to what would be predicted with a closed system model where total carbon rather than alkalinity is input. Since the well water is initially supersaturated, the simulation forces an immediate and drastic precipitation at the soil surface to achieve equilibrium (Ca decreases to below 0.8 $mmol_cL^{-1}$). Calcium concentrations also decrease with depth due to calcite precipitation (because alkalinity > Ca). The pH is 8.5 at the surface and increases to 8.8 with depth.

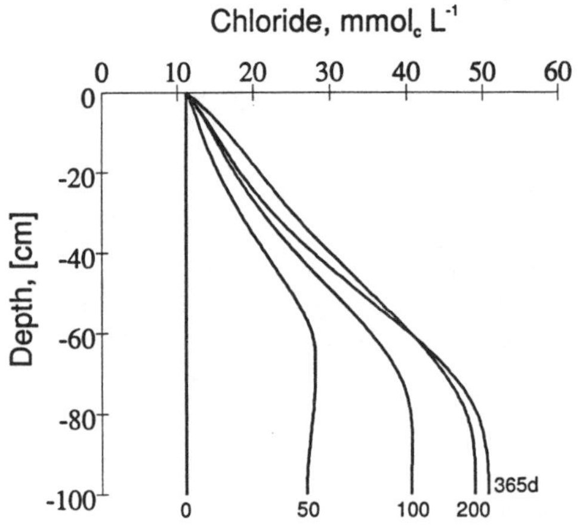

Figure 1 (right). Chloride concentration with time and depth.

Figure 2 (below). (a) Calcium and (b) alkalinity concentrations with depth and time assuming calcite equilibrium and atmospheric CO_2.

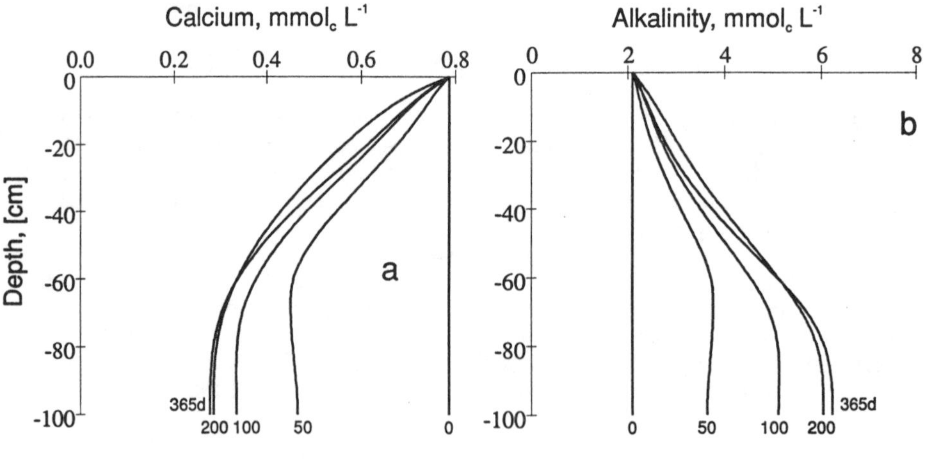

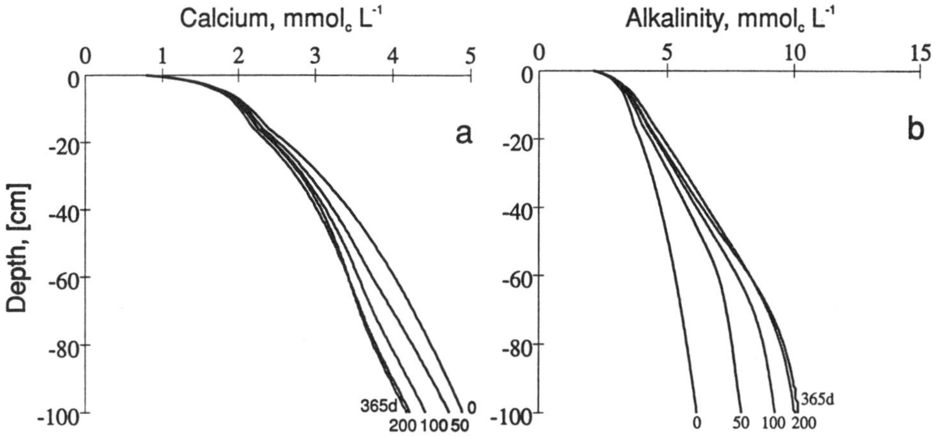

Figure 3. (a) Calcium and (b) alkalinity concentrations with depth and time assuming calcite equilibrium and CO_2 increasing linearly from atmospheric CO_2 at the surface to 0.02 atm (20 kPa) at 100 cm.

Figure 3 shows the Ca and alkalinity distributions with time and depth using the assumption of atmospheric CO_2 at the surface, increasing linearly with depth to a value of 0.02 atm CO_2 at 100 cm. Calcium concentrations now increase with depth, in contrast to Figure 2 where Ca concentrations decreased with depth. Calcium concentrations shown in Figure 3a are also much greater at the bottom of the profile than the concentrations shown in Figure 2a. Increasing Ca and alkalinity result from the increase in CO_2 with depth. The pH decreased from 8.5 at the surface to 7.35 at the bottom of the profile. Figure 4 shows the Ca, alkalinity, pH and pIAP for $(Ca^{2+})(CO_3^{2-})$ assuming a linear increase in CO_2 from atmospheric at the surface to 0.02 atm at 100 cm and a kinetic model for calcite precipitation/dissolution. Ca concentrations are greater throughout the profile as compared to the earlier simulations, especially in the shallow zones where supersaturation is greatest. As discussed earlier these represent realistic calcite saturation values. The pH was 8.9 at the surface and decreased to 7.4 at the bottom of the profile.

Figure 5 shows the Ca, alkalinity, pIAP for $(Ca^{2+})(CO_3^{2-})$, and CO_2 distribution with depth and time using the kinetic model and the CO_2 production transport model in UNSATCHEM (described in Šimůnek and Suarez, 1993b), with a CO_2 production term of 0.42 cm^3/cm^2/day. Calcium and alkalinity concentrations in the shallow and intermediate depths are greater than the concentrations in the earlier simulations. This increase is caused by the CO_2 distribution shown in Figure 5, with a CO_2 maxima near the surface. In contrast, the earlier simulations utilized a fixed CO_2 increasing linearly with depth. The cycles in the CO_2 distribution reflect the cyclic pattern of CO_2 increases associated with the infiltration events. The pIAP values show a rapid shift from supersaturation to undersaturation just below the surface, again related to the rapid increase in CO_2.

The effect of soil type on CO_2 concentration and solution composition is demonstrated by the simulation shown in Figure 6. In this instance we again used the kinetic model for calcite and the CO_2 production/transport option. The only difference is that we now infiltrate the same amount of water into a silt loam with a K_s of 6 cm/day instead of into a loam soil. The Ca and alkalinity concentrations shown in Figure 6 are much greater than those shown for the other simulations. In this simulation water was

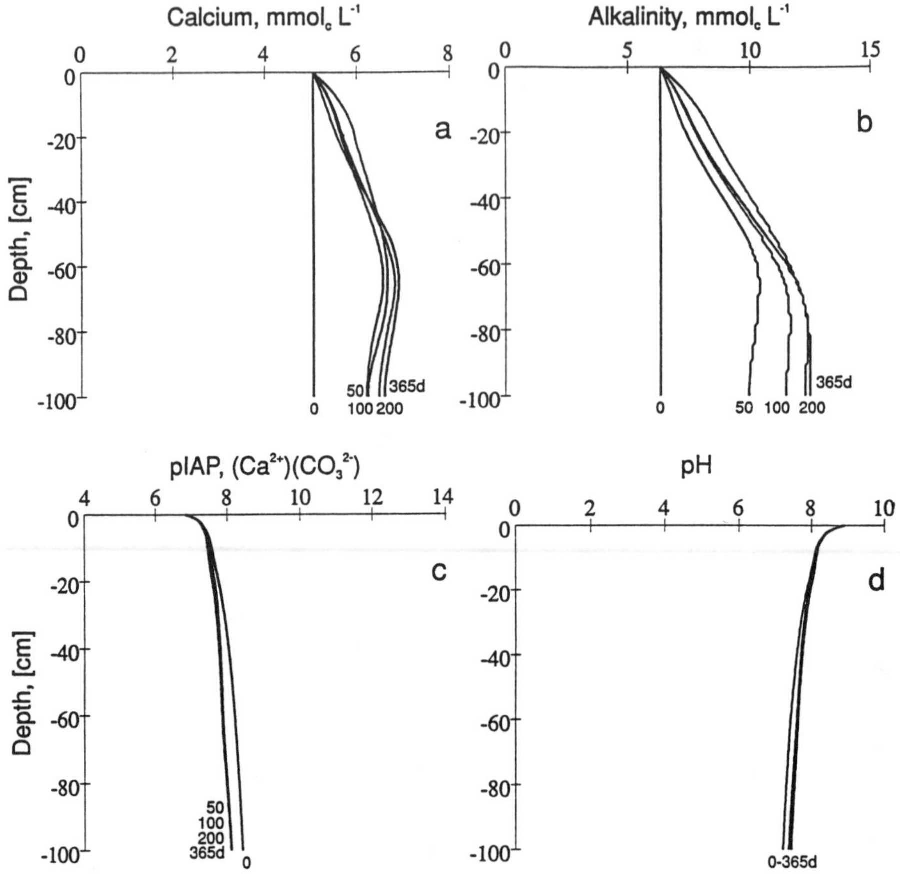

Figure 4 (above). (a) Calcium and (b) alkalinity concentrations and (c) pIAP (Ca^{2+})(CO_3^{2-}) and (d) pH with depth and time assuming a calcite kinetic model and CO_2 increasing linearly from atmospheric CO_2 at the surface to 0.02 atm. (20 kPa) at 100 cm.

Figure 5 (below). (a) Calcium and (b) alkalinity with depth and time assuming a calcite kinetic model and CO_2 predicted by the production/transport submodel. [Figure 5 continued next page.]

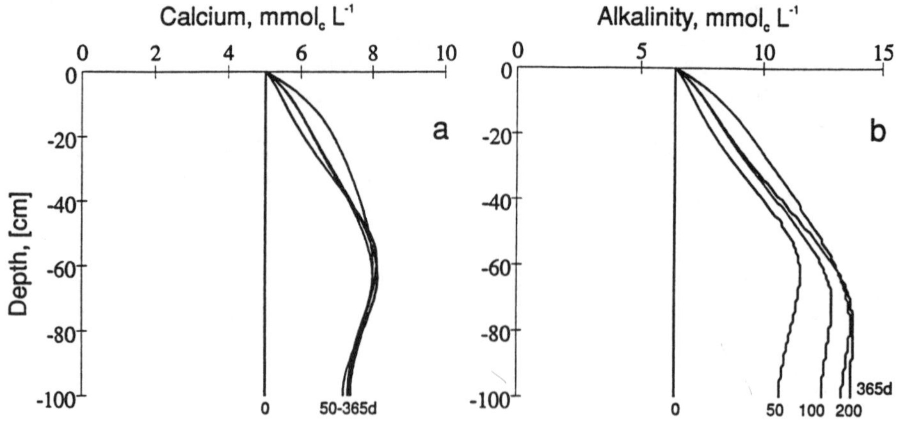

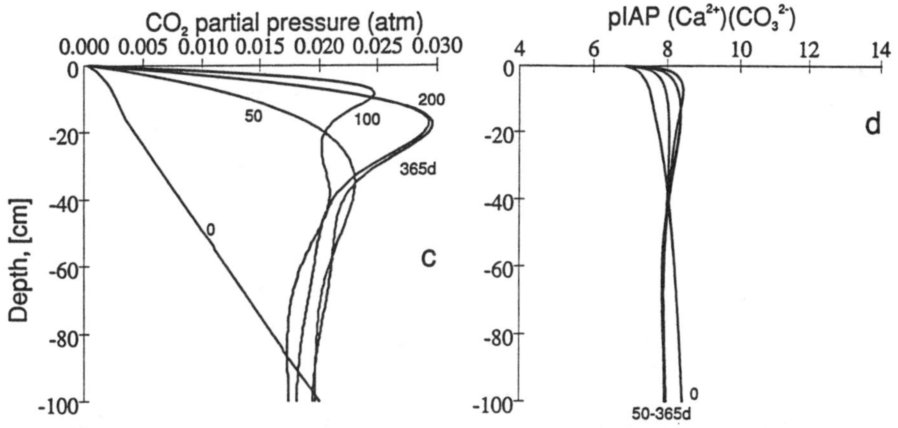

Figure 5 (above). (c) CO_2 partial pressure and (d) pIAP $(Ca^{2+})(CO_3^{2-})$ with depth and time assuming a calcite kinetic model and CO_2 predicted by the production/transport submodel.

Figure 6 (below). (a) Calcium and (b) alkalinity, (c) CO_2 partial pressure and (d) pH with depth and time assuming a calcite kinetic model and CO_2 predicted by the production/transport submodel. This simulation assumed infiltration into a silt loam soil.

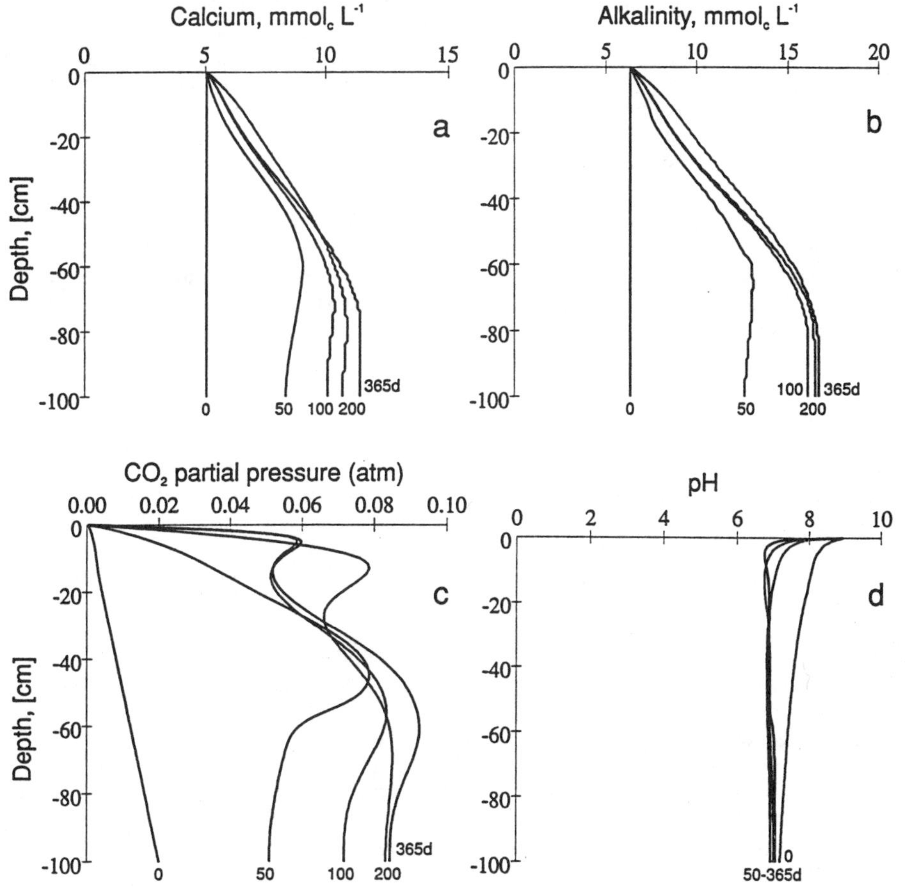

ponded on the surface during the infiltration and resulted in greater water content at the surface. This in turn resulted in decreased CO_2 diffusion out of the soil and thus increased the soil CO_2, as shown in Figure 6. Soil pH was also much lower, as can be seen from the comparison of Figure 6 with the data given for the earlier simulations. These simulations demonstrate the interaction of biological processes and soil physical properties on the chemical composition in the vadose zone. Predictions resulting from kinetic simulation of calcite reactions and dynamic, process-based simulation of soil CO_2 production and transport, rather than arbitrary model input assumptions, result in vastly different solute compositions and pH.

In the second example we assume that water is ponded continuously on a sodic soil, without rootwater uptake. In the absence of adverse chemical effects the soil has a saturated hydraulic conductivity of 60.5 cm/d, θ_r=0.000, θ_s=0.48, n=1.592 and α=0.015022 cm^{-1}. The cation exchange capacity was 200 mmol$_c$ kg^{-1}. The charge fraction on the exchanger was as follows: Ca, 0.2; Mg, 0.2; Na, 0.6. The initial soil water composition (before equilibration with the exchanger phase) expressed in mmol$_c$L^{-1} is: Ca= 0.2; Mg, 0.2; Na, 4.8; Cl, 4.6; alkalinity, 0.4; and SO_4, 0.0. The composition of the infiltrating water was as follows, expressed in mmol$_c$L^{-1}: Ca= 1.5; Mg, 0.5; Na, 2.0; Cl, 1.0; alkalinity, 0.5; and SO_4, 2.5.

In the first simulation of example 2 we do not consider the effect of solution chemistry on hydraulic conductivity. Figure 7a shows the infiltration of the tracer into the soil. After 0.6 d the front has moved almost to the bottom of the profile. The infiltration rate into the unsaturated soil exceeds the saturated hydraulic conductivity and then decreases to the 60.5 cm/d rate as the soil becomes saturated. As shown in Figure 7b it takes approximately 100 d for the soil solution Ca concentration to reach the concentration of the infiltrating water. The "retardation" of the Ca front is primarily due to Ca exchange with Na.

In the second simulation we utilize the functions described earlier which account for the chemical effects on hydraulic conductivity. The initial hydraulic conductivity is 2.4 $\times$ 10^{-5} cm/d, which increases to only 0.0128 cm/d as water saturation is achieved. As shown in Figure 8a after 5 years the tracer concentration in the profile has not yet stabilized to the infiltrating concentration. This delay is due to the reduced hydraulic conductivity. Note that the infiltration front of the tracer shown in Figure 8a is not as sharp as the front shown in Figure 7a, this is attributed to the increased importance of diffusion at slow infiltration rates. The Ca concentration profiles, shown in Figure 8b indicate that it takes 238 y for the Ca front to reach the 80 cm depth. Essentially all of the exchange occurred in the last 30 y. The infiltration rate increased rapidly in this time frame, as the increased Ca concentration allowed for increased infiltration which in turn allows for increased exchange.

The second set of simulations, shown in Figure 7 and Figure 8, demonstrate the importance of chemical effects on infiltration, as well as the effects of infiltration on solution chemistry. Such simulations are not only useful for describing the reclamation of a sodic agricultural soil but, are also directly applicable to waste disposal problems. Modeling contaminant transport likely represents the single largest use of chemical transport models. Sodium saturated smectites (bentonite) have been widely used in waste disposal and waste holding ponds. Such ponds have often been constructed without consideration of the effect of the waste solution on the physical properties of the restrictive clay layer. Consistent with field observations we predict that the layers can fail catastrophically due to conversion of the clay from Na-saturated to Ca saturated, causing a rapid increase in infiltration of the waste. This failure can occur after a variable

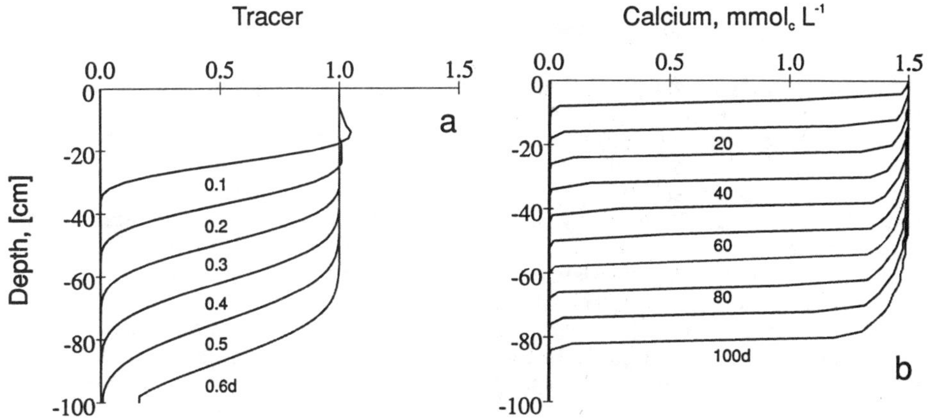

Figure 7. (a) Tracer and (b) Ca concentrations with depth and time assuming no effect of chemical composition on hydraulic properties. Simulation is infiltration of ponded water into a soil with a saturated hydraulic conductivity of 60.5 cm/d.

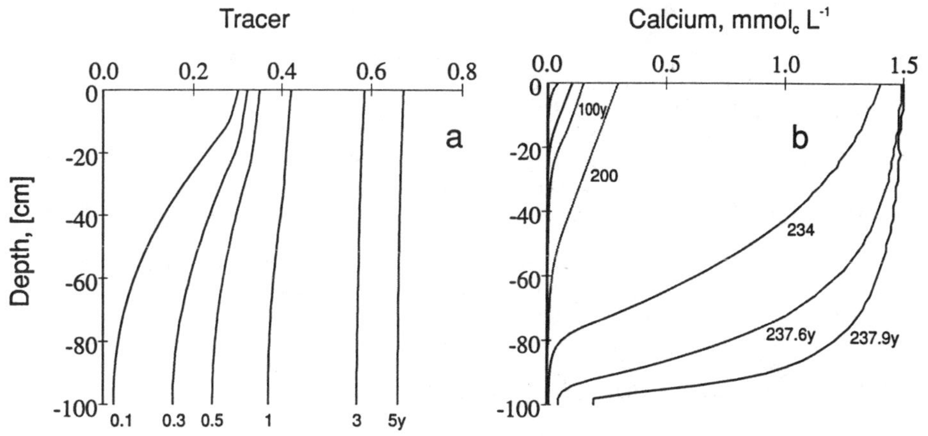

Figure 8. (a) Tracer and (b) Ca concentrations with depth and time assuming that the chemical composition of the solution affects the hydraulic properties. Simulation is infiltration of ponded water into a soil with a (optimal) saturated hydraulic conductivity of 60.5 cm/d.

number of years, depending on the initial hydraulic properties, layer thickness and solution composition. Models which attempt to simulate such processes must consider the chemical effects on soil physical properties.

FUTURE DEVELOPMENTS

Improvements in simulating solute transport in natural systems will require improved representations of the underlying chemical and physical processes. Kinetic expressions are lacking for realistic representation of most chemical processes under field conditions. At present there are no suitable representations for predicting changes in redox nor sufficient rate data for redox reactions. Representation of water flow using the Richards' equation is often not satisfactory for describing flow in natural environments. Many field scale simulations require three-dimensional representations of water flow. Computer codes which treat three-dimensional problems have not yet been coupled to solute transport and solution chemistry. Also needed are improved methods for estimating relevant flow, transport, and chemical parameters and their spatial distributions as input to the models. Finally we note that progress in model development is hindered by the severe lack of detailed field experiments which are needed to critically evaluate the existing models.

REFERENCES

Allison JD, Brown DS, Novo-Gradic KJ (1990) MINTEQA2/PRODEFA2, a geochemical assessment model for environmental systems: Version 3.0 Office Res Dev, USEPA, Athens, GA

Amrhein C, Suarez DL (1988) The use of a surface complexation model to decribe the kinetics of ligand-promoted dissolution of anorthite. Geochim Cosmochim Acta 52:2785-2793

Bresler E (1972) Interacting diffuse layers in mixed mono-divalent ionic systems. Soil Sci Soc Am Proc 36:891-896

Bresler E (1973) Simultaneous transport of solute and water under transient unsaturated flow conditions. Water Resour Res 9:975-986

Brooks RH, Corey AT (1964) Hydraulic properties of porous media. Hydrol Paper 3. 27 pp Colo State Univ, Fort Collins, Colo

Bryant SL, Schechter RS, Lake LW (1986) Interactions of precipitation/dissolution waves and ion exchange in flow through permeable media. AIChE J 32(5):751-764

Busenberg E, Plummer LN (1982) The kinetics of dissolution of dolomite in CO_2-H_2O systems at 1.5 to 65 C and 0 to 1 ATM P_{CO2}. Am J Sci 282:45-78

Cardon GE, Letey J (1992) Plant water uptake terms evaluated for soil water and solute movement models. Soil Sci Soc Am J 32:1876-1880

Carnahan CL (1990) Coupling of precipitation-dissolution reactions to mass diffusion via porosity changes. *In (ed) Melchior, DC, Bassett, RL,* Chemical Modeling of Aqueous System II. ACS Symposium Series 416. Am Chem Soc, Washington, DC

Cederberg GA, Street RL, Leckie JO (1985) A groundwater mass transport and equilibrium chemistry model for multicomponent systems. Water Resour Res 21(8):1095-1104

Chou L, Wollast R (1985) Steady-state kinetics and dissolution mechanisms of albite. Am J Sci 285:963-993

Davis JA, James RO, Leckie JO (1978) Surface ionization and complexation at the oxide/water interface. I. Computation of electrical double layer properties in simple electrolytes. J Colloid Interface Sci 63: 480-499

Davis JA, Leckie JO (1978) Surface ionization and complexation at the oxide/water interface. II. Surface properties of amorphous iron oxyhydroxide and adsorption of metal ions. J Colloid Interface Sci 67: 90-107

Davis JA, Leckie JO (1980) Surface ionization and complexation at the oxide/water interface. III. Adsorption of anions. J Colloid Interface Sci 74:32-43

Dudley LM, Hanks RJ (1991) Model SOWACH: Soil-plant-atmosphere salinity management model (User's manual) Utah Agric Expt Station Res Report 140, Logan UT

Feddes RA, Kowalik PJ, Zaradny H (1978) Simulation of Field Water Use and Crop Yield, Simulation Monograph. Pudoc, Wageningen, The Netherlands

Forster R (1986) A multicomponent transport model. Geoderma 38:261-278

Förster R, Gerke H (1988) Integration von Modellen des Wasser- und Stofftransports sowie physikochemischer Wechselwirkungen zur Analyse von Agrar-Ökosystemen, Verhandlungen der Gesellschaft für Ökologie, Band XVIII, Essen

Freijer JI, Leffelaar PA (1996) Adapted Fick's law applied to soil respiration. Water Resour Res 32:791-800

Furrer G, Stumm W (1986) The coordination chemistry of weathering: I Dissolution kinetics of δ Al_2O_3 and BeO. Geochim Cosmochim Acta 50:1847-1860

Gaber HM, Inskeep WP, Comfort SD, Wraith JM (1995) Nonequilibrium transport of atrazine through large intact soil cores. Soil Sci Soc Am 59:60-67

Gaudet JP, Jegat H, Vachaud G, Wierenga PJ (1977) Solute transfer, with exchange between mobile and stagnant water, through unsaturated sand. Soil Sci Soc Am J 41(4):665-671

Gerke HH, van Genuchten M Th (1993) A dual-porosity model for simulating the preferential movement of water and solutes in structured porous media, Water Resour Res 29(2):305-319

Gilmore E, Rogers JS (1958) Heat units as a method of measuring maturity in corn. Agron J 50:611-615

Goldberg S (1993) Chemistry and mineralogy of boron in soils p 3-44. *In* UC Gupta (ed) Boron and its role in crop production. CRC Press, Ann Arbor

Goldberg S, Glaubig RA (1985) Boron adsorption on aluminum and iron oxide minerals. Soil Sci Soc Am J 49:1374-1379

Griffioen J (1993) Multicomponent cation exchange including alkalinaization/acidification following flow through sandy sediment. Water Resour Res 29(9):3005-3019

Gureghian AB (1981) A two-dimensional finite-element solution for the simultaneous transport of water and multisolutes through a non-homogeneous aquifer under transient saturated-unsaturated flow conditions. Sci Total Environ 21:329-337

Harned HS, Davis Jr R (1943) The ionization constant of carbonic acid and the solubility of carbon dioxide in water and aqueous salt solutions from 0 to 50 C. J Am Chem Soc 653:2030-2037

Hutson JL, Cass A (1987) A retentivity function for use in soil water simulation models. J Soil Sci 38:105-113

Hutson JL, Wagenet RJ (1992) LEACHM:Leaching Estimation And Chemistry Model. ver 3 Dept of Soil Crop and Atmospheric Sciences, Res Series No 92-3, Cornell Univ, Ithaca, NY

Huyakorn PS, Kool JB, Wu YS (1991) VAM2D- Variably saturated analysis model in two dimensions, Version 5.2 with hysteresis and chained decay transport, Documentation and user's guide. NUREG/CR-5352, Rev 1. US Nucl Reg Comm, Washington, DC

Inskeep WP, Bloom PR (1985) An evaluation of rate equations for calcite precipitation kinetics at pCO_2 less than 0.01 atm and pH greater than 8. Geochim Cosmochim Acta 49:2165-2180

Inskeep WP, Bloom PR (1986) Kinetics of calcite precipitation in the presence of water-soluble organic ligands. Soil Sci Soc Am J 50:1167-1172

Jennings AA, Kirk DJ, Theis TL (1982) Multicomponent equilibrium chemistry in groundwater quality mdels. Water Reour Res 18(4):1089-1096

Keren R, Gast RG, Bar-Yosef B (1981) pH-dependent boron adsorption by Na-montmorillonite. Soil Sci Soc Am J 45:45-48

Kirkner DJ, Jennings AA, Theis TL (1985) Multisolute mass transport with chemical interaction kinetics. J Hydrol 76:107-117

Kirkner DJ, Reeves H (1988) Multicomponent mass transport with homogeneous and hetergeneous chemical reactions: Effect of the chemistry on the choice of numerical algorithm, 1: Theory. Water Resour Res 24:1719-1729

Lasaga AC (1979) The treatmeant of multi-component diffusion and ion pairs in diagenetic fluxes. Am J Sci 279:324-346

Lebron I, Suarez DL (1996) Calcite nucleation and precipitation kinetics as affected by dissolved organic matter at 25°C and pH >7.5. Geochem Cosmochim Acta 60:2765-2776

Liu CW, Narasimhan TN (1989) Redox-controlled multiple species reactive chemical transport, 1. Model development. Water Resour Res 25:869-882

Logan SH, Boyland PB (1983) Calculating heat units via a sine function. J Am Soc Hort Sci 108:977-980

Mangold DC, Tsang C (1991) A summary of subsurface hydrological and hydrochemical models, Reviews

of Geophysics 29(1):51-79
Massmann J, Farrier DF (1992) Effects of atmospheric pressure on gas transport in the vadose zone. Water Resour Res 28:777-791
Marshall TJ (1958) A relation between permeability and size distribution of pores. J Soil Sci 9:1-8
McNeal BL (1968) Prediction of the effect of mixed-salt solutions on soil hydraulic conductivity. Soil Sci Soc Am Proc 32:190-193
Means JL, Crear DA, Duguid JO (1978) Migration of radioactive wastes: Radionuclide mobilization by complexing agents. Science 200:1477-1481
Milly PCD, Eagleson PS (1980) The coupled transport of water and heat in a vertical soil column under atmospheric excitation. Tech Rep no 258. RM Parsons Lab, Mass Inst of Technol Cambridge
Molz FJ (1981) Models of water transport in the soil-plant system. Water Resour Res 17(5):1245-1260
Morel F, Morgan J (1972) A numerical method for computing equilibria in aqueous chemical systems. Environ Sci Tech 6:58-67
Mualem Y (1976) A new model for predicting the hydraulic conductivity of unsaturated porous media. Water Resour Res 12:513-522
Narasimhan TN, White AF, Tokunaga T (1986) Groundwater contamination from an inactive uranium mill tailings pile. 2: Application of a dynamic mixing model. Water Resour Res 22:1820-1834
Nassar IN, Horton R, Globus AM (1992) Simultaneous transfer of heat, water, and solute in porous media: I Theoretical development. Soil Sci Soc Am J 56:1350-1356
Nielsen DR, van Genuchten M Th, Biggar JW (1986) Water flow and solute transport processes in the unsaturated zone. Water Resour Res 22(9):89S-108S
Nkedi-Kizza P, Biggar JW, van Genuchten M Th, Wierenga PJ, Selim HM, Davidson JM, Nielsen DR (1983) Modeling tritium and chloride 36 transport through an aggregated oxisol, Water Resour Res 19(3):691-700
Parker JC, Jardine PM (1986) Effects of heterogeneous adsorption behavior on ion transport, Water Resour Res 22:1334-1340
Patwardhan AS, Nieber JL, Moore ID (1988) Oxygen, carbon dioxide, and water transfer in soils: Mechanism and crop response. Transactions of ASAE 31:1383-1395
Plummer LN, Wigley TM, Parkhurst LD (1978) The kinetics of calcite dissolution in CO_2 systems at 5° to 60°C and 0.0 to 1.0 atm CO_2. Am J Sci 278:179-216
Rao PSC, Davidson JM, Selim HM (1979) Evaluation of conceptual models for describing nonequilibrium adsorption-desorption of pesticides during steady flow in soils. Soil Sci Soc Am J 43: 22-28
Raats PAC (1974) Steady flows of water and salt in uniform soil profiles with plant roots. Soil Sci Soc Am Proc 38:717-722
Reardon EJ (1981) K_d's - Can they be used to describe reversible ion sorption reactions in contaminant migration? Ground Water 19:279-286
Robbins CW, Wagenet RJ, Jurinak JJ (1980a) A combined salt transport-chemical equilibrium model for calcareous and gypsiferous soils. Soil Sci Soc Am J 44:1191-1194
Robbins CW, Wagenet RJ, Jurinak JJ (1980b) Calculating cation exchange in a salt transport model. Soil Sci Soc Am 44:1195-1200
Rubin J (1990) Solute transport with multisegment, equilibrium controlled reactions: A feed forward method. Water Resour Res 26:2029-2025
Russo D (1986) Simulation of leaching of a gypsiferous-sodic desert soil. Water Resour Res 22:1341-1349
Russo D (1988) Determining soil hydraulic properties by parameter estimation: On the selection of a model for the hydraulic properties. Water Resour Res 24:453-459
Russo D, Bresler E (1977) Analysis of the saturated-unsaturated hydraulic conductivity in a mixed sodium-calcium soil solution. Soil Sci Soc Am J 41:706-710
Selim HM, Davidson JM, Mansell RS (1976) Evaluation of a two-site adsorption-desorption model for describing solute transport in soils. paper presented at Proceedings, Summer Computer Simulation Conference, Nat Sci Foundation, Washington, D.C
Shainberg I, Otoh H (1968) Size and shape of montmorillonite particles saturated with Na/Ca ions. Israel J Chem 6:251-259
Shani U, Dudley LM, Hanks RJ (1992) Model of boron movement in soils. Soil Sci Soc Am J 56:1365-1370
Šimůnek J, Suarez DL (1993a) UNSATCHEM-2D Code for Simulating two-dimensional variably saturated water flow, heat transport, carbon dioxide production and transport, and multicomponent solute

transport with major ion equilibrium and kinetic chemistry, US Salinity Laboratory Res Report No 128. US Salinity Laboratory USDA, Riverside, CA

Šimůnek J, Suarez DL (1993b) Modeling of carbon dioxide transport and production in soil: 1. Model development. Water Resour Res 29:487-497

Šimůnek J, Suarez DL (1994) Two-dimensional transport model for variably saturated porous media with major ion chemistry, Water Resour Res 30(4):1115-1133

Šimůnek J, van Genuchten M Th (1994) The CHAIN-2D code for simulating the two-dimensional movement of water, heat, and multiple solutes in variably-saturated porous media. US Salinity Laboratory Res Report 136. US Salinity Laboratory USDA, Riverside, CA.

Sposito G, Mattigod SV (1977) GEOCHEM: A computer program for the calculation of chemical equilibria in soil solutions and other natural water systems. Dept of Soil and Environ Sci, Univ California, Riverside

Stankovich JM, Lockinton DA (1995) Brooks-Corey and van Genuchten soil-water-retention models. J Irrg and Drain Engrg, ASCE 121:1-7

Stumm W, Morgan JJ (1981) Aquatic chemistry: An introduction emphasizing chemical equilibria in natural waters. John Wiley & Sons, New York

Suarez DL (1977a) Ion activity products of calcium carbonate in waters below the root zone. Soil Sci Soc Am J 41:310-315

Suarez DL (1977b) Magnesium, carbonate, and silica interactions in soils. US Salinity Laboratory Annual Report, USDA, 120 pp

Suarez DL (1985) Prediction of major ion concentrations in arid land soils using equilibrium and kinetic theories. *In* D DeCoursey: Proc of the Natural Resour. Modeling Symposium, Pingree Park, Co, 1983

Suarez DL, Dudley L (1996) Hydrochemical considerations in modeling water quality within the vadose zone. *In* Guitjens J, Dudley L (eds) Agroecosystems: Sources, control and remediation of oxyanions. Am Assoc Adv Sci

Suarez DL, Rhoades JR, Lavado R, Grieve CM (1984) Effect of pH on saturated hydraulic conductivity and soil dispersion. Soil Sci Soc Am 48:50-55

Suarez DL, Šimůnek J (1994). Modeling equilibrium and kinetic major ion chemistry with CO_2 production/transport coupled to unsaturated water flow. *In* Gee G, Wing R (ed.) In situ remediation: Scientific basis for current and future technologies. Part 2. p 1215-1246. Thirty-third Hanford symposium on health and the environment. Battelle Press, Columbus

Suarez DL, Šimůnek J (1993) Modeling of carbon dioxide transport and production in soil: 2. Parameter selection, sensitivity analysis and comparison of model predictions to field data. Water Resour Res 29:499-513

Suarez DL, Šimůnek J (1996) UNSATCHEM: Unsaturated water and solute transport model with equilibrium and kinetic chemistry. Soil Sci Soc Am J, In review

Suarez DL, Wood JD (1993) Predicting Ca-Mg exchange selectivity of smetitic soils. Agron Abst p 236. Am Soc Agron, Madison WI

Sverdrup HU, Warfvinge P (1988) Weathering of primary silicate minerals in the natural soil environment in relation to a chemical weathering model. Water Air Soil Poll 38:397-408

Thorstenson DC, Pollack DW (1989) Gas transport in unsaturated zones: Multicomponent systems and the adequacy of Fick's laws. Water Resour Res 25:477-507

Truesdell AH, Jones BF (1974) Wateq, a computer program for calculating chemical equilibria of natural waters. J Res U S Geol Surv 2:233-248

Tseng P, Sciortino A, van Genuchten M Th (1995) A partitioned solution procedure for simulating water flow in a variably saturated dual-porosity medium. Adv Water Resour 18:335-343

Valocchi AJ (1984) Describing the transport of ion-exchanging contaminants using an effective K_d approach. Water Resour Res 20:499-503

Valocchi AJ, Street RL, Roberts PV (1981) Transport of ion-exchanging solutes in groundwater: Chromatographic theory and field simulation. Water Resour Res 17(5):1517-1527

van Genuchten M Th (1980) A closed-form equation for predicting the hydraulic conductivity of unsaturated soils. Soil Sci Soc Am J 44:892-898

van Genuchten M Th (1987) A numerical model for water and solute movement in and below the root zone, Unpublished Research Report, US Salinity Laboratory, USDA, ARS, Riverside, CA

van Genuchten M Th, Leij FJ, Yates SR (1991) The RETC code for quantifying the hydraulic functions of unsaturated soils. EPA/600/2-91/065

van Genuchten M Th, Šimůnek J (1996) Evaluation of pollant transport in the unsaturated zone. *Proc Regional Approaches to Water Pollution in the Environment.* NATO Advanced Research Workshop, Liblice, Czech Republic , In press

van Genuchten M Th, Wagenet RJ (1989) Two-site/two-region models for pesticide transport and degradation: Theoretical development and analytical solutions. Soil Sci Soc Am J 53:1303-1310

van Genuchten M Th, Wierenga PJ, O'Connor GA (1977) Mass transfer studies in sorbing porous media, III. Experimental evaluation with 2,4,5,-T. Soil Sci Soc Am J 41:278-285

Wagenet RJ, Hutson JL (1987) LEACHM: A finite-difference model for simulating water, salt, and pesticide movement in the plant root zone. Continuum 2, New York State Resources Institute, Cornell University, Ithaca, NY

Walsh MP, Bryant SL, Schechter RS, Lake LW (1984) Precipitation and dissolution of solids attending flow through porous media. AIChE Jour 30(2):317-328

Wang JSY (1991) Flow and transport in fractured rocks, *Reviews of Geophysics* 29, supplement, US Natl Rep to Int. Union of Geodedy and Geophysics 1987-1990, pp 254-262

White N, Zelazny LW (1986) Charge properties in soil colloids. p 39-81. *In*: DL Sparks (ed.) Soil physical chemistry. CRC Press, Boca Raton, Florida

Wollast R, Mackenzie FT, Bricker OP (1968) Experimental precipitation and genesis of sepiolite at earth-surface conditions. Am Mineralogist 53:1645-1662

Yeh GT, Huff DD (1985) FEMA: A finite element model of material transport through aquifers, Rep ORNL-6063, Oak Ridge Natl Lab Oak Ridge, TN

Yeh GT, Tripathi VS (1989) A critical evaluation of recent developments in hydrogeochemical transport models of reactive multichemical components. Water Resour Res 25:93-108

Yeh GT, Tripathi VS (1991) A model for simulating transport of reactive multispecies components: Model development and demonstration. Water Resour Res 27:3075-3094

Zimmerman RW, Chen G, Hadgu T, Bodvarsson GS (1993) A numerical dual-porosity model with semianalytical treatment of fracture/matrix flow. Water Resour Res 29:2127-2137

Chapter 6

REACTIVE TRANSPORT IN HETEROGENEOUS SYSTEMS: AN OVERVIEW

A. F. B. Tompson and K. J. Jackson

Lawrence Livermore National Laboratory
Earth Sciences Division, P.O. Box 808
Livermore, California 94551U.S.A.

INTRODUCTION

Over the past decade or two, there has been a tremendous amount of research interest in the role and impact of geologic heterogeneity within the fields of contaminant hydrology and petroleum engineering. Heterogeneity is a catch word for the ubiquitous spatial variability in subsurface formation materials, their physical and chemical properties, and the myriad processes that occur within them. Physical heterogeneity is commonly associated with preferential flow and contaminant migration pathways, hydraulically inaccessible zones into which mobile constituents may only diffuse, accelerated mixing processes, and larger-scale dispersion phenomena. Chemical heterogeneity (as produced by nonuniform distributions of reactive minerals) can affect the overall mobility and species configurations within contaminant mixtures in the subsurface. Both forms of heterogeneity may inhibit or otherwise affect the impact of in-situ remediation technologies, ranging from traditional pumping methods to more advanced schemes that introduce chemical or microbial reagents into the subsurface, or petroleum recovery processes. Fine-scale heterogeneity (on the order of meters for a site comprising kilometers) reinforces the need to describe these systems as three-dimensional, and it is difficult, if not impossible, to fully characterize them on a continuous spatial basis.

From the perspective of contaminant hydrology, the impact of geologic heterogeneity is difficult to consider in practice when designing in-situ remediation procedures, estimating long-term health risks, or assessing the role of natural or accelerated contaminant attenuation processes in achieving "closures" of contaminated sites. Instead, oversimplified conceptualizations of system behavior are used as the basis for engineering design studies upon which overall health risks, cleanup projections, and economic costs are evaluated. Although this approach is usually "justified" in some sense by the lack of characteristic geologic data, it does nothing to scientifically address or quantify the uncertainties implied by the paucity and variability of information usually available. This can lead to unreliable and over-engineered remedial solutions that are unnecessarily expensive. Unfortunately, practical techniques for dealing with the important effects of heterogeneity in multiphase systems are few and far between.

This chapter will deal with the influence of spatially variable material properties on flow and reaction phenomena in natural subsurface systems. Medium heterogeneity has long been recognized as an important factor affecting the flow of multiphase fluid mixtures and the dispersion or dilution of aqueous chemical plumes in aquifers and petroleum reservoirs. Here, we will review some of the more important issues in this area from an engineering and hydrological perspective and explore their relevance and connection to systems involving reactions between aqueous and solid phase components.

0275-0279/96/0034-0006$05.00

SOME BACKGROUND FROM A HYDROLOGIC PERSPECTIVE

Typical setting

Figure 1. Cross-sectional view of a typical road cut showing nonuniform mix of soil and rock materials.

Figure 1 shows a cross-sectional photo of the earthen materials visible in a typical road cut. There is obviously a large degree of variation in the soil and rock materials in terms of their type, chemical composition, and spatial distribution. The associated "bulk" properties of the medium, whether they describe hydrologic flow, chemical, or mechanical behavior, will vary correspondingly. Figure 2 shows the spatial variation of hydraulic conductivity and medium porosity as measured along a transect within a drinking water source aquifer in Illinois. Here we see that the conductivity of the medium varies over four orders of magnitude over distances of a couple hundred feet.

Figure 3 shows histograms of over 200 hydraulic conductivity measurements estimated from well tests made within a one-square mile areal portion of an alluvial aquifer in Northern California, showing a similar degree of variability. Heterogeneity is a common feature of most subsurface environments such as fractured rock formations, alluvial riverflow deposits, and even "uniform" sandy aquifers, and may involve many different characteristic length scales. In quantitative terms, this has been documented in some detail with respect to aquifer flow properties (e.g. Bakr et al., 1978; Hoeksema and Kitanidis, 1985; Gelhar, 1986, 1993; Mackay et al., 1988; Boggs et al., 1990; Teutsch and Kobus, 1990; Leblanc et al., 1991; Gelhar et al., 1992; Ptak and Teutsch, 1994) and more recently for chemical properties of aquifer materials (Sheppard and Thibault, 1990; Bishop et al., 1991a,b; Robin et al., 1991; Jussel et al., 1994a; Smith et al., 1996); it is still the subject of much research. Clearly, heterogeneity is the rule rather than the exception.

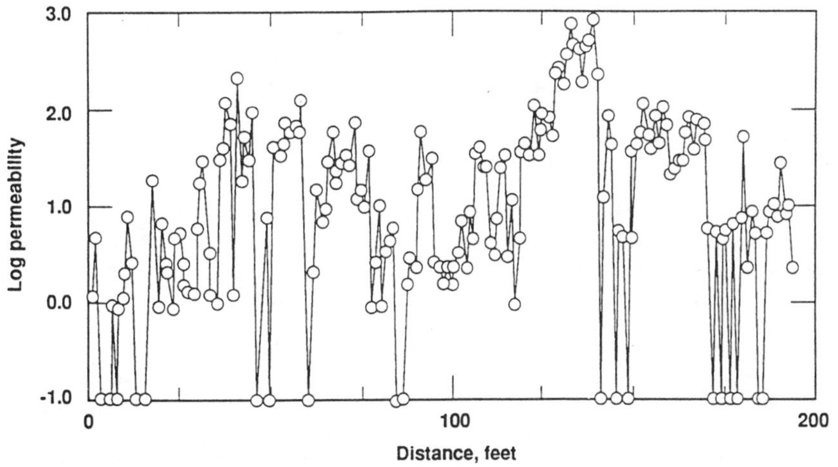

Figure 2. Spatial variation of hydraulic conductivity as measured along a transect within an aquifer in Illinois (after Bakr et al., 1978).

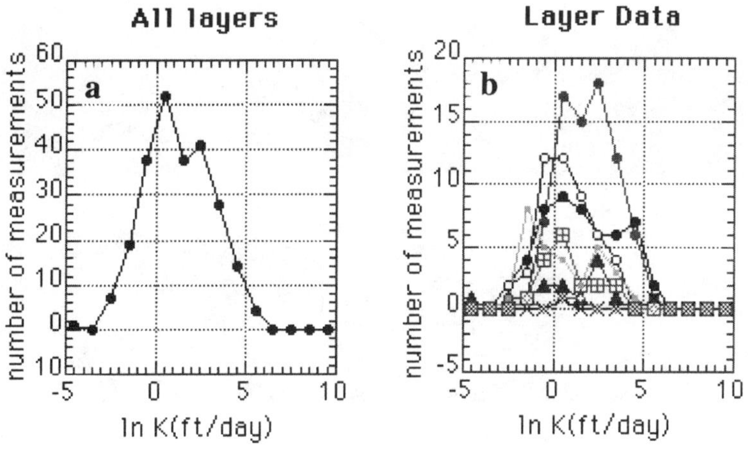

Figure 3. Histogram of over 200 hydraulic conductivity measurements made within a one-square mile areal portion of an alluvial aquifer beneath the Lawrence Livermore National Laboratory in northern California: (a) all 240 data points; (b) data segregated according to specific hydrostratigraphic unit from which they were measured.

Hydrologic impacts of heterogeneity

One of the principal impacts of heterogeneity is that it can lead to significant variation in fluid flow velocities over relatively short distances and promote the creation of preferential flow or channeling pathways in the subsurface. In a typical saturated aquifer, the mean groundwater velocity at a point in the medium is given by the familiar Darcy's Law,

$$\varphi \mathbf{v} = -K\nabla h \tag{1}$$

Here, K(**x**) and φ(**x**) are the hydraulic conductivity (L/T) and porosity of the medium, respectively, **v**(**x**,t) is the mean flow velocity (L) and h(**x**,t) is the hydraulic head (L), a combined measure of pressure and gravitational potential (Bear, 1972, 1979). When we say "mean flow velocity," we refer to an average pore velocity in some representative volume around the point **x**. We describe the system from a continuum viewpoint and do not explicitly recognize pore scale quantities; rather, we express their impacts at the continuum level in terms of material properties like K and φ.

The spatial gradient in head (known as the hydraulic gradient) is the main driving force for groundwater flow. If measurements of K at two nearby locations vary by an order of magnitude, then, for a similar hydraulic gradient, so will the velocity magnitudes. This can lead to some interesting behaviors.

Figures 3a and 3b show histograms of hydraulic conductivity data (in a logarithmic format) measured from well tests in the saturated alluvial materials beneath the Lawrence Livermore National Laboratory (LLNL) in California. Clearly, there is a wide variation in the values that have been measured. Figure 3a shows a histogram of all 200+ data points; Figure 3b shows how these data that pertain to 8 specific hydrostratigraphic layers that have been identified in the upper 500 feet or so of the saturated aquifer (Blake et al., 1995).

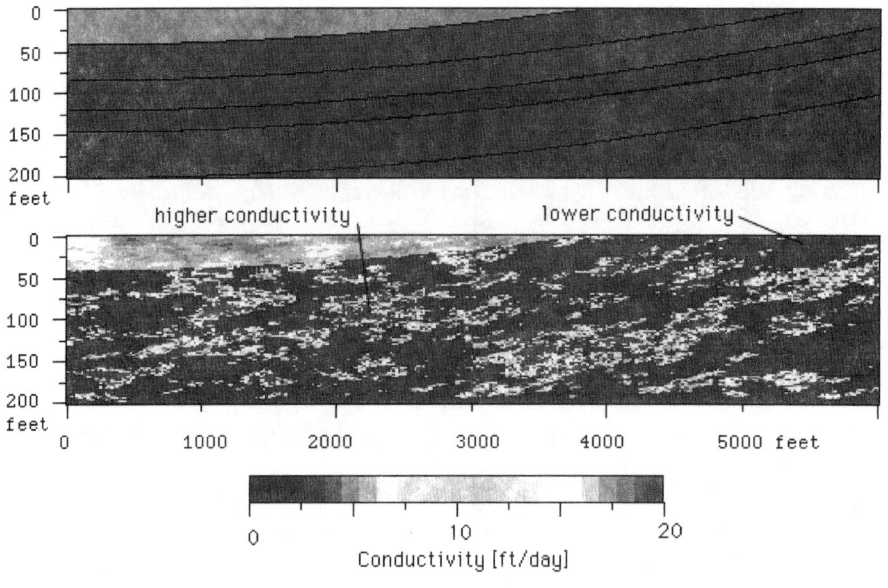

Figure 4. Two representations of the hydraulic conductivity distribution along a 6,000 ft-long vertical cross-section beneath the Lawrence Livermore National Laboratory, both derived from the same distributions shown in Figure 3b: (a, top): geometric mean conductivity specified as a constant in each hydrostratigraphic layer; (b, bottom): statistical "realization" of a nonuniform conductivity distribution, as sampled from the full distribution in Figure 3b and constrained by a model of spatial correlation (see text).

Figures 4a and 4b show two representations of the hydraulic conductivity distribution along a 6,000 ft-long vertical cross-section beneath LLNL that have been derived from the same distributions shown in Figure 3b. In Figure 4a, the hydrostratigraphic layers are considered to have distinct and uniform features characterized by constant conductivity values, shown here as the geometric mean of the appropriate distribution in Figure 3b. Here, much of the data in the distribution was "thrown away" in the averaging process, and we see a representation with very little variation in the property distribution. In Figure 4b, a statistical "realization" of a nonuniform conductivity distribution is shown in each layer, as sampled from the full distributions in Figure 3b and constrained by a model of spatial correlation (e.g. Tompson et al., 1989). Here, the main apparent features are the highs and lows in the values (that are consistent in a distributional sense with the data in Fig. 3b), and the bedding pattern provided by the correlation model (see below).

Across each of the sections in Figure 4, a mean hydraulic gradient was specified between the two ends to produce a steady, two-dimensional groundwater flow field in each section (from the right to the left). This was accomplished with a numerical model based upon (1) and a conservation of mass equation of the form (Bear, 1972, 1979)

$$\nabla \cdot (\varphi \mathbf{v}) = 0 \tag{2}$$

Within each section, two dilute, nonreactive tracer "pulses" were released and allowed to migrate with the flow. The distribution of tracer satisfies the chemical balance of mass,

$$\frac{\partial(\varphi c)}{\partial t} + \nabla \cdot (\varphi \mathbf{v} c) + \nabla \cdot (\varphi \mathbf{J}) = 0 \tag{3}$$

This expression just relates the change in solute mass in some incremental volume that has accrued over some time period to the net solute mass that has entered or left the volume via flow, diffusive, or dispersive processes. The concentration c($\mathbf{x}$,t) represents the average aqueous volumetric concentration (M/L^3, where M signifies mass or moles) in some representative volume around $\mathbf{x}$. The quantity $\mathbf{J}(\mathbf{x},t) \approx -(D\mathbf{I} + \mathbf{D}(\mathbf{v})) \cdot \nabla c$ is a combined diffusive / dispersive solute flux, where D is the molecular medium diffusivity, $\mathbf{I}$ is the unit tensor, and $\mathbf{D}$ is a velocity-dependent dispersion tensor (both L^2/T), the latter typically being much larger in magnitude than D (Bear, 1972, 1979).

Figure 5 shows the distribution of the "pulses" some 30 years after release in both flow fields. In Figure 5a (corresponding to Fig. 4a), they are confined to small, highly concentrated areas similar to their original size, and have mainly been displaced by the relatively uniform flow field. Although local medium scale dispersion has been included to represent solute mixing and dispersal as it moves through the porous medium (Bear, 1972, 1979), but its apparent effect is small. In Figure 5b, the tracer mass has been significantly dispersed and diluted further by the nonuniform velocity field corresponding to Figure 4b. The tracer migration illustrates the variability in flow pathways as induced by the nonuniform conductivity field.

Heterogeneity clearly impacts the degree to which tracers or other localized aqueous components are transported through the system. Given only a handful of conductivity or tracer concentration "measurements" from Figures 4b and 5b, it would be difficult to reconstruct the full distributions we started with or produced in the simulation. For example, it would be difficult, in general, to ascertain the maximum concentration in Figure 5b or where we would expect to find more or less of the tracer mass given only a few data points. In contaminant hydrology, we may be concerned with identifying where and at what levels the tracer "contaminants" go, where we should install pumps to extract contaminant mass most quickly, how well injected biomass or other remedial agents are distributed so as to contact or "mix" with the contaminants, or when we may assert the maximum concentrations fall below some critical action limit. This is all made more difficult by the hydraulic heterogeneity of the system.

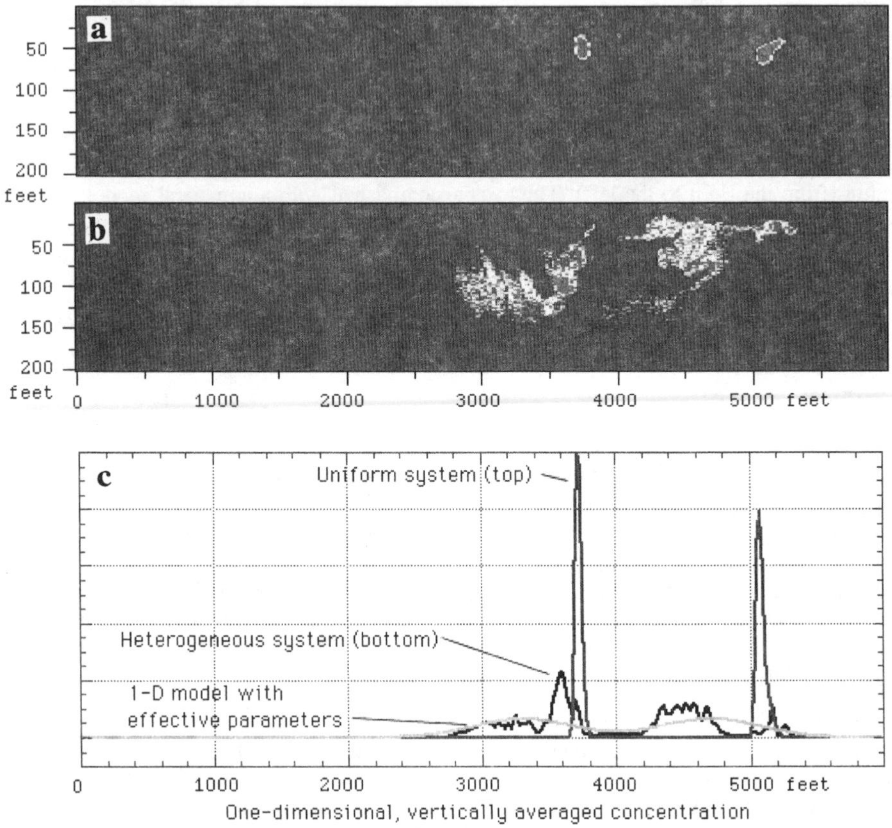

Figure 5. Distribution of two tracer "pulses" released into the two-dimensional groundwater flow fields in the cross-sections of Figure 4. In each section, flow was induced from the right to the left through the specification of an average hydraulic gradient across the section. In (a), the pulses are confined to small, highly concentrated areas similar to their original size, having mainly been displaced in the relatively uniform flow field. In (b), the tracer mass has been significantly dispersed and diluted by the nonuniform velocity field. In (c), vertically-averaged (1D) profiles from the uniform and heterogeneous cases are shown, as well as a prediction from a 1D model based upon an enhanced macrodispersion coefficient.

Describing and measuring spatial heterogeneity

There are many *reasons* and *ways* for describing (or modeling) heterogeneity. One of the main reasons is that we can never hope to fully measure the properties or other characteristics of an aquifer in sufficient detail to be able to forecast fluid migration and chemical transport at the level of detail shown in Figures 4b and 5b. This is due to the disparity in scales between the problem we need to analyze (aquifer scale, hundreds of meters to kilometers) and the heterogeneity scale (less than a meter to tens of meters), the ways in which we may gather or measure information on aquifer properties (hydraulic well tests, borehole geology and geophysical logs, seismic methods, hydrostratigraphy, etc.), and time and cost issues. If heterogeneity cannot be measured in fine detail, yet its impacts on flow or migration processes are believed to be significant, we must find a way to assess these impacts based on the data and other interpretive abilities we have. Many investigators have approached this problem by first developing detailed models or other representative "realizations" of geologic heterogeneity that recreate (in some approximate fashion) the degree of variability and geologic structure present in the system, as indicated by data (like Fig. 2 or Fig. 3) or other interpretive frameworks.

When we speak of a heterogeneity "model," we refer to an approximate spatial representation of the variability of a bulk property or geologic material of a formation, the overall quantitative features of this representation (length scales, statistical characteristics), and various relationships between material character, material properties, and material behavior. It is important to recognize that the hydraulic conductivity of a saturated medium has traditionally been the property of greatest interest because of the range of values it spans in many systems (e.g. Figs. 2 and 3) and its fundamental importance in controlling the flow field and chemical migration patterns (e.g. Eqns. 1-3). Medium porosity and other physical properties tend to have less of an influence. In unsaturated or multiphase flow systems, the distribution and variation of relative permeability and capillary pressure-saturation behavior with soil type becomes equally important in controlling fluid flow. Further issues associated with chemical properties of the medium and their variation will be addressed in a later section.

For reference, we present below a synopsis of the three techniques for modeling heterogeneity that have become rather common.

Correlated random fields. By far, this is one of the most simple and common models used. It is a member of a family of geostatistical models based upon the use of Spatial Random Functions (SRFs) (Deutsch and Journel, 1992; Isaaks and Srivastava, 1989; Journel and Huijbregts, 1978). Here, the spatial distribution of some relevant medium property (such as hydraulic conductivity) is considered to be random and spatially correlated within some geologic system. The approximate random character is used to represent the distribution of values that may be identified from a data set (i.e. Fig. 2 or Fig. 3); the correlation is used to reflect the spatial persistence of a property value in different directions, as a surrogate measure of aquifer material distribution. For example, it is common to see the distribution of the log-hydraulic conductivity expressed as

$$\ln K(\mathbf{x}) \approx F + f(\mathbf{x}) \tag{4}$$

where F is the mean of a measured distribution of ln K values (as in Fig. 3), and f(**x**) is a perturbation component, where the mean of the perturbation term squared is the variance of the distribution, $\langle f^2 \rangle = \sigma^2_f$. Spatial persistence is modeled with a correlation function C(**x**,**x**') that can be used to describe to what extent values of f or ln K at points **x** and **x**' are "similar." Mathematically, this function is defined as the mean of the product of f at two different locations,

$$C(\mathbf{x},\mathbf{x}') = \langle f(\mathbf{x}) f(\mathbf{x}') \rangle \approx \langle f(\mathbf{x}) f(\mathbf{x}+\mathbf{r}) \rangle \approx C(\mathbf{r}) \tag{5}$$

where $\sigma^2_f = C(\mathbf{x},\mathbf{x}) \approx C(0)$. The latter part of (5) is an approximation used in systems considered to be "stationary," where the degree and spatial structure of heterogeneity is considered to be uniform within the system of interest. In a stationary system, spatial correlation is only a function of the distance and direction between two points (as embodied in the separation vector **r**), and the mean and variance quantities are considered constant.

In general, correlation can be difficult to measure because of the paucity of data, especially closely-spaced data. Typically, a set of conductivity data will be analyzed using (4) to ascertain the mean (F) of the ln K distribution and whether it makes sense to consider this mean representative of an entire region or whether the system is trended (or nonstationary in the mean). The perturbation data (f) are then analyzed in pairs to detect correlation in different directions. The results can be fit to a mathematical model for stationary C(**r**), such as an exponential decay model

$$C(\mathbf{r}) = \sigma_f^2 \exp\left(-\left((r_x/\lambda_x)^2 + (r_y/\lambda_y)^2 + (r_z/\lambda_z)^2\right)^{1/2}\right) \tag{6}$$

where the λ terms represent correlation length scales [L] in three orthogonal directions. At large separation distances (often up to tens of meters), no correlation exists. In practice, the data are usually fit to an alternative representation called the variogram, defined by

$$\gamma(\mathbf{x},\mathbf{x}') = \frac{1}{2}\left\langle (f(\mathbf{x}) - f(\mathbf{x}'))^2 \right\rangle \approx \sigma_f^2 - C(\mathbf{r}) \tag{7}$$

Fitting data to a variogram model does not require an initial assumption of stationarity; rather, we can ascertain whether this can be assumed by looking at whether γ rises as a function or separation to a fixed "sill" value, whence the latter equivalence in (7) becomes valid, or whether it rises unbounded or to different fixed values, indicating changing degrees of variability with increasing spatial scale (Fig. 6).

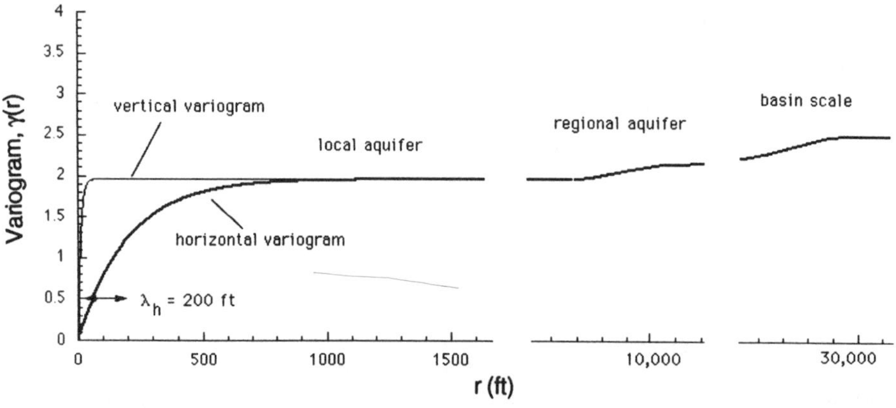

Figure 6. Typical variograms used to represent spatially correlated heterogeneity in the vertical and horizontal directions using the correlated random field model. The point variance (σ^2) of the ln(K) distribution is 1.96. Correlation scales (λ) in the horizontal and vertical directions are equal to 200 ft and 10 ft, respectively. As viewed over the first 1,500 ft, the variograms correspond to a stationary medium. At larger scales, variations representing larger geologic features begin to affect the variogram structure (See text and Gelhar, 1993).

This model is typically applied to a measured *parameter* of the system (e.g. hydraulic conductivity), as opposed to an explicit material or physical feature (such as a bed of silty sand or gravel). As such, the parameter value becomes a surrogate for the material at a given location. Because spatial correlation in this model is independent of the parameter value, it applies equally well to all values of the parameter and, hence, to all associated materials. Thus, the spatial structure and length scales attributed to a gravel bed (high conductivity) will necessarily be the same at those for a silty clay zone (low conductivity). In addition, the degree to which measured parameters correspond to specific physical materials or features will depend on how they are determined and what scale of "instrument" is used to measure them. Hydraulic conductivity, for example, can be measured from small, 3-inch soil cores taken in a drilling process, estimated from the response of larger-scale pumping tests in completed wells, or inferred from other geophysical observations. Thus, the scale of the measurements may affect how well particular physical features they refer to are resolved (Cushman, 1990; Shad and Teutsch, 1994). It is wise to seek consistency in scale when gathering and interpreting data for geostatistical analyses.

Example. The image in Figure 4b has been constructed from the random field model (4-7), the data in Figure 3b, and a baseline hydrostratigraphic layer interpretation (Tompson et al., 1989; Blake et al, 1995). Intrinsically, it is a single "realization" of the system (out of an infinite number of possibilities) that conforms to the data distribution and correlation model. Although it was not done here, it is possible to force a given realization to match specific measured data points in the generation process; this process is called "conditioning" (e.g. Deutsch and Journel, 1992).

According to the histogram data in Figure 3b, the degree of ln K variability (σ_f) within the layers shown in Figure 4b ranges between 1.4 and 2.3. These (dimensionless) numbers tend to reflect the composition of the layers, with the smaller values corresponding to more uniform (sandy) layers and the larger values representing units with a broader distribution of materials (gravels and sands to clays and silts). These values tend to be in agreement with data taken in many other aquifers (e.g. Table 6.1 in Gelhar, 1993), although closer scrutiny of Gelhar's table and related discussion shows there to be a dependence between the magnitude of the variability, the way in which the individual conductivity data were measured (e.g. size of the well screen, duration of the well test, etc.), and the overall length scale of the region from which the data were gathered.

The correlation length scales used in Figure 4b were 200 feet in the horizontal direction and 10 feet in the vertical direction. The larger degree of correlation was specified in the horizontal direction in order to approximate a natural bedding pattern. When correlation differs as a function of direction, the model (6) or (7) is said to be anisotropic (Fig. 6). The magnitudes of these correlation scales are not unreasonable for this type of region, although the range of correlation scales typically observed or measured in aquifers tends to be very large (e.g. Table 6.1 in Gelhar, 1993). This is again a result of the overall length scale of the problem (as there actually may be a hierarchy of different length scales, Cushman, 1990, and Fig. 6) and the size or scale of the individual conductivity measurements used to create the database. Measured horizontal correlation scales, for instance, range from a foot or two in small (500 ft) experimental sand aquifers to well over 3000 feet in some regional or basin scale aquifers (e.g. Gelhar, 1993, Fig. 6.8). In this sense, the type of structural element represented by a feature the size of a correlation scale may differ accordingly (see, e.g. Philips and Wilson, 1989).

Other SRF methods. The random field approach has been used to a great extent because it is amenable to mathematical manipulation within theoretical stochastic models of fluid flow and solute transport (e.g. Dagan, 1989; Gelhar, 1993). These models attempt to theoretically describe the effective or composite flow and transport behavior in heterogeneous systems without resorting to direct and detailed flow simulation, as was used in Figure 4b. Other variations of the SRF approach have been employed to broaden the applicability of the method, even though they are more appropriate for numerical flow simulation and less well suited for inclusion in theoretical models. The so-called binary or indicator models (Journel and Alalbert, 1990; Deutsch and Journel, 1992) involve the identification of discrete "indicators" to represent distinct classes of soils or other geologic materials. A "realization" of a geologic system will include an array of indicators to describe the spatial distribution of the indicator (soil) classes. Ideally, specific medium property values are then associated with each indicator to complete the realization. Because this approach provides a discrete representation of property values, it is important to carefully define the indicator classes ahead of time to allow the full distribution of properties to be included.

These techniques are fundamentally based on individual indicator variograms (as above) that represent the spatial continuity of each indicator class. Hence, it is possible to assign different models of variability with different length scales and principal axis orientations to different indicators (or soil classes). From a structural point of view, this approach offers more generality than the random field model. "Length scales" for indicator classes may be obtained from direct variogram measurement (as above), geologic interpretation of specific depositional environments, or both. In addition, most indicator algorithms are naturally conditional and can be modified to use both "soft" and "hard" forms of data. In the absence of sufficient property data, "soft information" can be used in lieu of "hard data" (using weakly defined cross-correlations between properties) to improve the overall conditioning and simulation process (Journel and Alalbert, 1990; Copty and Rubin, 1995). From a structural point of view, this approach offers more generality than the random field model.

Depositional and other geometric models. Depositional models of geologic heterogeneity seek to reproduce the patterns, structure, and distribution of geologic materials in by mimicking the processes that created them or recreating geometrical patterns and shapes of material bodies. For example, Webb (1994), Scheibe and Freyberg (1995), and Webb and Anderson (1996) discuss methods to simulate the depositional processes associated with braided streambed fluvial environments. Lake and Carroll (1986) and Lake et al. (1991) discuss a number of analogous approaches used within the oil industry. The issue is again focused on recreating elements of reality in a realization that are critical in reproducing important hydrologic (or petroleum flow) behavior. As with the indicator approaches, depositional models rely on simulating classes or types of media to which physical or chemical properties are later associated, although the concept of conditioning realizations to measured point data may not be possible in some of these approaches.

Dealing with physical heterogeneity

Given an approximate description of heterogeneity based upon one of these methods, what next? There are basically two camps of activity: those that work in the realm of "homogenization" theories and the like (e.g. stochastic models) and those focused on detailed numerical simulation (and, of course, those that combine both).

Homogenization. The concept of "homogenization" involves finding ways to describe and predict large-scale "bulk" behavior in nonuniform systems in terms of "effective" properties that are a function of an underlying structure and degree of heterogeneity. Given an idealized model of heterogeneity (such as a random field model in Eqns. 4-7) characterized by some fundamental length scale, the object would be to reliably predict the mean flow rates or effective solute migration and dispersion in such systems over much larger spatial scales on a simple, reliable basis.

For example, to estimate the net horizontal flow rate through the cross-section in Figure 4b under a known horizontal hydraulic gradient, one would normally multiply the magnitude of the gradient by an "average" hydraulic conductivity. Given the data distribution in Figure 3 and some interpretation of correlation structure (Fig. 6), how is the correct "average" chosen? If the geometric mean conductivity ($K_G = e^F$) is specified as a constant in each layer of the cross-section (Fig. 4a), then the apparent rate of plume migration is smaller than that in the heterogeneous case (compare results in Figs. 5a and b) — in other words, the net flow rate through the "average" system is too low. Stochastic theories of the sort advanced by Dagan (1989), Gelhar (1993), and others aim to predict a more appropriate "effective" horizontal conductivity $\hat{K}_h > K_G$ that would reproduce the correct flow rate under the same conditions. This effective property is determined in terms of the statistical variability and correlation structure of the medium. For example, Gelhar (1993) suggests that the effective horizontal conductivity in individual, statistically stationary units of the type comprising the stratified system in Figure 4b should be of the form $\hat{K}_h \approx K_G \exp(\sigma_f^2(1/2 - f(e))$, where f(e) is a complicated function of $e = \lambda_3 / \lambda_1$, the anisotropy ratio.

Consider, now, the overall dispersion and dilution of solute mass in Figure 5b, as induced by the nonuniform flow field. Here, the mass is largely dispersed and diluted, especially when compared with the same distributions induced by the more uniform field. This is also shown in Figure 5c, where we have reduced the previous 2D problems to their simpler one-dimensional counterparts by integrating the concentrations over the vertical dimension. The heterogeneous solution is still largely dispersed in comparison with the uniform case. Thus, for the simple 1D problem, how can the dispersive effect be described and reproduced? As above, stochastic theories attempt to estimate the degree of macroscopic dispersion as function of the statistical variability and correlation structure of the medium. This is usually expressed in terms of an effective dispersion tensor, $\mathbf{D}(s) = \mathbf{A}(s)|\mathbf{v}|$, whose principal components are related to local-scale dispersivities, α, the ln K variance, the correlation structural length scales, and even the net displacement, s(t), of the solute cloud. The appropriate mathematical form of the macrodispersivity tensor (**A**) will differ according to the problem dimensionality, overall problem scale, and the size of the solute cloud, and its overall applicability is still the subject of considerable debate (e.g. Dagan, 1989; Tompson and Gelhar, 1990; Bellin, et al., 1992; Gelhar et al, 1992; Gelhar, 1986, 1993; Rajaram and Gelhar, 1993; Tompson et al., 1996b). For example, Gelhar et al. (1992) discuss the increasing size of apparent field-scale dispersivity with increasing problem scale, and its relationship to increasing large scales of heterogeneity, such as those indicated in the variogram of Figure 6.

A third curve in Figure 5c represents a simple 1D simulation in which the effective velocity and dispersion behavior are incorporated. In this case, however, the appropriate forms of the effective parameters were determined from analysis of behavior observed in the heterogeneous simulation of Figure 5b, as opposed to the theoretical results. This was done mainly because multiple layers with different stochastic representations are included in this system, through which the expanding plumes migrate simultaneously. Although there is reasonable agreement in the basic features of the 1D solute distribution, this approach cannot provide the detail or reflect the local variability that the heterogeneity imparts.

Direct simulation. Direct simulation refers to solving heterogeneity problems at a very fine level of detail using numerical computer simulation. It is analogous to the procedure used in developing Figure 5b from Figure 4b, and allows basic material heterogeneities to be spatially resolved such that their subsequent impacts on flow and transport can be more carefully approximated. In this sense, uncertainties about the representation of effective behavior can be reduced by increasing the computational effort and property resolution. The approach is designed to combine sparse and variable hydraulic data and an integrated geologic interpretation onto a detailed probabilistic representation of the material property distribution that may allow multiple Monte Carlo simulations to be performed (Bellin et al., 1992, Essaid and Hess, 1993).

Increasingly, large-scale detailed simulation is being used to complement theoretical and experimental analyses of behavior in heterogeneous systems. Intensive calculations have been used, for example, to validate the accuracy of scaling or homogenization theories, benchmark the performance of intermediate or field scale experiments, or test and simulate the efficacy of specific remediation schemes within nonuniform environments (Ababou et al., 1989; Cole and Foote, 1990; Tompson and Gelhar, 1990; Macquarrie and Sudicky, 1990; Polmann et al., 1991; Schäfer and Kinzelbach, 1992; Tompson, 1993; Jussel et al., 1994b; Webb and Anderson, 1996; Thiele et al., 1996; Tompson et al., 1996a,b). Modeling capabilities that address behavior in three dimensions, resolve important small-scale property variations, incorporate specific remediation technologies or processes, or are capable of multiple Monte Carlo simulations are typically required. Such problems may have upwards of 10^8 computational nodes and may require solution approaches on massively parallel computers (Dougherty, 1991; Ashby et al., 1994).

THE CONCEPT OF CHEMICAL HETEROGENEITY

Reactions in porous media

In general, the composition of the groundwater in a natural formation will include a large number of dissolved constituents representative of the mineralogy of the geologic materials, atmospheric or other surface water interactions, and the occasional mix of "contaminants" that invade the subsurface from the surface. Principal types of reactions that may occur in an aqueous groundwater system include precipitation or dissolution of minerals in the mineral matrix, various forms of sorption and ion exchange between dissolved constituents and nearby mineral surfaces, formation of colloids, or numerous forms of complexation, hydrolysis, or redox reactions within the aqueous phase itself.

The impacts of reactions in the subsurface manifest themselves in many ways, and are often inextricably linked to the fluid flow itself. Over geologic time scales, we can speak of the development of "reaction fronts" and other phenomena associated with the nature and development of the soil or rock matrix, as based upon the existence of groundwater flow, very slow reactions, and a lot of time. Over shorter time scales, we may be concerned with the overall form and fate of liquid contaminant mixtures that enter the subsurface, move with the groundwater, and chemically interact with the solid or liquid phase constituents in the water or mineral matrix.

We will be focusing much of our discussion on the latter case, although many of the concepts used can be applied in a much broader sense. In the field of contaminant hydrology, the interest is usually in how reactions affect the migration or dispersive rates of contaminants, how they serve to remove contaminant mass from the aqueous system, and how the aqueous geochemistry may be subsequently modified by contaminant movements. Figure 7 shows the areal distribution of aqueous bromide and lithium at different times, as measured in a tracer test conducted at Cape Cod, MA (LeBlanc et al., 1991). Known amounts of each chemical were simultaneously injected into the shallow groundwater, allowed to migrate, and measured in a downgradient array of monitoring wells. Several distinct plumes were formed. As a result of solid phase sorption reactions, we clearly see different rates of plume migration and different plume shapes evolve during the test. After 461 days, the lithium plume has traveled only as far as the bromide plume at day 237, is much narrower, and shows some minor sharpened-front behavior at its leading edge (that is to say, the peak concentration in the plume seems closer to its advancing edge). Lithium was shown in this test to sorb onto the sandy medium materials according to a Langmuir isotherm, as shown in Garabedian et al. (1988, p. 124), and this form of isotherm has been shown to produce exactly this behavior (Tompson, 1993).

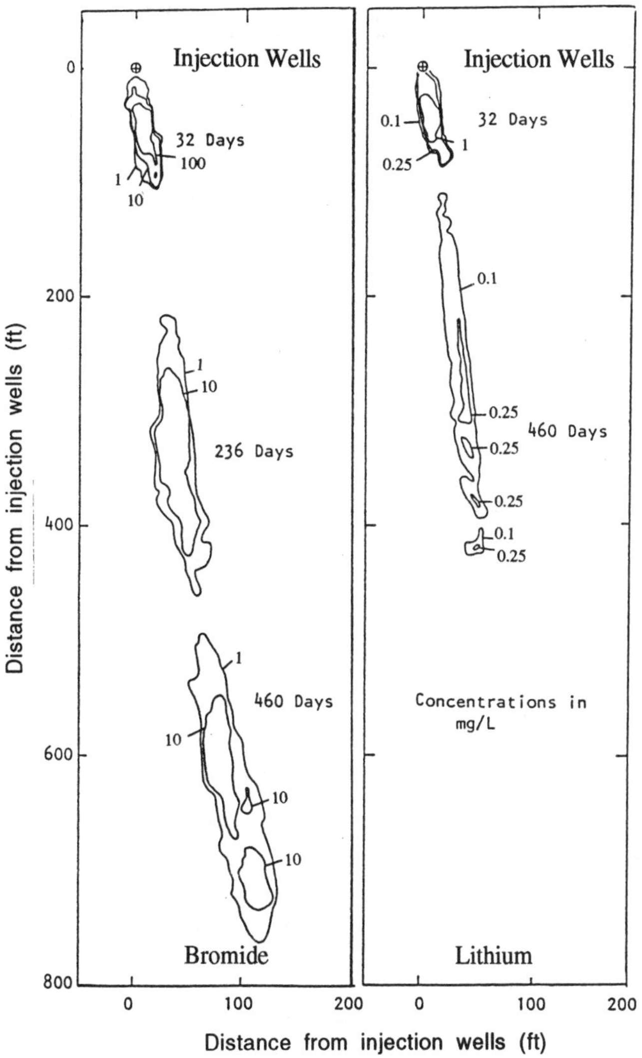

Figure 7. Spatial (areal) distribution of bromide and lithium plumes released as part of a tracer test in Cape Cod, Massachusetts (from Garabedian et al., 1988).

Reactions and heterogeneity

The notion of "chemical heterogeneity" may be loosely defined in terms of a spatial nonuniformity in the chemical properties or reactive mineral abundance in a porous medium

(Barber et al., 1992). Variation in the natural (as opposed to contaminant) geochemical composition of groundwater may also be regarded as a "signature" of a more fundamental mineralogical heterogeneity. The relevant properties or mineral of interest will depend on the problem at hand, as will the overall impacts. In a natural system (without contaminants or other "third-party" constituents), chemical heterogeneity may lead to different aqueous mineral compositions in different parts of a formation, as a function of the material distribution and flow regime. In a contaminated aquifer, the mobility or species composition of a plume may become dependent of the underlying structure of chemical heterogeneity. Let us consider two examples.

Example 1: Chemical heterogeneity and the aqueous geochemistry

Suppose the image in Figure 4b corresponded to the abundance and distribution of quartz sand in a groundwater system at low temperature. Suppose also that the surrounding material is devoid of silica. Let us consider three scenarios:

Scenario 1. If there were *no* groundwater flow and *no* molecular diffusion, then the concentration of silica in solution at a point would be in equilibrium with the sand at the same point. Hence, the spatial abundance of aqueous silica would correspond directly to the spatial distribution of sand—its nonuniformity or spatial abundance would trace back to the chemical heterogeneity of the system.

Scenario 2. Given the previous configuration, suppose we now allow horizontal flow to occur uniformly through the system, without any spatial variability—the hydraulic conductivity is considered constant. In this case, self-contained "material parcels" of water will move in order along horizontal streamlines. Depending on the streamline path and sand distribution, "silica-rich" water parcels may blindly flow into zones containing little or no quartz, or "silica-poor" water parcels may move into the quartz-rich areas (Thiele et al., 1996). If we assume that dissolved silica remains in solution and silica dissolution occurs fast enough (with respect to the flow velocity) to reach equilibrium, then the silica-poor packets will become richer in silica content as they move through the system. Ultimately, aqueous silica will be prevalent through the entire system in a very short time (after only a few "pore volumes" of fluid have passed through). Hence, the "signature" of chemical heterogeneity in the aqueous system is removed by the flow.

Scenario 3. Suppose we again begin with the configuration of Scenario 1 and allow flow to occur within the system. This time, we assume the hydraulic conductivity is directly related (or positively cross-correlated) to the occurrence of quartz sand. That is, where we have sand, we have high permeability, and vice versa. In this case, zones of preferential flow may develop between the zones of sand; hydraulically inaccessible zones may be created elsewhere outside of the sands. In other words, a set of streamlines (or streamtubes) will be established that link the zones of high permeability together (an effectively isolate them from lower permeability zones. In this case, the "silica-rich" water will preferentially remain in the sandy zones, while "silica-poor" water remains in the slower flow zones. Some cross-mixing can occur, but this will depend on the conductivity contrast between the different materials, the spatial structure of the material distribution, and the kinds of diffusion and dispersive mixing processes that occur.

The point in all of this is that the natural aqueous geochemistry may retain strong or weak signatures of any underlying chemical heterogeneity, as a function of the flow system, its associated physical heterogeneity, or the chemical components of interest (e.g. silica, carbonate, pH, or any other component whose abundance is controlled by chemical reaction with the rock matrix). In general, material parcels of water will translate down streamlines and their composition will chemically respond and adjust to the minerals that the parcels encounter (Thiele et al., 1996). In more complicated systems, this could lead to establishing an intricate chemical dynamical system associated with flow, reaction, and heterogeneity (Lichtner, 1988; Ortoleva, 1994). It is entirely possible that the water sampling process itself may cloud interpretations if solutions of different composition are mixed during collection (e.g. along the well bore of a fully penetrating groundwater well).

Example 2: Chemical heterogeneity and contaminant mobility

Sorption and retardation. As a second simple example, we will consider the migration of a single aqueous "contaminant" introduced into the groundwater and how certain reactions affect its mobility through retardation effects. First, let us review the concept of retardation and where it comes from.

The contaminant concentration distribution will satisfy a more general version of (3):

$$\frac{\partial(\varphi c)}{\partial t}+\nabla\cdot(\varphi \mathbf{v}c)+\nabla\cdot(\varphi \mathbf{J})=-\varphi R \tag{8}$$

The term on the right represents the *net* rate of aqueous contaminant *loss* [M/L^{3}T] due to sorption, precipitation, or other interactions with adjacent solid phase minerals, aqueous complexation, hydrolysis, or redox reactions, or any form of decomposition, such as bacteria mediated degradation or radioactive decay. For our present purposes, suppose only one form of interaction is present—that of a general binary sorption or ion exchange reaction onto a specific mineral phase (m) that may be nonuniformly distributed in the system. In this case, we assume R represents the net reaction rate resulting from a forward (aqueous to sorbed phase) reaction and a reverse (sorbed to aqueous phase) reaction.

The average abundance of the sorbed contaminant on the mineral surfaces will be denoted by an areal concentration, $s(\mathbf{x},t)$ [M/L^2] that satisfies

$$\frac{\partial(a_m^s s)}{\partial t}= \varphi R \tag{9}$$

where $a_m^s(\mathbf{x},t)$ is the specific surface area of mineral (m) per unit bulk volume of porous medium (L^{-1}). Because of the binary nature of the reaction, the net rate of aqueous mass loss must equal the net rate of sorbed phase production, so that the total contaminant mass (aqueous + sorbed) is conserved by (8) + (9):

$$\frac{\partial(\varphi c + a_m^s s)}{\partial t}+\nabla\cdot(\varphi \mathbf{v}c)+\nabla\cdot(\varphi \mathbf{J}) = 0. \tag{10}$$

Here, $\varphi c + a_m^s s \equiv \rho$ is the total chemical mass per unit bulk volume of porous medium [M//L^3].

If the net rate of chemical mass exchange between the liquid and mineral are considered fast with respect to the flow rate, then a local chemical equilibrium may be established such that

$$s = \Im(c, pH,\ldots) \tag{11}$$

This means that the sorbed and aqueous concentrations adjacent to a mineral surface are related according to a fixed isotherm. The function $\Im$ is used to approximate the specific mechanisms in the sorption process, whether they are dictated from thermodynamic mass action considerations or other empirical relationships, as well as any dependency on background geochemical conditions like pH or temperature. Strictly speaking, relationships like (11) are usually derived or justified in terms of idealized solutions (in a pore) near a perfect (i.e. clean) mineral surface; the use of macroscopic or average pore concentrations like s or c is an approximation. Equation (11) is usually called a sorption isotherm (Weber et al., 1991). Inserting (11) into (10) yields

$$\frac{\partial(\varphi c \Re)}{\partial t} + \nabla \cdot (\varphi \mathbf{v} c) + \nabla \cdot (\varphi \mathbf{J}) = 0 \tag{12}$$

where $\Re = 1 + a_m^s \Im(c,...)/\varphi c$ is a so-called retardation factor. When the temperature, pH, and all chemistry except the concentration of the component of interest are fixed, a "linear" isotherm may result and imply that $\Im \approx kc$, where k is related to intrinsic thermodynamic quantities independent of the bulk medium characteristics. In this case,

$$\Re \approx 1 + a_m^s k / \varphi \tag{13}$$

and the dependence on medium properties and more fundamental properties is separate and clear.

When the physical and mineralogical conditions in a formation are fixed, it is easy to see how $\Re$ will be constant and imply a uniform "retardation" in the migration ($\mathbf{v}/\Re$) and dispersive ($\mathbf{J}/\Re$) rates of an aqueous contaminant plume (as in Fig. 7). In the absence of the sorbing mineral ($a_m^s = 0$), the reaction and retardation effect will disappear altogether and (12) reduces to (3). In reality, the abundance of the mineral phase, and the overall reaction or retardation effect, are more likely to vary with position as a function of the physical and mineralogical composition of a formation. (Furthermore, in reality, the linearity of (11), the validity of the LEA, and many other assumptions may be suspect, but we will not question these assumptions now).

Chemical heterogeneity and sorption. Within geohydrology, most attempts at understanding the impacts of chemical heterogeneity on chemical transport have focused on the spatial variability of $\Re$, or its first cousin, K_d. In practical settings, it is most convenient to measure an apparent distribution coefficient K_d [L^3/M] in batch sorption experiments in which the fraction, ω, of contaminant mass sorbed per total solid phase mass is related to the aqueous concentration, c, via $\omega = K_d c$. The distribution coefficient becomes, in fact, a composite measure of the thermodynamic property, k, bulk medium density, ρ_b, the mineral surface area, a_m^s,

$$K_d = a_m^s k / \rho_b . \tag{14}$$

In terms of K_d, $\Re$ takes on the more familiar definition

$$\Re \approx 1 + \rho_b K_d / \varphi . \tag{15}$$

There are a number of instances where variations of K_d have been cataloged as a function of formation composition (Sheppard and Thibault, 1990; Bishop et al., 1991a,b). In Figure 8, results documented in Bishop et al. (1991a,b) show how measured K_d values for trichloroethlyene (TCE) and perchloroethlyene (PCE) from alluvial sediments beneath LLNL vary as a function of raw BET surface area (compiled by D. Hammond, pers. comm., 1994). These data show an approximate linear relationship suggested by (14). Figure 9 shows how the same results vary (on average) with organic content and lithology. Because the lithology categories are plotted in order of decreasing hydraulic conductivity, we see an approximate trend relating increased sorptive capacity to decreased permeability.

Indeed, Figure 10 shows this relationship more clearly for the same set of LLNL data (Bishop et al., 1991a,b; D. Hammond, pers. comm., 1994).

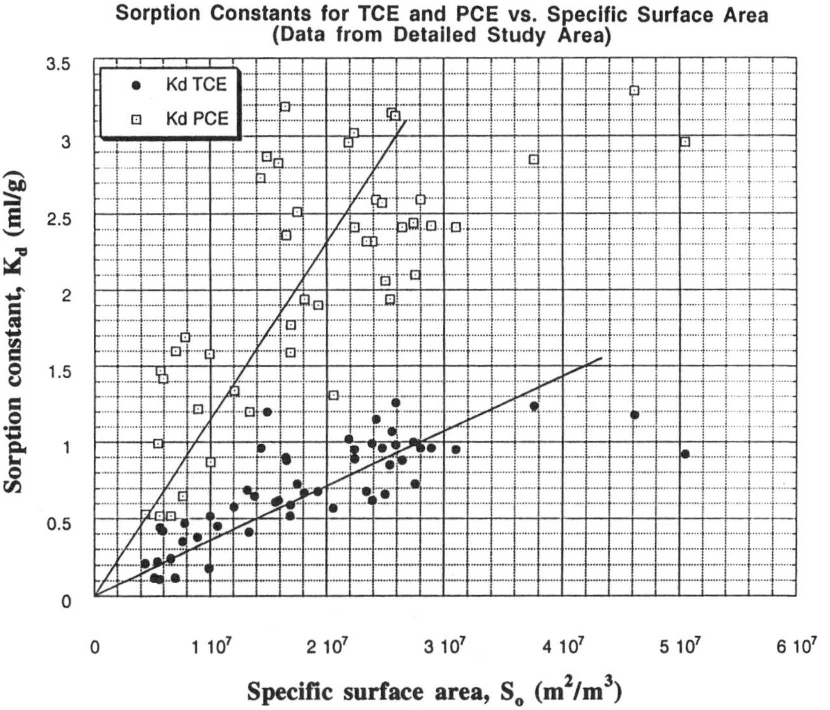

Figure 8. Variation of measured K_d values for TCE and PCE with BET surface area as measured in sediments beneath the Lawrence Livermore National Laboratory (D. Hammond, pers. comm., 1994; Bishop et al., 1991).

A number of investigators have used the fact that K_d may be inversely correlated with ln K in support of theoretical or computational studies designed to evaluate the impact of spatially variable retardation on plume migration and dispersive spreading rates (e.g. van der Zee and van Riemsdijk, 1987; Garabedian et al., 1988; Dagan, 1989; Valocchi, 1989; Chrysikopoulos et al., 1990; Kabala and Sposito, 1991; Robin et al., 1991; Bosma et al., 1993; Tompson, 1993; Burr et al., 1994). Numerous heuristic and theoretical arguments have been proposed to relate sorption (or retardation) to specific surface area (as in 13-14) as well as hydraulic conductivity (or permeability) to specific surface area, porosity, and grain-size distribution (Garabedian et al., 1988; Dullien, 1991; Tompson, 1993; Burr et al., 1994; Smith et al., 1996). For example, Smith et al. (1996) have proposed

$$\ln a^s \approx A_1 + A_2\varphi + \frac{A_3}{\varphi} - 0.5 \ln K \qquad (16)$$

for well sorted sands, where the A_i are constants, and a^s is the *total* specific surface area of the medium (L^{-1}). This relationship has been validated for a number of data sets based upon sandy materials (Smith et al.,1996); Tompson et al. (1996a) have used values of A_i equal to -37.7, 54.2,, and 6.6, respectively, for a fine, well-sorted sand. Assuming a^s_m = a^s and that the bulk density and porosity are constant, (12) and (16) suggest further that

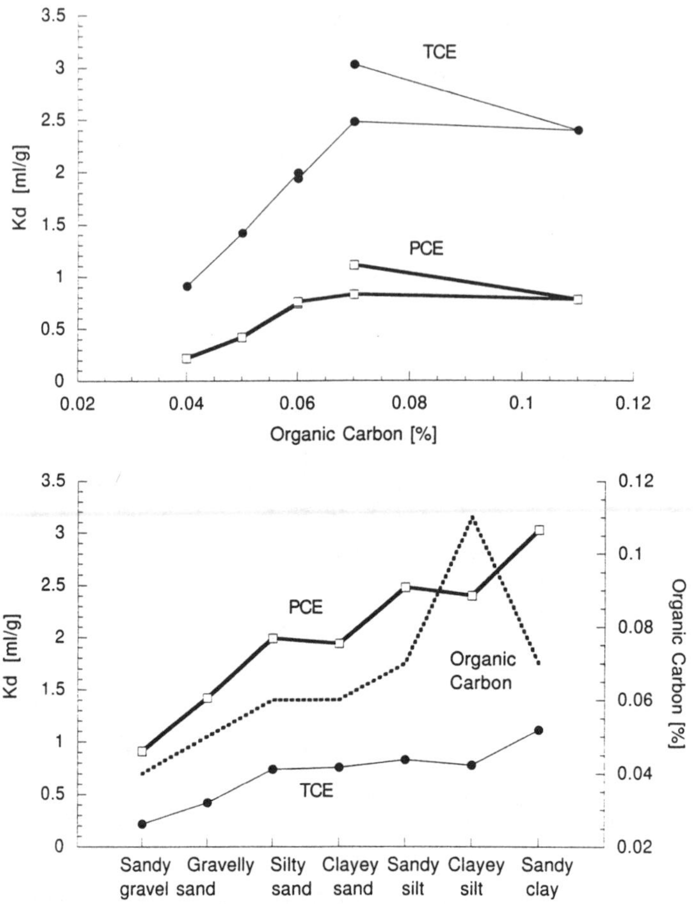

Figure 9. Average variation of K_d values for TCE and PCE with (top) organic carbon content, and (bottom) lithology, as measured in sediments beneath the Lawrence Livermore National Laboratory (after Bishop et al., 1991a,b). Because the sediment types are arranged in order of decreasing hydraulic conductivity, an inverse correlation of K_d with hydraulic conductivity may be inferred.

ln K_d is also negatively correlated to ln K with the same -0.5 slope. The data in Figure 10 are a good confirmation of this concept, even though the slope factor is closer to -0.2. The data measured by Robin et al. (1991) are less conclusive, however, although Smith et al. (1996) suspect there may be more of a "signal" in the data if variations with porosity are also taken into account.

The difference between the data slope in Figure 10 and the model slope in (16) may be due to the fact that (16) is a poor model for the mixed alluvial sediments at LLNL. Sorption is considered here to occur onto weakly abundant organic matter in the sands and silts, and possibly within microfractures of numerous other minerals. Figure 11 shows another set of data relating measured K_d values for toluene to surface area at three locations near LLNL (Bishop et al., 1991a,b; D. Hammond, pers. comm., 1994). The "clean" and "Pre" curves represent two sets of scattered data taken from two locations several hundred

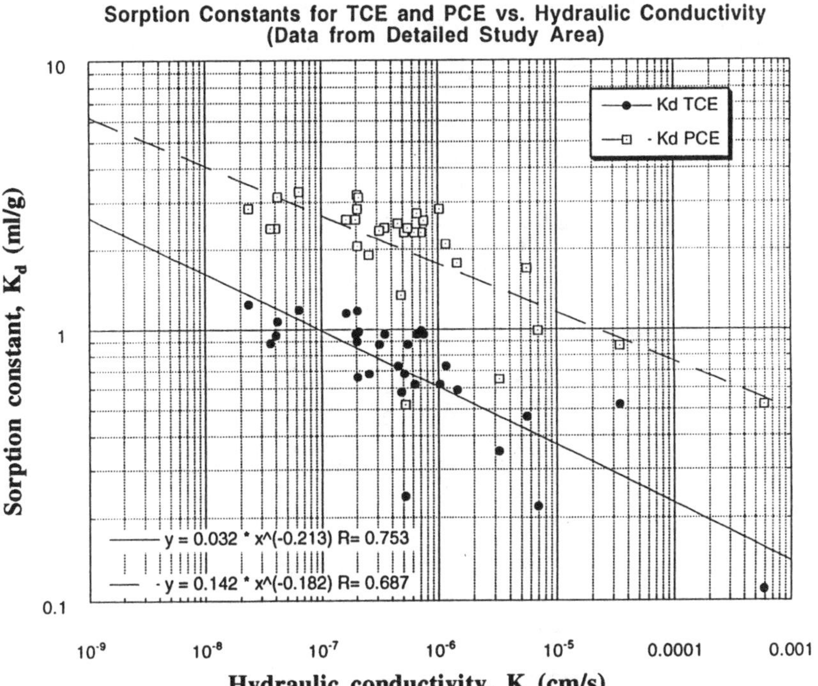

Figure 10. Variation of measured K_d values for TCE and PCE with hydraulic conductivity as measured in sediments beneath the Lawrence Livermore National Laboratory ((D. Hammond, pers. comm., 1994; Bishop et al., 1991a,b).

feet apart. There is a lot of scatter in the data, but the same type of trend seen in Figure 9 is seen here. The "Post" curve represents a new set of data taken at the same location as the "Pre" data, albeit after an in-situ steam stripping operation was conducted (Newmark et al., 1995). Here, we see the sorptive capacity of the system and the general distribution of surface area has been reduced by the stripping procedure. In this case, it appears that the sorption may have been associated with fine grained sediments or other organic material that were swept out in the operation.

Chemical heterogeneity impacts. Many of the theoretical analyses cited above have focused on how the bulk mobility and dispersive spreading rates of sorbing groundwater contaminants are affected by heterogeneity in derived parameters such as K_d or $\Re$. Most of these have assumed K_d or $\Re$ to be randomly distributed (in addition to ln K), and possibly cross-correlated with porosity and ln K, as in (14-16). The principal results involve the identification of "effective" retardation and dispersive coefficients as a function of their underlying statistical variability and cross-correlation.

On average, the results tend to indicate that effective retardation is reduced when cross-correlation between ln K_d and ln K is negative (as in Fig. 10), increased when it positive, and unchanged from the spatial arithmetic average of $\Re$ in the absence of correlation. This is understandable in the sense that sorption will be reduced in zones of higher flow rate, lowering, on average, the apparent bulk retardation effect (and vice versa, of course). Furthermore, the apparent degree of dispersive spreading can also be affected in

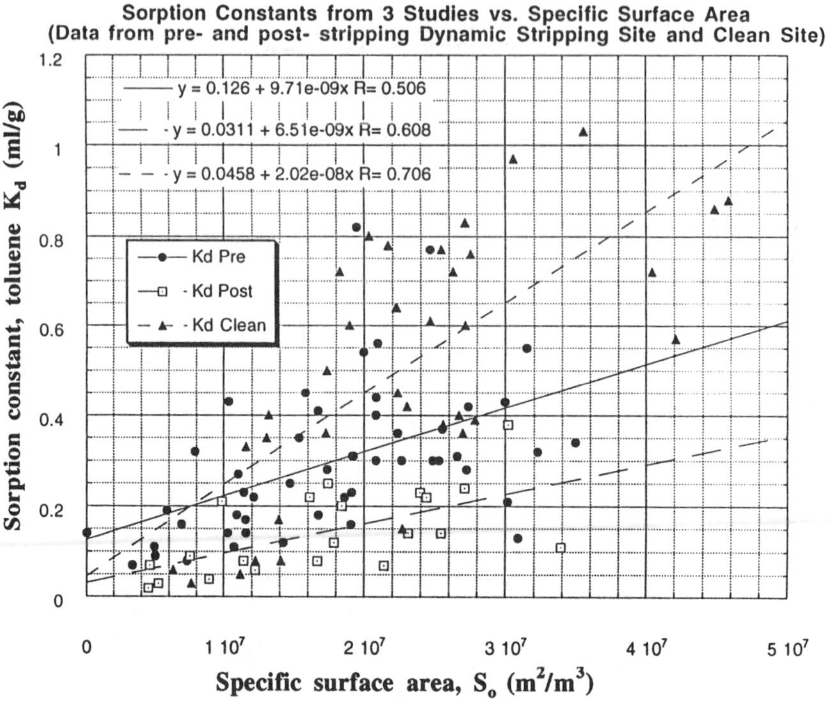

Figure 11. Variation of measured K_d values for toluene with BET surface area as measured at three locations in sediments beneath the Lawrence Livermore National Laboratory (D. Hammond pers. comm., 1994; Bishop et al., 1991a,b). "Pre" and "Clean" refer to measurements on samples collected from two locations separated by several hundred feet. "Post" refers to measurements made on samples taken very close to the "Pre" sampling spot, but after a gasoline steam stripping operation was performed. It is believed that the steam stripping removed a number of the fine and other organic materials onto which sorption occurred.

the same way. Garabedian et al. (1988) predicted enhanced bulk dispersive effects under negative correlation and very high positive correlation, and a regime of reduced dispersion under small positive correlation. The degree of change in retardation or dispersion is predicted to be significant in many situations of practical relevance, and less so in a number of others. Although many of these effects have been observed in numerical simulation experiments (e.g. Bosma et al., 1993; Tompson, 1993; Burr et al., 1994), little has been done in the field (through characterization of chemical heterogeneity or analysis of chemical migration) to document their existence and overall importance further.

LOOKING AT THE ROLE OF MULTICOMPONENT SYSTEMS

A more complicated example system

Clearly, the foregoing analyses and discussions have focused on the most simple forms of reaction in porous media involving isolated (single) constituents undergoing simple linear equilibrium reactions. For many hydrologic contamination problems, these sorts of conceptualizations serve a useful "engineering" purpose. Nevertheless, the relevant processes associated with contaminant migration can be much more complicated, and could involve multiple interacting components in contaminant mixtures, competitive sorption and ion exchange reactions, precipitation and dissolution phenomena, and so forth, all spanning a range of reaction rates that may simultaneously require both kinetic and equilibrium treatment.

Here we will review a recent analysis focused on the impacts of physical and chemical heterogeneity on the migration of a model co-contaminant mixture of uranium and citric acid (Tompson et al., 1996a). As reviewed by Riley et al. (1992), the interest in radionuclide and organic acid contaminant mixtures is motivated by their ubiquity in many U.S. Department of Energy facilities. The study reviewed here was designed to look at the role of physical and chemical heterogeneity in simple idealized saturated sand formation. The sand was assumed to be interlaced with a post-depositional hydrous-ferric oxide (HFO) mineral phase, such as goethite. Post depositional metal oxide coatings can be a significant cause of reactivity in a variety of subsurface systems (Smith and Jenne, 1991), and their spatial distribution can be very sporadic and nonuniform (Fig. 12). For comparison, analogous studies of cobalt-EDTA mixtures in the presence of metal oxide coatings are also in progress (Szescsody et al., 1994). Ultimately, the interest will be in how the presence of reactive media (e.g. goethite), its overall abundance and distribution (e.g. chemical heterogeneity), and its relationship to the hydraulic properties of the system (e.g. cross correlation) will influence the mobility, dispersive nature, and, in this case, the speciation patterns of the model contaminant mixture.

Figure 12. Cross-sectional view of a cut through a coastal sand formation in Virginia, showing various depositional patterns and interlaced iron oxide banding patterns (courtesy Annette Schafer, Idaho National Engineering Laboratory).

Approximate geochemical model

Although the interest here is to move toward a more complicated contaminant system, we will still keep many things simplified in order that the entire simulation process be kept tractable. We will consider a simple saturated sand aquifer system at fixed temperature, ionic strength and pH. Portions of the sand grains may be covered with a reactive goethite coating. Our "model" contaminant will involve an aqueous mixture of uranyl (UO_2^{2+}) and citric acid $(H_3)citrate$ that is introduced into the system. The mixture will be considered sufficiently dilute such that the ambient geochemical conditions in the sand remain unchanged by the presence and motion of the contaminants. However, these conditions may still influence the aqueous speciation and solid phase partitioning of the contaminants. The simulations described below assume a fixed aquifer pH of 5.

Within the aqueous phase, citric acid will dissociate into several hydrolysis species according to

$$H_n citrate^{n-3} \Leftrightarrow H^+ + H_{n-1} citrate^{n-4} \tag{17}$$

where n = 3, 2, or 1, while uranium will participate in hydrolysis reactions of the form

$$xUO_2^{2+} + yH_2O \Leftrightarrow (UO_2)_x(OH)_y^{2x-y} + yH^+ \tag{18}$$

In addition, the uranyl and $citrate^{n-3}$ ligand are expected to form an aqueous complex of the form [Shock, 1990]

$$aUO_2^{2+} + b\ citrate^{3-} \Leftrightarrow (UO_2)_a(citrate)_b^{2a-3b} \tag{19}$$

Formation of complexes with other citrate hydrolysis ligands will not be considered here. With concentrations of carbonate, magnesium, and calcium consistent with a groundwater at a constant pH of 5, the important aqueous species, as determined from calculations based on MINTEQ (Felmy et al., 1984), are defined by (x,y) = (1,0), (1,1), and (a,b) = (1,1).

Sorption of uranium and citrate onto goethite will be represented by surface complex reactions (Dzombak and Morrell, 1990). Assuming that the goethite is the only reactive mineral phase (i.e. uncoated quartz sands are assumed to be nonreactive) and considering only a single type of sorption site, these reactions will be described by

$$\equiv FeOH^0 + UO_2^{2+} \Leftrightarrow \equiv FeO(UO_2)^+ + H^+ \tag{20}$$

$$\equiv FeOH^0 + citrate^{3-} + H^+ \Leftrightarrow \equiv Fe_2 citrate^{2-} + H_2O \tag{21}$$

$$\equiv FeOH^0 + UO_2 citrate^- + H^+ \Leftrightarrow FeO(UO_2)citrate^0 + H_2O \tag{22}$$

Here the nomenclature follows Dzombak and Morel (1990), and the symbol $\equiv FeOH^0$ designates bonds at the surface of a hydrous ferric oxide solid and represents $[Fe(OH)_3]_n$. The stoichiometry of reaction (20) is consistent with the adsorption of uranyl on HFOs (Dicke and Smith, 1995). The stoichiometries of reactions (21) and (22) are currently unknown and are the subject of ongoing experimental evaluation. Dzombak and Morel (1990) represent the adsorption of trivalent anions by the use of up to four surface complexes with varying numbers of hydrogens. However, at constant pH, these multiple complexes can be treated using a single reaction. In addition, equilibria between H^+, $\equiv FeO^-$, $\equiv FeOH^0$, and $\equiv FeOH_2^+$ will also occur. However, at a fixed pH (e.g. 5), the ratio of the abundances of the $\equiv FeO^-$, $\equiv FeOH$, and $\equiv FeOH_2^+$ surface species are fixed by equilibrium considerations, allowing the abundance of $\equiv FeOH_2^+$ and the uncomplexed sites to be represented in terms of the abundance of $\equiv FeOH^0$.

The chemical model used contains two aqueous uranium hydrolysis species, four aqueous citrate species, one aqueous uranium citrate complex, and four surface species (protonated, uranium, citrate, and uranium-citrate). At a given location in the porous medium, partitioning among the aqueous and surface species will be affected by the abundance of goethite (or other HFOs) and the relative concentration of the species present.

The specific goethite surface area (average goethite surface area per unit volume of porous medium) is considered to be related to the total specific mineral surface area via

$$a_m^s(\mathbf{x}) = f(\mathbf{x}) \cdot a^s(\mathbf{x}) \tag{23}$$

where the fractional goethite surface area, $f(\mathbf{x})$, is bounded by 0 and 1. Both a_g^s and f can be used to indicate the macroscopic abundance of goethite at a point $\mathbf{x}$ in the system.

Equilibrium speciation. We assume that macroscopic flow processes are small enough to allow local equilibrium to be achieved by the complexing and sorption reactions (Bahr and Rubin, 1987). We also assume that the mass action expressions corresponding to (17-22) along with a balance of sorption sites, can be expressed in terms of average, macroscopic concentrations of the relevant species, as measured representative elementary volumes (REVs) of the porous medium (Hassanizadeh and Gray, 1979). Error in this representation will be minimized to the extent that conditions in the pores remain "well mixed" by local flow and diffusion processes (Murphy et al. 1989) and to the extent that goethite surface area is uniformly distributed in an REV.

Table 1. Aqueous and sorbed species considered in the uranium-citric acid problem. "Macroscopic concentrations" refer to bulk concentrations associated with a point or representative elementary volume in the porous medium. The "f" terms are derived from the hydrolysis and dissociation equilibria associated with uranyl and citric acid, and are constant for fixed pH.

Aqueous species	**Macroscopic Concentration** (M/L³)	**Species cross reference**
Total aqueous uranium	$c_u(\mathbf{x},t)$	$[UO_2^{2+}]+[UO_2(OH)^+]$
Total aqueous citrate	$c_c(\mathbf{x},t)$	$[H_3citrate]+[H_2citrate^-]+$ $[Hcitrate^{2-}]+[citrate^{3-}]$
Aqueous uranyl	$c_u(\mathbf{x},t)\cdot f_u$	$[UO_2^{2+}]$
Aqueous citrate	$c_c(\mathbf{x},t)\cdot f_c$	$[citrate^{3-}]$
Aqueous uranyl-citrate	$c_{uc}(\mathbf{x},t)$	$[UO_2-citrate^-]$
Sorbed or solid species	**Areal Macroscopic Concentration** (M/L²)	**Species cross reference**
Sorbed uranyl	$s_u(\mathbf{x},t)$	$[\equiv FeO(UO_2)^+]$
Sorbed citrate	$s_c(\mathbf{x},t)$	$[\equiv Fecitrate^{2-}]$
Sorbed uranyl-citrate	$s_{uc}(\mathbf{x},t)$	$[\equiv FeUO_2citrate^0]$
Free (unsorbed) site concentration	$s_g(\mathbf{x},t)$	$[\equiv FeOH^0]$

Mass-action expressions for complexing and sorption Reactions (19-22) are given by

$$Q_{uc} = \frac{c_{uc}}{c_c f_c c_u f_u} \tag{24}$$

$$Q_{su} = \frac{s_u \cdot 10^{-pH}}{s_g c_u f_u} \tag{25}$$

$$Q_{sc} = \frac{s_c \cdot 10^{pH}}{s_g c_c f_c} \tag{26}$$

$$Q_{suc} = \frac{s_{uc} \cdot 10^{pH}}{s_g c_{uc}} \tag{27}$$

where the concentrations are defined in Table 1 and the *Q*s are conditional equilibrium constants appropriate for fixed temperature, pressure, and fluid composition that incorporate thermodynamic activity coefficients. The f_u and f_c terms are used to represent the influence of the uranyl hydrolysis and citric acid dissociation reactions in the formulation, and remain constant for fixed pH (see Table 1 and Tompson et al., 1996a). These expressions may be combined with three molar balance equations of the form

$$\rho_u = \varphi(c_u + c_{uc}) + a_g^S(s_u + s_{uc}) \tag{28}$$

$$\rho_c = \varphi(c_c + c_{uc}) + a_g^S(s_c + s_{uc}) \tag{29}$$

$$\rho_s = a_g^s \Gamma = a_g^s(s_g + s_u + s_c + s_{uc}) \tag{30}$$

to determine the abundance of c_u, c_c, c_{uc}, s_u, s_c, and s_{uc}. The quantities $\rho_u(\mathbf{x},t)$ and $\rho_c(\mathbf{x},t)$ represent the macroscopic concentrations (M/L^3) of total-uranium or total-citrate, respectively, per bulk volume of medium. They will comprise the two principal unknown variables in the transport formulation below. Equation (30) is a macroscopic site balance, where $\rho_s(\mathbf{x},t)$ is the total sorption site concentration available on the goethite per bulk volume of the medium (M/L^3) and Γ is the intrinsic molar surface site density on goethite (M/L^3) Hsi and Langmuir (1985) have $\Gamma = 3 \times 10^{-6}$ mmol/cm^2 (Table 1). Aqueous and sorbed species considered in the uranium-citric acid problem. "Macroscopic concentrations" refer to bulk concentrations associated with a point or representative elementary volume in the porous medium. The "f" terms are derived from the hydrolysis and dissociation equilibria associated with uranyl and citric acid and are constant for fixed pH.

Sorption onto unoccupied sites has been assumed to be independent of conditions at other nearby sites. More specific (phenomenological) models can be used to account for effects arising from inter-site interactions (e.g. Dzombak and Morel, 1990). Because of the constraint embodied in (30) and the fact that only a single type of sorption site is considered, the Reactions (20-22) will exhibit coupled, Langmuir-like sorption behavior.

At any location or time, known values of $\rho_u(\mathbf{x},t)$, $\rho_c(\mathbf{x},t)$, and $\rho_s(\mathbf{x},t)$ may be inverted to obtain the equilibrium values of c_u, c_c, c_{uc}, s_u, s_c, and s_{uc} as a function of f_u, f_c, and the pH. To do this, the nonlinear set of Equations (28-30) must be inverted and solved for c_u, c_c, and s_{uc} subject to the constraints implied by (24-27). The nonlinear system takes the form

$$\mathbf{A}(c_u, c_c, s_g) \cdot \mathbf{c} = \mathbf{r} \tag{31}$$

where $\mathbf{c} = \{c_u, c_c, s_g\}$, $\mathbf{r} = \{\rho_u, \rho_c, \rho_s\}$, and $\mathbf{A}$ is implicitly dependent on the equilibrium constants, pH, as well as a_g^s and φ. Subsequently, values of c_{uc}, s_u, s_c, and s_{uc} are found by back substitution into (24-27). Recall that the notion of "chemical heterogeneity" refers here to spatial variability in the values of a_g^s and φ—this implies that the inversion process may yield different results for $\mathbf{c}$ at different locations even if $\mathbf{r}$ remains unchanged.

Preliminary calculations using MINTEQ (Felmy et al., 1984) have been conducted to estimate three of the four constants in (24-27) at pH 5. The results of Hsi and Langmuir (1985) and Mesuere and Fish (1992) (assuming sorption of citrate is qualitatively similar to that of oxalate) were used to estimate:

$$Q_{uc} f_u f_c \approx 1.7 \times 10^6 \text{ cm}^3\text{/mmol} \tag{32}$$

$$Q_{su} f_u \cdot 10^{pH} \approx 3.9 \times 10^3 \text{ cm}^3\text{/mmol} \tag{33}$$

$$Q_{sc} f_c \cdot 10^{-pH} \approx 1.2 \times 10^3 \text{ cm}^3\text{/mmol} \tag{34}$$

Although there are currently no values of Q_{suc} available to describe the formation of $\equiv FeUO_2citrate^0$, its sorption is not expected to be strong; hence, we will assume $Q_{suc} = 0$.

Transport formulation and simulations

Transport simulations will focus on solving two composite balance equations for total-uranium and total-citrate in the saturated sand system, with the assumption that the sorption and complexing reactions may be treated in an equilibrium fashion. At the macroscopic level, the relevant transport equations are given by (Tompson et al 1996a)

$$\frac{\partial \rho_u}{\partial t} + \nabla \cdot \left(\frac{\mathbf{v}\rho_u}{\Re_u}\right) - \nabla \cdot \left(\varphi \mathbf{D} \cdot \nabla \frac{\rho_u}{\varphi \Re_u}\right) = 0 \tag{35}$$

$$\frac{\partial \rho_c}{\partial t} + \nabla \cdot \left(\frac{\mathbf{v}\rho_c}{\Re_c}\right) - \nabla \cdot \left(\varphi \mathbf{D} \cdot \nabla \frac{\rho_c}{\varphi \Re_c}\right) = 0 \tag{36}$$

and are analogous to Equation (12) above for the one-component case. These equations can be used to evaluate the spatial distribution of total uranium and total citrate as a function of time, given appropriate "initial" conditions for the total concentrations and retardation factors. The entire species distribution can be subsequently evaluated at any time by inverting (31) as discussed above. The retardation factors in this case,

$$\Re_u = \frac{\rho_u}{\varphi(c_u + c_{uc})} \tag{37}$$

$$\Re_c = \frac{\rho_c}{\varphi(c_c + c_{uc})} \tag{38}$$

are concentration-dependent and will tend to increase with decreasing values of c_u and c_c.

The dependence of $\Re_u$ and $\Re_c$ on concentration will create differential retardation effects that can promote the formation of sharpened reaction fronts (e.g. van der Zee and van Riemsdijk, 1987; Tompson, 1993), such as the apparent effect shown in the lithium plume in Figure 7. This can be seen further by substituting (24-27), (28-29), and (32-34) into (37) and (38) and recalling that $Q_{suc} \approx 0$:

$$\Re_u \approx 1 + \frac{a_g^s s_g \cdot [3.9 \times 10^3 cm^3 / mmol]}{\varphi(1 + c_c \cdot [1.7 \times 10^6 cm^3 / mmol]} \tag{39}$$

$$\Re_c \approx 1 + \frac{a_g^s s_g \cdot [1.2 \times 10^3 cm^3 / mmol]}{\varphi(1 + c_u \cdot [1.7 \times 10^6 cm^3 / mmol]} \tag{40}$$

At lower concentrations, the sorption site constraint (30) will become insignificant such that $s_g \approx \Gamma \approx$ constant. In this case, the "leading" Langmuir behavior will emerge, although it will remain nonlinear and concentration dependent for values of c_u and c_c much greater than 10^{-7} mmol/cm^3. Given small enough concentrations, or sufficient time for dispersion and dilution processes to produce them, the ratios defined by (39) and (40) should eventually become "constant" or dependent only on local medium conditions.

Simulation strategy. The simplified geochemical model and transport formulation for the uranyl-citric acid mixture has been used to develop a series of numerical simulations depicting the migration, retardation, and speciation of various uranyl-citric acid mixtures through a 30 m-long cross section in a hypothetical heterogeneous sandy aquifer, much in the same way that was shown earlier in Figures 4 and 5.

The cross-section used for all simulations (except one) is shown in Figure 13—heterogeneity in the hydraulic conductivity distribution is again represented by a correlated random field with an anisotropic correlation structure to reflect horizontal bedding patterns. Flow in this system was induced from the left to the right and was determined from a computational flow model.

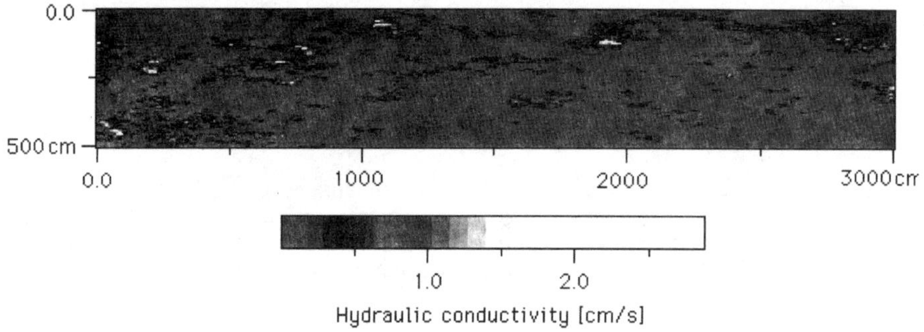

Figure 13. Approximate distribution of hydraulic conductivity in a 30 m-long cross-section through a hypothetical sand aquifer, as used in the uranyl-citrate migration simulations. This representation was developed from an anisotropic random field representation (4-7) as discussed in the text. Flow was later induced through this region from the left to the right.

Heterogeneity in the goethite abundance was specified in two ways—(i) via direct correlation of specific surface area to hydraulic conductivity, as in (16), and (ii) through specification of the fractional goethite surface area in (23) to represent "partial coverage" or specific banding patterns of goethite, as shown in Figure 12. Further details of all of these specifications can be found in Tompson et al. (1996).

Basic configuration. Our "basic problem configuration" consists of the cross section and associated variability in hydraulic conductivity, as shown in Figure 13, as well as the groundwater flow field determined in this system. The sand was assumed to be uniformly coated with goethite such that f = 1. The specific surface area of the sand was assumed correlated with ln K via (16) with a slope factor of -0.5 and an average value of 1,500 cm^{-1}. The range of surface area values is controlled by the variance of ln K distribution which was nominally set to $\sigma_f^2 = 1$, an average value for a sand formation. Further details of all of the specifications for this and all other configurations can be found in Tompson et al. (1996).

Two transport problems were initially conducted in this configuration (Figs. 14 and 15). In Problem 1, 50 mmol of a nonreactive tracer, 50 mmol of total uranium, and 50 mmol of total citrate (corresponding to initial concentrations of 0.0067, 0.002, and 0.002 $mmol/cm^3$, respectively) were released in the flow field and allowed to migrate in the groundwater flow and interact with the goethite. The first image in Figure 14 shows the tracer distribution after 8×10^6 s (93 days) of migration. The plume is well developed, evenly dispersed about its center of mass, and shows some effect of the hydraulic heterogeneity. The fourth and fifth images in Figure 14 show the aqueous-uranyl-citrate and aqueous-uranyl distributions at the same time. These plumes are significantly retarded, much thinner and less dispersed, asymmetrically distributed with "sharp front" features, and show again the effect of hydraulic heterogeneity. They also show how the complex is much more abundant, as dictated from equilibria concerns. The second image shows the tracer at an earlier time when it has traveled about the same distance as the uranyl and uranyl-citrate plumes. The qualities and relative differences in these results are remarkably similar to those seen in the Cape Cod tracer test images (Fig.7). This is consistent with the observation that the lithium in the tracer test was shown to have Langmuir sorption properties, in the same sense as the uranyl-citrate mixture has here.

In Problem 1-U, we released only 50 mmol of total uranium into the same system (no citrate). After the same 93 days, we find the uranyl distribution (shown in the third panel of Fig. 14) to be largely immobile. This results from the lack of citrate which serves as a mobilizing complex.

In Figure 15, we show the total mass of all aqueous and sorbed species in the system as a function of time for Problems 1 and 1-U. For Problem 1, the net mass of sorbed species increases as the total mass migrates and disperses through the system. Plume migration and dispersal promote plume dilution which, in turn, shifts the complexing and sorption equilibria towards more sorbed species. For Problem 1-U, we find most of the uranyl in a sorbed (immobile) state.

Modified correlation. In the next sequence of simulations (Problems 2-5), we deviate from the basic configuration by modifying the correlation function used to relate the specific sand surface area to the hydraulic conductivity, while still maintaining f = 1. These problems are identical to Problem 1 in terms of the initial chemical release and flow field configuration (Problem 3 only is based upon a uniform flow field). In Figure 16, the five panels show the aqueous uranyl-citrate complex after 93 days in Problems 1 (for reference) and 2-5.

In Problem 2, the specific surface area was specified as a constant throughout the domain, equal to the arithmetic mean of that used in the basic configuration. In Problem 3, the hydraulic conductivity was additionally kept constant at an appropriate "effective" value to maintain the same fluid flow rate through the system. In Problems 4 and 5, the surface area correlation (16) was used with modified slope factors of -0.3 and +0.3, respectively,

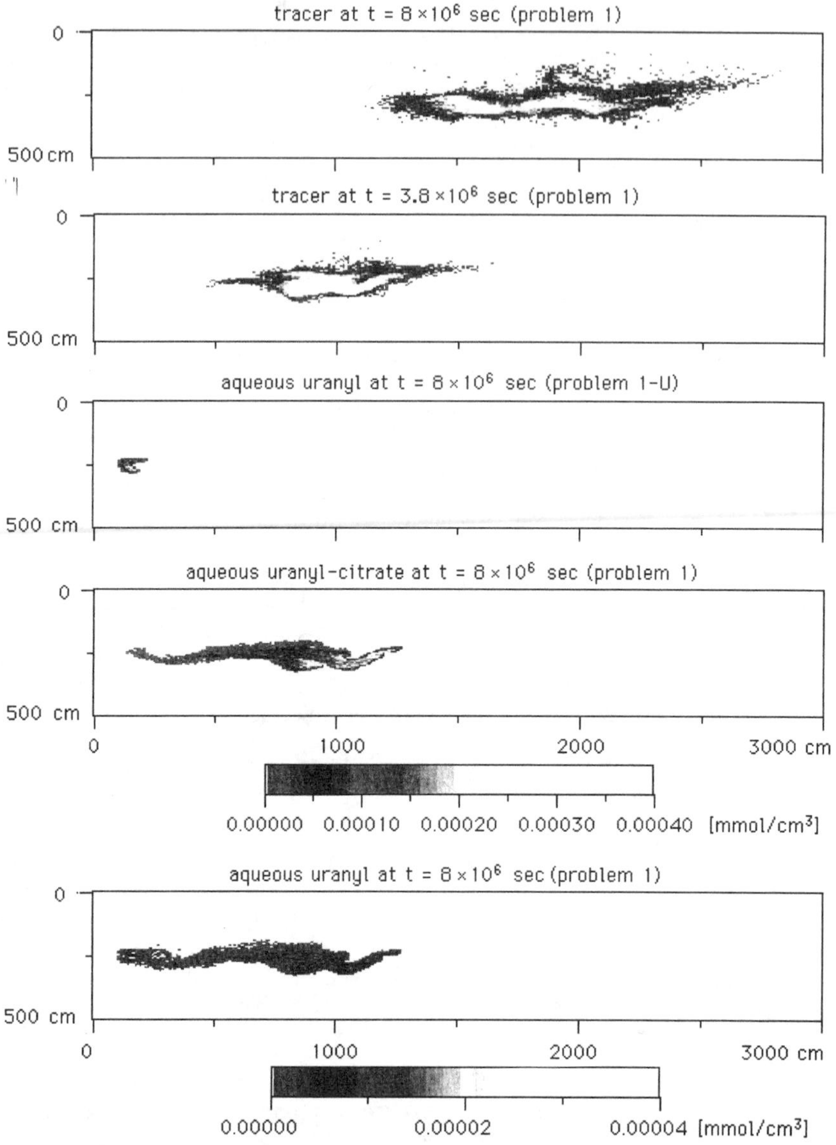

Figure 14. Snapshots of aqueous tracer, uranyl, and uranyl-citrate complex distributions for transport Problems 1 and 1-U (see text).

and the heterogeneous conductivity distribution. In Problems 2, 4 and 5, the apparent effects of these correlation modifications on the migration and dispersion facets of the plumes are hardly noticeable; aside from the uniform hydraulic nature in Problem 3, its plume is still similar to the others. These similarities are also shown in the total-mass plots calculated for aqueous uranyl-citrate, aqueous uranyl, and sorbed uranyl shown in Figures 17, 18 and 19.

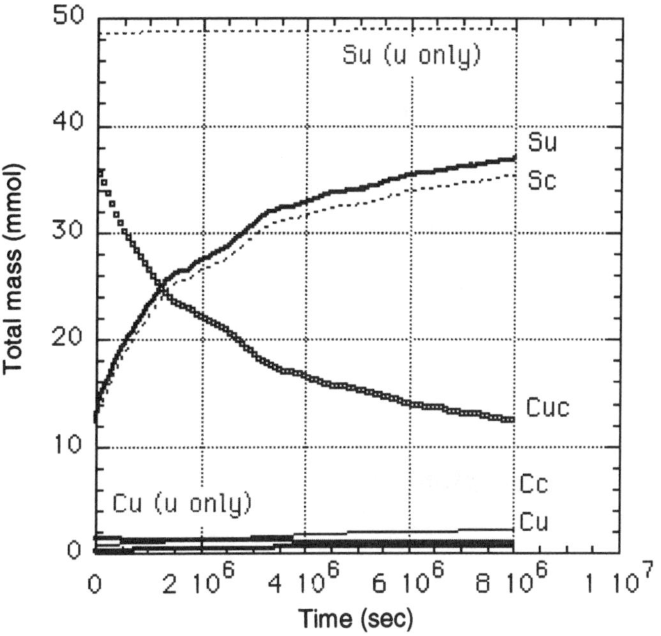

Figure 15. Total mass of aqueous and sorbed species in the system as a function of time for Problems 1 and 1-U.

Modified goethite distribution. In this sequence of simulations (Problems 6-9), we deviate from the basic configuration by modifying the fundamental abundance of goethite by specifying patterns and distributions according to the function f. The fundamental physical heterogeneity and surface area correlation used in the basic case are retained. Figure 20 shows five basic patterns of the f distribution used in these problems. The first corresponds to a uniform value of f = 1, as used in Problems 1-5; the second corresponds to a case where f is uniformly equal to 0.1 (Problem 6), the third through fifth correspond to cases where f = 1 in a light or heavy banded pattern, or a block pattern, and 0 elsewhere (Problems 7-9).

In Figure 21, the five panels show the aqueous uranyl-citrate complex after 93 days in Problems 1 (for reference) and 6-9. Clearly, the lack of geoethite in the f = 0.1 and banded cases (panels 2-4) tends to produce less sorption and more plume mobility. In the last panel, the plume organization in the blocked goethite area is the same as in Problem 1, but in the non-goethite area, the plume has become much more mobile. These same features can be discerned in the total mass plots in Figures 17, 18 and 19.

Modified source composition. Figures 22 and 23 show results from three additional problems in which the composition of the initial release has been modified with respect to Problems 1-9. The first panel in each figure shows the distribution of aqueous uranyl-citrate and aqueous citrate from Problem 1 (for reference).

In Problem 1-S (panel 2 in each figure), the initial composition of total uranium and total citrate was shifted to 37.5 and 62.5 mmol, respectively, forcing a stoichiometric overabundance of citrate (initial concentrations were proportionally modified). This configuration was used to run Problem 1 over again. Although the panel-two plumes in

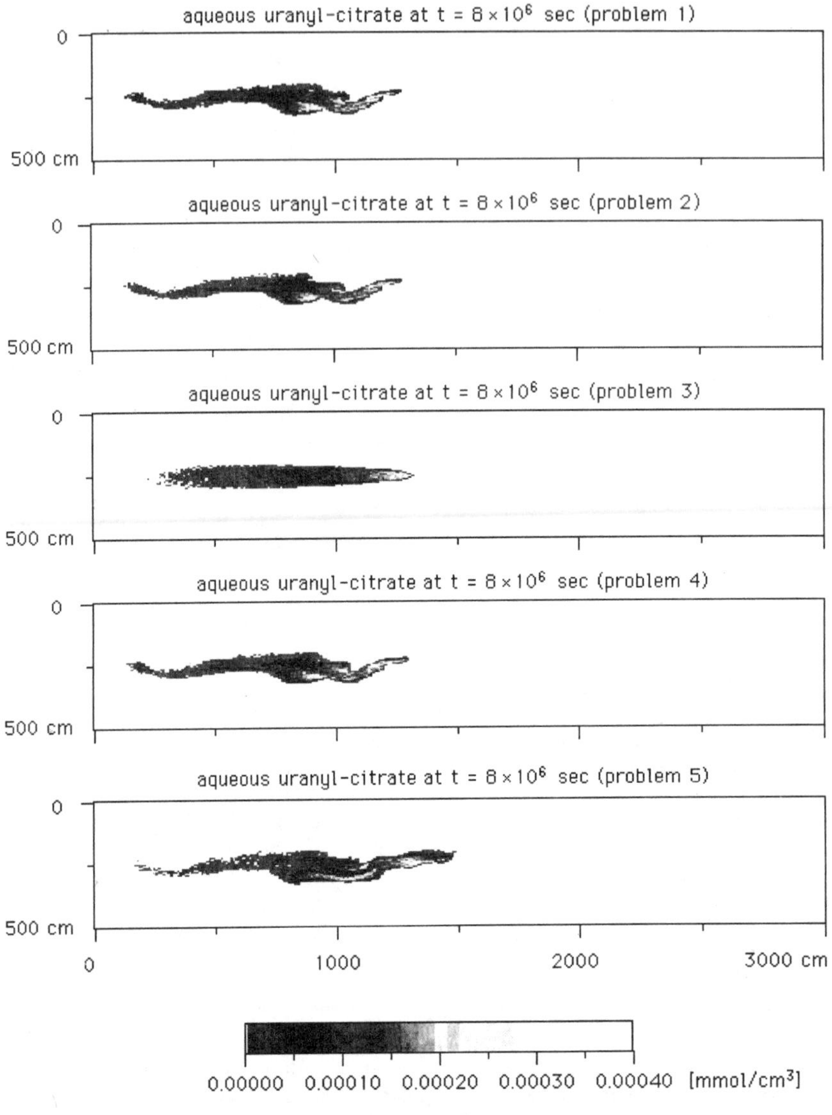

Figure 16. Snapshots of aqueous uranyl-citrate distribution for transport problems 1-5 (see text).

these figures look "similar", the excess citrate (that can find no uranyl to complex with) forms a highly concentrated, immobile anchor near the citrate plume tail.

In Problem 8-S (panel 3 in each figure), the modified initial composition was used to run Problem 8 over again. In this case, the excess citrate is much more mobile, owing to the general reduction in goethite associated with the banding pattern. In fact the banding pattern is faintly reflected in the citrate concentration distribution. This result and its difference from that in Problem 1-S exemplify the importance and impacts of chemical heterogeneity.

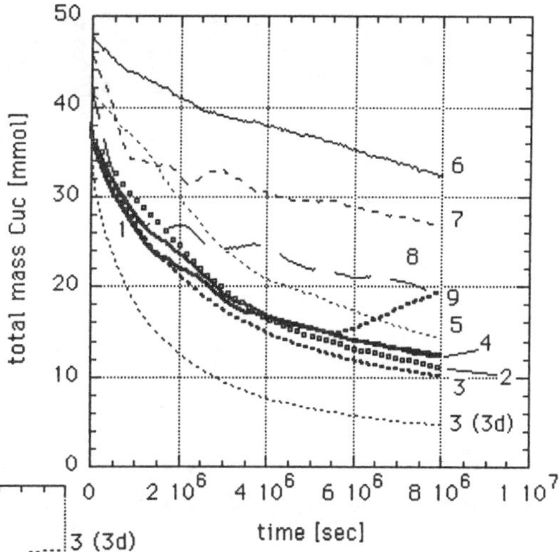

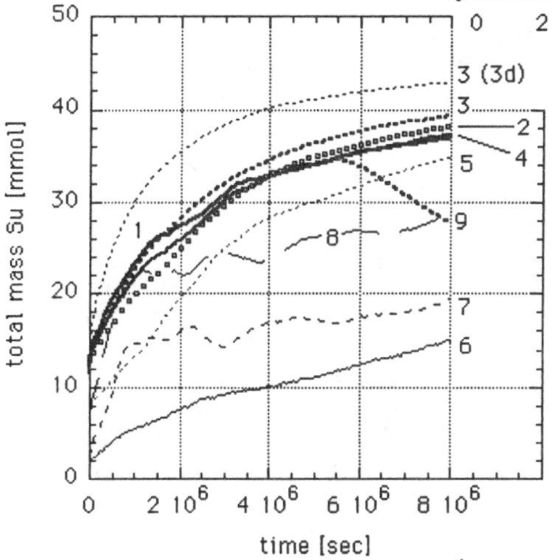

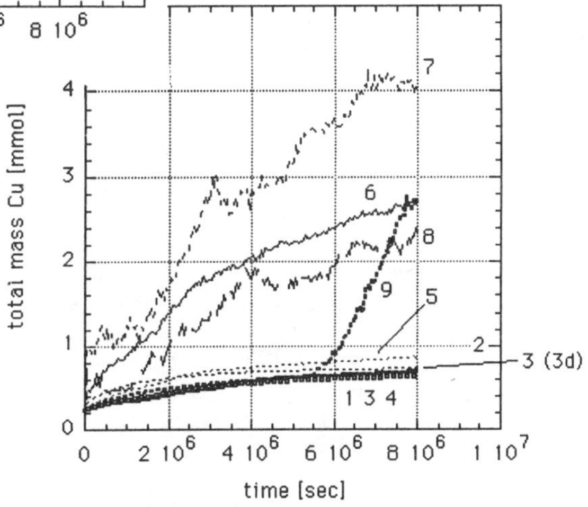

Figure

17 (top). Total mass of aqueous uranyl-citrate in the system as a function of time for Problems 1-9.

18 (middle). Total mass of aqueous uranyl in the system as a function of time for Problems 1-9.

19 (bottom). Total mass of sorbed uranyl in the system as a function of time for Problems 1-9.

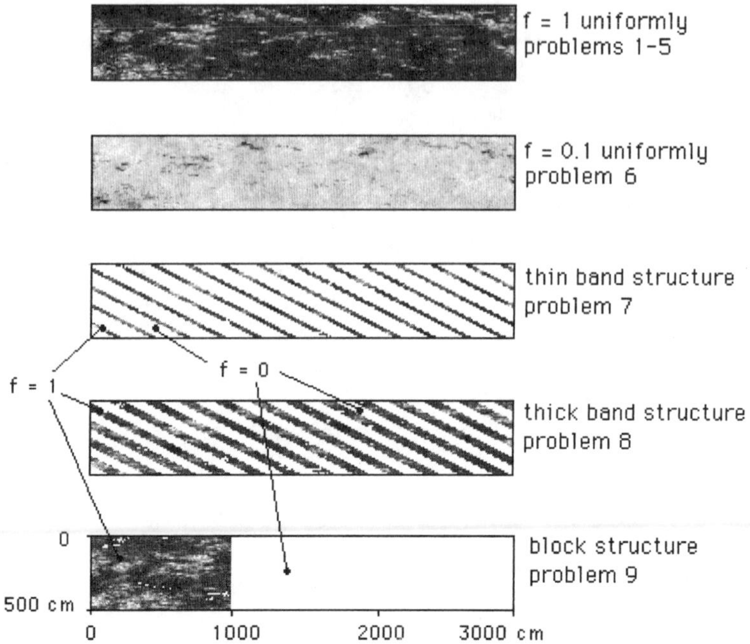

Figure 20. Spatial distribution of goethite surface area as represented by the fraction, f, in Problems 1-5 (panel 1, f = 1), Problem 6 (panel 2, f = 0.1), and Problems 7-9 (panels 3-5, f = 1 in various banded or block patterns).

In Problem 1-3x (panel 4 in each figure), initial composition and concentrations used in Problem 1 were increased by a factor of three (150 mmol of total uranyl and total citrate), and the problem was rerun. Even though the stoichiometry was kept "even," the distribution of aqueous uranyl-citrate changed because limitations in available sorption sites (30) did not allow a proportional increase in sorbed mass; hence overall mobility was increased.

A final remark

Although this multicomponent example was formulated on a rather simple geochemical basis, the results seem to show a great deal of rich behavior with respect to the mobility, speciation patterns and dispersive characteristics of the contaminant mixtures as they move through the system. These features are influenced not only by the general reactive nature of the system (stoichiometric balance in mixture composition, sorption site constraints, etc.), but the physical and chemical heterogeneity of the system. In particular, the abundance and distribution of goethite, as described by banding or block patterns that mimic real pattern observations (Fig. 12) seem to be quite influential in migration behavior, much more so than controls related to variable surface area alone. Over longer time frames, of course, goethite distribution itself may be modified, either from iron dissolution, additional geochemical precipitation, or even microbiologic activity (e.g. Ferris et al., 1989).

A FIELD EXAMPLE INVOLVING BIOREMEDIATION

In situ bioremediation of aquifers contaminated with volatile organic compounds has received increased attention in recent years, as a possible alternative to the more standard pump-and-treat remediation options (Wilson et al., 1986; Lee et al., 1988; Baker and Herson, 1991). Current efforts to develop techniques for bioremediation of contaminated

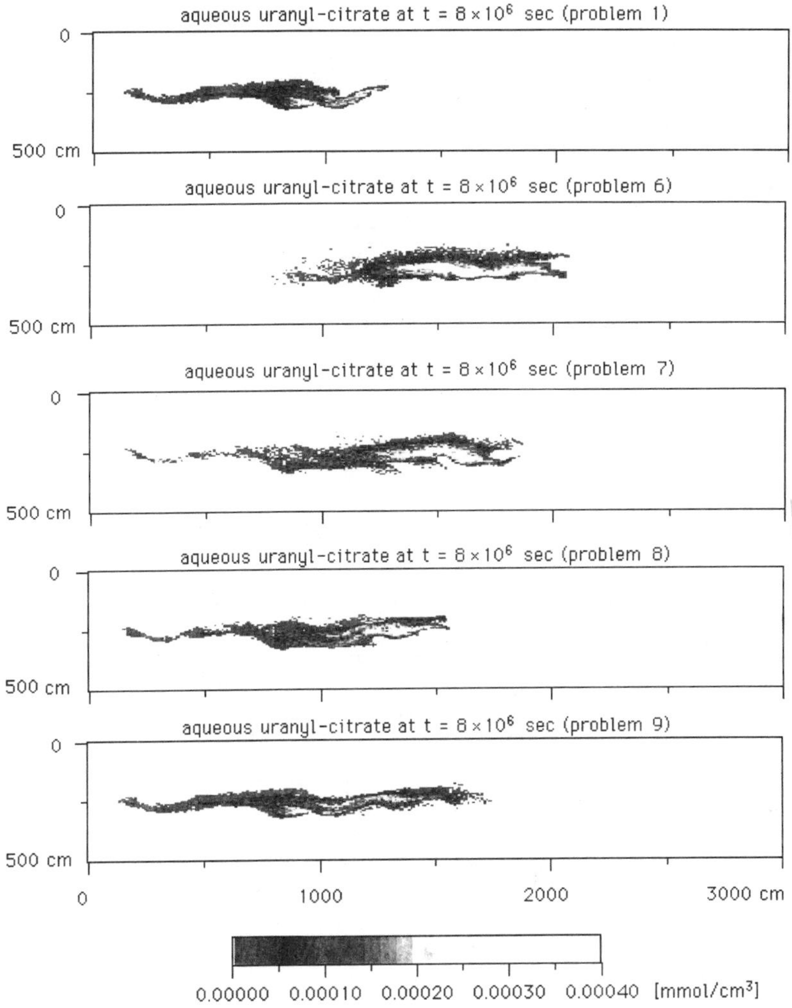

Figure 21. Snapshots of aqueous uranyl-citrate distribution for transport Problems 1 and 6-9 (see text).

aquifers center around two contrasting approaches: biostimulation and bioaugmentation. Field implementation of either approach is profoundly influenced by media heterogeneities and the success of any field-scale application of bioremediation hinges on the degree to which it is possible to successfully circumvent the engineering challenges presented by natural variations in the physical and chemical properties of aquifers.

Biostimulation

Biostimulation is an attractive alternative because of its simplicity and promise of low cost. In biostimulation, a suite of nutrients and electron acceptors—such as oxygen—is injected into the subsurface to increase the population of indigenous, contaminant-degrading microorganisms. These microorganisms are usually found attached to the solid

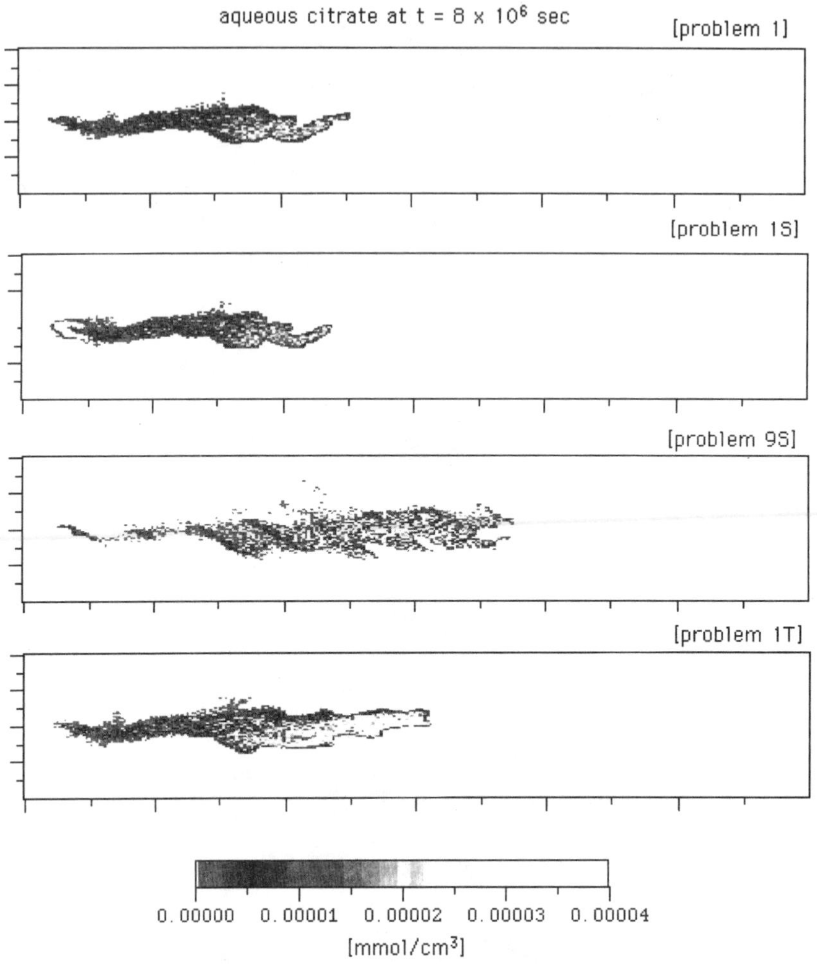

Figure 22. Snapshots of aqueous citrate distribution for transport Problems 1, 1S, 9S, and 1T (see text).

matrix, although they can detach and move with the groundwater as well. The injected compounds are generally inexpensive and the surface operations are straight-forward to implement. The biostimulation process has three main aspects: (1) selecting an injection package that is appropriate for the indigenous microbial community, the subsurface environment, and the contaminant(s) of interest; (2) injecting the package into the groundwater system to increase the indigenous bacterial population densities over an area sufficiently large to ensure good potential for degradation, and (3) achieve contact between the stimulated microbial community and the contaminated fluids. One of the chief difficul ties in this process is that injected fluids tend to displace contaminated groundwaters away from what eventually becomes the "biostimulated" zone. If injection ceases and the contaminated water is drawn back under natural or artificial conditions, some degradation may occur, although at a diminished efficiency because the nutrients that support the enhanced microbial activity are no longer available to the bacteria.

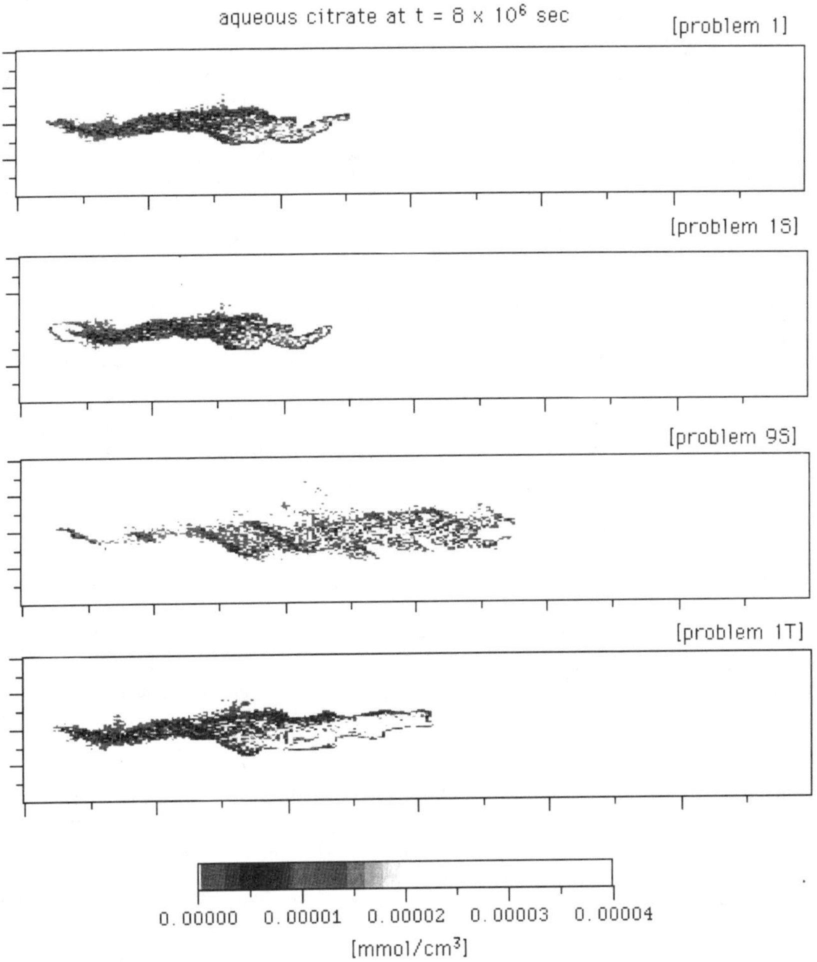

Figure 23. Snapshots of aqueous uranyl-citrate distribution for transport Problems 1, 1S, 9S and 1T (see text).

Extensive biostimulation field tests conducted at Moffet Field, California successfully demonstrated the first two points (Roberts et al., 1990; Semprini et al., 1990; Semprini et al., 1991; Hopkins et al., 1993) by achieving an approximately 90% TCE degradation in a 1 ppm contaminant stream at the end of a 60-day interval in which phenol-degrading microorganisms were stimulated. Achieving contact between contaminant and the stimulated population was not tested in this experiment, because the contaminants were injected along with the nutrient package—thereby forcing contact. Much poorer results were obtained for a TCE degradation experiment when methanotrophs were stimulated (Roberts et al., 1990; Semprini et al., 1990; Semprini et al., 1991). In part, the reason for this hinged on the fact that contaminated groundwater was displaced past the growth media.

Because of this displacement and mixing issue, biostimulation is most applicable in aquifers where a significant fraction of the contaminant mass is relatively insoluble or sorbed to the solid media. This contaminant immobility can result in more efficient mixing between contaminant, injected fluid, and biomass, and has the potential to yield substantial biodegradation (Wilson et al., 1994). However, it should be pointed out that biostimulation field studies conducted without a tracer in the injection package are ambiguous because of the inability to discriminate between biodegradation and dilution (Nelson et al., 1990; Kinsella and Nelson, 1994). Additional challenges in field-scale engineering of a biostimulation remediation strategy include competitive inhibition (Semprini et al., 1991), preferential growth near the injection well (Peyton et al., 1995), and a limited ability to select the bacterial strains that are stimulated to grow.

Bioaugmentation

In the resting-state in situ bioaugmentation approach (Taylor et al., 1993), naturally-occurring bacteria are grown in surface bioreactors, separated from their growth medium, re-suspended in an aqueous solution that is devoid of added growth nutrients, and then injected into the subsurface. During the injection phase a portion of the bacteria attaches to the aquifer material and forms a fixed-bed biofilter. After the microbial injection phase is completed, subsequent ambient or induced flow delivers contaminated groundwater to the biofilter region, resulting in contaminant biodegradation.

The key to this approach is that nutrient injection over large areas, and the subsequent complications, is not required. The degree of biodegradation depends on the flux of contaminants, the attached population density, and the contaminant residence time in the biofilter, just as in any chemical reactor. Because these "resting state" bacteria are not growing or multiplying, the emplaced biofilter will eventually expire because the injected, non-dividing bacteria have both a finite biotransformation capacity for degrading contaminants and a finite period of longevity during which the bacteria will remain metabolically active in the resting state (Taylor et al., 1993; Tompson et al., 1994; Taylor and Hanna, 1995).

Thus in order to continue in situ bioremediation by this process, regular replenishment of the bacterial population by re-injection is required. However, this approach avoids many of the problems associated with the biostimulation approach; especially those associated with the displacement of mobile contaminant species, biofouling due to enhanced bacterial growth near injector wells, and the limitations imposed by the relatively low solubility of some components of the nutrient package. A barrier to its successful implementation is the difficulty of achieving bacterial contact with highly retarded or immobile contaminants.

Bioaugmentation field test

A recent field test of the resting-state bioaugmentation approach (Duba et al., 1996) used a naturally occurring species of methanotrophic bacteria (Methylosinus trichosporium OB3b) together with a "huff-and-puff" implementation in a trichloroethene (TCE) plume at the Chico Municipal Airport in northern California. TCE is the sole groundwater contaminant in this plume, with concentrations generally less than 2000 ppb. Contaminated groundwater is restricted to a 15 m thick, partially confined aquifer, with a depth to groundwater of about 27 m. Aquifer materials consist of highly heterogeneous volcaniclastic sediments of the Pliocene Tuscan Formation (Harwood et al., 1981). Suitability, treatability, and groundwater tracer experiments led to selection of a single well near the center of the plume for the test.

Results of multicomponent conservative tracer tests confirmed the existence of a highly heterogeneous subsurface environment that presented a difficult challenge for implementing this in situ biodegradation method and provided a basis for determining the extent of biodegradation achieved during the test. In the same way that in situ solutes are dispersed and preferentially distributed by heterogeneity (Fig. 5), the injected bacteria will be similarly affected. In this experiement, aquifer heterogeneity dramatically affected the subsurface distribution of injected bacteria and governed the dimensions of the emplaced biofilter region. Instead of the roughly spherical biofilter that would be formed in a homogeneous aquifer (Fig. 24), the tracer test indicated large variations in hydrologic properties in both the horizontal and vertical directions. A computer simulation of the tracer injection/withdrawal process was conducted using the NUFT code (Nitao, 1995) with the assumption of a homogeneous aquifer. Comparison of the measured evolution of the conservative tracer (bromide ion) concentration at the two monitoring wells with the calculations (Fig. 25) confirms that the aquifer behaves as a highly heterogeneous unit in the injection horizon. Several key reservoir characteristics can be inferred from computer modeling of the tracer data: (1) there are azimuthal and depth-related variations in hydraulic conductivity over the 1 m (or less) distance between the injection/withdrawal well and the monitoring wells suggested by both the relative amplitude and phasing of the bromide pulses; (2) relative to the homogeneous model, breakthrough at two monitoring wells located 1 m on either side of the injection well is essentially instantaneous, suggesting at least one thin, high-conductivity zone; (3) log-normal analysis (Chesnut, 1994) of pump-test data suggests that the variance of the ln K distribution (σ_f^2) is close to 1, indicating significant heterogeneity.

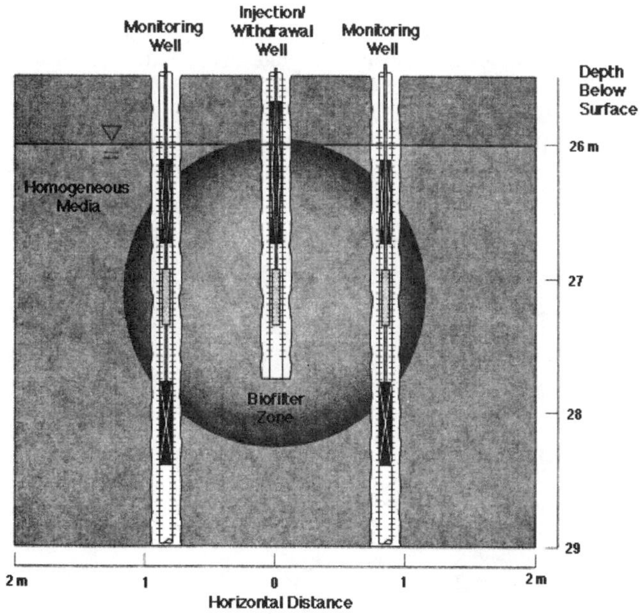

Figure 24. Highly idealized cross-section of a nearly spherical microbial filter as it would appear if emplaced in a homogeneous aquifer (adapted from Duba et al., 1996). The biofilter radius is about 1.1 m. Two monitoring wells are placed about 1 m on either side of the injection well. Depth below the ground surface is indicated by the scale to the right.

Based on these tracer test results, it can be assumed that portions of the aquifer with greater conductivity preferentially accepted the injected fluid, which resulted in a biofilter of large radial extent in these zones (Fig. 26). Less of the injected material flowed into zones of lower conductivity, which led to a biofilter of relatively small radial extent. Unfortunately, no unique picture of the shape of the biofilter can be defined by the data provided by tracer and pump tests. However, these results indicate that the microbial filters technique may prove to be a practical technique for remediation of groundwater plumes contaminated with chlorinated ethenes, but they also emphasize that interpretation of the test results are impossible without an adequate understanding of the effects of media heterogeneities.

SUMMARY

We have attempted to review some issues concerning the role and impact of geologic heterogeneity on the migration and reaction of subsurface fluids and dissolved chemical components. We have chosen to emphasize aspects associated with contaminant hydrogeology as it is a most timely and generally illustrative problem. In terms of reactions and other chemical phenomena in the subsurface, their true dimension cannot be understood without an honest regard for the fluid flow processes and their relationship to heterogeneity. Contaminant migration, dilution, speciation, retardation, and other reaction and mobility issues will all be affected by both physical and chemical heterogeneity. Thus, to the extent that these processes need to be understood, manipulated, or otherwise exploited, the need to characterize and quantify the impacts of heterogeneity will be ever-present.

ACKNOWLEDGMENTS

This work was conducted under the auspices of the U.S. Department of Energy (DOE) by Lawrence Livermore National Laboratory under contract W-7405-Eng-48. It was supported in part by the Subsurface Science Program of the Office of Health and Environmental Research of the DOE through subcontracts C94-160801 and C95-175509 from the Idaho National Engineering Laboratory, Idaho Falls, ID, and the DOE Office of Technology Development (Environmental Restoration Research and Development Program). Contributions of Robert Smith and Annette Schafer of the Idaho National Engineering Laboratory are gratefully acknowledged. We thank William Glassley for his lucid review.

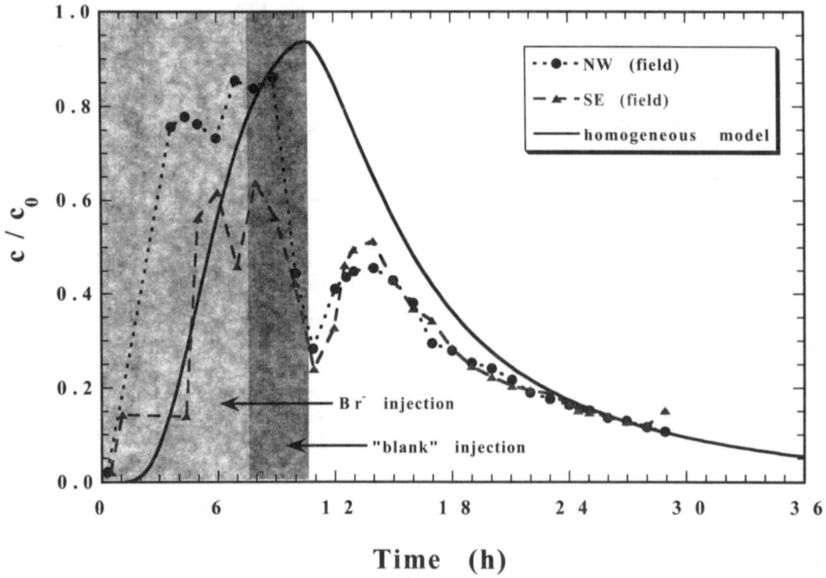

Figure 25. Bromide concentrations measured in the two monitoring wells (Fig. 22) observed during the tracer test, compared with those predicted from a computer simulation assuming a homogeneous aquifer.

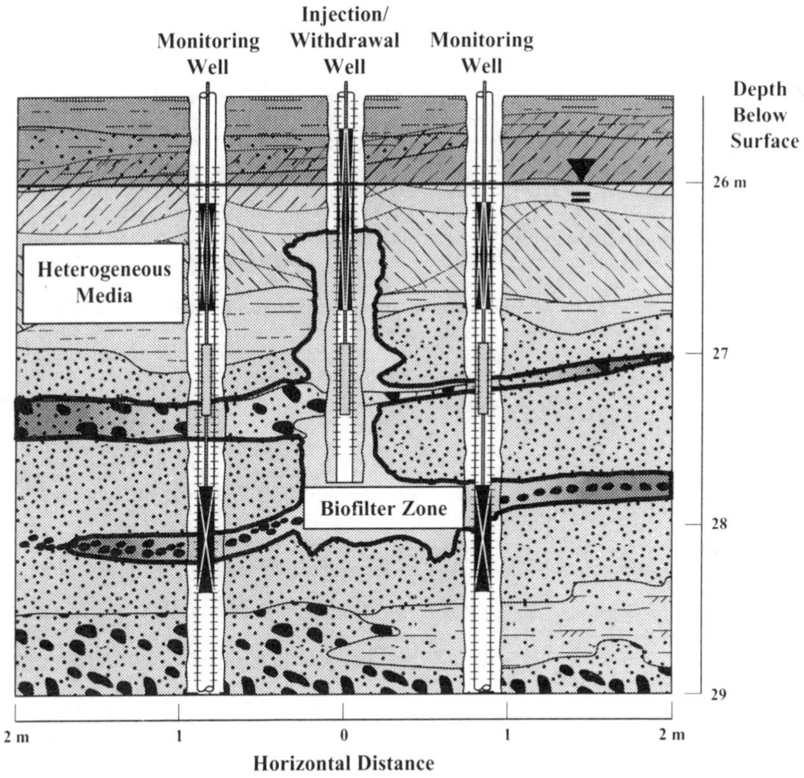

Figure 26. Schematic of a more realistic biofilter established in a heterogeneous geologic media.

REFERENCES

Ababou R, McLaughlin DB, Gelhar LW, Tompson AFB (1989) Numerical simulation of saturated flow in randomly heterogeneous porous media. Transport in Porous Media 4:549-565

Ashby SF, Bosl WJ, Falgout RD, Smith SG, Tompson AFB, Williams TJ (1994) A numerical simulation of groundwater flow and contaminant transport the CRAY T3D and C90 supercomputers. Proc 34th Cray User Group Conf, October 10-14, 1994, Tours, France; (also Lawrence Livermore Nat'l Lab, Livermore CA, UCRL-JC-118635)

Bahr JM, Rubin J (1987) Direct comparison of kinetic and local equilibrium formulations for solute transport affected by surface reactions. Water Resources Res 23:438-452

Baker KH, Herson DS (1991) In situ bioremediation of contaminated aquifers and subsurface soils, Geomicrobiol 8:133-146

Bakr AA, Gelhar LW, Gutjahr Al L, MacMillan JR (1978) Stochastic analysis of spatial variability in subsurface flows 1. Comparison of one- and three-dimensional flows, Water Resources Res 14:263-271

Barber II, LB, Thurman EM, Runnells DD (1992) Geochemical heterogeneity in a sand and gravel aquifer: Effect of sediment mineralogy and particle size on the sorption of chlorobenzenes, J Contaminant Hydrol 9:35-54

Bear J (1972) Dynamics of Fluids in Porous Media. Elsevier

Bear J (1979) Hydraulics of Groundwater, McGraw Hill

Bellin A, Salandin P, Rinaldo A (1992) Simulation of dispersion in heterogeneous porous formations—Statistics, 1st order theories, convergence of computations, Water Resources Res 28:2211-2227

Bishop D, Rice D, Rogers L, Webster-Scholten C (1991a) Comparison of field-based distribution coefficients (Kd's) and retardation factors (R's) to laboratory and other determinations of Kd's, Lawrence Livermore Nat'l Laboratory, Livermore, CA (UCRL-AR-105002)

Bishop D, Rice D, Nelson SC (eds) (1991b) Dynamic Stripping Project Characterization Report, Lawrence Livermore Nat'l Laboratory, Livermore, CA (UCRL-AR-108707)

Blake RG, Maley MP, Noyes CM (1995) Hydrostratigraphic Analysis—The Key to Cost Effective Ground Water Cleanup at Lawrence Livermore National Laboratory, Lawrence Livermore Nat'l Laboratory, Livermore, CA (UCRL-JC-120614)

Boggs JM, Young SC, Benton DJ, Chung YC (1990) Hydrogeologic characterization of the MADE site, EPRI EN-6915, Res project 2484-5, Interim report, Electric Power Research Inst, Palo Alto, CA, July, 1990

Bosma WJP, Bellin A, van der Zee SEATM, Rinaldo A (1993) Linear equilibrium adsorbing solute transport in physically and chemically heterogeneous porous formations: 2. Numerical results, Water Resources Res 29:4031-4043

Burr DT, Sudicky EA, Naff RL (1994) Nonreactive and reactive solute transport in three-dimensional heterogeneous porous media: Mean displacement, plume spreading, and uncertainty. Water Resources Res 30:791-815

Chesnut DA (1994) Dispersivity in heterogeneous permeable media. Proc 5th Annual Int'l Conf High Level Nuclear Waste Management, Las Vegas, NV

Copty N, Rubin Y (1995) A stochastic approach to the characterization of lithofacies from surface seismic and well data. Water Resources Res 31:1673-1686

Chrysikopoulos CV, Kitanidis PK, Roberts PV (1990) Analysis of one- dimensional solute transport through porous media with spatially variable retardation factor. Water Resources Res 26:437-446

Cole CR, Foote HP (1990): Multigrid Methods for Solving Multi-Scale Transport Problems, in Dynamics of Fluids in Hierarchical Porous Media, J. Cushman, ed., Academic Press

Cushman JH (1990) An introduction to hierarchical porous media. In: Dynamics of Fluids in Hierarchical Porous Media. Cushman JH (ed) Academic Press

Dagan G (1989) Flow and Transport in Porous Formations, Springer-Verlag

Deutsch CV, Journel AG (1992) GSLIB: Geostatistical Software Library and User's Guide. Oxford University Press

Dougherty DE (1991) Hydrologic applications of the connection machine CM-2. Water Resources Res 27:3137-3147

Dicke CA, Smith RW (1995) Double diffuse layer modeling of actinide adsorption on hydrous iron oxides. Idaho Nat'l Engineering Laboratory. (in review)

Duba AG, Jackson KJ, Jovanovich MC, Knapp RB Taylor RT (1996) TCE remediation using in situ, resting-state bioaugmentation. Environ Sci Tech 30:1982-1989

Dullien FAL (1992) Porous Media. Fluid Transport and Pore Structure (2nd edition), Academic Press

Dzombak DA, Morel FM (1990) Surface Complex Modeling, Hydrous Ferrous Oxide, Wiley Interscience.

EPA (1993) Drinking Water Regulations and Health Advisories, Office of Water, U S Environmental Protection Agency, Washington DC, May, 1993

Essaid HI, Hess KM (1993) Monte Carlo simulations of multiphase flow incorporating spatial variability of hydraulic properties, Groundwater 31:123-134

Felmy AR, Girvin DC, Jenne EA (1984) MINTEQ: A computer program for calculating aqueous geochemical equilibria, U S Environmental Protection Agency, report EPA-600/3-84-032, available as report TIS PB84-157148, Nat'l Technical Information Service, Springfield, Virginia

Ferris FG, Schultze S, Witten TC, Fyfe WS, Beveridge TJ (1989) Metal interactions with microbial biofilms in acidic and neutral pH environments, Appl Environ Microbiol 55:1249-1257

Garabedian S, Gelhar LW, Celia MA (1988) Large-scale dispersive transport in aquifers: field experiments and reactive transport theory, Tech. Report 315, R M Parsons Laboratory, Dept Civil Engineering, Massachusetts Inst Technology, Cambridge MA

Gelhar LW (1986) Stochastic subsurface hydrology: from theory to applications, Water Resources Res 22:135s-145s

Gelhar LW (1993) Stochastic Subsurface Hydrology. Prentice Hall

Gelhar LW, Welty C, Rehfeldt KR (1992) A Critical Review of Data on Field-Scale Dispersion in Aquifers. Water Resources Res 28:1955-1974

Harwood DS, Helley EJ, Doukas MP (1981) Geologic Map of the Chico Monocline and Northeastern Part of the Sacramento Valley, California. Map I-1238, U S Geological Survey, Reston, Virginia

Hassanizadeh SM, Gray WG (1979) General conservation equations for multi-phase systems: Averaging procedures. Adv Water Resources 2:131-144

Hoeksema RJ, Kitanidis PK (1985)Analysis of the spatial structure and properties of selected aquifers. Water Resources Res 21:563-572

Hopkins GD, Munakata J, Semprini L, McCarty PL (1993) Trichloroethlyene concentration effects on pilot field scale in situ groundwater bioremediation by phenol-oxidizing microorganisms. Environ Sci Tech 27:2542-2547

Hsi C-KD, Langmuir D (1985) Adsorption of uranyl onto ferric oxyhydroxides: application of the surface complexation site-bonding model. Geochem Cosmochim Acta 49:1931-1941

Isaaks EH, Srivastava RM (1989) Applied Geostatistics. Oxford University Press

Journel AG, Huijbregts ChJ (1978) Mining Geostatistics. Academic Press

Journel AG, Alabert F (1990) New method for reservoir mapping J Petrol Tech, p 212-218

Jussel P, Stauffer F, Dracos T (1994a) Transport modeling in heterogenous aquifers, 1. Statsitcal description and numerical generation of gravel deposits. Water Resources Res 30:1803-1817

Jussel P, Stauffer F, Dracos T (1994b) Transport modeling in heterogenous aquifers, 2. 3-dimemsional transport model and stochastic tracer experiments.Water Resources Res 30:1819-1831

Kabala ZJ, Sposito G (1991) A stochastic model of reactive solute transport with time-varying velocity in a heterogeneous aquifer. Water Resources Res 27:341-350

Kinsella JV, Nelson MJK (1994) In situ bioremediation: Site characterization, system design, and full-scale field remediation of petroleum hydrocarbon- and trichloroethylene-contaminated groundwater. In: Bioremediation Field Experience. Flathman PE, Jerger CE, Exner JH (eds) CRC Press, p 413-428

Lake LW, Carroll HB Jr (1985) Reservoir Characterization, Academic Press

Lake LW, Carroll HB Jr, Wesson TC (1991) Reservoir Characterization II, Academic Press

Leblanc DR, Garabedian SP, Hess KM, Gelhar LW, Quadri RD, Stollenwerk KG, Wood WW (1991) Large-scale natural gradient tracer test in sand and gravel, Cape Cod, Massachusetts, 1. Experimental Design and observed tracer movement. Water Resources Res 27:895-910

Lee MD, Thomas JM, Borden RC, Bedient PB, Ward CH, Wilson JT (1988) Biorestoration of aquifers contaminated with organic compounds. CRC Critical Rev Environ Control 18:29-89

Lichtner PC (1988) The quasi-stationary state approximation to coupled mass transport and fluid-rock interaction in a porous medium, Geochim Cosmochim Acta 52:143

Mackay DM, Freyberg DL, Roberts PV, Cherry JA (1986) A natural gradient tracer experiment on solute transport in a sand aquifer, 1. Approach and overview of plume movement. Water Resources Res 22:2017-2029

Macquarrie KTB, Sudicky EA (1990) Simulation of biodegradable organic contaminants in ground water, 2, Plume behavior in uniform and random flow fields, Water Resources Res 26:223-239

Mesuere K, Fish W (1992) Chromate and oxalate adsorption on goethite. 1. Calibration of surface complexation models. Environ Sci Tech 26:2357-2364

Murphy W, Oelkers E, Lichtner P (1989) Surface reaction versus diffusion control of mineral dissolution and growth rates in geochemical processes. Chem Geol 78:357-380

Nelson, M. J., J. V. Kinsella, and T. Montoya, In-situ biodegradation of TCE contaminated groundwater, Environmental Progress, 9(3), 190-196, 1990.

Newmark RL, Aines RD (1995) Summary of the LLNL Gasoline Spill Demonstration—Dynaminc Underground Stripping Project. Lawrence Livermore Nat'l Laboratory. Livermore, CA (UCRL-ID-120416)

Nitao JJ (1995) Reference Manual for the NUFT Flow and Transport Code, Version 1.0. Lawrence Livermore Nat'l Laboratory, Livermore, CA (UCRL-IC-113520)

Ortoleva PJ (1994) Geochemical Self-Organization. Oxford University Press

Payton BM, Truex MJ, Skeen RS, Hooker BS (1995) Design of an in situ carbon tetrachloride bioremediation system. In: Bioremediation of Chlorinated Solvents. Hinchee RE, Leeson A, Semprini L (eds) Battelle Press, p 111-116

Phillips FM, Wilson JL (1989) An approach to estimating hydraulic conductivity spatial correlation scales using geological characteristics. Water Resources Res 25:141-143

Polmann D, McLaughlin D, Luis S, Gelhar L, Ababou R (1991) Stochastic modeling of large scale flow in heterogeneous, unsaturated soils. Water Resources Res 27:1447-1458

Ptak T, Teutsch G (1994) Forced and natural gradient tracer tests in a highly heterogeneous porous aquifer—instrumentation and measurements. J Hydrology 159:79-104

Rajaram H, Gelhar LW (1993) Plume scale dependent dispersion in heterogeneous aquifers, 2. Eularian analysis and 3-dimensional aquifers. Water Resources Res 29:3261-3276

Riley RG, Zachara JM, Wobber FJ (1992) Chemical Contaminants on DOE Lands and Selection of Contaminant Mixtures for Subsurface Science Research. U S Dept of Energy Report DOE/ER-0547T

Roberts PV, Hopkins GD, Mackay DM, Semprini L (1990) A field evaluation of in situ biodegradation of chlorinated ethenes: Part 1, Methodology and field site characterization. Ground Water 28:591-604

Robin MJL, Sudicky E, Gillham R, Kachanoski R (1991) Spatial variability of strontium distribution coefficients and their correlation with hydraulic conductivity in the Canadian Forces Base Borden aquifer. Water Resources Res 27:2619-2632

Schäfer W, Kinzelbach W (1992) Stochastic modeling of in-situ bioremediation in heterogeneous aquifers. J Contaminant Hydrol 10:47-73

Scheibe TD, Freyberg DL (1995) Use of sedimentological information for geometric simulation of natural porous media structure. Water Resources Res 31:3259-3270

Semprini L, Hopkins GD, Roberts PV, Grbiç-Galiç D, McCartyPL (1991) A field evaluation of in situ biodegradation of chlorinated ethenes: Part 3, studies of competitive inhibition. Ground Water 29:239-250

Semprini L, Roberts PV, Hopkins GD, McCarty PL (1990) A field evaluation of in situ biodegradation of chlorinated ethenes: Part 2, Results of biostimulation and biotransformation experiments. Ground Water 28:715-727

Shad H, Teutsch G (1994) Effects of the investigation scale on pumping test results in heterogeneous porous aquifers. J Hydrology 159:61-77

Shock EL (1990) Thermodynamic Data for Metal-Organic Acid Complexes at Groundwater Conditions. Final Report for EG&G Idaho Consultant Agreement C90-103054

Sheppard MI, Thibault DH (1990) Default soil solid/liquid partition coefficients for four major soil types: A compendium. Health Physics 59:471-482

Smith RW, Jenne EA (1991) Recalculation, evaluation, and prediction of surface complexation constants for metal adsorption on iron and manganese oxides. Environ Sci Tech 25:525-531

Smith RW, Schafer AL, Tompson AFB (1996) Theoretical relationships between reactivity and permeability for monomineralic porous media. In: Proc XIXth Symp on the Scientific Basis for Nuclear Waste Management. Murphy WM, Knecht DF (eds) Materials Research Society

Szescsody JE, Zachara JM, Bruckhart PL (1994) Adsorption-dissolution reactions affecting the distribution and stability if Co(II)-EDTA in iron-oxide-coated sand. Environ Sci Tech 28:1706-1716

Taylor RT, Duba AG, Durham WB, Hanna ML, Jackson KJ, Jovanovich MC, Knapp RB, Knezovich JP, Shah NN, Shonnard DR Wijesinghe AM (1993) In situ bioremediation of trichloroethylene-contaminated water by a resting-cell methanotrophic microbial filter. Hydrol Sci J 38:323-342

Taylor RT, Hanna ML (1995) Laboratory treatability studies for resting-cell in situ microbial filter bioremediation. In: Bioaugmentation for Site Remediation. Hinchee RE, Fredrickson J, Alleman RE (eds) Battelle Press, p 15-29

Teutsch G, Kobus H (1990) The environmental research field site Horkheimer-Insel-research program, instrumentation, and first results. J Hydraulic Res 28:491-501

Thiele MR, Batycky RP, Blunt MJ, Orr FM Jr (1996) Simulating flow in heterogeneous media using streamtubes and streamlines. SPE Reservoir Engineering, p 5-12, February, 1996

Tompson AFB, Ababou R, Gelhar LW (1989) Implementation of the three-dimensional turning bands random field generator. Water Resources Res 25:2227-2243

Tompson AFB, Gelhar LW (1990) Numerical simulation of solute transport in randomly heterogeneous porous media. Water Resources Res 26:2541-2562

Tompson AFB (1993) Numerical simulation of chemical migration in physically and chemically heterogeneous porous media. Water Resources Res 29:3709-3726

Tompson AFB, Knapp RB, Hanna ML, Taylor RT (1994) Simulation of TCE migration in a porous medium under conditions of finite degradation capacity. Adv Water Resources 17:241-249

Tompson AFB, Schafer AL, Smith RW (1996a) Impacts of physical and chemical heterogeneity on co-contaminant transport in a sandy porous medium,. Water Resources Res 32:801-818

Tompson AFB, Falgout RD, Smith SG, Bosl WJ, Ashby SF (1996b) Analysis of Subsurface Contaminant Migration and Remediation Using High Performance Computing (in review) Adv Water Resources (also, Lawrence Livermore Nat'l Laboratory, Livermore, CA, UCID-JC-124650, `http://www-ep.es.llnl.gov/www-ep/esd/sstrans/tompson/AWR96/`)

Valocchi AJ (1989) Spatial moment analysis of the transport of kinetically and sorbing solutes through stratified aquifers. Water Resources Res 25:273-279

van der Zee SEATM, van Riemsdijk WH (1987) Transport of reactive solute in spatially variable soil systems. Water Resources Res 23:2059-2069

Webb E (1994) Simulating the three-dimensional distribution of sediment units in braided stream deposits. J Sedimentary Res B64:219-231

Webb E, Anderson M (1996) Simulation of preferential flow in three-dimensional, heterogeneous conductivity fields with realistic internal architecture. Water Resources Res 32:533-545

Weber WJ, McGinley PM, Katz LE (1991) Sorption phenomena in subsurface systems: concepts, models, and effects on contaminant fate and transport. Water Resources Res 25:499-528

Wilson JT, Armstrong JH, Rafai HS (1994) A full-scale field demonstration on the use of hydrogen peroxide for in situ bioremediation of an aviation gasoline-contaminated aquifer. In: Bioremediation Field Experience. Flathman PE, Jerger DE, Exner JH (eds) CRC Press, p 333-360

Wilson JT, Leach LT, Hensen MJ Jones JN (1986) In situ biorestoration as a ground water remediation technique, Ground Water Monit Rev 6:56-64

Yeh GT, Tripathi VS (1991) A model for simulating transport of reactive multispecies components—model development and demonstration. Water Resources Res 27:3075-3094

Chapter 7

MICROBIOLOGICAL PROCESSES IN REACTIVE MODELING

Bruce E. Rittmann and Jeanne M. VanBriesen

Department of Civil Engineering
Northwestern University
2145 Sheridan Road
Evanston, IL 60208 U.S.A.

INTRODUCTION

Microbiological reactions are becoming increasingly vital in models of reactive transport in the subsurface (NRC, 1990). Relatively recent discoveries that microorganisms are present throughout the subsurface mean that microbiologically catalyzed reactions probably are affecting the fate of organic and inorganic contaminants. Furthermore, manipulations of the underground environment for the purpose of increasing microbiological activity offer the potential to clean-up the contamination via engineered bioremediation (NRC, 1993).

Bringing microbiological reactions fully into reactive modeling is not a trivial undertaking. Historically, modelers have utilized one of four methods: first order models, instantaneous reaction models, Monod and dual Monod models, and biofilm models. Table 1 indicates recent modeling work with each method.

First-order models generally model substrate loss as dependent upon the concentration of the substrate and an empirically determined first order rate constant. Instantaneous reaction models consider biodegradation to be an instantaneous reaction between the organic electron donor and oxygen as an electron acceptor. This modeling method assumes that transport of electron donor and acceptor, rather than microbial kinetics, is the controlling factor for the rate of the degradation reaction. Although first order and instantaneous modeling methods are suitable for certain conditions in reactive modeling, they ignore the fact that the reactions are catalyzed by living organisms. Monod and biofilm models consider the presence and activity of the living organisms that catalyze the reactions and represent a more mechanistic modeling approach. Furthermore, biologically catalyzed reactions frequently affect or are affected by other kinds of reactions. These close connections often control the overall fate of contaminants and the success or failure of bioremediation schemes.

In this manuscript, we provide a proper foundation for incorporating microbiological reactions into reactive transport models. First, we describe the primary reactions catalyzed by microorganisms, explain why these reactions occur, and systematically present the types of kinetic expressions needed to represent the reactions in a reactive model. Second, we identify other types of chemical reactions that are affected by the primary microbiological reactions. The key here is how to link these other reactions directly to the primary microbiological reactions. Finally, we present several modeling examples in which we properly connect the different reactions and predict interesting and important findings about what controls the fate of organic and inorganic components present in a subsurface environment. Although this manuscript does not discuss the transport aspects of modeling, it presents the reactions needed and how they are inter-related when microbiology is included in any model.

0275-0279/96/0034-0007$05.00

Table 1. Recent biodegradation modeling

Modeling Method	Reference
First Order	Frind et al, 1990
	Bolten et al, 1993
	Fry et al, 1993
	McNab and Narasimhan 1994
	Baik and Lee, 1994
	Fry and Istok, 1994
Instantaneous Reaction	Borden and Bedient, 1987
	Rifai and Bedient, 1990
Monod and Dual Monod	Moltz et al, 1986
	Srinivasan and Mercer, 1988
	Widdowson et al, 1988
	Celia et al, 1989
	Kindred and Celia, 1989
	MacQuarrie et al, 1990
	Kinzelbach et al, 1991
	Bae and Rittmann, 1996b
Biofilm	Moltz et al, 1986
	Widdowson et al, 1988
	Rittmann and McCarty, 1981
	Bouwer and Cobb, 1987
	Baveye and Valocchi, 1989
	Taylor and Jaffe, 1990 a-d
	Odencrantz et al, 1990
	Zysett et al, 1994

MICROBIOLOGICAL REACTIONS

Although modelers often are driven to include microbial reactions because of their interest in the fate of biodegradable contaminants, they cannot ignore that the catalysts for the biodegradation reactions are alive. Just like human beings, bacteria need to have food supplies and a hospitable environment if they are to grow, sustain themselves, and accomplish useful tasks (to humans and themselves). Therefore, an adequate model for microbial reactions should explicitly include the microorganisms themselves and the factors that allow them to exist and function.

Primary metabolism

The most fundamental reason that microorganisms catalyze chemical reactions is to make more of themselves. Called biomass synthesis, this basic growth event requires that the cells extract from their environment

(1) nutrients (like C, N, P, O, and H) to provide the elemental building materials of new biomass,

(2) electrons to reduce the elements to the proper reduction state, and

(3) energy to fuel the reduction step and the assembly of the building materials into sophisticated and highly organized cells.

Fortunately for the microorganisms, they usually can obtain the electrons, energy, and many of the nutrients from a single material, the *electron-donor primary substrate*. The term *primary substrate* indicates that the material is involved in supplying the needed nutrients and energy for synthesis. The term *electron donor* indicates that it is the source of the electrons.

In primary-substrate utilization, the electron-donor substrate is oxidized in a set of electron transfers (usually 2 electrons per step) that form reduced electron carriers inside the cell. Reduced nicotinanaide adenine dinucleotide (NADH) is the most common example of a reduced electron carrier (Stryer, 1988). An organic substrate that is mineralized is completely oxidized such that all of its electrons are transferred to NADH molecules, while the carbon is released as inorganic carbon in the form of carbonate—H_2CO_3, HCO_3^-, and CO_3^{2-}.

These facets of primary-donor oxidation are illustrated with acetate, one of the most commonly found substrates in nature. In Reaction R1, acetate undergoes a two-electron oxidation that releases inorganic carbon (H_2CO_3, which is the same as H_2O+CO_2) and reduces NAD^+ to NADH.

$$CH_3COOH + NAD^+ + 2\ H_2O \rightarrow CH_3OH + NADH + H^+ + H_2CO_3 \qquad \text{(R1)}$$

[The microorganisms actually carry out this reaction using a more elaborate set of intermediates via the tricarboxylic acid cycle, but the ultimate result is as shown in R1 (Stryer, 1988).] Reaction R2 shows the mineralization of acetate to produce 2 H_2CO_3 and 4 NADH molecules per acetate molecule.

$$CH_3COOH + 4\ NAD^+ + 4\ H_2O \rightarrow 2\ H_2CO_3 + 4\ NADH + 4H^+ \qquad \text{(R2)}$$

Once the bacteria have the NADH molecules from oxidation of the electron-donor substrate, they can invest those electrons in two ways. The first option, called respiration, involves transferring the electrons from NADH to an extracellular electron acceptor in a reaction that generates free energy that can be captured. The most important electron-acceptor primary substrates are O_2, NO_3^-, SO_4^{2-}, CO_2, and Fe^{3+}. Reactions R3 and R4 illustrate these respiratory electron transfers for O_2 and NO_3^-:

$$O_2 + 2\ NADH + 2\ H^+ \rightarrow 2\ H_2O + 2\ NAD^+ + \text{free energy} \qquad \text{(R3)}$$

$$NO_3^- + 2.5\ NADH + 3.5\ H^+ \rightarrow 0.5\ N_2 + 2.5\ NAD^+ + 3\ H_2O + \text{free energy} \qquad \text{(R4)}$$

For the fraction of the electrons that participate in respiration, an overall, stoichiometric reaction can be written by combining donor and acceptor reactions. For example, the aerobic mineralization of acetate combines R2 and 2 times R3:

$$CH_3COOH + 2\ O_2 \rightarrow 2\ H_2CO_3 + \text{free energy} \qquad \text{(R5)}$$

The free energy generated in reactions like R5 is captured in high-energy phosphate-ester bonds of adenosine triphosphate (ATP). The means for that capture via the proton-motive force is beyond the scope of this manuscript, but is described in microbiology and biochemistry texts (e.g. Brock et al, 1994; Stryer, 1988). The cells can spend the ATP-stored energy for a variety of purposes, including mobility, osmotic regulation, and repair of cellular macromolecules. However, a major sink for the stored energy is synthesis of new biomass. Here, ATP and NADH are invested to create the highly organized (i.e. low entropy) and significantly reduced macromolecules that constitute biomass. These macromolecules include proteins, polysaccharides, nucleic acids, and lipids.

Represented simply, the synthesis of new biomass from investments of electrons, energy, and environmentally available nutrients is

$$10\ NADH + 5\ H_2CO_3 + NH_4^+ + 9\ H^+ + \text{free energy} \rightarrow C_5H_7O_2N + 10\ NAD^+ + 13\ H_2O \qquad (R6)$$

In R6, $C_5H_7O_2N$ is a simple elemental formula for biomass and has 20 electrons invested in its carbon. For the electrons originally in acetate that are invested in biomass, R6 can be combined with 2.5 times R2 to give

$$2.5\ CH_3COOH + NH_4^+ + \text{free energy} \rightarrow C_5H_7O_2N + 3\ H_2O + H^+ \qquad (R7)$$

The electrons originally present in acetate are proportioned out between R3, the energy generating redox reaction, and R7, the cell synthesis reaction. The proportioning depends on the relative energy yield of R3 compared to the energy requirements of R7. The fraction of electrons going to synthesis, f_s°, is large (e.g. > 0.5) when R3 releases a large amount of free energy for each electron flowing to respiration, while the free-energy costs of R7 are small. (This proportioning has been addressed systematically and quantitatively by McCarty (e.g. Christensen and McCarty, 1975); the quantification details are not addressed here.) For most organic electron donors, f_s° is between 0.5 and 0.7 when O_2 or NO_3^- is the electron acceptor. Decreases to the energy yield in R3 and/or increases to the energy costs in R7 can make f_s° small. Inorganic electron donors, other electron acceptors, and autotrophy (CO_2 fixation for synthesis) can drive f_s° to values less than 0.05. Reactions involving such low f_s° values are still possible, but lead to very slow growth rates for the organisms. Since the mineralization of the organic contaminant is accomplished by the microorganisms, slow growth leads to fewer organisms and less degradation.

In many instances of organic contamination, the organic contaminant can serve as the electron-donor primary substrate (Rittmann et al., 1994). This is quite convenient, because utilization of the target contaminant results in the growth of more bacteria able to carry out the contaminant degradation reaction, which is very useful to humans. This good result occurs only when the bacteria also have their electron-acceptor primary substrate for energy generation and nutrients for biomass synthesis. If any of these critical "other" materials is missing, biomass synthesis will not increase the numbers of degrading bacteria.

Special status of oxygen

Dissolved oxygen can play two key roles in microbiological reactions. The first role—as an electron-acceptor primary substrate—was described in the previous section. In that role, O_2 is noteworthy because it yields the greatest amount of free energy when it accepts electrons during respiration. Hence, having dissolved oxygen available helps make f_s° large and spurs the most growth of new biomass.

In some cases, dissolved oxygen has a second role—a direct cosubstrate in critical "activation" reactions for aliphatic and aromatic hydrocarbons. Two good examples of this activation role are given by Reactions R8 for toluene and R9 for nitrilotriacetic acid (NTA):

$$C_6H_5CH_3 + O_2 + NADH + H^+ \rightarrow C_6H_5CH_2OH + H_2O + NAD^+ \qquad (R8)$$

$$N(CH_2COOH)_3 + O_2 + NADH + H^+ \rightarrow (HOOCCH_2)_2N(CHOHCOOH) + H_2O + NAD^+ \qquad (R9)$$

Both reactions are monooxygenations, in which O_2 and NADH are direct cosubstrates for toluene or NTA. Toluene and NTA are oxidized by two electrons, and an -OH is substituted for an -H. The O in -OH comes from O_2. The other O in O_2 is reduced with 2 electrons from NADH and ends up in H_2O. By inserting -OH groups, monooxygenation makes the molecules more susceptible to subsequent dehydrogenation and hydroxylation reactions, which are the standard 2-electron oxidations that produce reduced electron carriers.

Secondary utilization

Biodegradation of a contaminant does not always require that the bacteria be growing through primary utilization of that contaminant. As long as capable bacteria are present and nothing is inhibiting their reactions, biodegradation is possible through *secondary utilization*, which is defined as biodegradation of a substrate that yields no or negligible energy to support synthesis (Rittmann et al., 1994). When the bacteria do not gain energy from utilization (i.e. no primary substrate is available), they either never exist or die away.

In many cases, a target contaminant can serve as a primary substrate, but its concentration is less than the minimum concentration, S_{min}, capable of providing enough energy to sustain the bacteria (Rittmann et al., 1994). These sub-S_{min} concentrations can occur for low-solubility compounds or after considerable biodegradation has reduced the concentration. Secondary utilization of such a trace-level contaminant can take place if other primary substrates are simultaneously present at super-S_{min} concentration (Namkung et al., 1983) or if the target (or other) primary substrate was available in the recent past and grew substantial biomass (Rittmann and Brunner, 1984).

A special case of secondary utilization is called *cometabolism*, for which the target compound is incapable of being a primary substrate at any concentration (Rittmann et al., 1994). In general, cometabolism occurs as a "fortuitous" reaction in which an enzyme incidentally catalyzes a reaction with the target contaminant, even though its normal substrate is a different, albeit structurally related, compound. The products of cometabolic reactions often accumulate, because they are not good substrates for the enzymes catalyzing subsequent reactions.

Kinetics

Whether the contaminant is biodegraded as a primary or secondary substrate, its kinetics are essential for reactive-transport modeling. The most widely used kinetic formulation is the classical Monod relationship, written here for an electron-donor:

$$r_d = -q_{md} \frac{S_d}{K_d + S_d} X \qquad \text{(E1)}$$

in which r_d = the rate of accumulation of an electron-donor substrate, $M_dL^{-3}T^{-1}$; q_{md} = maximum specific utilization rate of the donor, $M_dM_x^{-1}\,T^{-1}$; K_d = donor's half-maximum-rate concentration, M_dL^{-3} ; S_d = concentration of the donor substrate, M_dL^{-3} ; and X = concentration of the biomass active in degrading the donor substrate, M_xL^{-3}.

The Monod expression is a hyperbolic function that simplifies to a zero-order expression in S_d [i.e. $r_d = -q_{md}X$] when $S_d \gg K_d$ and to a first-order expression in S_d

[i.e. $r_d = \frac{-q_{md}X}{K_d} S_d$] when $S_d \ll K_d$.

In any case, r_d is first order with respect to X. This means that the concentration of active biomass must be known or predicted if the rate of degradation of the donor is to be computed. Simplistic first-order decay rates ($r_d = -k_1S_d$) ignore the reality that active microorganisms must be present to catalyze the reactions.

Because a significant portion of the electrons from a donor substrate are ultimately transferred to the primary electron-acceptor substrate, the electron acceptor also can control the kinetics of donor utilization. Detailed kinetic and physiological studies by Bae and Rittmann (1996a,b) showed that the best way to represent the limitation by an acceptor is through the multiplicative-Monod formulation, also sometimes called double-Monod:

$$r_d = -q_{md}\frac{S_d}{K_d + S_d}\frac{S_a}{K_a + S_a}X \tag{E2}$$

in which K_a = acceptor's half maximum-rate concentration, M_aL^{-3}; and S_a = concentration of the acceptor substrate, M_aL^{-3}. K_a values for acceptors that serve only in respiration tend to be quite small, usually much less than 1 mg/l, but K_a usually is larger when O_2 is also used as a cosubstrate for monooxygenation reactions (Malmstead et al., 1995). Equation E2 gives a low utilization rate when either S_d or S_a is low, and it declines very much when both are low together, a situation termed dual limitation (Bae and Rittmann, 1996a).

Additional complexity can be added to kinetic models when the system includes inhibitory compounds, mixed substrates, multiple bacterial species, sequential degradation reactions, or other features particular to a contamination situation. The modifications necessary to consider specific cases are beyond the scope of this paper.

Active biomass

When the active biomass is being grown and sustained by the contaminant's biodegradation (most contaminants serve as electron donors and are oxidized), the biomass synthesis is linked to donor-substrate utilization:

$$r_{syn} = -Yr_d \tag{E3}$$

in which r_{syn} = rate of accumulation of newly synthesized active biomass, $M_xL^{-3}T^{-1}$; and Y = true yield coefficient, $M_xM_d^{-1}$. The true yield is a thermodynamically controlled coefficient expressing the fraction of primary electron donor electrons that are invested into new biomass; thus Y is proportional to f_s^o. For example, when f_s^o = 0.64 e^- eq to synthesis/e^- eq total donor for aerobic degradation of NTA, then

$$Y = \left(0.64\frac{e^- eqcells}{e^- eqdonor}\right) x \left(\frac{113gcells}{20e^- eqcells}\right) x \left(\frac{18e^- eqNTA}{191gNTA}\right) = \frac{0.341gcells}{gNTA}$$

where cells are represented by $C_5H_7O_2N$ (MW = 113 g/mole, 20 e^-eq in C) and NTA is $C_6H_9O_6N$ (MW = 191 g/mole, 18 e^-eq in C). For nitrate respiration, (but with N in NTA as the N source for cell synthesis), an f_s^o of 0.62 gives

$$0.62\left(\frac{e^- cells}{e^- eqdonor}\right) x \left(\frac{113gcells}{20e^- eqcells}\right) x \left(\frac{18e^- eqNTA}{191gNTA}\right) = \frac{0.330gcells}{gNTA}$$

New synthesis is not the only phenomenon affecting active biomass. For several reasons, active biomass decays or is lost. This is represented mathematically by

$$r_{decay} = -bX \tag{E4}$$

in which r_{decay} = rate of accumulation of active biomass due to loss processes, $M_xL^{-3}T^{-1}$; and b = biomass-decay coefficient, T^{-1}. Generally, the main loss phenomenon is endogeneous respiration, in which the biomass is oxidized, the primary acceptor is reduced, and energy is generated to fulfill so-called maintenance needs, such as mobility, regulating osmotic pressure, and repairing macromolecules. Thus, the modeling assumes that the primary-donor substrate supplies energy and electrons for new biomass synthesis, while endogeneous respiration supplies energy and electrons for maintenance. This division surely is not absolute in reality, but it has proven a very useful representation.

The net rate of biomass accumulation is the synthesis less the maintenance, or

$$r_{net} = r_{syn} + r_{decay} = -Yr_d - bX \tag{E5}$$

When the donor and acceptor concentrations are high enough, $-r_d$ is large, making r_{net} positive. On the other hand, very low substrate concentrations drive $-r_d$ to near zero, and r_{net} is negative. Thus, active biomass can be in net growth or net loss, depending on $-r_d$.

Electron-acceptor substrate

The utilization of the electron-acceptor substrate occurs in two ways. First, the electrons extracted from the electron-donor substrate, but not invested in biomass, go directly to the primary electron acceptor for energy generation (also for activation). In units of electron equivalents,

$$f_e^{\,o} = 1 - f_s^{\,o} \tag{E6}$$

in which $f_e^{\,o}$ = fraction of donor electrons sent to the acceptor. Second, the electrons removed from the biomass during endogeneous respiration go to the primary acceptor for energy generation to support maintenance. We call that fraction f_m, or the fraction of electrons originally in the primary donor, synthesized into biomass, and finally sent to the acceptor for maintenance respiration. Thus, the total flow of electrons to the acceptor (f_e) is

$$f_e = f_e^{\,o} + f_m = 1 - f_s^{\,o} + f_m \tag{E7}$$

The rate of utilization of the acceptor depends on the rates of donor utilization and endogeneous maintenance respiration by means of appropriate stoichiometric conversion. These conversions can be carried out by using a direct proportioning from rate equations E3 and E4:

$$r_a = f_e^{\,o}\alpha_{a/d}r_d - \alpha_{a/x}bX \tag{E8}$$

in which r_a=rate of accumulation of the acceptor (it has a negative sense), $M_aL^{-3}\,T^{-1}$; $\alpha_{a/d}$ and $\alpha_{a/x}$ are stoichiometric factors needed to achieve the proper units. For example, if NTA is the donor and expressed as mg NTA/l, the biomass is represented as $C_5H_7O_2N$, and the acceptor is O_2, then $\alpha_{a/d}$ = 0.753 mg O_2/mg NTA and $\alpha_{a/x}$ = 1.42 mg O_2/mg cells. When the acceptor is NO_3^--N, we have $\alpha_{a/d}$ =0.264 mg NO_3^--N/mg NTA and $\alpha_{a/x}$ = 0.496 mg NO_3^--N/mg cells.

The stoichiometric (α) values are obtained for O_2 by combining the appropriate acceptor half reaction, written for one electron equivalent (i.e. R3+4 for O_2, or R4+5 for NO_3^-), with Reaction R10 for synthesis and Reaction R11 for the oxidation of NTA:

$$0.25\ H_2CO_3 + 0.05\ NH_4^+ + 0.45\ H^+ + 0.5\ NADH \rightarrow 0.05\ C_5H_7O_2N + 0.65\ H_2O + 0.5\ NAD^+ \quad \text{(R10)}$$

$$0.0556\ C_6H_9O_6N + 0.5\ NAD^+ + 0.667\ H_2O \rightarrow 0.3336\ H_2CO_3 + 0.5\ NADH + 0.\ 0556\ NH_4^+ + 0.445\ H^+ \quad \text{(R11)}$$

The $\alpha_{a/d}$ values are obtained by combining (R3+4 or R4+5) with R11.

$$0.0556\ C_6H_9O_6N + 0.25\ O_2 + 0.0556\ H^+ + 0.166\ H_2O \rightarrow 0.333\ H_2CO_3 + 0.0556\ NH_4^+ \quad \text{(R12)}$$

$$0.0556\ C_6H_9O_6N + 0.2\ NO_3^- + 0.2556\ H^+ + 0.0665\ H_2O \rightarrow 0.333\ H_2CO_3 + 0.0556\ NH_4^+ + 0.1\ N_2 \quad \text{(R13)}$$

Thus, for O_2,

$$\alpha_{a/d} = \frac{0.25 moles O_2}{0.0556 moles NTA} x \frac{32 g O_2}{mol O_2} x \frac{mol NTA}{191 g NTA} = 0.753 \frac{g O_2}{g NTA} \quad \text{(E9)}$$

And, for NO_3^-

$$\alpha_{a/d} = \frac{0.2 mol NO_3^-}{0.0556 mol NTA} x \frac{14 g N}{mol NO_3^-} x \frac{mol NTA}{191 g NTA} = 0.264 \frac{g NO_3^- - N}{g NTA} \quad \text{(E10)}$$

The $\alpha_{a/x}$ values are obtained in a similar manner by combining (R3+4) or (R4+5) with R10 written in the reverse direction to represent decay of biomass:

$$0.05\ C_5H_7O_2N + 0.25\ O_2 + 0.05\ H^+ + 0.15\ H_2O \rightarrow 0.25\ H_2CO_3 + 0.05\ NH_4^+ \quad \text{(R14)}$$

$$0.05\ C_5H_7O_2N + 0.2\ NO_3^- + 0.25\ H^+ + 0.05\ H_2O \rightarrow 0.25\ CO_2 + 0.05\ NH_4^+ + 0.1\ N_2 \quad \text{(R15)}$$

For O_2

$$\alpha_{a/x} = \frac{0.25 mole O_2}{0.05 mole cells} x \frac{32 g O_2}{mole O_2} x \frac{mole cells}{113 g cells} = 1.42 \frac{g O_2}{g cells} \quad \text{(E11)}$$

For NO_3^-

$$\alpha_{a/x} = \frac{0.2 mole NO_3^-}{0.05 mole cells} x \frac{14 g N}{mole NO_3^-} x \frac{mole cells}{113 g cells} = 0.496 \frac{g NO_3^- - N}{g cells} \quad \text{(E12)}$$

Finally, by combining f_s times R10 for synthesis with f_e times R12 or R13 for donor oxidation, we can write the complete reactions for cell synthesis using the electron-donor primary substrate (NTA here) and the chosen electron acceptor.

$$0.055\ C_6H_9O_6N + 0.0875\ O_2 + 0.023\ H^+ + 0.07\ H_2O \rightarrow 0.032\ C_5H_7O_2N + 0.17\ H_2CO_3 + 0.023\ NH_4^+ \quad \text{(R16)}$$

$$0.055\ C_6H_9O_6N + 0.075\ NO_3^- + 0.098\ H^+ + 0.032\ H_2O \rightarrow 0.031\ C_5H_7O_2N + 0.175\ H_2CO_3 + 0.023\ NH_4^+ + 0.038\ N_2 \quad \text{(R17)}$$

These complete reactions are used for biodegradation modeling. An advantage of writing out the full stoichiometric reactions for synthesis (R16 and R17) and endogeneous respiration (R14 and R15) is that the stoichiometry for other materials becomes evident. For example, R16 shows that 0.023 equivalents of strong acid (H^+) are consumed for each 0.055 mole of NTA mineralized aerobically for cell growth and energy generation. This converts to consumption of $\alpha_{H+/d}$ = 0.418 H^+ eq/mole NTA, or 0.00219 H^+ eq/g NTA. With NO_3^- as the electron acceptor, the acid consumption (from R17) is $\alpha_{H+/d}$ = 1.78 H^+ eq/mole NTA, or 0.015 H^+ eq/g NTA. Thus, nitrate respiration consumes much more acid than does aerobic respiration. Likewise, aerobic and nitrate respiration of biomass consume $\alpha_{H+/X}$ = 1H^+ eq/mol $C_5H_7O_2N$ and 5 H^+ eq/mole $C_5H_7O_2N$, respectively. Thus, the consumption rate of strong acid is directly proportional to $-r_d$ and bX, just as are O_2 and NO_3^--N consumption rates, and they are quantified through stoichiometric (α) conversions.

Creating mass balance equations

In the subsurface, most microorganisms are attached to the solid medium (Harvey et al., 1984; Rittmann, 1993), which means that most of the bacteria do not transport with the advecting water. Whereas a simple one-dimensional transport model with reaction for substrates includes advection and dispersion terms for the substrates, the model should not include transport for the attached bacteria. Mathematically, this translates to nonsteady-state mass balances of

$$\frac{\partial S_d}{\partial t} = u\frac{\partial S_d}{\partial x} - D_H\frac{\partial^2 S_d}{\partial x^2} + r_d \quad \text{(E13)}$$

$$\frac{\partial S_a}{\partial t} = u\frac{\partial S_a}{\partial x} - D_H\frac{\partial^2 S_a}{\partial x^2} + f_e^o\alpha_{a/d}r_d - \alpha_{a/x}bX \quad \text{(E14)}$$

$$\frac{\partial X}{\partial t} = -Yr_d - bX \quad \text{(E15)}$$

in which t = time, T; x = longitudinal distance along the flow path, L; u = actual longitudinal velocity, LT^{-1}; and D_H = hydrodynamic dispersion coefficient, L^2T^{-1}. When sources and flow velocity are stable, Equations E13, E14 and E15 can go to steady state, in which the left side equals zero.

In many instances, researchers use suspended-growth, batch reactors to determine kinetic characteristics in the laboratory. Often the batch reactors have no solid media. Modeling of batch reactors is much different from modeling transport in the subsurface. Most critically, batch reactors have no advection, and the mass-balances are inherently nonsteady state. The mass balance equations are

$$\frac{dS_d}{dt} = r_d \quad \text{(E16)}$$

$$\frac{dS_a}{dt} = f_e^{\,o}\alpha_{a/d} r_d - \alpha_{a/x} bX \tag{E17}$$

$$\frac{dX}{dt} = -Yr_d - bX \tag{E18}$$

Other mass balances, such as on acidic hydrogen or nutrients, can be added as needed. The rate terms are based on stoichiometry, as discussed in the previous section.

Whether for transport in porous media or reaction with no transport in a batch reactor, the relevant mass-balance equations must be solved simultaneously to yield S_d, S_a, X, and other concentrations with time and location (subsurface transport scenario only). In general, numerical techniques are required, since the equations are highly nonlinear.

Macroscopic versus biofilm modeling

A very important distinction for microbiological modeling in the subsurface is whether to treat the microbial reactions as "macroscopic" versus "biofilm". Baveye and Valocchi (1989) defined a macroscopic model as one in which all organisms in a control volume are exposed to the substrate concentrations prevailing in that volume's bulk-liquid. This is in contrast to a biofilm setting, in which mass-transport resistances can create concentration gradients; some organisms are exposed to substrate concentrations much less than in the bulk liquid.

Biofilms are layerlike aggregates of microorganisms attached to solid surfaces. Because the subsurface is noteworthy for having copious surface area and bacteria attached to those surfaces, it may seem *a priori* to fall into the biofilm category. However, this conclusion is not always justified. To have significant concentration gradients set up inside or just outside a biofilm, the accumulation of biomass per unit surface area and the substrate flux to the biofilm need to be sufficiently large. When the biofilm accumulation is sparse, as often is the case in the subsurface, significant concentration gradients do not occur. Odencrantz (1992) systematically investigated this question using the steady-state biofilm model of Rittmann and McCarty (1980). He determined that most realistic subsurface scenarios do not support enough biofilm accumulation to generate significant differences between the results predicted by macroscopic versus biofilm models. Likewise, Wood et al (1994) presented an analysis of the applicability of single phase (macroscopic) and multiple phase (biofilm) models for the description of microbial kinetics. They conclude that when the mass transfer rate between mobile and immobile regions is fast compared to the microbial reaction rate, the single phase model is sufficiently predictive. This condition would be expected in systems with the sparse biofilm accumulation typical of the subsurface.

Because we anticipate few subsurface circumstances for which the added complications of a full biofilm model are justified, we use a strictly macroscopic model in the examples that follow. The reader should keep in mind, however, that our macroscopic formulation allows the bacteria to be attached to solid surfaces in the subsurface, as shown by E15. Thus, using a macroscopic model does not imply that the cells are suspended; instead, it means that we do not need to account for concentration gradients to the attached bacteria.

CHEMICAL REACTIONS RELATED TO SUBSURFACE MICROBIOLOGY

In addition to the direct microbial reactions discussed in the previous section, chemical reactions can have a profound effect on microbiological processes and must be considered in reactive transport modeling. Reactions that influence microorganisms include reactions taking place in the aqueous "bulk" phase, as well as those taking place at the interfaces between the bulk phase and the solid phase of the subsurface materials. Both types of reactions are introduced here, and their impacts on microbiological processes are discussed. Finally, the methodology for incorporating modeling reactions and linking microbiological and chemical reaction modeling is presented.

Acid, base and complexation reactions

Acid/base reactions involve the transfer of protons among species and are extremely rapid. Many acid-base reactions occur in natural systems, and these reactions affect or control the pH of natural waters. For example, acetic acid dissociates to form the acetate ion and a proton in water:

$$C_2H_3O_2H \leftrightarrow H^+ + C_2H_3O_2^- \qquad \text{(R18)}$$

This reaction releases a proton and can decrease the pH in the system. It also releases the acetate ion for other reactions, including complexation.

Complexation reactions involve formation of soluble compounds consisting of one or more central atoms (usually metals) covalently bonded to one or more attached species called ligands. For example, the acetate ion can form complexes with metals in solution:

$$C_2H_3O_2^- + Co^{2+} \leftrightarrow Co(C_2H_3O_2)^+ \qquad \text{(R19)}$$

This reaction decreases the free Co^{2+} and acetate ion concentrations. Very strong complexors can sequester biologically required metal ions (e.g. Fe^{3+}) or toxic metal ions (e.g. Cu^{2+}). Although some complexation reaction are slow, most complexation reactions are fast compared to microbiological reactions and can be assumed to go to equilibrium.

Species that are acids or bases often also participate in complexation. In general, ligands are bases, such as the acetate ion. Central metals, such as Co^{2+}, often are Lewis acids, reacting with water to form a variety of soluble hydroxide complexes:

$$Co^{2+} + H_2O \leftrightarrow Co(OH)^+ + H^+ \qquad \text{(R20)}$$

$$Co^{2+} + 2\,H_2O \leftrightarrow Co(OH)_2^{o} + 2\,H^+ \qquad \text{(R21)}$$

$$Co^{2+} + 3\,H_2O \leftrightarrow Co(OH)_3^- + 3\,H^+ \qquad \text{(R22)}$$

Interactions between bulk phase reactions and biodegradation

The interaction between these bulk-phase, equilibrium-controlled reactions and biodegradation reactions can significantly affect system pH and biological activity. Many electron donors and acceptors required by microorganisms for growth can be acids or bases, and they can participate in complexation reactions[1]. The distribution of the acids and bases is controlled by system pH and acid/base and complexation reactions. For example,

[1] A well studied *exception* involves hydrocarbon biodegradation. Inorganic bulk reactions have little impact on the kinetics of hydrocarbon degradation (Baedecker et al, 1993), and degradation of hydrocarbons usually does not generate strong acids or bases. Thus, integrating chemical reaction modeling with biological reaction modeling is less important for the special case of hydrocarbons (Borden and Bedient, 1987).

acetic acid is an electron donor to the cells and an acid, while the acetate ion is a complexation reactant. pH and system parameters control the distribution of acetate in aqueous forms (HAc, Ac^-, $CoAc^+$, etc., where Ac^- represents the acetate ion). Of great importance here is that some aqueous forms are biologically degradable, but others are not. Therefore, the distribution of acetate (and many other organic chemicals) among its various chemical species can have a significant impact on the possibility of degradation in the system and on the rate of that degradation.

When degradation takes place, it decreases the total amount of acetate in the solution and may result in changes in the aqueous distribution of the remaining acetate, depending on which acetate species is degraded. For example, if only HAc is degradable, an acid form is lost during degradation; this requires a redistribution from other (basic) forms to re-establish thermodynamic equilibrium. Thus, the degradation of only HAc should result in an increase in pH in order to compensate for the loss of basic acetate to replenish degraded acetic acid. The increase in pH shifts the ratio of HAc/Ac^- to a lower value and decreases the amount of degradable HAc available for the microorganisms. The decrease in HAc available for degradation, represented by a decrease in S_d in E1 or E2, decreases the rate of the degradation reaction.

A second interaction between biodegradation and acid/base reactions involves the direct chemical participants in the degradation reaction: the reactants and products. Consumption (or production) of acidic hydrogen in biodegradation reactions is not unusual. For example, acetic acid as a primary electron donor is oxidized in the following half reaction:

$$0.125\ CH_3COOH + 0.5\ H_2O \rightarrow 0.25\ H_2CO_3 + e^- + 1.0\ H^+ \qquad \text{(R23)}$$

This half reaction consumes a weak acid (acetic acid, HAc) and produces a weak acid (carbonic acid, H_2CO_3), a strong acid (hydrogen ion, H^+), and an electron. These effects are compounded by the necessary coupling of the donor reaction (R23) with an acceptor reaction. Reaction R5, written previously for aerobic oxidation of acetic acid, shows no net production or consumption of acidic hydrogen. However, when NO_3^{2-} or Fe^{3+} are electron acceptors, the following represent balanced donor oxidations:

$$CH_3COOH + 1.6\ NO_3^{2-} + 1.6\ H^+ \rightarrow 2\ H_2CO_3 + 0.8\ N_2 + 0.8\ H_2O \qquad \text{(R24)}$$

$$CH_3COOH + 4\ H_2O + 8\ Fe^{3+} \rightarrow 2\ H_2CO_3 + 8\ H^+ + 8\ Fe^{2+} \qquad \text{(R25)}$$

R24, using NO_3^{2-} as the electron acceptor substrate, *consumes* 1.6 acid equivalents per mole of acetic acid mineralized. R25, using Fe^{3+} as the electron acceptor substrate, *generates* 8 acid equivalents per mole of acetic acid mineralized. Thus, biological degradation of acetate with oxygen as an acceptor does not affect system pH via acidic hydrogen, but degradation with nitrate as an acceptor consumes H^+ and can cause the pH to rise, while degradation with Fe^{3+} as the acceptor produces H^+ and can cause pH to fall.

The most significant impact of complexation on biodegradation is in the redistribution of the degradable electron donor into multiple complexes. As with redistribution into different acid/base species, these new complexes may be more or less degradable than the parent compound. For example, nitrilotriacetic acid (NTA) is believed to be biologically degraded in the $HNTA^{2-}$ form (Bolton et al., 1996). However, equimolar solutions of Co and NTA have a speciation that is 95% $CoNTA^-$ and 5% $HNTA^{2-}$ at pH 6. As $HNTA^{2-}$ is degraded and its concentration decreases, the ratio of total Co to total NTA increases, and the proportion in HNTA drops even lower. Thus, the degradation rate is reduced in two ways: the decrease in the overall NTA concentration due to degradation and the decrease in the fraction of that NTA that is degradable.

Competing chelates have significant effects on this speciation-induced control of degradation. For example, the addition of a small amount of the stronger complexor, such as ethylenediaminetetraacetic acid (EDTA), increases the rate of degradation of NTA. The EDTA complexes the Co, freeing the NTA for distribution among acid species and increasing the concentration of $HNTA^{2-}$ in solution. As another example, increased rates of NTA degradation were observed in metal-NTA systems containing the Tris buffer compared to phosphate buffer (Firestone and Tiedje, 1975). Since Tris is a weak chelator, its complexation with the metal frees the NTA for distribution among acid species and increases the concentration of $HNTA^{2-}$ in solution

Modeling bulk phase reactions

The modeling of acid/base and complexation reactions has traditionally utilized the local equilibrium assumption (LEA) (e.g. Morel and Morgan, 1972). LEA states that equilibrium formulations are valid as long as the chemical reaction is reversible and the rate of chemical reaction is rapid compared to rates of other system processes (Rubin, 1983). The extremely rapid nature of all acid/base and most complexation reactions in aqueous solution suggests that LEA will be valid in most cases. Therefore, we use an equilibrium formulation to model these reactions. The solution algorithm for an LEA system involves the simultaneous solution of mass action and mass balance equations.

When the pH is specified and fixed, the set of equations created for mass action and mass balance can be solved either analytically for simple systems or numerically (for example, with a Newton-Raphson iterative technique) for complex systems. An example involves a Co-NTA system having two components, Co^{2+} and NTA^{3-}, and two mass balance equations:

$$C_{TCO} = [Co^{2+}] + [CoOH^{+}] + [Co(OH)_2^{o}] + [Co(OH)_3^{-}] + [CoNTA^{-}] + [Co(NTA)_2^{4-}] + [CoOHNTA^{2-}] \qquad (E19)$$

$$C_{TNTA} = [NTA^{3-}] + [HNTA^{2-}] + [H_2NTA^{-}] + [H_3NTA] + [CoNTA^{-}] + 2[Co(NTA)_2^{4-}] + [CoOHNTA^{2-}] \qquad (E20)$$

These equations are formed by summing the concentrations of all possible forms of Co and NTA in the system. For Co, we consider the cobalt hydroxide complexes shown in R20 through R22 and complexes formed between NTA^{3-} and Co^{2+} , as shown below:

$$Co^{2+} + NTA^{3-} \leftrightarrow CoNTA^{-} \qquad (R26)$$

$$Co^{2+} + 2\,NTA^{3-} \leftrightarrow Co(NTA)_2^{4-} \qquad (R27)$$

$$Co^{2+} + NTA^{3-} + H_2O \leftrightarrow CoOHNTA^{2-} + H^{+} \qquad (R28)$$

This final reaction is a complexation and an acid-forming reaction. For NTA, we consider these complexation reactions, as well as the acidic nature of NTA. NTA is a triprotic acid, and the following reactions occur:

$$H_3NTA \leftrightarrow H^{+} + H_2NTA^{-} \qquad (R29)$$

$$H_2NTA \leftrightarrow H^+ + HNTA^{-2} \tag{R30}$$

$$HNTA \leftrightarrow H^+ + NTA^{-3} \tag{R31}$$

For each of these reactions, a mass-action equation can be written in the form K= Π\{products\}/Π\{reactants\} , where \{\} refers to the activity of the species. These mass action equations are defined by their equilibrium constants: K_a for an acid dissociation reaction and K_f for a complex-formation reaction. For example, the mass-action equations for H_3NTA and for CoNTA are written as:

$$K_a = \frac{\{H^+\}\{H_2NTA^-\}}{\{H_3NTA\}} \tag{E21}$$

$$K_f = \frac{\{CoNTA^-\}}{\{Co^{2+}\}\{NTA^{3-}\}} \tag{E22}$$

Rearrangement of the mass action equations in terms of a single form allows substitution into the mass balance equations. For example, E22 is rearranged as:

$$\{CoNTA^-\} = K_f\{Co^{2+}\}\{NTA^{3-}\} \tag{E23}$$

When all mass action equations are written and rearranged in terms of only the components, Co^{2+} and NTA^{3-} , these equations are substituted into the mass balance equations. However, the mass balance equations are written in terms of concentrations, while the mass action equations are written in terms of activities. In order to integrate these two types of equations, we use an activity coefficient, defined as gamma (γ) by the equation:

$$\{CoNTA^-\} = \gamma[CoNTA^-] \tag{E24}$$

The activity coefficients depend on the solution's ionic strength and the ion's valence (Snoeyink and Jenkins, 1980). For the purposes of this discussion, the ionic strength is assumed to be low, and the activity coefficients (γs) that allow conversion of activities to concentrations are taken as equal to one. With this assumption, the mass balances are rewritten in terms of concentrations of the components, Co^{2+} and NTA^{3-}.

$$C_{TCO} = [Co^{2+}] + [Co^{2+}][OH^-]K_{h1} + [Co^{2+}][OH^-]^2K_{h2} + [Co^{2+}][OH^-]^3K_{h3} + [Co^{2+}][NTA^{3-}]K_{f1} + [Co^{2+}][NTA^{3-}]^2K_{f2} + [Co^{2+}][OH^-][NTA^{3-}]K_{f3} \tag{E25}$$

$$C_{TNTA} = [NTA^{3-}] + \frac{[H^+][NTA^{3-}]}{K_{a3}} + \frac{[H^+]^2[NTA^{3-}]}{K_{a3}K_{a2}} + \frac{[H^+][NTA^{3-}]}{K_{a3}K_{a2}K_{a1}} + [Co^{2+}][NTA^{3-}]K_{f1} + 2[Co^{2+}][NTA^{3-}]^2K_{f2} + [Co^{2+}][OH^-][NTA^{3-}]K_{f3} \tag{E26}$$

For fixed pH, these two mass balance equations can be solved simultaneously for the concentration at equilibrium of NTA^{3-} and Co^{2+}. From these two concentrations, all other species are computed from the mass action equations.

The restriction of fixed pH is reasonable when the water contains a high buffer intensity. On the other hand, many environmental systems are not highly buffered. Many degradation reactions produce or consume acidic hydrogen, and bacteria can be quite sensitive to shifts in pH. For instance, kinetic parameters for degradation and bacterial growth depend on the prevailing pH. A fixed-pH model does not consider the impact of these reactions on the system pH and all the reactions influenced by the changing pH. Therefore, we need to be able to compute the effect of changing pH on biodegradation reactions and the effect of biological reactions on the pH.

When the pH is not fixed, there are two approaches to equilibrium modeling. First, one can assume that acidic hydrogen behaves in solution as any other cation and add it as a component. A mass balance is generated as discussed above, and solution to the coupled equation set proceeds as before. This method has been widely used in modeling (Lichtner, 1985) and is mathematically and chemically sound. Alternatively, acidic hydrogen can be treated as a unique component, and a specific mass balance for acidic hydrogen—called a proton condition—can be generated. Either the proton condition or the mass balance on acidic hydrogen is required for accurately describing acid/base effects on pH, and one or the other is necessary to determine the equilibrium pH. The advantage to the proton condition is realized when dealing with systems in which kinetic reactions are changing the total acidic hydrogen in the system over time. We employ the proton condition, which is described in this section, followed by an example where its advantage is illustrated.

In common usage, creation of the proton condition requires knowledge of the acid / base status of the species added to make the initial solution. The added species present are considered the *reference levels*. The proton condition is created relative to these reference levels: All species with more protons than the reference level are placed on the left side of the equation, and all species with fewer protons than the reference level are placed on the right side of the equation. The units for all terms in the proton condition are proton equivalents per liter, with the equivalent concentration always equaling the molar concentration multiplied by the difference in protons from the reference level. This "bookkeeping" ensures that, regardless of the changes that take place in the speciation of components, the number of acid equivalents produced (left side) is matched by the production of balancing conjugate base equivalents (right side). In effect, the proton condition is a mass balance on acidic hydrogens. While the pH in a system may change dramatically, protons cannot be created or destroyed in the system in acid-dissociation reactions; they can only be exchanged among aqueous species.

In order to avoid the necessity of recreating a proton condition for each problem that could be considered, we standardize the reference levels for the proton condition. If the actual addition differs from this reference level, the proton condition is adjusted using a method called *equivalent additions*.

Generating a proper proton condition with a standardized reference level is best understood with an example. Consider the example discussed above: a solution containing NTA and Co. The reference levels for the proton condition are a matter of choice; however, solution to the coupled mass balance and proton condition equations is simplified by choosing uncomplexed forms for the reference level. As shown in derivation of equations E25 and E26 above, the uncomplexed NTA^{3-} and the uncomplexed Co^{2+} were chosen as components. These are also chosen, along with water in its molecular form (H_2O), as reference levels for the proton condition.

Compared to these reference species, the following species have more protons: H_3NTA, H_2NTA^-, $HNTA^{2-}$, and H^+ Compared to the reference species, the following species have fewer protons: OH^-, $CoOH^+$, $Co(OH)_2^o$, $Co(OH)_3^-$, and $CoOHNTA^{2-}$. Additionally, $CoNTA^-$ and $Co(NTA)_2^{4-}$ can form; however, these complexes are at the same proton reference level as uncomplexed Co^{2+} and NTA^{3-}.

The following equation is the proton condition when all species are added at their reference level and no other acid or base species are added. Note that all species with more protons are placed on the left side of the equation, those with fewer protons are placed on the right side, and a multiplier (in equivalents per mole) is used for each species to represent how many protons difference there is between the species and the reference level.

$$3[H_3NTA] + 2[H_2NTA^-] + [HNTA^{2-}] + [H^+] = [OH^-] + [CoOH^+] + 2[Co(OH)_2^o] + 3[Co(OH)_3^-] + [CoOHNTA^{2-}] \quad (E27)$$

The model easily can be adjusted to consider situations in which Co or NTA was added to the system in a form other than its reference level. For example, if the system begins with $CoOHNTA^{2-}$ as the added component, then the proton condition in E27 is incorrect, because the added form of NTA—$CoOHNTA^{2-}$—is more basic than NTA^{3-}. Each of the components is one proton more in excess when $CoOHNTA^{2-}$ is the reference level instead of NTA^{3-}, the structured reference level. Although this inconsistency could be overcome by recreating the proton condition with $CoOHNTA^{2-}$ as the reference level, a simpler approach, and the one we use, is an *equivalent addition*. Given the large number of acid/base species that could be added, the equivalent addition method is the easiest and most systematic approach.

The utility of equivalent addition comes about because the difference between the actual starting species and the fixed reference level is known. All the changes to the proton condition are in the same direction, and the effect of rewriting the condition is to change the "sum" on the left side of the equation. Since the numeric amount of the difference between the reference levels can be computed, an equivalent addition in the known change to the "sum" achieves the same result as rewriting the entire proton condition. A term BASE allows for this addition and is added to the left side of the equation:

$$3[H_3NTA] + 2[H_2NTA^-] + [HNTA^{2-}] + [H^+] + BASE = [OH^-] + [CoOH^+] + 2[Co(OH)_2^o] + 3[Co(OH)_3^-] + [CoOHNTA^{2-}] \quad (E28)$$

This BASE term can be used to represent increases to either the right or the left by controlling the sign of the value entered. When the added species is more basic than the reference level, BASE is positive; when the added species is more acidic than the reference level, BASE is negative. For addition of $CoOHNTA^{2-}$ instead of NTA^{3-} (the added species is more basic than the standard reference level), BASE is positive and has a value equal to the $C_{TCo} \times 1$ eq/mole. If NTA^{3-} were the actually added species for NTA, then BASE = 0, and E28 equals E27.

An example shows the full procedure using equivalent additions. Consider proton condition as written in E28 to be used for a system that started with the following additions:

10^{-3} M $CoOHNTA^-$

10^{-5} M $Co(OH)_3^-$

10^{-4} M $CoNTA^-$

To utilize the standard proton condition (E28), an equivalent addition in terms of the reference species (NTA^{3-}, Co^{2+}, H_2O and strong acid or base) must be created. Since the

$CoOHNTA^{2-}$ is one proton more basic than the reference level of NTA^{3-}, but at the reference level of Co^{2+}, its equivalent addition is 10^{-3} NTA^{3-}, 10^{-3} Co^{2+}, and 10^{-3} strong base. Since the $Co(OH)_3^-$ is three protons more basic than the reference level of Co, its equivalent addition is 10^{-5} Co^{2+} and 3×10^{-5} strong base. The $CoNTA^-$, although complexed, is at the same proton reference level as uncomplexed Co^{2+} and NTA^{3-} and does not require modification. The complete equivalent addition is therefore:

$$10^{-3} + 10^{-5} + 10^{-4} \text{ M Co} = 1.11 \times 10^{-3} \text{ M } Co^{2+}$$

$$10^{-3} + 10^{-4} \text{ M NTA} = 1.1 \times 10^{-3} \text{ M } NTA^{3-}$$

$$10^{-3} \text{ strong base from the } CoOHNTA^{2-} + 3 \times 10^{-5} \text{ strong base from the } Co(OH)_3^- = 1.03 \times 10^{-3} \text{ eq /L BASE}$$

The NTA and Co have been adjusted to appear in their reference levels for the existing proton condition. Only the strong base must be numerically considered in altering the proton condition. Therefore, the equivalent addition solution requires a BASE term of 1.03×10^{-3} M.

In addition to eliminating the need to rewrite the proton condition for each new problem formulation, the equivalent-addition methodology also permits consideration of kinetic reactions in the system that produce or consume acidic hydrogen. These reactions change the total amount of acidic equivalents present and can be tracked using the same BASE term. For example, aerobic, primary utilization of NTA consumes 0.418 moles of acid per mole of NTA degraded (R16). A rate of loss of acid is computed from the rate of NTA biodegradation (e.g. E2), and this rate is incorporated into the next call of the equilibrium routine by augmenting the BASE term in the proton condition.

Additional chemical reactions in the subsurface

While beyond the scope of this paper, additional chemical reactions in the subsurface affect and are affected by biological reactions. Surface reactions of sorption, precipitation, and dissolution affect solution pH and equilibrium speciation. Through these effects, as well as by direct sequestration of degradable components, these reactions affect biological degradation. Subsurface models incorporate these reaction types as either equilibrium or kinetically controlled. Interactions between these reactions and biological reactions should be considered in a parallel manner to the interactions with aqueous phase complexation and acid/base reactions discussed in this paper.

MODELING EXAMPLES

Several examples involving the biodegradation of NTA illustrate the methods outlined in earlier sections. Stoichiometric reactions were created for oxidation of NTA with each potential donor electron acceptor substrate (R16 and R17). Biomass decay reactions also were created (R14 and R15). The stoichiometry of these reactions is used along with kinetic parameters for species known to degrade the target electron donors to compute a rate of substrate utilization following E2. This substrate utilization rate is linked directly to biomass synthesis through E5 and to rates for additional reactants through rate equations formatted as E8.

The dependence of biodegradation reactions on the chemical form of the substrate presents an important modeling challenge. While data suggest that certain bacterial species favor particular forms of electron donors (e.g. Bolton et al., 1996), different bacterial species may have different preferences. It is also possible that different chemical forms of a substrate lead to different biodegradation rates for the same species of bacteria (Bolton et al., 1996). Thus, our model allows us to specify degradable and nondegradable forms of each substrate. The concentrations of all substrate forms as received from the equilibrium speciation are stored; however, only the degradable forms are added to the total substrate available for degradation, S_d utilized in E2.

For each time step, the model computes the speciation for the NTA system; determines the S_d, S_a, and X for the timestep; and links these with kinetic parameters and the stoichiometry given by R16 or R17 and R14 or R15 to determine rates of change for all components in the system.

As an example, the model computes the following rates for each time step for the obligate aerobic NTA degrading organism, *Chelatobacter heintzii* :

$$r_{NTA} = -q_{md} \frac{S_{NTA}}{K_{NTA} + S_{NTA}} \frac{S_{O2}}{K_{O2} + S_{O2}} X \quad \text{rate of donor utilization} \qquad \text{(E29)}$$

$$r_{net} = r_{syn} + r_{decay} = -Yr_{NTA} - bX \quad \text{rate of biomass accumulation} \qquad \text{(E30)}$$

$$r_{O2} = 1.59 r_{NTA} + 5bX \quad \text{rate of oxygen utilization} \qquad \text{(E31)}$$

$$r_{H+} = 0.418 r_{NTA} + 1bX \quad \text{rate of acidic hydrogen utilization} \qquad \text{(E32)}$$

$$r_{CO2} = 3.11 r_{NTA} + 5bX \quad \text{rate of } CO_2 \text{ production} \qquad \text{(E33)}$$

$$r_{NH4+} = 0.418 r_{NTA} + 1bX \quad \text{rate of } NH_4^+ \text{ production} \qquad \text{(E34)}$$

These rates are utilized to update the total concentrations of each component. The model then respeciates all bulk-liquid components and complexes and repeats the biodegradation computations for the next time step. A standard explicit finite difference scheme is employed that is exact for a sufficiently small time step.

Table 2. Equilibrium and kinetic parameters for NTA degredation by *Chelatobacter heintzii*

Parameter	Value Used	Source of Parameter
pK_a for H_3NTA	1.6	Sillen and Martell (1964)
pK_a for H_2NTA	3.0	Sillen and Martell (1964)
pK_a for HNTA	10.3	Sillen and Martell (1964)
pK_f for $CoOH^+$	4.3	Sillen and Martell (1964)
pK_f for $Co(OH)_2^o$	5.1	Sillen and Martell (1964)
pK_f for $Co(OH)_3^-$	10.5	Sillen and Martell (1964)
pK_f for $Co(OH)_4^{2-}$	32.3	Sillen and Martell (1964)
pK_f for $CoNTA^-$	10.6	Sillen and Martell (1964)
pK_f for $Co(NTA)_2^{4-}$	14.4	Sillen and Martell (1964)
pK_f for $CoOHNTA^{2-}$	14.5	Sillen and Martell (1964)
Cell Yield	0.45 g cells/ g THOD	computed from stoichiometry
Maximum specific rate of substrate utilization	4.9 g THOD/g-cells-day	Egli (1988)
Monod half maximum rate concentration for donor (NTA)	1.1 x 10^{-4} g COD/L	Alder (1990)
Monod half maximum rate concentration for acceptor (Oxygen)	0.20 mg/L	Siegrist et al (1989)
Endogeneous decay constant	0.05 1/day	Odencrantz (1992)

The model is utilized for biodegradation and linked-equilibrium modeling for a variety of cases that illustrate the interactions between biodegradation rates and system pH. As indicated in previous discussions, the pH of the system controls aqueous speciation of the primary electron donor. The concentration of degradable forms of the electron donor control the rate of degradation. Biodegradation reactions produce and consume acidic hydrogen, further altering the pH and affecting the re-speciation of the donor.

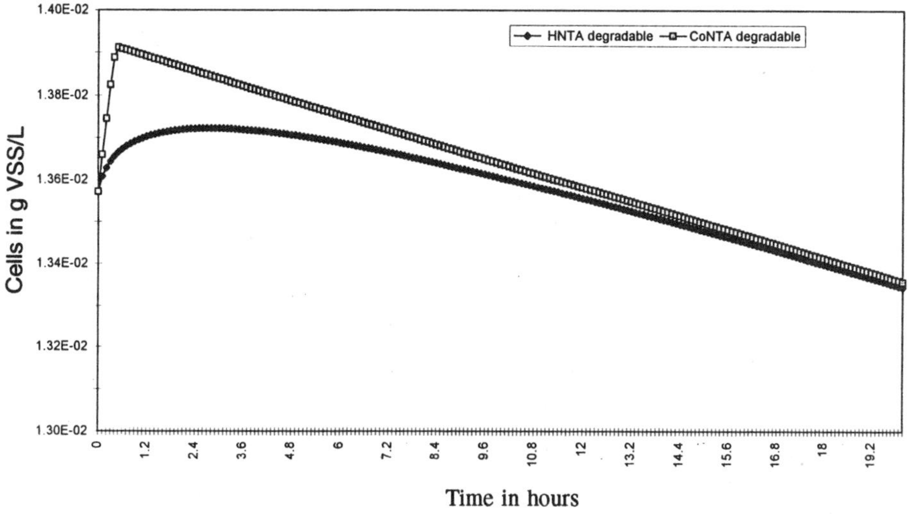

Figure 1. Effect of degradable species on cell growth during degradation on NTA at a fixed pH = 6.

First, we consider the impact of system pH on biodegradation reactions in batch experiments. As a case study, we consider the degradation of nitrilotriacetic acid (NTA) by *Chelatobacter heintzii*. The kinetic parameters for this system were collected from the literature and are given in Table 2 (above), along with equilibrium parameters for the Co/NTA system. For this system, there is significant research to suggest that not all aqueous forms are degradable (Bolton et al., 1996). The predominance of the degradable form in aqueous solution can significantly affect the degradation predicted. Figures 1 and 2 show cell growth and loss of NTA in a system in which *Chelatobacter* utilizes either $HNTA^{2-}$ or $CoNTA^{-}$. The system is modeled for initially equimolar NTA:Co (5.23×10^{-6} M) at pH 6. The initial organisms are at 10^{9} CFU/ml, and the dissolved oxygen is constant at 1 mg/L. Speciation for the initial conditions has 95% of the NTA present as $CoNTA^{-}$ and 4.75% present as $HNTA^{2-}$. When $CoNTA^{-}$ is directly degradable by *Chelatobacter heintzii*, we observe rapid cell growth (Fig. 1) and a rapid loss of NTA (Fig. 2). NTA is 99% degraded in approximately 0.6 hours. However, when $HNTA^{2-}$ is directly degradable by the organisms, cell growth is more gradual (Fig. 1), and NTA loss is slower, being 99% complete by day 20 (Fig. 2). These results demonstrate that the degradable form of the electron donor can play a dramatic role in the kinetics of biological degradation. While cells grow and decay and NTA is fully degraded in each case, the time to complete degradation is strongly controlled by what species is degradable. When the non-dominant species is the only one degradable, the rate of degradation slows, and cell growth is reduced.

The effect of pH on speciation and on biodegradation can also be explored by considering degradation of NTA at different initial pH values. For the same starting

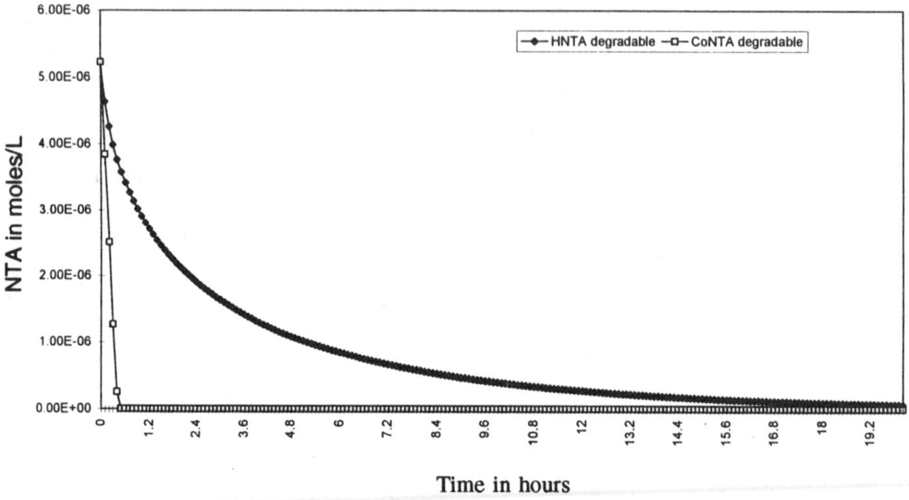

Figure 2. Effect of degradable species on loss of NTA during degradation on NTA at a fixed pH = 6.

conditions, NTA degradation was modeled at pH 6, 8, and 10 for the case considering $HNTA^{2-}$ as the only degradable form. Speciation predicts $HNTA^{2-}$ will comprise 5% of the NTA at pH 6, 3.8% at pH 8, and 0.1% at pH 10. Figure 3 indicates cell growth for pH 6, but only decay for pHs 8 and 10. Figure 4 shows rapid (compared to pH 8 and 10) utilization of NTA at pH 6, with 99% mineralization by day 20. At pH 8, however, NTA declines more slowly, reaching 24% degraded by day 20. For pH 10, virtually no NTA is degraded in 20 days. Again, the system pH controls the amount of $HNTA^{2-}$ in solution. When this is the form *Chelatobacter heintzii* is capable of degrading, high pH profoundly slows the rate of degradation and prevents net biomass growth.

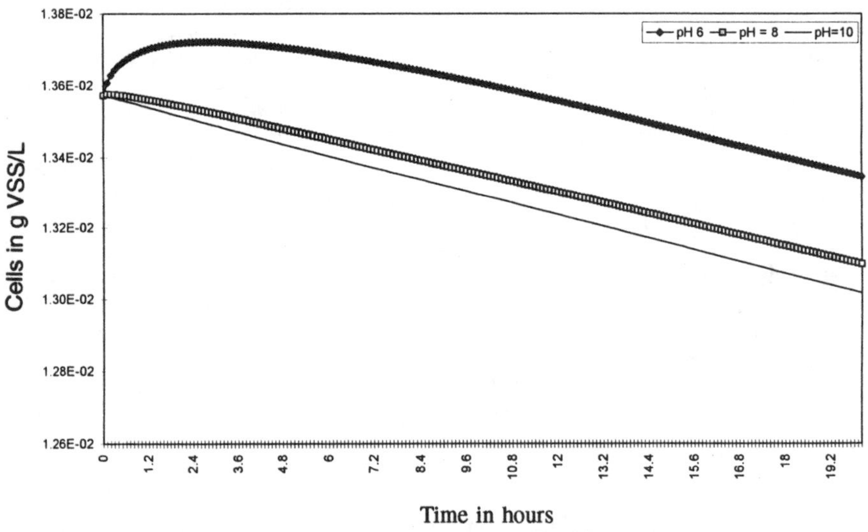

Figure 3. Effect of system pH on cell growth during degradation on NTA as HNTA.

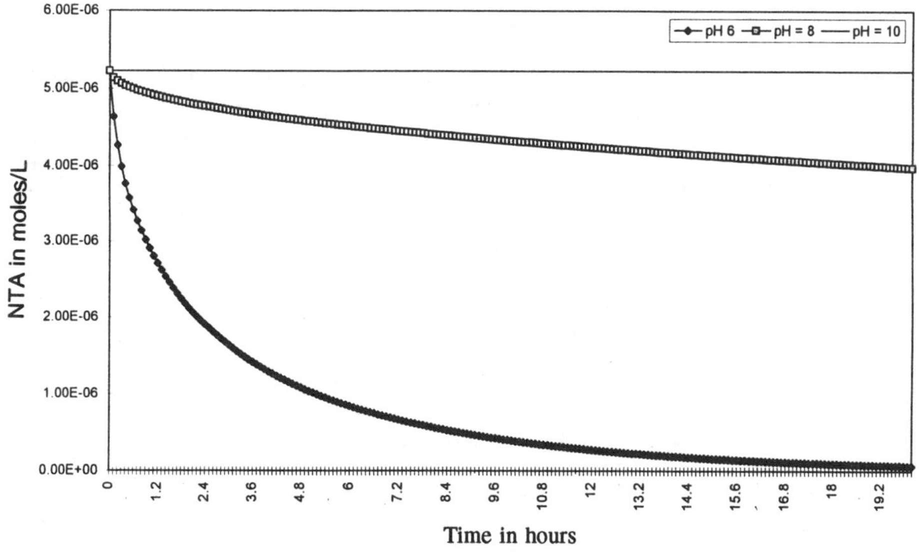

Figure 4. Effect of system pH on loss of NTA during degradation on NTA as HNTA.

Finally, the degradation of NTA utilizes acidic equivalents and, therefore, can affect the system pH. The model was run without fixing the pH to determine the effect of the degradation reaction on acidic hydrogen in the system. Beginning with slightly higher NTA concentration (5.23×10^{-5}) and the same other initial conditions as above and with pH starting at 6, modeling (Fig. 5) predicts a rise in pH over time to 8.5 after 20 days. For this system, the changing pH has only a slight effect on the speciation of the degradable form, HNTA, because the $pK_{a,2}$ and $pK_{a,3}$ values are 2.5 and 10.3, which are outside the pH range of 6.0 to 8.5. Therefore, the rate of the substrate utilization is only reduced slightly by the increase in pH for this example (not shown).

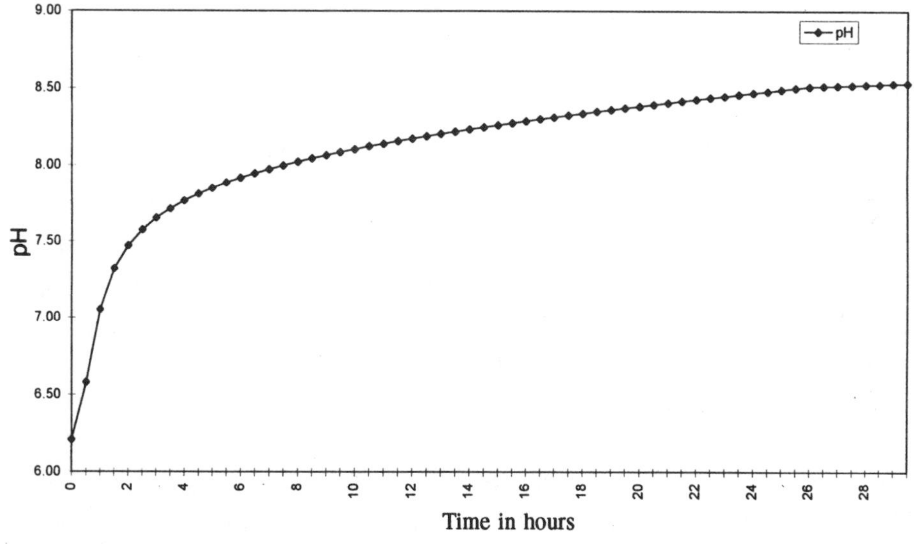

Figure 5. Effect of degradation of NTA on system pH.

CONCLUSIONS

In this presentation, we detail the basic components of microbiological modeling required for inclusion with subsurface transport. Our modeling approach emphasizes that the microorganisms' most fundamental phenomena is to gain electrons and energy to synthesize and maintain themselves. By tracking the acquisition of electrons from the primary electron donor substrate and their uses—to reduce environmental materials to make cells and to generate energy—we are able to determine the full stoichiometry of the mineralization and cell growth reaction. This full stoichiometry allows prediction not only of the loss of the primary electron donor substrate, but also prediction of the impacts of biodegradation on concentrations of the microorganisms, the electron acceptor substrate, acidic hydrogen, and inorganic carbon.

The degradation reactions are affected by and affect nonbiological reactions, particularly including complexation and acid/base reactions. We explicitly link these equilibrium controlled reactions to the kinetically controlled biodegradation reactions.

Several examples show how complexation and acid/base reactions can dramatically alter the biodegradation rate of organic substrates when the different acid/base or complexed species have different biodegradation kinetics. Furthermore, we demonstrate how microbial reactions affect the system pH, speciation, and biodegradation rates.

ACKNOWLEDGMENTS

This work is supported by a grant from the Co-Contaminant Chemistry subprogram within the Subsurface Science Program of the Department of Energy's Office of Health and Environmental Research.

REFERENCES

Alder AC, Siegrist H, Gujer W, Giger,W (1990) Behavior of NTA and EDTA in biological wastewater treatment. Water Res 24:6:733-742

Bae W, Rittmann BE (1996a) Responses of intracellular cofactors to single and dual limitation. Biotech Bioengineer 49:690-699

Bae W, Rittmann BE (1996b) A structured model of dual-limitation kinetics. Biotech Bioengineer 49:683-689

Baedecker MJ, Cozzarelli IM, Eganhouse RP, Siegel DI, Bennett PC (1993) Crude oil in a shallow sand and gravel aquifer III. Biogeochemical reactions and mass balance modeling in anoxic groundwater. Appl Geochem 8:569-586

Baik MH, Lee KJ (1994) Transport of radioactive solutes in the presence of chelating agents. Annals of Nuclear Energy 21: 81-96

Baveye P, Valocchi AJ (1989). An evaluation of mathematical models of the transport of reacting solutes in saturated soils and aquifers. Water Resources Res 25:1413-1421

Bolton H. Jr., Girvin DC, Plymale AE, Harvey SD, Workman DJ (1996) Degradation of metal nitrilotriacetate Complexes by *Chelatobacter heintzii*. Environ Sci Technol 30:931-938

Bolton H Jr., Li SW, Workman DJ, Girvin DC (1993) Biodegradation of synthetic chelates in subsurface sediments from the southeast coastal plain. J Environ Quality 22:125-132

Borden RC, Bedient PB (1987) Transport of dissolved hydrocarbons influenced by oxygen-limited biodegradation, 1, Theoretical development. Water Resources Res 22:1973-1982

Brock TD, Madigan MT, Martinko JM, Parker J (1994) Biology of Microorganisms, 7th edn. Prentice-Hall, Englewood Cliffs, NJ

Bouwer EJ, Cobb GD (1987) Modeling of biological processes in the subsurface. Water Sci Tech 19:769-779

Celia MA, Kindred JS, Herrera I (1989) Contaminant transport and biodegradation 1. A numerical model for reactive transport in porous media. Water Resources Res 25:1141-1148

Christensen DR, McCarty PL (1975). Multiprocess biological treatment model. J Water Pollution Control Federation 47:2652-2664

Egli T (1988). (An)Aerobic breakdown of chelating agents used in household detergents. Microbiol Sci 5:36-41

Firestone MK, Tiedje JM (1975) Biodegradation of metal-NTA complexes by a pseudomonas species: Mechanism of reaction. Appl Microbiol 29:758-64

Frind EO, Suynisveld WHM, Strebel O, Boettcher J (1990) Modeling of multicomponent transport with microbial transformation in ground water: the Fuhrberg case. Water Resources Res 26:1707-1719

Fry VA, Istok JD (1994) Effects of rate limited desorption on the feasibility of in situ bioremediation. Water Resources Res 30:2413-2422

Fry VA, Istok JD, Guenther RB (1993) An analytical solution to the solute transport equation with rate limiting desorption and decay. Water Resources Res 29:3201-3208

Harvey RW, Smith RL, George L (1984). Effect of organic contamination upon microbial distributions and heterotrophic uptake in Cape Cod, MA aquifer. Appl Environ Microbiol 48:1197-1201

Kindred JS, Celia MA (1989) Contaminant transport and biodegradation 2. Conceptual model and test simulations. Water Resources Res 25:1149-1159

Kinzelbach W, Schafer W, Herzer J (1991) Numerical modeling of natural and enhanced denitrification processes in aquifers. Water Resources Res 27:1123-1135

Lichtner PC (1985) Continuum model for simultaneous chemical reactions and mass transport in hydrothermal systems. Geochim Cosmochim Acta 49:779-800

MacQuarrie KTB, Sudicky EA, Frind EO (1990) Simulation of biodegradable organic contaminants in ground water 1. Numerical formulation in principal direction. Water Resources Res 26:207-222

Malmstead MJ, Brockman F, Valocchi AJ, Rittmann BE (1995). Modeling biofilm biodegradation requiring cosubstrates: The quinoline example. Water Sci Tech 31:71-84

McNab WW Jr, Narasimhan TN (1994) Modeling reactive transport of organic compounds in groundwater using a partial redox disequilibrium approach. Water Resources Res 30:2619-2635

Moltz FJ, Widdowson MA, Benefield LD (1986) Simulation of microbial growth dynamics coupled to nutrient and oxygen transport in porous media. Water Resources Res 22:1207-1216

Morel F, Morgan J (1972) A Numerical method for computing equilibria in aqueous chemical systems. Environ Sci Technol 6:58-667

Namkung E, Stratton R, Rittmann BE (1983) Predicting removal of trace organic compounds by biofilms. J Water Pollution Control Federation 55:1366-1372

NRC = National Research Council (1990) Ground Water Models: Scientific and Regulatory Applications. National Academy Press, Washington, DC

NRC = National Research Council (1993) In Situ Bioremediation: When Does It Work? National Academy Press, Washington, DC

Nortemann B (1992) Total degradation of EDTA by mixed cultures and a bacterial isolate. Appl Environ Microbiol 58:2:671-676

Odencrantz JE (1992) Modeling the Biodegradation Kinetics of Dissolved Organic Contaminants in a Saturated Heterogeneous Two-Dimensional Aquifer. PhD dissertation, Dept. of Civil Engineering, University of Illinois, Urbana, IL

Odencrantz, JE, Valocchi, AJ, Rittmann, BE (1990) Modeling two-dimensional solute transport with different biodegradation kinetics. Paper presented at Petroleum Hydrocarbons and Organic Chemicals in Groundwater, Am Petrol Inst, Houston, TX, Oct 31 to Nov. 2, 1990

Rifai HS, Bedient PB (1987) Comparison of biodegradation kinetics with an instantaneous reaction model for groundwater. Water Resources Res 26:637-645

Rittmann BE (1993) The significance of biofilms in porous media. Water Resources Res 29:2195-2202

Rittmann BE, Brunner CW (1984) The nonsteady-state-biofilm process for advanced organics removal. J Water Pollution Control Federation 56: 874-880

Rittmann BE, McCarty PL (1980) A Model of steady-state-biofilm kinetics. Biotech Bioengineer 22:2343-2357

Rittmann BE, McCarty PL (1981) Substrate flux into biofilms of any thickness. J Environ Eng 107:831-849

Rittmann BE, Seagren E, Wrenn BA, Valocchi AJ, Ray C, Raskin L (1994) In Situ Bioremediation, 2nd edn, Noyes Publishers, Park Ridge, NJ

Rubin J (1983) Transport of Reacting Solutes in Porous Media: Relation Between Mathematical Nature of Problem Formulation and Chemical Nature of Reactions. Water Resources Res 19:1231-1252

Siegrist H, Alder A, Gujer W, Giger W (1989) Behaviour and modeling of NTA degradation in activated sludge systems. Water Science Tech 21:315-324

Sillen LG, Martell AE (1964) Stability Constants of Metal-Ion Complexes. Special Publ 17, The Chemical Society, London

Snoeyink VL, Jenkins D (1980) Water Chemistry. John Wiley & Sons , New York

Srinivasan P, Mercer JW (1988) Simulations of biodegradation and sorption processes in ground water. Ground Water 26:475-487

Stryer L (1988). Biochemistry, 3rd edn, W H Freeman & Co, San Francisco, CA

Taylor SW, Jaffe PR (1990a) Biofilm growth and the related changes in the physical properties of a porous medium 1. Experimental investigation. Water Resources Res 26:2153-2159

Taylor SW, Jaffe PR (1990b) Biofilm growth and the related changes in the physical properties of a porous medium 2. Permeability. Water Resources Res 26:2161-2169

Taylor SW, Jaffe PR (1990c) Biofilm growth and the related changes in the physical properties of a porous medium 3. Dispersivity and model verification. Water Resources Res 26:2171-2180

Taylor SW, Jaffe PR (1990d) Substrate and biomass transport in a porous medium. Water Resources Res 26:2181-2194

Widdowson MA, Motz FJ, Benefield LD (1988) A numerical model for oxygen and nitrate based respiration linked to substrate and nutrient availability in porous media. Water Resources Res 24:1553-1565

Wood BD, Dawson CN, Szecsody JE, Streile GP (1994) Modeling contaminant transport and biodegradation in a layered porous media system. Water Resources Res 30:1833-1845

Zyssett AF, Stauffer F, Dracos T (1994) Modeling of reactive ground water transport governed by biodegradation. Water Resources Res 30:2423-2434

Chapter 8

BIOGEOCHEMICAL DYNAMICS IN AQUATIC SEDIMENTS

Philippe Van Cappellen

School of Earth and Atmospheric Sciences
Georgia Institute of Technology
Atlanta, Georgia 30332 U.S.A.

Jean-Francois Gaillard

Department of Civil Engineering
Northwestern University
Evanston, Illinois 60208 U.S.A.

INTRODUCTION

The sedimentary column contains the record of past environmental conditions at the earth's surface. Sediments are not passive recorders, however. Deposited organic and inorganic particulate matter participates in a variety of chemical and microbial reactions that lead to the disappearance of certain constituents, the formation of residual organic phases, or the appearance of new minerals. The accurate interpretation of the sedimentary record depends on our ability to distinguish post-depositional changes from the primary signals registered at the water-sediment interface.

The biogeochemical transformations occurring in sediments cause chemical mass transfers with the water column (Fig. 1). As a result of the microbial oxidation of deposited organic matter, dissolved oxidants are transported into the sediment, while opposite fluxes return inorganic nutrient species to the bottom waters. Consequently, the productivity and chemical composition of terrestrial and marine surface waters may be tightly coupled to the activity of benthic microbial populations. On geologic time scales, this coupling profoundly influences the chemical evolution of the oceans and atmosphere (Berner, 1989).

Aquatic surface sediments are also inhabited by macro-invertebrates and bottom-dwelling fish. The macrofauna markedly affect physical sediment properties, and may significantly enhance particulate and solute transport fluxes (e.g. Matisoff, 1995). Overall, reactive transport in surface sediments is characterized by a major involvement of biological activity. While recent studies have provided definite evidence for microbially-mediated chemical gradients in subsurface systems (e.g. Chapelle, 1993), aquatic sediments offer easily accessible porous media in which to study the interactions between chemical reactions, physical transport and biological activity, on spatial scales ranging from millimeters to meters.

Much of what we know about the biogeochemical dynamics of surface sediments has been derived from measured concentration profiles of pore water and solid sediment species. In order to interpret the profiles, however, we must be able to separate the effects of coupled sedimentological, biological and geochemical processes. Laboratory and field experiments can be designed to probe individual transport or reaction processes. In particular, the mechanisms, kinetics and/or equilibrium states of individual biogeochemical reactions can be investigated in the laboratory, while tracers can be used to parameterize transport in field-scale experiments. Reactive transport models describe the interaction of

0275-0279/96/0034-0008$05.00

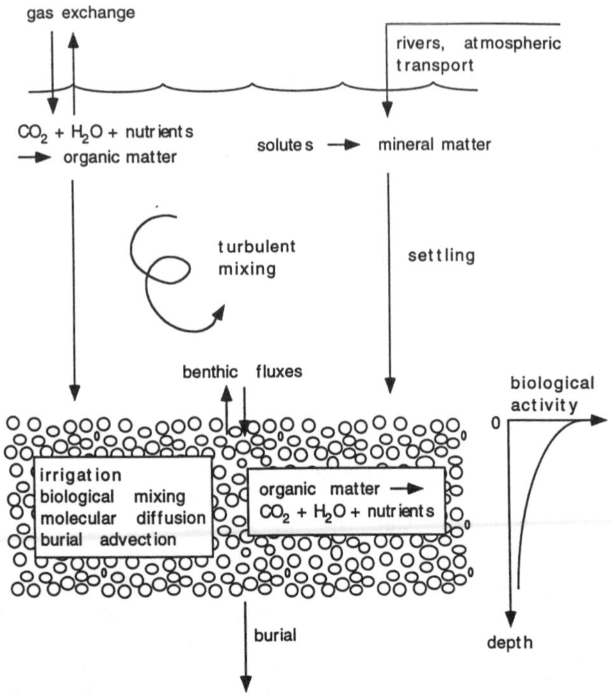

Figure 1. Aquatic surface sediments are formed by the deposition of organic and mineral matter from the water column. They represent active reaction-transport systems where the deposited materials undergo extensive biogeochemical processing prior to their ultimate preservation in the sedimentary column. The diagram emphasizes the open nature of surface sediments, as well as the central role of biological activity.

biogeochemical reactions and transport processes, and thus provide the link between the field measurements and laboratory studies.

Theoretical models that focus on chemical mass transfers occurring shortly after deposition of sediments are referred to as early diagenetic models (Berner, 1980). These models offer quantitative tools to investigate the environmental and biogeochemical factors controlling the chemical dynamics of aquatic sediments. They can be used to quantify the importance of various reaction pathways and sedimentary sinks, and allow one to calculate elemental budgets and benthic exchange fluxes.

In this chapter, the treatment of aquatic sediments as reactive transport systems is reviewed. After a short introduction to sediment biogeochemistry and a historical review of early diagenetic modeling, a detailed discussion on the mathematical representation of transport fluxes and rates of biogeochemical reactions is presented. The expressions for fluxes and rates are then tied together in the mass conservation equations. Numerical methods of solution are only briefly discussed, as they are treated extensively elsewhere in this volume.

AQUATIC SEDIMENTS: BACKGROUND

Aquatic sediments as porous media

Aquatic sediments are porous media formed by the deposition from water of organic and mineral particulate matter produced in the overlying water body, or derived from a more distant source (Fig. 1). Sediments differ significantly from the water column in terms of their transport properties. The presence of a solid matrix eliminates large-scale fluid flow and turbulent mixing. Advective velocities of particulate matter also change from relatively fast settling velocities to slower sediment accumulation rates. Reduced transport rates and longer particulate residence times promote the establishment of steep pore water and solid sediment concentration gradients below the water-sediment interface.

Most freshly deposited aquatic sediments have fairly high total porosities, typically ranging from 60 to 95% (Manheim, 1970; Ullman and Aller, 1982). Resistivity profiles often reveal a decrease of the porosity by as much as 10% from the water-sediment interface down to depths of a few centimeters. At greater depths, the porosity decreases much more slowly. To account for compaction effects, porosity depth profiles are usually fitted to exponential functions of the form (e.g. Rabouille and Gaillard, 1991),

$$\phi(x) = \phi_\infty + (\phi_0 - \phi_\infty)\exp(-\alpha_\phi x) \tag{1}$$

where x is depth below the water-sediment interface, ϕ_0 is the porosity at the water sediments interface, ϕ_∞ is the asymptotic porosity value at depth, and α_ϕ is the porosity depth attenuation constant. Note that, strictly speaking, Equation (1) assumes a steady-state porosity distribution. Non-steady state compaction effects have not, to the authors' knowledge, been investigated in the context of early diagenetic models.

The porous structure of well-sorted sandy sediments can be represented by simple models based on the regular packing of spheres or rods (Bear, 1972). These models can be used to calculate matrix and transport properties, such as the specific surface area, porosity and tortuosity. They do not apply to fine-grained, organic-rich surface sediments, however. Inert tracer breakthrough experiments performed in the senior author's laboratory on undisturbed core sections of surficial aquatic sediments show that, in contrast to coarse detrital sediments, the pore lengths in fine-grained sediments are orders-of-magnitude larger than the mean grain size. This is consistent with a loose network of small sediment particles with large pore spaces in between (Engelhardt, 1977). The breakthrough curves further confirm the absence of stagnant pore water or a dual porosity, in all marine and freshwater sediments studied. Thus, for unconsolidated sediments, the effective porosity equals the total porosity.

Aquatic sediments as biogeochemical reactors

Aquatic sediments are open systems which exchange matter with the overlying water column and the underlying sediment repository (Fig. 1). Of particular importance is the influx of particulate organic matter produced in the surface waters. For shallow-water sediments located within the photic zone, organic matter may also originate from photosynthesis right at the water-sediment interface (Revsbech and Jørgensen, 1983). The net supply of organic matter to the sediment represents the ultimate source of energy sustaining the benthic populations of micro- and macroorganisms. The major changes affecting pore water and solid sediment chemistry can be traced back, directly or indirectly, to the supply and degradation of organic matter.

Because the organic matter is supplied at the water-sediment interface, the latter is also the site of the most intense heterotrophic activity. Indicators of biological activity, e.g. measurements of ATP (adenosine triphosphate), cell abundance, metabolic rates or macrofaunal mixing intensity, are generally highest in the topmost sediment layer and decrease with depth. In parallel, the organic carbon concentration usually decreases with depth below the water-sediment interface, unless the sediment is vigorously mixed or experiences a non-steady state supply of organic matter at the water-sediment interface.

Equally important, however, is the gradual drop in reactivity of the organic matter (Van Cappellen et al., 1993). This is due to the fact that natural organic matter is a mixture of many different compounds. As the heterotrophic organisms preferentially degrade the more readily hydrolizable substrates, the residual organic matter tends to become progressively more refractory. As a result of the decrease in abundance and reactivity of the organic matter, the net rate of organic carbon oxidation is usually found to slow down rapidly with depth below the water-sediment interface (e.g. Aller and Mackin, 1989; Klump and Martens, 1989; Mackin and Swider, 1989; McNichol et al., 1988; Canfield et al., 1993b) (Fig. 2).

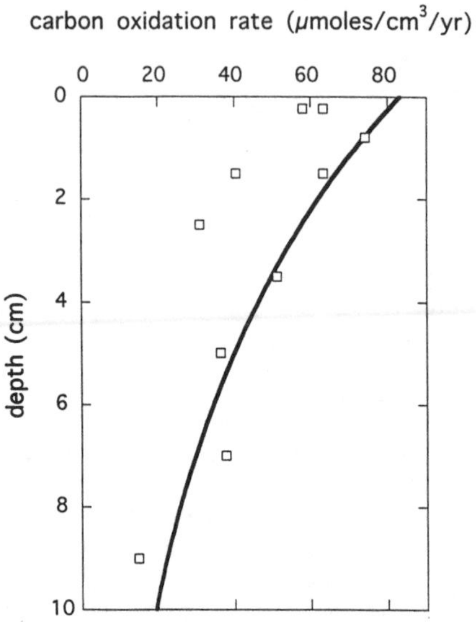

Figure 2. Depth distribution of net organic carbon oxidation in a nearshore marine sediment (Skagerrak, Denmark). The symbols represent measured rates, the solid line an exponential fit to a subset of the data. A generally decreasing trend of the organic carbon oxidation rate with depth is observed in most surface sediments (data from Canfield et al., 1993b).

The metabolic oxidation of organic matter consumes oxidants and produces reduced species. Hence, sediments tend to be net sinks of oxidized chemical species (O_2, NO_3, SO_4), while reduced products are returned to the water column. Under favorable conditions, for example shallow, turbulent waters, volatile products of organic matter degradation (e.g. CH_4, N_2, N_2O, H_2S) may escape to the atmosphere. Heterotrophic activity in sediments also releases dissolved organic matter and mineral nutrients to the water column, hence linking biological productivity of surface waters to biogeochemical processes in sediments. This benthic-pelagic coupling is particularly strong in productive, shallow-water environments, for instance estuaries and coastal areas (e.g. Wollast, 1993).

The oxidation of organic matter may lead to precipitation of authigenic minerals phases which can be buried permanently in the sediment column. The preservation of these solid phases, including sulfides, carbonates, silicates and phosphates, offers important clues for the reconstruction of paleodepositional conditions (see, e.g. Huckriede and Meischner, 1996). Detrital mineral phases that survive weathering and transport by rivers or wind are for the most part unreactive upon incorporation into aquatic sediments. The oxyhydroxides of iron and manganese are an important exception. These phases are stable under oxygenated conditions encountered in the water column, however, they may be reduced after burial or mixing below the aerobic surface layer of a sediment (Canfield, 1989).

In addition to organic matter, biological activity also supplies biogenic mineral phases to sediments, principally calcium carbonate and amorphous silica (opal). Biogenic opal is thermodynamically unstable in most surface waters, and dissolution starts immediately after death of the biomineralizing organism. In sediments containing biogenic opal, the dissolution process causes a build-up of the concentration of silicic acid with increasing depth below the water-sediment interface (e.g. Rabouille et al., 1996). In biosiliceous oozes, the dissolution process continues until the pore waters reach equilibrium with the dissolving biosiliceous debris. When the sediment receives a significant deposition flux of detrital material, reprecipitation of silicic acid and soluble aluminum may induce the formation of authigenic alumino-silicates (Mackin and Aller, 1984; Mackin, 1987; Van Cappellen and Qiu, 1996). The silicic acid concentration in the pore waters may then tend toward a steady-state between opal dissolution and authigenic silicate formation, rather than to thermodynamic equilibrium.

The saturation state of sediment pore waters with respect to calcium carbonate is linked to the early diagenesis of organic matter. Depending on the dominant pathway of organic matter oxidation, $CaCO_3$ can either dissolve or precipitate. Aerobic respiration results in the complete oxidation of organic carbon to CO_2 and hence promotes the dissolution of deposited biogenic carbonate debris (e.g. Archer et al., 1989; Cai et al., 1995). Anaerobic respiration pathways on the other hand generate bicarbonate. This will counter dissolution and, under the right circumstances, may cause the precipitation of authigenic $CaCO_3$ (Gaillard et al., 1989).

From the preceding discussion, the reader will have gathered that aquatic sediments are extremely dynamic and complex biogeochemical systems. Their overall state of non-equilibrium is due mainly to the supply of organic matter, which can be viewed as the external source of "metabolic" energy driving the biogeochemical transformations. To a lesser degree, disequilibrium is also maintained by the transfer of thermodynamically unstable mineral phases from the water column. Overall, the quantitative description of chemical dynamics in early diagenetic systems demands a kinetic modeling approach which accounts for the coupling of reaction rates to transport rates.

Spatial and temporal scales of early diagenesis

Diagenetic reactions continue over the entire thickness of the sedimentary column. The most rapid changes, however, take place in close proximity to the water-sediment interface. This is also the region of sediment which is physically mixed by macrofauna or bottom currents, and which exchanges solutes with the water column. Early diagenetic models focus on this most reactive part of the sediment. Quantitative estimates of the vertical extent of the zone of early diagenesis can be derived from the magnitudes of the main transport and reaction parameters (Table 1).

Table 1. Typical values of main transport and reaction parameters in marine sediments:
ω: sedimentation rate; D_b: particle mixing coefficient; k_C: apparent first-order rate constant of organic matter degradation. Also given are the characteristic depth (x_{ed}) and time (t_{ed}) scales for early diagenesis in the different depositional settings.

	Pelagic sediments	Hemipelagic sediments	Shelf/coastal sediments
water depth (m)	6000-3000	3000-200	$\leq$ 200
ω (cm yr^{-1})	0.001	0.02	0.1
D_b (cm^2 yr^{-1})	0.1	1.5	10
k (yr^{-1})	0.001	0.02	0.1
x_{ed} (cm)	10	9	10
t_{ed} (yr)	10,000	450	100

Given the importance of organic matter degradation as the primary process driving chemical change in sediments, the reactivity of particulate organic matter is a major parameter controlling the thickness of the zone of early diagenesis. In the simplest kinetic model, the reactivity is parameterized by a first-order reaction rate constant for organic carbon degradation, k_C (Berner, 1980). The characteristic length scale of early diagenesis can then be estimated from ω/k_C or $(D_b/k_C)^{1/2}$, where ω is the sediment accumulation rate, and D_b the sediment mixing coefficient (Van Cappellen and Wang, 1995). The first formula applies when sediment advection is the main process of vertical particle transport below the water-sediment interface, the second when random mixing by macrofauna (bioturbation) or bottom currents dominates.

Bioturbation usually dominates particulate transport in the upper 5 to 20 cm of sediments, unless conditions are particularly inhospitable for macrofauna, e.g. when the bottom waters are anoxic or dysoxic (e.g. Tromp et al., 1995). Using the values for D_b and k_C given in Table 1, characteristic depths of early diagenesis on the order of 10 cm are calculated for widely different depositional environments. This reflects the rough correlations that exists between the intensity of particle mixing by benthic macrofauna, and the supply flux plus reactivity of sedimentary organic matter (Tromp et al., 1995).

While the main chemical changes associated with organic matter oxidation are confined to the first 10 to 100 cm of sediment, over a broad range of depositional conditions, the time scales involved are quite variable. An estimate of the characteristic duration of early diagenesis is obtained by dividing the characteristic depth scale by the sediment accumulation rate (Table 1). Thus, the compositional profiles observed in a one meter core from the deep-sea reflect the interaction between transport and reaction processes covering a time span on the order of 10^4 years, compared to 100 years or less in a similar core retrieved in a shallow-water environment. However, the extent of early diagenetic processing is not proportionally larger in the deep-sea core, because of the much lower average reactivity (k_C) of the organic matter (Table 1).

Trends in field studies

The development of increasingly sophisticated transport-reaction models for surface sediments is driven in part by the need to describe and utilize the detailed biogeochemical data sets that are being collected. These data sets are characterized not only by the measurement of growing numbers of chemical species, but also by the improved spatial resolution with which the distributions are being measured. Spectacular advances in microsensor technology make it now possible to routinely measure concentrations of pore water species on submillimeter scales (e.g. Revsbech and Jørgensen, 1986; Cai and Reimers, 1993; Jensen et al., 1993; Cai et al., 1995; Brendel and Luther, 1995) (Fig. 3).

Closer attention is also being paid to the speciation and reactivity of solid-bound constituents. Chemical extraction techniques are applied to follow the redistribution of elements among their different solid pools during early diagenesis (e.g. Canfield, 1989; Huerta-Diaz and Morse, 1992; Ruttenberg, 1992; Kostka and Luther, 1994; Raiswell et al., 1994), while kinetic experiments are used to directly assess the reactivity of sedimentary phases (e.g. Aller and Mackin, 1989; Van Cappellen, 1996). The measurements of depth distributions of chemical concentrations and properties are further being complemented by benthic chamber experiments which measure solute exchange fluxes at the water-sediment interface (e.g. Jahnke and Christiansen, 1989; Jahnke, 1990).

The interpretation of pore water and solid sediment profiles in sediments is often complicated by the existence of competing reaction pathways. Many redox transformations, for example, can proceed via inorganic and microbial pathways. One novel way to address

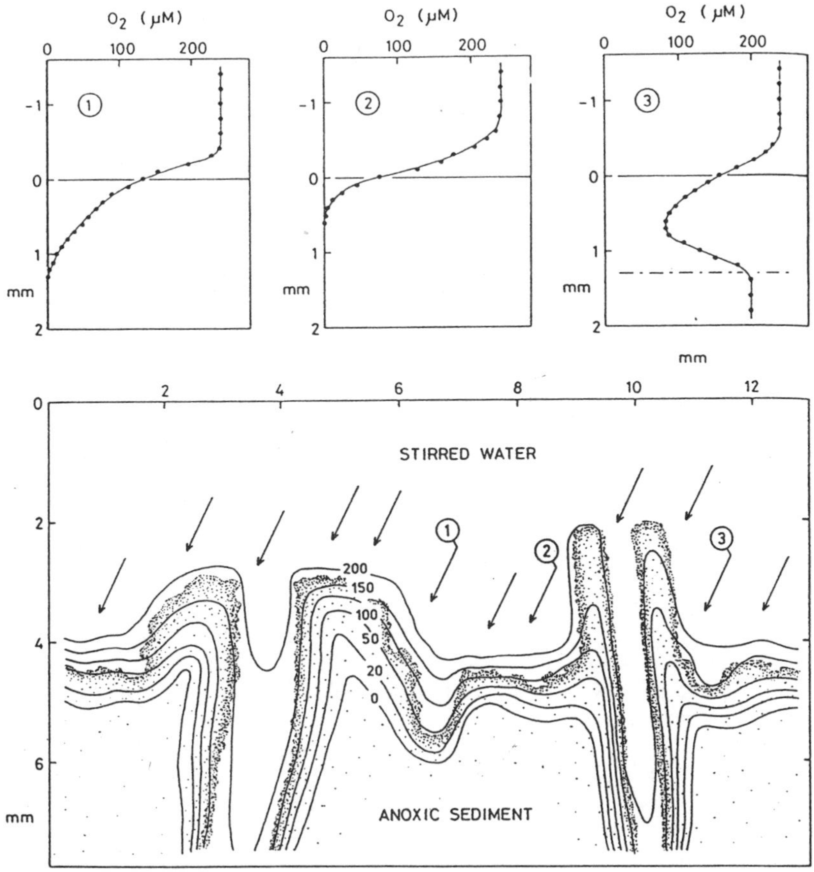

Figure 3. Oxygen concentration field at the water-sediment interface of an organic-rich shallow water environment determined with microelectrodes. The tubular structures are burrows. Arrows indicate positions of microprofiles, three of which are shown as the top figures (from Jorgensen and Revsbech, 1985).

this type of problem in the field is to deploy nucleic acid probes that are designed to detect the presence and activity of specific microbial populations (e.g. Amman et al., 1990, 1991; Poulsen et al., 1993; DiChristina and DeLong, 1993; Ramsing et al., 1996). The combined determination of the distributions of molecular probe signals and chemical concentrations promises to greatly enhance our ability to unravel the complex interactions between the biological and inorganic components of sediments.

It is worth emphasizing that the observational data are collected on a variety of spatial scales. Thus, while certain pore water constituents can be measured with microelectrodes of a few micrometer in diameter mounted on micromanipulators (Fig. 3), the vertical resolution for solid sediment constituents is usually on the order of several millimeters at best. This disparity of observational scales has implications for the validation of reactive transport models by field data (see below).

EARLY DIAGENETIC MODELING

Historical perspective

The application of continuum models in early diagenesis was introduced and popularized by Robert A. Berner. His book "Early Diagenesis: A Theoretical Approach" (Berner, 1980) provides the theoretical framework for formulating the differential equations describing mass conservation of solute, solid and adsorbed species in aquatic sediments. It discusses the reference frame, boundary conditions and approximations that have been adopted in most subsequent work. It also includes many examples of model applications published up to 1980, with an emphasis on the decomposition of organic matter and its effects on pore water and solid sediment composition. A review of more recent applications can be found in Van Cappellen et al. (1993).

In the first generation of early diagenetic models, the rate of organic matter degradation was described by first-order kinetics, the so-called one-G model, which assume that the net degradation rate depends only on the reactivity and concentration of metabolizable organic carbon (Berner, 1964, 1974). The resulting reactive transport equation is one relating the concentration of metabolizable organic carbon to time and the spatial variable(s). In one dimension, and assuming steady state, the one-G model predicts exponential decreases with depth of the bioavailable organic carbon content and its degradation rate. To a first approximation, this behavior is observed in many sediments (Fig. 2).

First-order degradation kinetics have been justified by assuming that the extracellular hydrolysis of the macromolecular particulate organic matter is the rate-limiting step in the overall degradation process (Gujer and Zehnder, 1983; Meyer-Reil, 1991). Recent work on the degradation pathways of specific macromolecules in sediments indicates that this hypothesis may not be generally valid (Arnosti et al., 1994). The one-G model has also been criticized for being unable to account for the progressive decrease in organic matter reactivity with advancing diagenesis. Hence models have been developed where the total organic carbon is divided into a series of fractions, each with its own first-order rate constant for degradation (multi-G model), or models where the overall first-order rate coefficient is continuously decreasing as diagenesis progresses. Nonetheless, despite its approximate nature, the introduction of the one-G model was a crucial step in the development of reactive transport models for aquatic sediments, because it provided a tool to quantify the rate of the main process affecting the electron and proton balances in the sediment.

Once the distribution of the rate of carbon oxidation is known, it becomes possible to calculate the rates of consumption or production of the principal metabolic reactants and products. The depth distribution of pore water oxygen, for example, can be modeled by assuming that O_2 consumption is the result of aerobic respiration, hence, leading to a simple stoichiometric coupling between the rate of organic carbon oxidation and the rate of O_2 reduction (e.g. Emerson, 1985). When the organic matter flux at the water-sediment interface is sufficiently small, all metabolizable organic carbon will be oxidized before the pore waters run out of dissolved oxygen (organic carbon limitation). A simple transport-reaction model for the coupled early diagenesis of organic carbon and oxygen can then be build which includes a one-G representation for the rate of aerobic respiration. At steady-state, the model predicts that the oxygen concentration decreases with depth in the sediment, until it levels off at a constant asymptotic value after all bioavailable organic matter has been oxidized (Rabouille and Gaillard, 1991). By fitting this simple model to observed dissolved oxygen profiles, estimates can be obtained for the rate constant of aerobic respiration in oligotrophic depositional environments.

In more organic-rich sediments, pore water oxygen is depleted before all the metabolizable organic carbon is oxidized (oxygen limitation), and further degradation takes place via anaerobic pathways. Reduced byproducts such as ammonia, hydrogen sulfide, Fe(II), Mn(II) and methane, are then produced below the depth of oxygen penetration. Physical and macrofaunal processes will induce a net transport of the reduced species from the deeper anaerobic layers to the sediment surface. As a result, a fraction of the oxygen consumed in the surface layer of the sediment is diverted away from aerobic respiration, toward the reoxidation of the reduced species (e.g. Boudreau, 1991). Thus, the distribution of pore water oxygen is now coupled to the distributions of species other than organic carbon.

As a general rule, the fate of a given element in a sediment is coupled to that of many others. It is therefore not surprising that the tendency in recent years has been to include multiple reaction pathways and coupling of species in early diagenetic models (e.g. Boudreau, 1991; Rabouille and Gaillard, 1991; Boudreau and Canfield, 1993; Cai et al., 1995; Van Cappellen and Wang, 1995, 1996; Dhakar and Burdige, 1996; Wang and Van Cappellen, 1996). Mathematically, this leads to differential equations containing more than one dependent variable, or concentration. Numerical methods then become necessary to simultaneously solve the conservation equations of the various independent species.

The emphasis of applications of early diagenetic models has also evolved. In much of the earlier work, an analytical solution of a simple transport-reaction model was fitted to the measured depth profile of a given dissolved or solid-bound species. The match between model and data gave some confidence that the critical transport and reaction processes had been identified. In applications where the reaction kinetics of the species were well-known, e.g. in the case of radioactive decay, the model fit was used to derive values of unknown transport parameters. Many estimates of biologically-induced mixing rates, or biodiffusion coefficients, have been determined in this manner (e.g. Benninger et al., 1979; DeMaster and Cochran, 1982; many others). When the transport parameters were constrained, model fits were used to yield values for reaction parameters (for examples, see, Berner, 1980).

The systematic use of simple analytical models has undoubtedly led to a better quantitative understanding of the external and internal controls of early diagenesis (e.g. Toth and Lerman, 1977; Berner, 1978; Tromp et al., 1995). Despite many simplifying assumptions, analytical solutions have been used successfully to reproduce observed distributions of chemical species and reaction rates in sediments. There are however limitations to the use of simple analytical models.

Curve-fitting of an analytical solution to an observable does not in itself represent a validation of the transport-reaction model, or its underlying assumptions. To return to an earlier example, let us assume that the rate of oxygen reduction in a sediment is due only to aerobic respiration. We can then develop a simple one-dimensional reactive transport model with a one G-type rate expression representing the consumption of molecular oxygen. Such a model will frequently produce a satisfactory match with measured oxygen profiles. This is true, even in organic-rich settings where it has been conclusively demonstrated that oxygen reduction results mainly from the reoxidation of secondary reduced species (Jørgensen, 1982; Aller, 1990; Canfield, 1993; Wang and Van Cappellen, 1996).

The simple model fits the data, because of the strong effect of molecular diffusion which tends to erase features in the dissolved oxygen profile that might otherwise distinguish the different kinetic regimes of O_2 reduction. Hence, an averaging of the rates of the various reactions consuming O_2 reproduces the observed profile. Nonetheless, the simple oxygen model is inherently incapable of providing information on the relative contributions of different oxygen reduction pathways, and little insight is gained into the role of oxygen in the sedimentary redox cycles of elements such as nitrogen, sulfur, manganese or iron.

The newer multicomponent reactive transport models offer more realistic representations of the coupled biogeochemical dynamics in sediments. As a result, they are increasingly powerful as diagnostic and prognostic tools. Hence, model applications are shifting from the fitting of data to sensitivity studies and prediction. Because they incorporate the multiple reaction couplings of early diagenesis, the newer models are better equipped to simulate the collective behavior of reactive species and its response to changes in internal and external forcings.

The continuum approach

Nearly all quantitative descriptions of early diagenesis are based on a differential form of the principle of conservation of mass. These continuum models use effective transport and reaction parameters, that is, averages of the corresponding microscopic quantities over a representative elementary volume, or REV (Bear, 1972). This averaging is central to the continuum approaches for reactive transport in natural porous media used throughout this book. The REV must be large enough to yield relatively constant values for the effective properties, but small enough to reproduce their macroscopic variations at the scale of interest, say, that of a sediment core. Ideally, the same REV should apply to all properties.

Because, in general, the biogeochemical and physical properties of sediments vary much faster in the vertical than horizontal direction, most continuum models of early diagenesis are one-dimensional (1-D). The horizontal length scale of the REV must therefore integrate lateral heterogeneities. The data in Figure 3 help illustrate some of the caveats involved in applying the REV concept. The figure shows a detailed two-dimensional map of the dissolved oxygen concentration field near the water-sediment interface at a nearshore site, obtained with microelectrodes (Jørgensen and Revsbech, 1985). Inspection of the figure indicates that a horizontal length scale of about 5 mm is needed to obtain a statistically valid average oxygen concentration. In the vertical direction, however, the REV must be much thinner, at most 0.1 mm. Thus, the dissolved oxygen data suggests a REV on the order of $5 \cdot 5 \cdot 0.1 = 2.5$ mm^3 in the upper sediment.

In microelectrode studies, typically one or a few microprofiles of a given parameter are measured at a site. For a sediment-water interface such as the one shown in Figure 3, a small number of profiles may not be enough to define a meaningful average, however. In other words, a solute profile obtained with a microelectrode may provide information at a spatial scale that is smaller than that of the REV appropriate for a one-dimensional reactive transport model. In contrast, the measurement techniques of other properties, including the concentrations of most solid constituents, require samples that are much larger than 2.5 mm^3. Thus, frequently, the observables are measured at scales that do not agree with that of the REV of the model. This is an important point to keep in mind when comparing the output of a model calculation to field data. For example, a 1-D model for pore water oxygen is not expected to reproduce the features exhibited by microprofile no. 3 (Fig. 3). It could be fitted to the two other profiles (no. 1 and 2), but would predict a different rate of oxygen reduction in each case.

The REV varies spatially within a sediment. The vertical dimension of the REV is typically smallest near the water-sediment interface, because this is the region where the gradients in chemical, physical and biological properties are steepest. With increasing depth in the sediment, the vertical length scale of the REV generally increases. The REV also varies significantly from sediment to sediment. In settings where macrofaunal burrows are spaced much further apart than in the coastal deposit shown in Figure 3, multiple coring may be required to accurately calibrate a 1-D model.

In theory, it is possible to solve the 2-D conservation equation of molecular oxygen for the detailed morphology of the water-sediment interface shown in Figure 3. This would dramatically increase the computational work, but it is doubtful that much additional information of practical value would be obtained. Furthermore, the water-sediment interface in Figure 3 is not static. Its structure is constantly being reworked by bottom currents, sedimentation and biological activity. The solution of the full, time-dependent solution of the 2-D oxygen field may well be mathematically and computationally intractable with conventional continuum approaches. The alternative, besides using a laterally-averaged 1-D model, is to consider an idealized, steady-state representation of the burrowed surface (see below).

Recipe for a multicomponent early diagenetic model

The development of a multicomponent reactive transport model for early diagenesis begins with the identification of the biogeochemical processes to be included. Each process must be represented by a formal reaction stoichiometry. This may seem straightforward, but it is not. The majority of chemical transformations in aquatic sediments proceed via complex pathways, during which many intermediate species are produced and consumed. As a guideline, one may include only those dissolved or solid-bound species that react or accumulate appreciably on the time scale of early diagenesis. In other words, both the highly unreactive and highly reactive species are excluded, because they do not directly affect the overall chemical dynamics of the system. Nonetheless, the selection of the reaction stoichiometries and, therefore, the reactive species remains somewhat arbitrary. In addition to the personal bias of the modeler, it reflects the level of chemical detail required in a particular model application.

Each reaction process is represented in the model by a rate expression or, for fast reactions, an equilibrium constant. Most enzymatic processes, redox transformations, and mineral dissolution-precipitation reactions taking place in aquatic sediments proceed at characteristic rates that are sufficiently slow, relative to the transport time scales, as to require a kinetic description. Equilibrium conditions can be used for sufficiently rapid homogeneous complexation or sorption reactions.

The mass conservation equations of the reactive species are obtained by merging the reaction rate expressions and equilibrium constants with expressions for the solute and solid-bound transport fluxes. Because the rate expression (or equilibrium constant) of a reaction appears in the conservation equations of the reactants and products, the distributions of the species are mathematically coupled to one another. Some degree of coupling among species may also derive from ionic diffusion (see below). With appropriate initial and boundary conditions, the coupled conservation equations can then be solved simultaneously, using a numerical discretization scheme.

TRANSPORT PROCESSES IN AQUATIC SEDIMENTS

Various modes of transport affect surface sediments after deposition, including advection, diffusion, dispersion and mixing. Furthermore, transport can be continuous or episodic in nature. The distinction between the possible modes of transport entails the averaging of properties over different time scales. For instance, macrofaunal activity may induce a more or less continuous mixing of the pore fluid and sediment particles, while a resuspension event may cause the near-instantaneous homogenization of the surface sediment. Nonetheless, it may be impossible to separate the two types of mixing when observing a sediment which, over the course of its history, has been perturbed by both processes. Based on the observations, it may only be possible to model the effects of the two processes with a single average mixing coefficient. This points to the fact that mathematical models of sediment processes may only be valid on temporal and spatial domains where the average properties are relevant. This imposes de facto some basic constraints on the time and spatial functions that can be used to describe transport.

In this section, the major transport processes that have been represented in early diagenetic models are described. Because in most sediments the movement of water relative to the solid matrix is very slow, hydrodynamic dispersion can be neglected. As a result, the diffusion and dispersion coefficients that appear in the continuity equations of solutes and solids are independent of the advective velocity. This does not hold for sediments within a flow field, such as river bed sediments or lake sediments that are subjected to groundwater infiltration. Fluid flow transport of this type is discussed elsewhere in this volume. Here we are primarily concerned with molecular or ionic diffusion of solutes in sediments, biological mixing of bulk sediment, and irrigation due to infaunal activity (e.g., burrow flushing).

Ionic and molecular diffusion

Molecular diffusion refers to the flux of matter caused by random thermal motion. It is a continuous process which is driven by the existence of a gradient of chemical potential which, usually, can be translated into a gradient of concentration. Molecular diffusion deals primarily with the migration of neutral species, whereas ionic diffusion refers to the migration of charged species which are interacting electrostatically. Even for neutral species, the migration of one species can influence that of others via short range interactions and reaction (Cussler, 1995). In pore solutions, neutral species are present at low concentrations. Hence, solute-solute interactions are weak and the effects of co-migrating species can be neglected. That is, the solvent-solute interactions determine the rate of migration of the solute species. In the case of electrolyte solutions, solute-solute interactions strongly couple the migration of species. For example, the diffusion of NaCl in water results in a strict co-diffusion of Na^+ and Cl^-, despite the differences in their intrinsic diffusion coefficients. In this case, a more complex formalism is needed, namely that of multicomponent diffusion.

Molecular diffusion: The Stokes-Einstein equation. Diffusion coefficients in liquids can be estimated with the Stokes-Einstein equation (Cussler, 1995). In this model, the migration of a solute species is described as the movement of a hard sphere through a continuum. Consequently, it is only expected to be a good approximation when the solute is much larger than the solvent molecules. Typically, diffusion coefficients in water estimated by the Stokes-Einstein equation are accurate to about 20%.

The Stokes-Einstein equation expresses that the net velocity of a hard sphere is proportional to the drag force exerted on it. Because the motion is sufficiently slow, the drag force can be calculated from Stokes law. By equating the force setting the particle into motion to the gradient in chemical potential, and by assuming that the solution is ideal, the solute's diffusion coefficient, D, is given by:

$$D = \frac{k_B T}{6\pi\eta R_0} \tag{2}$$

where k_B is Boltzmann's constant, T is the absolute temperature, η is the viscosity of the solvent and R_0 is the radius of the solute.

Although the Stokes-Einstein equation leads to poor estimates of diffusion coefficients, it provides some important insight into the diffusion process. From an inspection of Equation (2), it is clear that diffusion coefficients increase with temperature and decrease with increasing viscosity. Temperature and viscosity effects can be condensed in the simple relationship:

$$D_1\eta_1T_2 = D_2\eta_2T_1 \tag{3}$$

where the subscripts refer to two different temperatures. It is important to note that viscosity variations due to temperature have a greater effect on diffusivities than does temperature alone, though the distinction between the effects is here rather subtle. It is also expected that the larger the molar volume of the species, the smaller its diffusion coefficient.

In principle, the application of the Stokes-Einstein equation is restricted to infinitely dilute solutions. With increasing concentration of the solute, its diffusion coefficient will change. This can be attributed to the effect of the movement of other molecules, i.e. their hydrodynamic wakes, on the drag force experienced by a given molecule. Non-ideal effects are usually formalized by the Nernst-Hartley equation,

$$D = D_0\left(1+\frac{\partial \ln\gamma}{\partial \ln C}\right) \tag{4}$$

where D_0 is the diffusion equation at infinite dilution, and γ and C are the activity coefficient and concentration of the solute. This equation holds for neutral molecules and for a single salt where D_0 is the mutual diffusion coefficient of the salt.

Fortunately, to palliate the lack of precise theoretical predictions, diffusion coefficients in water of simple neutral species and ions have been measured with great precision by a variety of methods (e.g. Robinson and Stokes, 1959; Cussler, 1995). The use of radioactive tracers has also provided a powerful means to measure diffusion coefficients in complex media, including pore waters (e.g. Applin and Lasaga, 1984). In most instances where a species is present at very small concentrations, the diffusion coefficients obtained through different methods (e.g. limiting equivalent conductivities, radiotracer flux measurements) are almost identical. The coefficient value in this limiting case is referred to as the tracer diffusion coefficient.

Ionic diffusion: the multicomponent approach. The preceding discussion considered the diffusion of a single species. Natural pore waters, however, are solutions composed of a large number of different molecules and ions, with variable intrinsic mobilities. Therefore, we have to envision the collective movement of charged and neutral particles. Let us assume that the neutral species are little affected by the presence of other species and migrate according to their tracer diffusion coefficients. Focusing on the charged species, a formalism of multicomponent diffusion can be used to derive theoretical expressions for the coupled ionic diffusion coefficients. This approach, with slight variations, has been applied to the calculation of pore water fluxes by McDuff and Ellis (1979), Lasaga (1979), Gaillard (1982) and Simonin et al. (1989).

The simplest way to introduce the theoretical development is to consider an example: the diffusion of a single strong electrolyte (NaCl) in water. It is known that the intrinsic diffusion coefficient is smaller for Na^+ than for Cl^-. Therefore, the Cl^- ions have a tendency to move faster than Na^+ ions. This creates, at the molecular scale, a charge separation. In return, this charge separation generates an electrical potential whose effect is to speed up the slower ions and slow down the faster ones. The result, at the macroscopic scale, is that the Na^+ and Cl^- ions migrate at the same velocity. The corresponding diffusion coefficient is the mutual diffusion coefficient of the salt.

The potential created by the intrinsic difference in ionic mobility, or diffusion potential, plays a central role in determining the rates of diffusion of ions in electrolyte solutions. To account for its effect, one needs to depict the electrostatic interactions between the ions, and use the rather cumbersome method of multicomponent diffusion. In some circumstances, this formalism may be of limited value because the coupling effects arising from the electrostatic interactions are minor. This is the case for the diffusion of most ions in marine sediments, where the presence of a large excess of background electrolyte, i.e. NaCl, cancels out the effects of small diffusion potentials. In freshwater sediments, however, multicomponent diffusion cannot so easily be dismissed.

In this section, a general equation for the diffusive transport of electrolytes is presented. Then, from the general expression, we show that simpler diffusion equations can be obtained when dealing with solutions with a rather constant background electrolyte. The derivations follow those in Gaillard (1982).

Because diffusion is a process occurring close to equilibrium, it is legitimate to use thermodynamic functions to calculate the forces acting on ions. One can even go further and, following Einstein's assumption, derive an expression for the force exerted on a single ion or molecule. This force should take into account the chemical forces, that is, the gradient in chemical potential, and the electrical forces, under the form of a diffusion potential:

$$F_i = -\nabla\mu_i + z_i eE \tag{5}$$

where μ_i is the chemical potential of species i, zi is the charge of the ion, and E is the diffusion potential created by the movement of all the ions.

Assuming that the ion is a hard sphere moving in the solvent, the balance of the force given by Equation (5) and the drag force exerted on the sphere leads to a net velocity given by Stokes law. According to the linear transport theory, one can relate the flux J_i of the species to its mobility u_i^0, and its concentration C_i, via

$$J_i = u_i^0 C_i F_i \tag{6}$$

Hence,

$$J_i = u_i^0 C_i(-\nabla\mu_i + z_i eE) \tag{7}$$

Because the aqueous solution must remain electroneutral at the macroscopic scale, the net flux of charges must be zero:

$$\sum_i^n z_i J_i = 0 \tag{8}$$

where the sum is taken over all ionic species. By combining Equations (7) and (8), we can calculate the diffusion potential:

$$E = \frac{\sum_k^n z_k u_k^0 C_k \nabla\mu_k}{e\left(\sum_j^n z_j^2 u_j^0 C_j\right)} \tag{9}$$

and, hence,

$$J_i = -u_i^0 C_i \nabla\mu_i + u_i^0 z_i C_i \frac{\sum_k^n z_k u_k^0 C_k \nabla\mu_k}{\sum_j^n z_j^2 u_j^0 C_j} \tag{10}$$

In order to derive an expression for the flux in terms of concentrations and concentration gradients, the chemical potential is expanded to:

$$\mu_i = \mu_i^0 + k_B T \ln(\gamma_i C_i) \tag{11}$$

where γ_i is the activity coefficient of species i. Differentiating Equation (11) gives

$$\nabla\mu_i = k_B T \nabla \ln\gamma_i + k_B T \frac{\nabla C_i}{C_i} \tag{12}$$

Combining Equations (10), (12), and the Nernst-Einstein relationship for the tracer diffusion coefficient $D_i^0 = k_B T u_i^0$, we arrive at the following general expression for the flux of species i,

$$J_i = -D_i^0 \left(\nabla C_i + C_i \nabla \ln\gamma_i - \frac{z_i C_i}{\sum_j^n z_j^2 C_j D_j^0} \left(\sum_k^n z_k D_k^0 (C_k \nabla \ln\gamma_k + \nabla C_k) \right) \right) \tag{13}$$

Activity coefficients can be estimated using various expressions, from the Debye-Hückel limiting law in dilute solutions to Pitzer's equations in concentrated brines. Here, in order to keep the expressions simple, an extended Debye-Hückel law is used to approximate departure from non-ideal behavior (e.g. Whitfield, 1973):

$$\ln\gamma_i = \frac{-A z_i^2 \sqrt{I}}{1 + b\sqrt{I}} + \sum_j^n \beta_{ij} C_j \tag{14}$$

where A and b are solvent parameters, β_{ij} are virial coefficients that account for specific interactions between the ions, and I, the ionic strength, is given by,

$$I = \frac{1}{2} \sum_j^n z_j^2 C_j \tag{15}$$

Equation (14) can be applied up to an ionic strength of about 1. After some algebraic manipulation, one obtains

$$\nabla \ln\gamma_i = \frac{-A z_i^2}{2\sqrt{I}\left(1 + b\sqrt{I}\right)^2} \sum_j^n z_j^2 \nabla C_j + \sum_j^n \beta_{ij} \nabla C_j \tag{16}$$

This expression can then be incorporated into the flux equation (Eqn. 13).

The flux of one ionic species under the coupling of all other diffusing species can be expressed under the form of Fick's law:

$$J_i = -\sum_j^n D_{ij} \nabla C_j \tag{17}$$

By identifying Equations (13) and (17), we obtain

$$D_{ij} = D_i^0 \left(\delta_{ij} + C_i \lambda_{ij} - \frac{z_i z_j C_i D_j^0}{\sum_k^n z_k^2 C_k D_k^0} - z_i C_i \lambda_{ij} \frac{\sum_k^n z_k C_k D_k^0}{\sum_k^n z_k^2 C_k D_k^0} \right) \tag{18}$$

where δ_{ij} is the kroenecker delta, and

$$\lambda_{ij} = \frac{-z_i^2 z_j^2 A}{2\sqrt{I}\left(1 + b\sqrt{I}\right)^2} + \beta_{ij} \tag{19}$$

Hence, in the case of a multicomponent solution of n species, one obtains the general transport equation, or Fick's second law:

$$\frac{\partial C}{\partial t} = \underline{D} \nabla^2 C \tag{20}$$

where **C** is the vector of the n species concentrations, and $\underline{D}$ is the matrix with elements D_{ij}.

It should be noted that, in the above treatment, each diffusing species, i.e. each individual chemical entity, needs to be considered, including free ions, ion-pairs and complexes. In pore waters where only a small number of species may be of interest, the multicomponent approach therefore renders the task of quantifying solute diffusion quite complex. The modeler will have to evaluate the cost/benefit of using the approach, depending on the objectives of her (his) study. Fortunately, under the right conditions, the full multicomponent description can be reduced to the simpler case of one-component diffusion.

Limiting cases. An examination of Equation (18) reveals that the D_{ij} values are functions of C_i. In the limit that the concentration C_i becomes vanishingly small, the only important term left in the diffusion matrix is $D_{ii} = D_i^0$. Consequently, the flux of a trace species is simply given by:

$$J_i = -D_i^0 \nabla C_i \tag{21}$$

Equation (21) also applies to ionic species present at higher concentrations, but diffusing in a constant background electrolyte. This is the case for most ions in seawater. Essentially, the uniform background electrolyte screens out the effects of the diffusion potential. In the absence of concentration gradients of the major ions, all terms on the right-hand-side of Equation (13), except $-D_i^0 \nabla C_i$, become very small. Therefore, little departure from tracer diffusion will arise from multicomponent diffusion and it is legitimate to use a simple Fickian formulation for the flux.

Multicomponent diffusion has been largely neglected in early diagenetic models of marine sediments. One case where this may not be valid is in sediments where dissolved sulfate, a major seawater anion, is depleted with depth due to bacterial sulfate reduction. The pore water sulfate gradient may then significantly affect the migration of other pore solutes. The importance of multicomponent diffusion and the coupling effects of ion-pairing in the early diagenesis of marine sediments has been emphasized by Lasaga (1979, 1981a).

In freshwater sediments, cross-coupling effects are nearly always important, because there is no background electrolyte, and because the early diagenetic reactions cause major changes in the chemical composition of the pore waters. These effects are magnified at the water-sediment interface, where the pore water compositional gradients are largest (e.g.

Gaillard, 1982; Carignan and Lean, 1991). Moreover, dissolved ions that are minor in the overlying waters may become major constituents in the pore waters, for example Fe^{2+} and NH_4^+. The diffusion matrices of these species may therefore change dramatically over short distances across the water-sediment interface. However, the consequences of multicomponent diffusion for reactive transport in freshwater sediments have yet to be investigated in a systematic way.

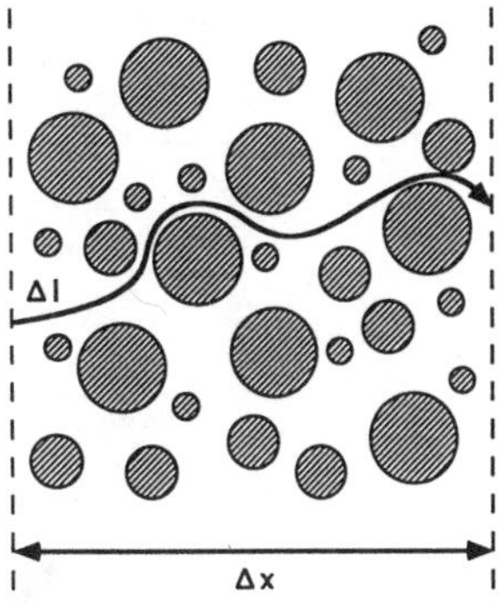

Figure 4. Convolute diffusion pathways in a sediment result in a reduction of the effective diffusion coefficient (see text).

Porosity and tortuosity. In a sediment, or for that matter any porous system, the presence of solid particles causes the diffusion paths of solutes to deviate from straight lines (Fig. 4). Consequently, the diffusion coefficients of solutes must be corrected for the tortuosity. The latter is defined as:

$$\theta = \frac{\Delta l}{\Delta x} \tag{22}$$

where Δl is the actual distance traveled by the solute per unit length Δx of the medium. Clearly, tortuosity must be ≥ 1.

Because, at the microscopic level, the diffusion process in a condensed phase can be visualized as a succession of elementary jump events, with a characteristic jump distance λ, the diffusion coefficient of a solute can be interpreted as (Lasaga, 1981b):

$$D = \Gamma \lambda^2 \tag{23}$$

where Γ is the jump frequency (in units of inverse time). From Equations (22) and (23), the following relationship between the diffusion coefficient in a porous medium and that in free solution can then be derived,

$$D_s = D\frac{(\Delta x)^2}{(\Delta l)^2} = \frac{D}{\theta^2} \tag{24}$$

where the subscript s refers to the coefficient in the porous medium, in this case the sediment. It can easily be seen that $D_s \leq D$.

Experimentally, the tortuosity of a sediment can be obtained by measuring the porosity, ϕ, and formation resistivity factor, F. The three quantities are related by (McDuff and Ellis, 1979):

$$\theta^2 = \phi F \tag{25}$$

The formation factor is determined by measuring the resistivity of the porous medium, relative to that of the free solution. Microelectrode technology currently allows one to obtain fine-scale profiles of the formation factor in surface sediments (Fig. 5). Frequently, however, this information is missing and we must then rely on empirical relationships that relate the formation factor to the sediment porosity.

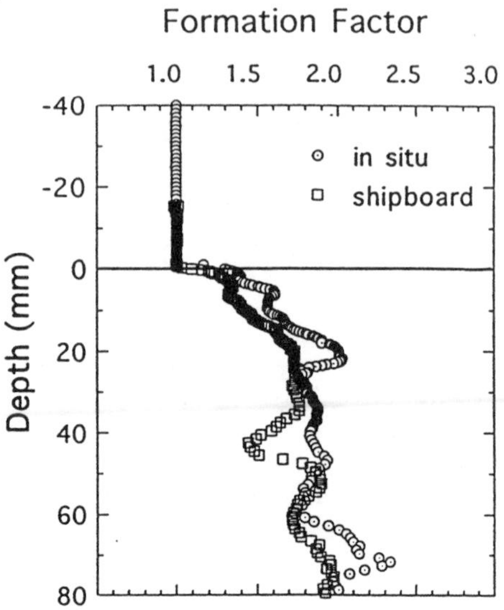

Figure 5. In situ and shipboard microprofiles of formation factor determined with a resistivity microelectrode. The position of the water-sediment interface is determined precisely from the rapid change in the formation factor. Differences between in situ and shipboard measurements may reflect disturbance during core recovery (from Cai et al., 1995).

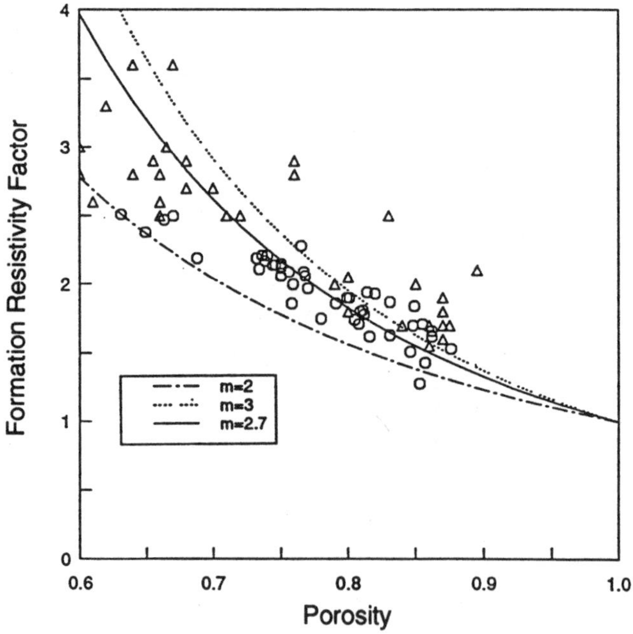

Figure 6. Functional dependence of resistivity forma-tion factor on sediment porosity. The lines represent Archie's law (Eqn. 26) for different values of the coefficient m (data compiled by Tromp, 1992).

As expected, the formation resitivity factor increases with decreasing porosity (Fig. 6), a relationship which is frequently described by Archie's law (Archie, 1942):

$$F = \phi^{-m} \tag{26}$$

where m is an empirical constant applicable to a given sediment type. For unconsolidated, porous (f ≥ 0.6) muds, an average value of m = 2.7±0.3 has been proposed (Andrews and Bennett, 1981; Ullman and Aller, 1982; Tromp, 1992). This suggests the following correction for diffusion coefficients in aquatic surface sediments:

$$D_s = D\phi^{1.7\pm0.3} \tag{27}$$

Boudreau (1996) recently reviewed a number of empirical tortuosity-porosity relationships. His analysis of the available data favors the following equation:

$$D_s = \frac{D}{1-\ln\left(\phi^2\right)} \tag{28}$$

For a typical porosity of a surficial deposit of $\phi = 0.85$, Equations (27) and (28) both predict a decrease in diffusivity due to tortuosity of about 25%.

An additional reduction of diffusion fluxes in porous media results from the partial obstruction of a cross-sectional area by solid particles. Based on a dimensional analysis, it can be shown that the correct form of Fick's first law for one-component diffusion through a sediment is given by (Berner, 1980):

$$J = \phi D_s \nabla C \tag{29}$$

where J is the flux expressed per unit total surface area of sediment. Thus, for a given gradient in concentration, the diffusion flux in a sediment of porosity $\phi = 0.85$ is about 36% smaller than in homogeneous solution.

Biological mixing

Bioturbation refers to the reworking of sediment by the activity of infaunal macro-invertebrates (for an excellent review, see, Matisoff, 1995). Given the wide variety of organisms (polychaetes, crustaceans, mollusks, etc.) and types of activities (burrowing, selective feeding, excretion, etc.) involved, one may expect bioturbation to be an extremely complex process. The simplest, and most popular, approach to deal with this complexity is to assume that, when considered over a sufficient period of time, transport resulting from bioturbation can be described as a random process. In a one-dimensional early diagenetic model, the intensity of sediment mixing can then be quantified by a so-called biodiffusion coefficient, D_b. Under these conditions, the governing 1-D equation for a solid constituent is (Berner, 1980),

$$\frac{\partial(1-\phi)C_i}{\partial t} = -\frac{\partial}{\partial x}\left(-D_b\frac{\partial(1-\phi)C_i}{\partial x} + \omega(1-\phi)C_i\right) + (1-\phi)\sum_j R_{ij} \tag{30}$$

where C_i is the concentration of the constituent per unit volume dry sediment, ω is the sediment advection rate (units of length total sediment per unit time), and R_{ij} is the rate of production of the constituent by reaction j.

Equation (30) has been fairly successful at reproducing the depth profiles of radionuclides in sediments. In fact, fitting of Equation (30) to radionuclide distributions has been the main source of numerical values for D_b found in the literature (for a compilation of D_b values in marine sediments, see Tromp et al., 1995). The principal controls on the intensity of biological mixing appear to be (1) the magnitude of the food supply (i.e. the deposition flux of organic detritus), (2) the bottom water oxygenation, and (3) the supply rate of detrital matter. Sediments that receive a higher flux of organic matter are capable of sustaining larger populations of macrofauna and, consequently, exhibit higher values of D_b (Van Cappellen et al., 1993). The depth and intensity of bioturbation, however, are also affected by the dissolved oxygen content of the bottom waters. Benthic macrofaunal activity slows down when the oxygen level at the water-sediment interface declines below 2 ml l^{-1}, and it virtually ceases at levels ≤ 0.2 ml l^{-1} (e.g. Rhoads and Morse, 1971; Baden et al., 1990; Tyson and Pearson, 1991). Bioturbation may also be limited by very high rates of detrital deposition which interfere with the establishment of a stable population of benthic infauna.

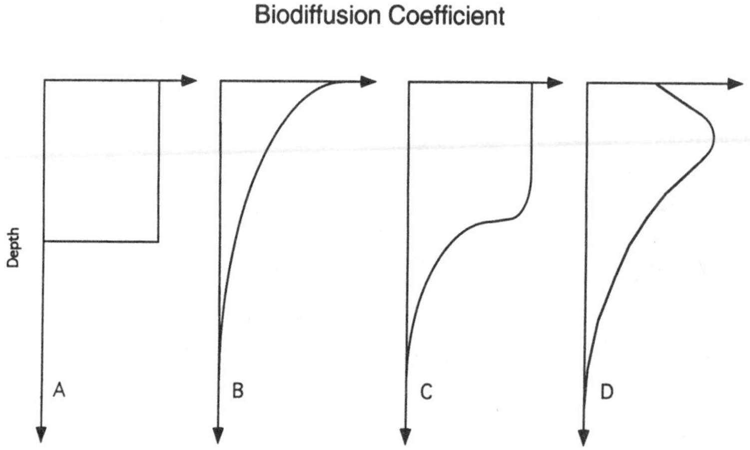

Figure 7. Theoretical depth distributions of the biodiffusion coefficient. (A) uniformly bioturbated layer overlying non-bioturbated sediment; (B) exponentially decreasing biodiffusivity; (C) gradual transition between uniformly bioturbated surface layer and non-bioturbated sediment; (D) sub-surface maximum of mixing intensity.

The main difference among biodiffusional models is the depth distribution assigned to D_b. The simplest model consists of a bioturbated layer characterized by a constant value of D_b, overlying a non-bioturbated layer (e.g. Guinasso and Schink, 1975). This and other vertical profiles of D_b that have been used in the literature are shown in Figure 7. One way to rationalize these distributions is to interpret the net biodiffusion coefficient as the product of an intrinsic biodiffusivity per average organism and the number density of organisms (e.g. Matisoff, 1982). The shape of the vertical profile of D_b should then parallel that of the vertical distribution of macrofaunal abundance. Unfortunately, data on population density of macrofauna in sediments are rather scarce. Furthermore, in order to obtain representative animal counts, the vertical resolution must be fairly coarse, as illustrated by the data in Figure 8. Even a visual inspection of the data reveals that at least three of the theoretical D_b profiles shown in Figure 7 (A, C, and D) would provide reasonable fits to the measured distribution of infaunal abundance (with perhaps a slight advantage for the D-type profile). Thus, it may be difficult to uniquely define the fine-scale distribution of D_b. Partly, this reflects the fact that the representation of biological mixing by a biodiffusion coefficient is only valid at rather large spatial and temporal scales.

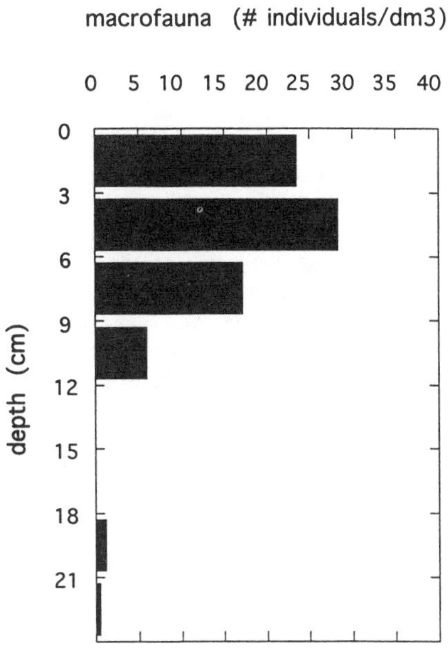

Figure 8. Depth distribution of average number density of macrobenthos in a Mediterranean sediment (1000 m water depth) (data from Gerino et al., 1993).

We can get a rough idea of the representative elementary volume (REV) for the biodiffusional model by considering the magnitude of the sampling volume which gives a statistically reproducible measurement of the density of the organisms responsible for mixing. Because of the sizes and abundances of the organisms, core slices of one to several centimeters thick are usually needed (Fig. 8). Therefore, we should not expect the biodiffusional model to provide a good description of biological transport at a vertical scale much smaller than 1 cm. With an estimated biodiffusion coefficient of 1 $cm^2\ yr^{-1}$ at the deep-sea setting of Figure 8, Einstein's equation, $x = (2D_bt)^{1/2}$, then predicts that the biodiffusion coefficient averages the mixing intensity over approximately one year at this site. The spatial and temporal scales over which biodiffusion averages biologically-induced transport will vary from site to site, depending on the types and abundances of organisms.

The large REV of the biodiffusion approach contrasts with the much smaller REV estimate based on the dissolved oxygen distribution discussed in an earlier section. Again, it is important to recognize the implications of this disparity in scales for model validation. Suppose we construct a multicomponent model that includes both pore water and solid sediment constituents, and apply it to sediments where solute transport is dominated by molecular diffusion. By including bioturbation under the form of a biodiffusion coefficient, we automatically build into the model an inherent difference in the spatial and temporal resolution with respect to solutes and solids. Hence, the ability of the model to reproduce observed profiles may be limited by the coarser resolution with which transport of the solids is simulated. This is particularly true when the profiles record processes occurring at time scales shorter than that for which the biodiffusion model is valid.

In addition to the limitations of the biodiffusion model noted above, the activities of certain infaunal organisms lead to transport that is inherently non-diffusive (Boudreau, 1986; Wheatcroft et al., 1990). Certain species of worms, for example, ingest sediment at depth in the sediment and redeposit it, after intestinal processing, at the water-sediment interface (e.g. Robbins et al., 1979; McCall and Fisher, 1980). This induces a non-local, "conveyor belt" transport that cannot be represented as a diffusional flux. An added complication arises from the fact that deposit feeders often preferentially ingest certain particles, hence creating different rates of reworking depending on particle type (e.g. Jahnke et al., 1986).

Boudreau (1986) and Boudreau and Imboden (1987) have presented a theoretical model to describe generalized, non-local transport in sediments. In the model, mixing at all scales is handled by exchange functions. The general exchange function $K(x_i, x_j, t)$ describes the time-dependent rate of transfer of sedimentary material from depth x_i to depth x_j. The application of this approach requires one to define the form of the exchange function at the site studied. Boudreau and Imboden have investigated the theoretical tracer profiles produced by a number of simple exchange functions, as well as a function describing burrow excavation and infilling. Controlled tracer experiments will be needed, however, to constrain the forms of exchange functions that apply to natural sediments. Meanwhile, an important function of the general mixing model of Boudreau and Imboden is that it provides a framework in which simpler models, such as the biodiffusion model, can be analyzed.

The models of biological mixing in sediments discussed so far are of a deterministic nature. However, given the partially random nature of macrofaunal activity, stochastic models that rely on the realization of a random function should provide a viable alternative. Notwithstanding a few studies (e.g. Jumars et al., 1981), this approach has yet to be fully explored.

Irrigation

The transfer flux of a solute across the water-sediment interface can be determined in two ways. First, one can measure the vertical pore water concentration profile of the solute and calculate the benthic flux, assuming one-component diffusion, with

$$J_0 = -\phi D_s \left.\frac{\partial C}{\partial x}\right|_{x=0} \tag{31}$$

where the parameter values and gradient on the right-hand side are evaluated at the water-sediment interface (x = 0). Second, one can place an enclosure on the sediment, a so-called benthic chamber, and monitor the build-up, or depletion, of the solute's concentration inside the chamber. From the known dimensions of the chamber, one can then convert the initial rate of change of the concentration into a flux in or out of the sediment (for technical details, see Jahnke and Christiansen, 1989).

When sediments are inhabited by macrofauna, a comparison of the two methods invariably yields lower estimates for the diffusional fluxes estimated with Equation (31) (Matisoff, 1995, and references therein). Given that the benthic chamber estimates are closer to the true fluxes, this indicates that solute transport mechanisms other than molecular diffusion are active. These mechanisms are collectively referred to as irrigation and are the result of infaunal activity (Aller, 1982). Burrows increase the surface area through which sediment and overlying water can exchange (Fig. 3), while tube-flushing actively injects overlying water at depth in the sediment. It should be noted that in certain environments enhanced solute transport can also be caused by abiotic mechanisms, e.g. wave-pumping in shallow waters.

One approach to modeling the effects of bioirrigation on pore water composition and benthic exchanges is to use an idealized representation of the burrowed water-sediment interface. Aller (1980) developed such a model for the case of permanent burrow dwellings

(Fig. 9). In this model, the burrows are vertical cylinders of fixed diameter. They are further assumed to be uniformly distributed and perfectly flushed, that is, the water composition along the burrow walls is equal to that of the overlying water. Taking advantage of the cylindrical symmetry, the three-dimensional conservation equation for the concentration of a dissolved species around a central burrow can be written as:

$$\frac{\partial \phi C_i}{\partial t} = -\frac{1}{r}\frac{\partial}{\partial r}\left(-D_s\phi\frac{\partial C_i}{\partial r}\right) - \frac{\partial}{\partial x}\left(-D_s\phi\frac{\partial C_i}{\partial x}\right) + \phi\sum_j R_{ij} \tag{32}$$

where the concentration C_i is in units of mass per unit volume solution, and r is the radial distance from the axis of the central burrow (Fig. 9). Equation (32) assumes that the sediment advection term can be neglected, and that diffusion can be approximated by Fick's law for one-component diffusion (Eqn. 29).

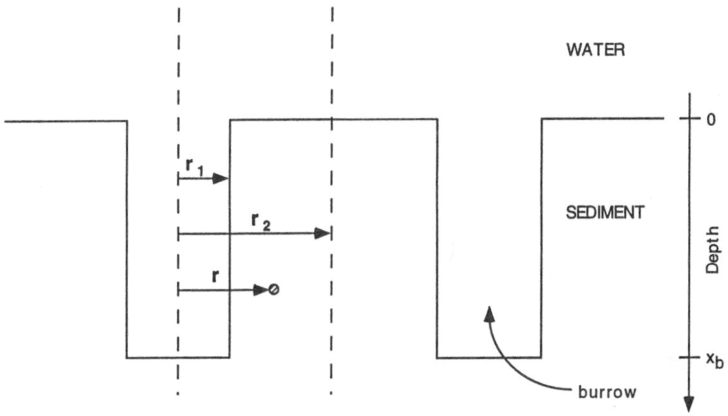

Figure 9. Idealized representation of burrowed water-sediment interface (compare with Fig. 3). The geometry of the interface is completely determined by the quantities r_1, r_2 and x_b (after Aller, 1980).

The concentrations predicted by Equation (32) can be compared to measured fine-scale concentration fields such as the one shown in Figure 3. More commonly, however, pore waters are extracted from whole core slices. The measured concentrations must then be compared to the theoretical values averaged spatially over the burrowed sediment (Aller, 1980). The radial diffusion model has been tested in microcosm studies with variable (but known) densities of burrowing animals (Matisoff, 1995). It usually performs well in the surficial irrigated zone of the sediment but, because of poor coupling to the underlying non-bioturbated zone, not as well at greater depths.

In principle, bioirrigation could be described by a general exchange function, similar to the one discussed in the context of biologically-enhanced particle transport. So far, however, the non-local models for irrigation that have been presented in the literature have been highly simplified. Emerson et al. (1984), for example, used a steady state non-local model where the intensity of water flushing is quantified by an irrigation exchange coefficient, α. The rate of addition/removal of a solute at depth x in the sediment, S(x), is then given by,

$$S(x) = \alpha(x)\left(C_0 - C(x)\right) \tag{33}$$

where C(x) is the concentration at depth x, C_0 is the concentration in the overlying water, and the exchange coefficient is expressed in units of inverse time.

Martin and Banta (1992) have used tracer distributions of pore water bromide and $^{222}Rn/^{226}Rd$ disequilibrium to constrain the depth distributions of α(x) in sediments of Buzzards Bay (Massachusetts, USA). Their results show generally decreasing trends of the irrigation intensity with depth. In some instances their data could be fitted to a simple exponential decay:

$$\alpha(x) = \alpha_0 \exp(-\alpha_1 x) \tag{34}$$

where α_0 and α_1 are positive constants. Wang and Van Cappellen (1996) have successfully used Equation (34) to reproduce the effects of irrigation on the pore water compositions of two nearshore marine sediments off the coast of Denmark. At a third site, however, a distribution of α with a subsurface maximum resulted in a better fit to the data.

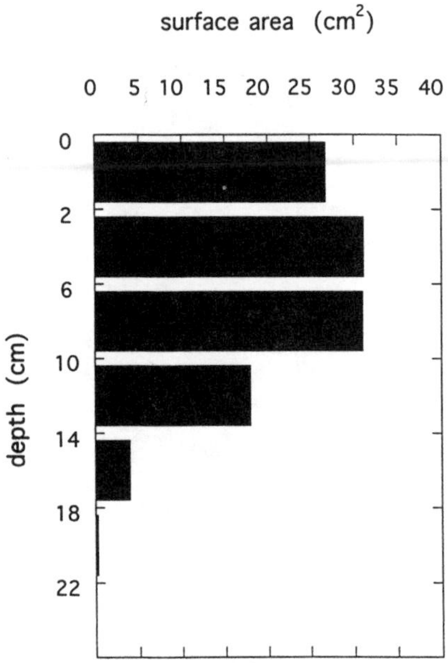

Figure 10. Depth distribution of the mean surface area of burrow walls. The surface areas were calculated from resin casts of burrows constructed by polychaetes in a controlled in vitro experiment. The casts were digitized to generate computer images from which the surface areas were determined. In this particular experiment, burrowing of the sediment increased the total water-sediment interface area by at least a factor of 3.2 (data from Gerino and Stora, 1991).

As with sediment reworking, more experimental data are needed to define the exchange functions for irrigation in natural sediments. Tracer experiments represent the traditional method for calibrating transport in geochemical models. Other approaches, however, may also provide useful information. For example, Figure 10 shows the depth distribution of the surface area of polychaete burrows (Nereis diversicolor) determined from cast images in an in vitro experiment (Gerino and Stora, 1991). Because, in principle, the intensity of exchange should be proportional to the amount of surface area of burrow walls, the data in Figure 10 suggest that, at least in this particular setting, Equation (34) may not be the best representation of the distribution of the irrigation exchange coefficient.

CHEMICAL PROCESSES IN AQUATIC SEDIMENTS

Many, if not most, chemical changes in aquatic sediments are the result of microbial metabolism. The degradation of organic matter by heterotrophic microorganisms is responsible for the establishment of the vertical zonation of redox conditions (Froelich et al., 1979). The respiratory and fermentative pathways of organic matter degradation are extremely complex and involve consortia of many different organisms (Laanbroek and Veldkamp, 1982). The concentrations of some of the important intermediate species (e.g. acetate, molecular hydrogen) have been measured in a number of studies (e.g. Novelli et al., 1988). Nonetheless, most early diagenetic models focus on the initial reactants (organic matter and oxidants) and the final products (inorganic nutrients and reduced metabolites), and therefore rely on overall reaction stoichiometries (Berner, 1980; Gaillard and Rabouille, 1992; Van Cappellen et al., 1993).

Chemically, the degradation of organic matter corresponds to a redox transformation. Organic carbon is oxidized while an oxidant is reduced. New chemical entities, including reduced pore water, adsorbed and solid species, are produced. Transport of these species along the redox gradient of the sediment will result in secondary redox reactions, many of which are also carried out by microorganisms. Chemolithotrophic bacteria enzymatically oxidize reduced sulfur, nitrogen, iron and manganese species for their energy supply (Brock et al., 1994). Benthic populations of nitrifying bacteria, for example, are responsible for the transformation of ammonia to nitrate, in the presence of oxygen (Henricksen and Kemp, 1988). In other cases, both microbial and chemical pathways are possible, as in the oxidation of Fe(II) and Mn(II) species (e.g. Emerson et al., 1982; Kepkay et al., 1984; Morgan et al., 1985; Moses et al., 1986).

In this section, we discuss rate expressions that can be used to represent chemical transformations in sediments, with an emphasis on microbially-mediated reactions. Despite our incomplete understanding of the mechanisms of many of the processes, it is nonetheless possible to formulate reasonable rate expressions based on fundamental principles of microbial and chemical kinetics.

Kinetics of organic matter degradation

Most kinetic treatments of organic matter degradation that have been used in early diagenetic models are based on the first-order kinetic (one-G) model introduced by Berner (1964). In this model the rate of organic carbon decomposition is given by,

$$R_C = -\frac{d[CH_2O]}{dt} = k_C[CH_2O]_m \tag{35}$$

where R_C corresponds to the net rate of organic carbon degradation, k_C is a first-order rate constant, $[CH_2O]$ is the concentration of particulate organic carbon, and the subscript m refers to the metabolizable fraction of the organic carbon.

The use of Equation (35) assumes the existence of two fractions of particulate organic matter. The metabolizable fraction is potentially degradable while the remainder is resistant to microbial attack, on the time scale of interest. The logical extension of this model is to break the metabolizable fraction down into subclasses of different reactivities (multi-G model) (Jørgensen, 1978). Based on a kinetic laboratory study, Westrich and Berner (1984) have argued for the existence in nearshore sediments of two distinct metabolizable fractions with first-order degradation constants that differ by nearly an order of magnitude.

As an alternative to multi-G models, one can adopt a continuum description for the reactivity of organic matter (Middelburg, 1989; Boudreau and Ruddick, 1991; Van Cappellen et al., 1993). The rate coefficient k_C in Equation (35) then progressively decreases with advancing diagenesis. One advantage of continuum models is that they no longer require a separation of the total particulate organic carbon into metabolizable and non-metabolizable fractions. The rate of organic matter degradation comes to a halt when the value of of the rate coefficient drops below a critical value. The remaining amount of organic carbon is then essentially undegradable. It can be shown that the critical value of k_C, beyond which further degradation becomes negligible, depends on the magnitudes of the transport parameters and the depth/time scale of interest (Van Cappellen et al., 1993).

The first-order rate dependence of the rate of organic carbon degradation is consistent with a degradation pathway in which the rate-limiting step is the initial hydrolysis of the particulate organic matter by extracellular enzymes (e.g. Gujer and Zehnder, 1983). The first step results in smaller, dissolved molecules that can be taken up by microorganisms for further intracellular processing. According to this model, the rate coefficient k_C is primarily a measure of the reactivity of the particulate organic matter toward extracellular enzymatic hydrolysis. It further assumes that the microbial populations rapidly adjust their production of extracellular enzymes to changes in availability of degradable organic matter. The rapid response of microbial activity to substrate availability explains why the microbial biomass, or the number of active microorganisms, do not appear explicitly in the rate expression (Van Cappellen et al., 1993).

Equation (35) and related expressions consider the decomposition of organic matter as a whole. In principle, however, it is possible to develop kinetic models for the degradation of individual groups of organic compounds. Such models are likely to become more common, as we gain a better understanding of the structure and degradation pathways of the different classes of molecules that make up the organic matter in sediments. A recent study by Arnosti et al. (1994) on the mineralization of polysaccharides in marine sediments suggests the following reaction scheme,

particulate organic matter ⟶ HMW dissolved polymers

⟶ LMW dissolved monomers ⟶ inorganic carbon

(HMW: high molecular weight, LMW: low molecular weight). Their experimental results show that the intermediate products may accumulate to fairly high levels, which is contrary to a simple rate-limitation by the initial hydrolysis of the particulate matter. This suggests that more realistic kinetic descriptions of the decomposition of organic matter could be developed by including rate expressions for the key intermediate steps in the mineralization process.

Monod kinetics

The rate of oxidation of organic matter by aerobic bacteria is independent of the concentration of dissolved molecular oxygen, as long as the latter remains higher than a critical level, typically around 3 to 10 µM (Devol, 1975; Nielsen et al., 1990). Below the critical level, the rate of aerobic respiration decreases with decreasing O_2 concentration, and vanishes when no O_2 is left. This behavior is characteristic of the utilization of a substrate by a microbial population. It is quantitatively described by the so-called Monod law. According to this formulation, the limitation of aerobic respiration by O_2 obeys,

$$R_{O_2} = R_{max} \frac{[O_2]}{K_{O_2} + [O_2]} \qquad (36)$$

where R_{max} is the maximum rate (i.e. when the supply of the limiting substrate is unrestricted), $[O_2]$ is the dissolved oxygen concentration, and K_{O_2} is the half-saturation concentration, or Monod constant.

Equation (36) is formally equivalent to the Michaelis-Menten rate equation for enzymatic reactions. A straightforward derivation of the Michaelis-Menten equation involves the calculation of the quasi-steady state rate of reaction of a substrate catalyzed by one enzyme, in a two-step process (Laidler, 1987). However, many more complicated enzyme-catalyzed reactions obey the Michaelis-Menten equation, indicating that adherence to the empirical rate Equation (36) does not imply a simple mechanism with only one intermediate substrate-enzyme complex (Laidler, 1987). In other words, Equation (36) provides a robust expression for representing the overall rate of consumption of a substrate in a complex microbial process, even when detailed information about the reaction pathway(s) is missing.

According to Equation (36), when $[O_2] << K_{O_2}$, the availability of oxygen is limiting and the rate of aerobic respiration exhibits a first-order dependence on $[O_2]$. When, on the other hand, $[O_2] >> K_{O_2}$, the aerobic bacteria are functioning at maximum capacity, that is, they cannot produce more enzymes to utilize the excess O_2. Oxygen saturation occurs, because the maximum level of enzymatic activity is set by the availability of the organic substrates that serve as the energy source to the bacteria. Under normal circumstances, saturation with respect to the organic substrates is never reached, because the microbial populations can always respond to an increase in the supply of organic substrates by producing more enzymes. The fact that the availability of energy substrates is always limiting also justifies the use of the first-order rate expression for organic carbon degradation, Equation (35). (Note: this is no longer the case when an external limiting agent, e.g. a toxic substance, is introduced in the system.) Furthermore, we can identify R_{max} in Equation (36) with the rate of organic carbon oxidation when O_2 is saturating. Hence,

$$R_{O_2} = k_{O_2}[CH_2O]\frac{[O_2]}{K_{O_2} + [O_2]} \tag{37}$$

where k_{O_2} is the first-order rate coefficient of carbon oxidation by aerobic respiration. Note that we dropped the subscript m from the concentration of particulate organic matter, $[CH_2O]$. Thus, all changes in the reactivity of the organic matter are incorporated into the variable rate coefficient k_{O_2}.

In other microbial reactions, saturation with both reactants or substrates may occur. Rate expressions for two-substrate enzymatic reactions have been derived for a number of different reaction mechanisms. These expressions can usually be reduced to the form (e.g. Laidler, 1987):

$$R_{AB} = \frac{k[A][B]}{K + K_A[A] + K_B[B] + [A][B]} \tag{38}$$

where k, K, K_A and K_B are constants. When there is a large excess of one of the substrates, the simple Michaelis-Menten expression is recovered. For example, when $[B] >> [A]$, then,

$$R_{AB} = \frac{k[A]}{K_B + [A]} \tag{39}$$

Another interesting limiting case of Equation (38) occurs when both substrates are present at low concentrations. The expression then approaches the bimolecular rate law:

$$R_{AB} = k^*[A][B] \tag{40}$$

where $k^* = k/K$.

Inhibition and competition

As long as molecular oxygen is present at sufficient levels in the pore waters of a sediment, all organic matter oxidation is carried out via aerobic respiration. Only when nearly all O_2 has been reduced will dentritifying bacteria be able to start oxidizing the particulate organic matter that remains (Nielsen et al., 1990). Oxygen thus inhibits denitrification, possibly because it blocks nitrate reductase, the enzyme responsible for the reduction of nitrate to the intermediate product nitrite (Brock et al., 1994).

A variety of mathematical equations have been proposed to describe the inhibition of a metabolic pathway (Humphrey, 1972). One way to express the inhibition of denitrification by O_2 is (Van Cappellen et al., 1993),

$$R_{NO_3} = k_{NO_3}[CH_2O]\frac{[NO_3^-]}{K_{NO_3}+[NO_3^-]}\frac{K_{O_2}^{in}}{K_{O_2}^{in}+[O_2]} \tag{41}$$

where K_{NO_3} is the half-saturation nitrate concentration, $K_{O_2}^{in}$ the constant of inhibition due to oxygen, and k_{NO_3} the first-order rate coefficient for denitrification. Note that Equation (41) also accounts for limitation of denitrification by nitrate availability.

An inspection of Equation (41) reveals that inhibition of denitrification by oxygen ceases only when $[O_2]$ drops significantly below the value of the inhibition constant. Obviously, a given microbial reaction may be acted upon by more than one inhibitor. In that case, the combined effect of all inhibitors can be described by a product of terms similar to the last one on the right-hand-side of Equation (41).

The poisoning of an enzyme system is only one mechanism whereby the activity of a microbial pathway can be suppressed. Another mechanism involves the competition of two groups of microorganisms for the same substrate. The relative rates of organic matter degradation by sulfate reduction and methanogenesis, for example, appears to be controlled by a competition for acetate (Lovley and Klug, 1983, 1986). The latter is an intermediate product of the anaerobic decomposition of organic matter, which is used as an energy substrate by both sulfate reducing and methane producing bacteria. When the levels of dissolved sulfate are high enough, acetate uptake by the sulfate reducers outcompetes that of methanogens. As the sulfate concentration drops, however, the activity of the sulfate reducers becomes limited by the availability of sulfate ($[SO_4] \leq K_{SO_4}$). Acetate uptake is now diverted to the methanogens, and methanogenesis takes over as the dominant degradation pathway for organic matter.

Because the competition between sulfate reduction and methanogenesis is fairly well understood, it is possible to develop a kinetic model with rate expressions for acetate production, plus acetate utilization by sulfate reducers and methanogens. In such a model, the suppression of methanogenesis is the result of the intrinsically faster kinetics of acetate uptake by sulfate reducers. Only when the latter become limited by the availability of their terminal electron acceptor can the rate of methanogenesis reach significant levels. In many instances, however, a detailed mechanistic understanding of the inhibitory or competitive interactions between microbial populations is not available. One can then always fall back on an empirical description such as Equation (41).

Another type of competition opposes microbial and abiotic reaction pathways. For example, the reduction of manganese and iron oxyhydroxides in sediments can be carried out by heterotrophs that couple the reduction of Mn(IV) and Fe(III) directly to the oxidation of organic matter (Lovley, 1987, 1991; Nealson and Meyers, 1992; Burdige, 1993;

DiChristina and DeLong, 1993; Canfield et al., 1993a,b). These organisms, however, must compete with numerous inorganic and organic reductants that can react inorganically with the oxyhydroxides (Berner, 1970; Postma, 1985; Stone, 1987; Wehrli et al., 1989; Luther et al., 1992; Sunda and Kieber, 1994). The most straightforward way to account for this type of competition is to represent each potential microbial and chemical pathway by a rate expression. As for any set of parallel reactions, the fastest pathway then dominates the overall transformation rate. This approach has been used by Wang and Van Cappellen (1996) to identify the environmental conditions which favor either dissimilatory (microbial) or non-dissimilatory (chemical) reduction of Fe and Mn in sediments.

A kinetic model for organic matter degradation

With the rate equations presented so far, it is possible to construct a reasonably realistic kinetic model which accounts for the principal chemical changes accompanying the degradation of sedimentary organic matter. The standard set of overall reaction stoichiometries that are used to represent the main pathways of degradation are (e.g. Van Cappellen and Wang, 1996):

aerobic respiration:

$$\begin{aligned}&(CH_2O)_x(NH_3)_y(H_3PO_4)_z + (x+2y)O_2 + (y+2z)HCO_3^- \\ &\longrightarrow (x+y+2z)CO_2 + yNO_3^- + zHPO_4^{2-} + (x+2y+2z)H_2O\end{aligned} \tag{42}$$

denitrification:

$$\begin{aligned}&(CH_2O)_x(NH_3)_y(H_3PO_4)_z + \left(\tfrac{4x+3y}{5}\right)NO_3^- \longrightarrow \\ &\left(\tfrac{2x+4y}{5}\right)N_2 + \left(\tfrac{x-3y+10x}{5}\right)CO_2 + \left(\tfrac{4x+3y-10z}{5}\right)HCO_3^- + zHPO_4^{2-} + \left(\tfrac{3x+6y+10z}{5}\right)H_2O\end{aligned} \tag{43}$$

dissimilatory manganese reduction:

$$\begin{aligned}&(CH_2O)_x(NH_3)_y(H_3PO_4)_z + 2xMnO_2 + (3x+y-2z)CO_2 + (x+y-2z)H_2O \\ &\longrightarrow 2xMn^{2+} + (4x+y-2z)HCO_3^- + yNH_4^+ + zHPO_4^{2-}\end{aligned} \tag{44}$$

dissimilatory iron reduction:

$$\begin{aligned}&(CH_2O)_x(NH_3)_y(H_3PO_4)_z + 4xFe(OH)_3 + (7x+y-2z)CO_2 \longrightarrow \\ &4xFe^{2+} + (8x+y-2z)HCO_3^- + yNH_4^+ + zHPO_4^{2-} + (3x-y+2z)H_2O\end{aligned} \tag{45}$$

sulfate reduction:

$$\begin{aligned}&(CH_2O)_x(NH_3)_y(H_3PO_4)_z + x/_2 SO_4^{2-} + (y-2z)CO_2 + (y-2z)H_2O \\ &\longrightarrow x/_2 H_2S + (x+y-2z)HCO_3^- + yNH_4^+ + zHPO_4^{2-}\end{aligned} \tag{46}$$

methanogenesis:

$$\begin{aligned}&(CH_2O)_x(NH_3)_y(H_3PO_4)_z + (y-2z)H_2O \\ &\longrightarrow x/_2 CH_4 + \left(\tfrac{x-2y+4z}{2}\right)CO_2 + (y-2z)HCO_3^- + yNH_4^+ + zHPO_4^{2-}\end{aligned} \tag{47}$$

where $(CH_2O)_x(NH_3)_y(H_3PO_4)_z$ is an idealized representation of the sedimentary particulate organic matter.

Reactions (42) to (47) are listed roughly in the sequence in which they occur with increasing depth in a sediment. The successive utilization of terminal electron acceptors has been rationalized in terms of the free energy yields of the overall reactions, which decreasing order $O_2 > NO_3 > Mn(IV) > Fe(III) > SO_4$ (Berner, 1980; Canfield, 1993). Thus, for given availabilities of the various oxidants, the ecosystem maximizes the amount of energy it can harvest from the degradation of the organic matter. The microbial populations, however, rely on kinetic strategies, e.g. inhibition and competition, to achieve their optimal "thermodynamic efficiency". If we adopt Equation (41) to describe the suppression of a given degradation pathway by the availability of more powerful oxidants, we obtain the following kinetic scheme for the decomposition of the organic matter:

$$R_{O_2} = k_{O_2}[CH_2O]\frac{[O_2]}{K_{O_2}+[O_2]} \tag{48}$$

$$R_{NO_3} = k_{NO_3}[CH_2O]\frac{[NO_3^-]}{K_{NO_3}+[NO_3^-]}\frac{K_{O_2}^{in}}{K_{O_2}^{in}+[O_2]} \tag{49}$$

$$R_{Mn} = k_{Mn}[CH_2O]\frac{[Mn(IV)]}{K_{Mn}+[Mn(IV)]}\frac{K_{O_2}^{in}}{K_{O_2}^{in}+[O_2]}\frac{K_{NO_3}^{in}}{K_{NO_3}^{in}+[NO_3^-]} \tag{50}$$

$$R_{Fe} = k_{Fe}[CH_2O]\frac{[Fe(III)]}{K_{Fe}+[Fe(III)]}\frac{K_{O_2}^{in}}{K_{O_2}^{in}+[O_2]}\frac{K_{NO_3}^{in}}{K_{NO_3}^{in}+[NO_3^-]}\frac{K_{Mn}^{in}}{K_{Mn}^{in}+[Mn(IV)]} \tag{51}$$

$$R_{SO_4} = k_{SO_4}[CH_2O]\frac{[SO_4^=]}{K_{SO_4}+[SO_4^=]}\frac{K_{O_2}^{in}}{K_{O_2}^{in}+[O_2]}\frac{K_{NO_3}^{in}}{K_{NO_3}^{in}+[NO_3^-]}\frac{K_{Mn}^{in}}{K_{Mn}^{in}+[Mn(IV)]}\frac{K_{Fe}^{in}}{K_{Fe}^{in}+[Fe(III)]} \tag{52}$$

$$R_{CH_4} = k_{CH_4}[CH_2O]\frac{K_{O_2}^{in}}{K_{O_2}^{in}+[O_2]}\frac{K_{NO_3}^{in}}{K_{NO_3}^{in}+[NO_3^-]}\frac{K_{Mn}^{in}}{K_{Mn}^{in}+[Mn(IV)]}\frac{K_{Fe}^{in}}{K_{Fe}^{in}+[Fe(III)]}\frac{K_{SO_4}^{in}}{K_{SO_4}^{in}+[SO_4^=]} \tag{53}$$

$$R_C = R_{O_2}+R_{NO_3}+R_{Mn}+R_{Fe}+R_{SO_4}+R_{CH_4} \tag{54}$$

where all the rates are expressed in mass equivalents of organic carbon. A similar kinetic model for organic matter degradation was recently developed by Boudreau (in press).

The above kinetic scheme will reproduce the sequence of organic matter decomposition pathways (42) to (47). The degree of overlap between successive organic matter degradation pathways will depend on the inhibition constants. Unfortunately, values of these constants in natural aquatic sediments have not been determined independently. From an ecological point of view, however, we expect that the interactions between the degradation pathways would result in an optimal use of the energy source. This implies that inhibition of a given degradation pathway can no longer be significant, once the previous pathway becomes limited by the availability of its terminal oxidant. Stated otherwise, once an oxidant concentration is no longer saturating, the pathway using the next most powerful oxidant should become active. This will be the case when the half-saturation constant of an oxidant is of the same order of magnitude as the corresponding inhibition constant. Thus, as a first approximation, we may assume $K_i \approx K_i^{in}$. Values of half-saturation concentrations applicable to natural aquatic sediments can be found in Van Cappellen and Wang (1995) and Wang and Van Cappellen (1996).

A kinetic scheme such as the one represented by Equations (48) to (53) creates a high degree of coupling between species distributions. For example, in order to calculate the rate at which reactive Fe(III) oxyhydroxides are being reduced by Reaction (45), at any given depth and time in a sediment, the concentrations of particulate organic matter, dissolved oxygen, dissolved nitrate, reactive Mn oxyhydroxides, and reactive Fe(III) oxyhydroxides must all be known (Eqn. 51). Therefore, when the kinetic scheme is implemented in a reactive transport model, the distributions of organic carbon and all the potential oxidants must be solved simultaneously. The result is a coupled model where the downward succession of organic matter degradation pathways, and therefore the characteristic redox zonation of the sediment, comes about as the natural consequence of the microbial kinetics.

A final remark about the kinetic model of organic matter degradation presented here concerns its relationship to the acid-base properties of the pore waters. Inspection of Reactions (42) to (47) reveals that the stoichiometries are not balanced by protons, but by bicarbonate and carbonic acid. Hence, the reaction stoichiometries describe the net production/consumption of pore water alkalinity, rather than that of H^+. This procedure avoids having to include H^+ as an independent species, with all the ambiguities surrounding the definition of its transport properties in aqueous media. The stoichiometries of all secondary reactions can be dealt with in a similar fashion (see Table 2). This leads to a pore water acid-base model where the total concentrations of weak acids (principally carbonic acid and hydrogen sulfide) and total alkalinity are treated as conservative parameters, that is, their distributions are described by continuity equations. When coupling these

Table 2. Some important secondary redox reactions in natural aquatic sediments. $\equiv S-H^0$ represents a hydrated site on a mineral or biological surface.

$$\equiv S\text{-}Mn^+ + \frac{1}{2}O_2 + HCO_3^- \longrightarrow \equiv S\text{-}H^0 + MnO_2 + CO_2 \tag{55}$$

$$Fe^{2+} + \frac{1}{4}O_2 + 2HCO_3^- + \frac{1}{2}H_2O \longrightarrow Fe(OH)_3 + 2CO_2 \tag{56}$$

$$\equiv S\text{-}Fe^+ + \frac{1}{4}O_2 + HCO_3^- + \frac{3}{2}H_2O \longrightarrow \equiv S\text{-}H^0 + Fe(OH)_3 + CO_2 \tag{57}$$

$$2Fe^{2+} + MnO_2 + 2HCO_3^- + 2H_2O \longrightarrow 2Fe(OH)_3 + Mn^{2+} + 2CO_2 \tag{58}$$

$$NH_4^+ + 2O_2 + 2HCO_3^- \longrightarrow NO_3^- + 2CO_2 + 3H_2O \tag{59}$$

$$H_2S + 2O_2 + 2HCO_3^- \longrightarrow SO_4^{2-} + 2CO_2 + 2H_2O \tag{60}$$

$$H_2S + 2CO_2 + MnO_2 \longrightarrow Mn^{2+} + S^o + 2HCO_3^- \tag{61}$$

$$H_2S + 4CO_2 + 2Fe(OH)_3 \longrightarrow 2Fe^{2+} + S^o + 4HCO_3^- + 2H_2O \tag{62}$$

$$FeS + 2O_2 \longrightarrow Fe^{2+} + SO_4^{2-} \tag{63}$$

$$CH_4 + 2O_2 \longrightarrow CO_2 + 2H_2O \tag{64}$$

$$CH_4 + CO_2 + SO_4^{2-} \longrightarrow 2HCO_3^- + H_2S \tag{65}$$

continuity equations to equilibrium expressions for the reversible proton-dissociation reactions of the weak acids, all acid-base properties, including the pH, can be calculated (for details, see, Van Cappellen and Wang, 1995, 1996).

Secondary reactions

Secondary chemical processes in sediments include redox transformations, acid-base reactions, homogeneous complexation reactions, sorption processes, dissolution and precipitation of mineral phases, colloidal phenomena, plus abiotic transformations of organic molecules. Here, we focus on the secondary redox reactions, and refer the reader to the vast chemical and geochemical literature dealing with the kinetic and equilibrium formulations of the other processes (standard references include, Berner, 1980; Lasaga, 1981b; Sposito, 1989; Morel and Hering, 1993; Brezonik, 1994; Stumm and Morgan, 1996).

Secondary redox reactions are driven by the appearance of reduced by-products during the degradation of organic matter. As shown by reactions (42) to (47), heterotrophic activity in a sediment may result in the formation of Mn^{2+}, Fe^{2+}, NH_4, H_2S, and CH_4. Additional species, e.g. dissolved organics and polysulfides, may be produced during the decay of particulate organic matter. All of the reduced species can be transported within the sediment under dissolved, adsorbed or solid forms. Reduced Fe(II) and sulfide, for example, can precipitate as iron sulfide phases. Bioturbation can then move the solid sulfides to the oxidized surface layer where they oxidatively dissolve, resulting in the reprecipitation of Fe(III) oxyhydroxides. The latter subsequently migrate downward, by mixing and advection, until they are reduced again via biotic or abiotic pathways. Thus, the combination of transport processes and redox transformations may lead to a continuous cycling of elements between their oxidized and reduced states.

Overall reaction stoichiometries that represent some of the major secondary redox reactions in sediments are presented in Table 2. It can be seen that, in general, competing pathways exist for the oxidation of a given reduced species. Thus, dissolved hydrogen sulfide that has not precipitated out as a solid sulfide phase may potentially react with Fe/Mn oxyhydroxides or with dissolved oxygen. Furthermore, multiple pathways may hide beneath the stoichiometric reactions presented in Table 2. The oxygenation of adsorbed Mn(II), for instance, may proceed abiotically (e.g. Diem and Stumm, 1984), or it may be carried out by specialized chemolithotrophic bacteria (e.g. Emerson et al., 1982).

The kinetics of homogenous abiotic redox reactions have been studied quite extensively. The rate laws can usually be written as:

$$R = k[\mathrm{Ox}][\mathrm{Red}] \tag{66}$$

where Ox and Red stand for the oxidant and reductant respectively (e.g. Larson and Weber, 1994). One example is the homogeneous oxygenation of dissolved Fe^{2+} whose kinetics have been shown to follow Equation (66) (Stumm and Lee, 1961). The rate coefficient, however, is not a true kinetic constant, but incorporates the effect of pH on the reaction rate.

Abiotic redox reactions in natural systems are frequently catalyzed by solid surfaces. Hence, in the presence of appropriate binding sites, the abiotic oxygenation of Fe^{2+} proceeds via a heterogeneous pathway where the rate is proportional to the amount of adsorbed Fe^{2+} (Wehrli, 1990). If few binding sites are available, saturation may occur and the rate will be independent of the concentration of the reductant, Fe^{2+}. In the presence of a large excess number of surface sites, the rate will exhibit a first-order dependence with respect to Fe^{2+}. That is, we observe a behavior very similar to that described by the Monod law for substrate utilization by microbial populations (Eqn. 36). This reflects the fact that the second term on the right-hand-side of Equation (36) is formally analogous to a Langmuir adsorption isotherm. Appropriate kinetic formulations for heterogeneous redox reactions are discussed in detail by Wehrli (1990).

Microbially-mediated pathways for secondary redox reactions can be treated as two-substrate enzymatic reactions. For example, the available data on the microbial oxidation of Mn^{2+} show that, at low reactant concentrations, the oxidation rate correlates positively with the concentrations of both dissolved Mn^{2+} and O_2 (Kepkay et al., 1984; Tebo and Emerson, 1985; Taylor, 1986). At higher levels, the rate becomes independent of $[O_2]$ and $[Mn^{2+}]$. This suggests that the kinetics could be modeled with an expression of the form of Equation (38).

The representation of secondary redox reactions in a reactive transport model by the full equations for heterogeneous or microbial kinetics is feasible, although fairly cumbersome. Fortunately, in natural sediments, secondary redox reactions are often confined to narrow redox fronts, where the concentrations of both the oxidant and reductant are fairly low (Fig. 11). This can be understood as follows. If a fast abiotic redox pathway exists, then the combination of rapid reaction kinetics and finite transport rates will force a narrow reaction zone (Fig. 11). If, because of kinetic hindrance, all the inorganic pathways are slow, a broader reaction zone could develop where high concentrations of the reactants would coexist. In a natural system, however, this would create a niche for microorganisms which could sustain their metabolism by catalyzing the thermodynamically favorable reaction. The ecosystem will therefore attempt to provide biotic catalysts for the reaction, rather than waste the chemical energy source.

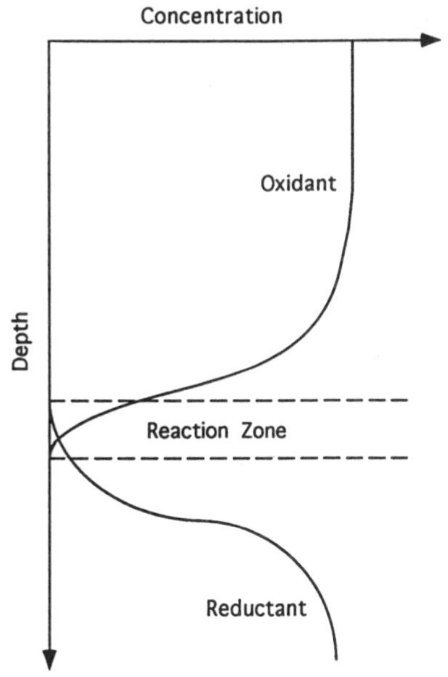

Figure 11. Reaction front of a secondary redox reaction. Rapid reaction kinetics, relative to the transport rates, results in a fairly narrow reaction zone where the reactants coexist at low concentrations.

From a kinetic point of view, the above limiting case means that the rate of reaction of a secondary reaction will often reduce to the simple bimolecular expression of Equation (66), whether the process is microbial or abiotic. That is, Equation (66) represents the simplest, yet kinetically justified, choice for representing secondary redox reactions in early diagenetic models.

Rate coefficients for secondary redox reactions under natural conditions are still poorly constrained. Abiotic reaction kinetics are usually determined in laboratory systems that bear little resemblance to actual sediments, while microbial studies frequently do not emphasize the collection of data that can be reduced into quantitative rate expressions. As an alternative to experimentation, reactive transport modeling can be used to extract kinetic information on reaction pathways directly from field data. Wang and Van Cappellen (1996) have applied this approach to estimate the rate coefficients of the reactions listed in Table 2. When comparing the estimates obtained for three separate coastal marine sediments, they found very similar values for those reactions involving only aqueous reactants. When solid or interfacial species act as reactants, they noted significant variations in the rate coefficients from one site to the other. This, however, is not entirely unexpected, given the highly variable reactivities and surface properties of sedimentary solids.

CONTINUITY EQUATIONS

The most general form of the continuity equation describing the reactive transport of a dissolved, interfacial or solid in a sediment is

$$\frac{\partial \hat{C}}{\partial t} = -\Sigma \nabla \bullet J + \Sigma S + \Sigma R \tag{67}$$

where $\partial C / \partial t$ is the Eulerian derivative of concentration with respect to time, $\hat{C}$ is the concentration of the constituent per unit volume total sediment (solids plus pore space), $\nabla \bullet J$ is the divergence of the local transport flux J (i.e. an advective or diffusive flux), S is the source strength of a non-local transport process, and R is the rate of change of the constituent's concentration due to a biogeochemical reaction.

In a multicomponent reactive transport model, one continuity equation must be solved for each independent species. As an example, let us construct the continuity equation for dissolved molecular oxygen. Transport of O_2 can occur by molecular diffusion, irrigation, bulk sediment mixing and pore water advection. Using the formulations presented in earlier sections, the one-dimensional continuity equation for O_2 is then:

$$\frac{\partial \phi [O_2]}{\partial t} = -\frac{\partial}{\partial x}\left(-D_s \phi \frac{\partial [O_2]}{\partial x} - D_b \frac{\partial \phi [O_2]}{\partial x} + v\phi [O_2] \right) + \phi \alpha \left([O_2]_0 - [O_2] \right) + \phi \Sigma R \tag{68}$$

where the concentration of O_2 is now in units of mass per unit volume pore water. The advective velocity, v, refers to fluid flow relative to the water sediment interface (x = 0). This velocity will be equal to the burial rate of the solid particles, ω, only in the special case of no compaction and no externally impressed flow (Berner, 1980). Note that in Equation (68) the transport parameters D_s, D_b, v and α can all be functions of time and depth.

If we consider a sediment below the photic zone, only reactions consuming O_2 take place. The reaction rate term in Equation (68) which accounts for aerobic respiration and secondary reactions (Table 2) is then given by

$$\begin{aligned} \Sigma R = &-\left\{ \frac{k_{56}[O_2][Fe^{2+}]}{4} + 2k_{59}[O_2][NH_4^+] + 2k_{60}[O_2][H_sS] + 2k_{64}[O_2][CH_4] \right\} \\ &-F\left\{ \frac{x+2y}{x}\left(k_{O_2}[CH_2O]\frac{[O_2]}{K_{O_2}+[O_2]} \right) + \frac{k_{55}[O_2][\equiv S-Mn^+]}{2} + \frac{k_{57}[O_2][\equiv S-Fe^+]}{4} + 2k_{63}[O_2][FeS] \right\} \end{aligned} \tag{69}$$

where bimolecular expressions are used for the secondary reactions reducing O_2. All the rate constants in Equation (69) are expressed in units of volume pore water per unit mass per unit time, except k_{O_2} which is in units of inverse time. The factor F converts between units of dissolved and particulate concentrations:

$$F = \frac{\rho(1-\phi)}{\phi} \tag{70}$$

where ρ is the dry sediment density.

As illustrated by Equation (69), reaction rate terms create a strong coupling among reactive species in a multicomponent model. This imposes stringent constraints on the validation of the model, because a unique set of reaction stoichiometries, rate expressions and rate coefficients is required to simultaneously reproduce the distributions of all the reactive species, dissolved and solid-bound. Any error in the formulation or parameterization of the rate expression of a reaction automatically propagates in the continuity equations of all the species involved in that reaction, hence reducing the likelihood that the model will satisfactorily reproduce the field observations. In general, the penalty for this type of error increases with the degree of coupling. Thus, when more reactive species are added to a reactive transport model, the increase in the number of adjustable parameters is compensated by a decrease in degree of freedom due to the reaction coupling of the species.

The continuity equations off the reactive species must be supplied with appropriate sets of boundary conditions. It is usually assumed that the concentrations of solutes at the water-sediment interface are equal to their bottom water values. For solid-bound constituents, the physically most natural coupling to the overlying water column is through the deposition flux. The appropriate flux-continuity condition is

$$J_0 = \left\{ -D_b \frac{\partial \rho(1-\phi)C}{\partial x} + \omega\rho(1-\phi)C \right\}\Bigg|_{x=0} \tag{71}$$

where J_0 is the deposition flux at the water-sediment interface, in mass per unit sediment surface area per unit time. One often imposes zero concentration gradients at the lower boundary, hence assuming that early diagenetic reactions cease at sufficiently great depth in the sediment.

The numerical solutions to coupled sets of continuity equations such as (68) are discussed in detail elsewhere in this volume. For one-dimensional early diagenetic models, finite differences are most frequently used. The main complication in the numerical solutions arises from the presence of non-linear rate terms, such as Equation (69). An efficient way to linearize rate terms in the discretized continuity equation of a given species is to expand the rate at spatial node l and time step m as follows,

$$R_l^m = R_l^{m-1} + \frac{\partial R}{\partial C}\bigg|_{l,m-1} \left(C_l^m - C_l^{m-1} \right) \tag{72}$$

where C is the concentration of the species. Full details on how to implement the above linearization scheme into an iterative method for solving 1-D finite difference equations are given in Van Cappellen and Wang (1996).

APPLICATION

Wang and Van Cappellen (1996) recently fitted a comprehensive, multicomponent model of reactive transport to extensive chemical and kinetic data sets collected by Canfield et al. (1993a,b) in three coastal marine sediments. The model consistently reproduced the field observations. From the model fits, values for reaction plus transport parameters were retrieved. These values were then used in extensive sensitivity analyses which highlighted the early diagenetic roles of transport processes, Fe and Mn deposition fluxes, and the total rate of organic carbon oxidation. By integrating the rate distributions of reaction pathways, detailed steady state budgets for the major redox elements were constructed. As an example of the results, Table 3 presents the calculated dissolved oxygen budgets at the three sites. The total oxygen uptake rates by the sediments are of the same order of magnitude. When considering the relative importance of the individual reaction pathways of oxygen reduction, however, large differences are observed between the sites. The steady state budget for iron at one of the sites (Fig. 12) further illustrates the level of detailed quantitative understanding about elemental cycling one can achieve by combining field data to reactive transport modeling.

Table 3. Model-derived, depth-integrated rates of O_2 reduction in three coastal marine sediments. Total rates of reduction are given in μmol cm^{-2} yr^{-1}. The other numbers correspond to the percentage contributions of the various pathways to total O_2 consumption (from Wang and Van Cappellen, 1996).

Pathway	Site S4	Site S6	Site S9
$C_{org} \rightarrow CO_2$	21.8 %	45.8 %	30.6 %
$Mn(II)_{adsorbed} \rightarrow MnO_2$	4.5 %	17.8 %	65.5 %
$Fe(II)_{aqueous} \rightarrow Fe(OH)_3$	1.2 %	6.6 %	0 %
$Fe(II)_{adsorbed} \rightarrow Fe(OH)_3$	34.9 %	15.2 %	0 %
$H_2S \rightarrow SO_4^{2-}$	<0.1 %	<0.1 %	0 %
$NH_4^+ \rightarrow NO_3^-$	2.7 %	1.8 %	3.9 %
$FeS \rightarrow SO_4^{2-}$	34.8 %	12.7 %	0 %
TOTAL	358.5	364.5	296.6

CONCLUSIONS

The past few years have witnessed the development of increasingly sophisticated models of early diagenesis. This trend can be explained by a combination of factors, including the availability of more complete and integrated data sets on aquatic sediments, a better understanding of the speciation and reaction pathways in early diagenetic systems, and the rapid evolution of personal computers and workstations. Multicomponent reactive transport models provide the quantitative tools that are needed to describe, interpret and systematize the currently available chemical and biological information on aquatic sediments.

Clearly, there is room for improvement. Overall, the mathematical treatment of biologically-induced transport in most models remains fairly crude. Additional data and experiments, combined with theoretical work, will be required to constrain the functions that describe biological transport at a variety of spatial and temporal scales. Similarly, detailed studies of the pathways and mechanisms of biogeochemical processes will be needed in order to produce more general rate models for chemical transformations in sediments. Hence, as shown by the most recent developments, progress in modeling will continue to be strongly linked to advances in field and experimental studies.

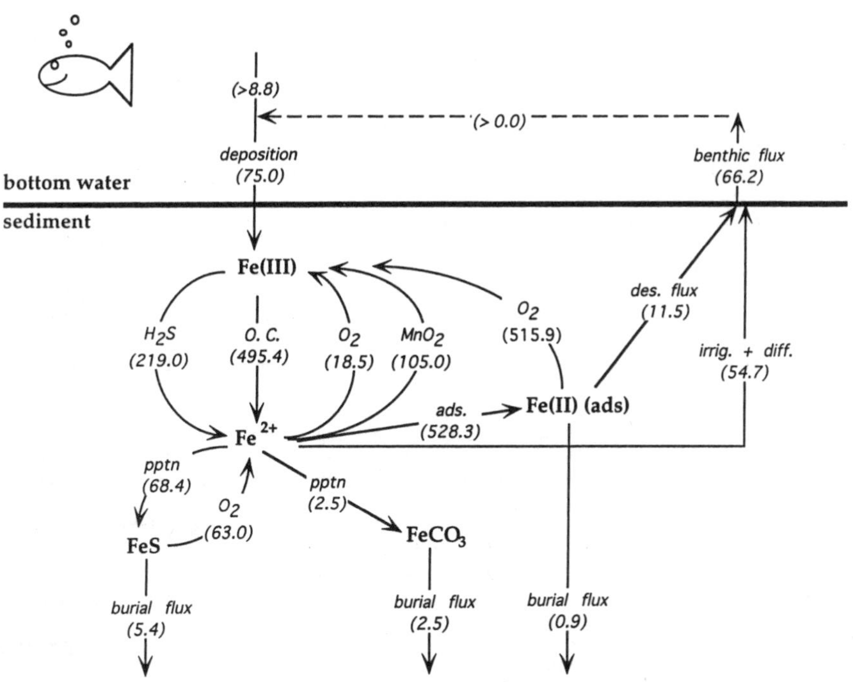

Figure 12. Iron cycling in a coastal marine sediment. The numbers are model-derived fluxes and depth-integrated reaction rates (in μmol cm^{-2} yr^{-1}). They were obtained by fitting a multicomponent reactive transport model to measured concentration profiles of pore water and solid sediment constituents. (O.C.: organic carbon; ads: adsorption; des: desorption; irrig: irrigation; diff: molecular/ionic diffusion; ppt.: precipitation.) From Wang and Van Cappellen (1996).

ACKNOWLEDGMENTS

Because they have inspired us in our work on early diagenesis, we thank Bob Berner, Yifeng Wang, Christophe Rabouille, Don Canfield, Ellery Ingall, Bernard Boudreau, Tracey Tromp, and Carl Steefel. The senior author also acknowledges the financial support of the U.S. Environmental Protection Agency and the Subsurface Science Program of the U.S. Department of Energy.

REFERENCES

Aller RC (1980) Quantifying solute distributions in the bioturbated zone of marine sediments by defining an average microenvironment. Geochim Cosmochim Acta 44:1955-1965

Aller RC (1982) The effects of macrobenthos on chemical properties of marine sediment and overlying water. In: Animal-Sediment Relations: The Biogenic Alteration of Sediments. PL McCall, MJ Tevesz (eds) p 53-102. Plenum Press, New York

Aller RC (1990) Bioturbation and manganese cycling in hemipelagic sediments. Phil Trans R Soc Lond A331, 51-68

Aller RC, Mackin, JE (1989) Open-incubation, diffusion methods for measuring solute reaction rates in sediments. J Marine Res 47: 411-440

Amann RL, Krumholz L, Stahl DA (1990) Fluorescent-oligonucleotide probing of whole cells for determinative, phylogenetic, and environmental studies in microbiology. J Bacteriol 172:762-770

Andrews D, Bennett A (1981) Measuring diffusivity near the sediment-water interface with a fine-scale resistivity probe. Geochim Cosmochim Acta 45:2169-2175

Applin KR, Lasaga AC (1984) The determination of SO_4^{2-}, $NaSO_4^-$, and $MgSO_4^0$ tracer diffusion coefficients and their application to diagenetic flux calculations. Geochim Cosmochim Acta 48:2151-2162

Archer DE, Emerson S, Reimers C (1989) Dissolution of calcite in deep-sea sediments: pH and O_2 microelectrode results. Geochim Cosmochim Acta 53:2831-2846

Archie GE (1942) The electrical resitivity log as an aid in determining some reservoir characteristics. Petrol Tech 1:55-62

Arnosti C, Repeta DJ, Blough NV (1994) Rapid bacterial degradation of polysaccharides in anoxic marine systems. Geochim Cosmochim Acta 58:2639-2652

Baden SP, Loo L-O, Pihl L, Rosenberg R (1990) Effects of eutrophication on benthic communities including fish: Swedish west coast. Ambio 19:113-122

Bear J (1972) Dynamics of Fluids in Porous Media. Dover Publishers New York

Benninger LK, Aller RC, Cochran JK, Turekian KK (1979) Effects of biological sediment mixing on the ^{210}Pb chronology and trace metal distribution in a Long Island Sound sediment core. Earth Planet Sci Lett 43:241-259

Berner RA (1964) An idealized model of dissolved sulfate distribution in recent sediments. Geochim Cosmochim Acta 28:1497-1503

Berner RA (1970) Sedimentary pyrite formation. Am J Sci 268:1-23

Berner RA (1974) Kinetic model for the early diagenesis of nitrogen, sulphur, phosphorus, and silicon in anoxic marine sediments. In: The Sea, volume 5 ED Goldberg (ed) p 427-450. Wiley, New York

Berner RA (1978) Sulfate reduction and the rate of deposition of marine sediments. Earth Planet. Sci. Lett. 37:492-498

Berner RA (1980) Early Diagenesis: A Theoretical Approach. Princeton Univ Press, Princeton, New Jersey

Berner RA (1989) Biogeochemical cycles of carbon and sulfur and their effect on atmospheric oxygen over Phanerozoic time. Paleogeogr Paleoclimatol Paleoecol 75:97-122

Boudreau BP (1986) Mathematics of tracer mixing in sediments: II. Nonlocal mixing and biological conveyor-belt phenomena. Am J Sci 286:199-238

Boudreau BP (1991) Modelling the sulfide-oxygen reaction and associated pH gradients in porewaters. Geochim Cosmochim Acta 55:145-159

Boudreau BP (1996) The diffusive tortuosity of fine-grained unlithified sediments. Geochim Cosmochim Acta 60:3139-3142

Boudreau BP (in press) A method-of-lines code for carbon and nutrient diagenesis in aquatic sediments. Computers and Geosciences

Boudreau BP, Imboden DM (1987) Mathematics of tracer mixing in sediments: III. The theory of nonlocal mixing within sediments. Am J Sci 287:693-719

Boudreau BP, Ruddick BR (1991) On a reactive continuum representation of organic matter diagenesis. Am J Sci 291:507-538

Boudreau BP, Canfield DE (1993) A comparison of closed- and open-system models for porewater pH and calcite-saturation state. Geochim Cosmochim Acta 57:317-334

Brendel P, Luther GW III (1995) Development of a gold amalgam voltammetric microelectrode for the determination of dissolved Fe, Mn, O_2, and S(-II) in porewaters of marine and freshwater sediments. Environ Sci Technol 29:751-761

Brezonik PL (1994) Chemical Kinetics and Process Dynamics in Aquatic Systems. Lewis Publ., Boca Raton, FL

Brock TD, Madigan MT, Martinko JM, Parker J (1994) Biology of Microorganisms. Prentice Hall, Englewood Cliffs, New Jersey

Burdige DJ (1993) The biogeochemistry of manganese and iron reduction in marine sediments. Earth-Science Rev 35:249-284

Cai W-J, Reimers CE (1993) The development of pH and pCO_2 microelectrodes for studying the carbonate chemistry of porewaters near the sediment-water interface. Limmol Oceanogr 38:1762-1773

Cai W-J, Reimers CE, Shaw T (1995) Microelectrode studies of organic carbon degradation and calcite dissolution at a California continental rise site. Geochim Cosmochim Acta 59:497-511

Canfield DE (1989) Reactive iron in marine sediments. Geochim Cosmochim Acta 53:619-632

Canfield DE (1993) Organic matter oxidation in marine sediments. In: Interactions of C, N, P, and S Biogeochemical Cycles and Global Change. R Wollast, FT Mackenzie, L Chou (eds) p 333-363. Springer-Verlag, Berlin

Canfield DE, Thamdrup B, Hansen JW (1993a) The anaerobic degradation of organic matter in Danish coastal sediments: Iron reduction, manganese reduction, and sulfate reduction. Geochim Cosmochim Acta 57:3867-3883

Canfield DE, Jørgensen BB, Fossing H, Glud R, Gundersen J, Ramsing NB, Thamdrup B, Hansen JW, Nielsen, LP, Hall POJ (1993b) Pathways of organic carbon oxidation in three continental margin sediments. Marine Geol 113:27-40

Carignan R, Lean, DRS. (1991) Regeneration od dissolved substances in a seasonally anoxic lake: The relative importance of processes occurring in the water column and in the sediments. Limnol Oceanogr 36:683-707

Chapelle FH (1993) Ground-Water Microbiology and Geochemistry. Wiley, New York

Cussler EL (1995) Diffusion: Mass Transfer in Fluid Systems. Cambridge University Press, Cambridge, UK

DeMaster DJ, Cochran JK (1982) Particle mixing rates in deep-sea sediments determined from excess ^{210}Pb and ^{32}Si profiles. Earth Planet Sci Lett 61:257-271

Devol AH (1970) Biological Oxidation in Oxic and Anoxic Marine Environments, Rates and Processes. PhD Dissertaion, Univ of Washington

Dhakar SP, Burdige DJ (1996) A coupled, non-linear, steady-state model for early diagenetic processes in pelagic sediments. Am J Sci 296:296-330

DiChristina TJ, DeLong EF (1993) Design and application of rRNA-targeted oligonucleotide probes for the dissimilatory iron- and manganese-reducing bacterium Shewanella putrefaciens. Appl Environ Microbiol 59:4152-4160

Diem D, Stumm W (1984) Is dissolved Mn^{2+} being oxidized by O_2 in absence of Mn-bacteria or surface catalysis? Geochim Cosmochim Acta 48:1571-1573

Emerson S (1985) Organic carbon preservation in marine sediments. In: The Carbon Cycle and Atmospheric CO_2: Natural Variations Archean to Present. ET Sundquist, WS Broecker (eds). Am Geophys Union Geophys Monogr Series 32:78-87

Emerson S, Kalhorn S, Jacobs L, Tebo BM, Nealson KH, Rosson RA (1982) Environmental oxidation rate of manganese (II): Bacterial Catalysis. Geochim Cosmochim Acta 46:1073-1079

Emerson S, Jahnke R, Heggie D (1984) Sediment-water exchange in shallow water estuarine sediments. J Marine Res 42:709-730

Engelhardt WV (1977) The Origin of Sediments and Sedimentary Rocks. Halstead Press, New York

Froelich PN, Klinkhammer GP, Bender ML, Luedtke NA, Heath GR, Cullen D, Dauphin, P, Hammond D, Hartman B, Maynard V (1979) Early oxidation of organic matter in pelagic sediments of the eastern equatorial Atlantic: Suboxic diagenesis. Geochim Cosmochim Acta 43:1075-1090

Gaillard J-F (1982) Comportement Geochimique du Fer et du Phosphore lors de leur Diagenese. Dissertation, University of Paris VI

Gaillard J-F, Rabouille, C (1992) Using Monod kinetics in geochemical models of organic carbon mineralization in deep-sea surficial sediments. In: Deep-Sea Food Chains and the Global Carbon Cycle. GT Rowe, V Pariente (eds) p 309-324. Kluwer, Dordrecht, The Netherlands

Gaillard J-F, Pauwels H, Michard G (1989) Chemical diagenesis in coastal marine sediments. Oceanol Acta 12:175-187

Gerino M, Stora G (1991) Analyse quantitative in vitro de la bioturbation induite par la Polychete Nereis diversicolor. C-R Acad Sci Paris 313:489-494

Gerino M, Stora G, Gontier G, Weber O (1993) Quantitative approach of bioturbation on continental margin. Ann Inst océanogr (Paris) 69:177-181

Guinasso NL, Schink DR (1975) Quantitative estimates of biological mixing rates in abyssal sediments. J Geophys Res 80:3032-3043

Gujer W, Zehnder AJB (1983) Conversion processes in anaerobic digestion. Water Sci Tech 15:127-16

Henriksen K, Kemp WM (1988) Nitrification in estuarine and coastal narine sediments. In: Nitrogen Cycling in Coastal Marine Environments, SCOPE 33. TH Blackburn, J Sørensen (eds) p 207-249. Wiley, Chichester, UK

Huckriede H, Meischner D (1996) Origin and environment of manganese-rich sediments within black-shale basins. Geochim Cosmochim Acta 60:1399-1413

Huerta-Diaz MA, Morse JW (1992) Pyritization of trace metals in anoxic marine sediments. Geochim Cosmochim Acta 56:2681-2702

Humphrey AE (1972) The kinetics of biosystems: A review. In: Chemical Reactor Engineering RF Gould (ed) Am Chem Soc Adv Chem Series 109:630-650

Jahnke RA (1990) Early diagenesis and recycling of biogenic debris at the seafloor, Santa Monica Basin, California. J Marine Res 48:413-436

Jahnke RA, Christiansen, MB (1989) A free-vehicle benthic chamber instrument for sea floor studies. Deep-Sea Res 36:625-637

Jahnke RA, Emerson SR, Cochran JK, Hirschberg DJ (1986) Fine scale distributions of porosity and particulate excess ^{210}Pb, organic carbon and $CaCO_3$ in surface sediments of the deep equatorial Pacific. Earth Planet Sci Lett 77:59-69

Jensen K, Revsbech NP, Nielsen LP (1993) Microscale distribution of nitrification activity in sediment determined with a shielded microsensor for nitrate. Appl Environ Microbiol 59:3287-3296

Jørgensen BB (1978) A comparison of methods for the quantification of bacterial sulfate reduction in coastal marine sediments. II. Calculation from mathematical models. Geomicrobiol J 1:29-47

Jørgensen BB (1982) Mineralization of organic matter in the sea bed—the role of sulfate reduction. Nature 296:643-645

Jørgensen BB, Revsbech, NP (1985) Diffusive boundary layers and the oxygen uptake of sediments and detritus. Limnol Oceanogr 30:111-122

Jumars PA, Nowell ARM, Self RFL (1981) A simple model of flow-sediment-organism interaction. Marine Geol 42:155-172

Kepkay PE, Burdige DJ, Nealson KH (1984) Kinetics of bacterial manganese binding and oxidation in the chemostat. Geomicrobiol J 3:245-262

Klump VJ, Martens CS (1989) The seasonality of nutrient regeneration in an organic-rich coastal sediment: Kinetic modeling of changing pore-water nutrient and sulfate distributions. Limnol Oceanogr 34:559-577

Kostka JE, Luther GW III(1994) Partitioning and speciation of solid phase iron in saltmarch sediments. Geochim Cosmochim Acta 58:1701-1710

Laanbroek HJ, Veldkamp H (1982) Microbial interactions in sediment communities. Phil Trans R Soc London B 297:533-550

Laidler KJ (1987) Chemical Kinetics. Harper Collins, New York

Larson RA, Weber EJ (1994) Reaction Mechanisms in Environmental Organic Chemistry. Lewis Publ, Boca Raton, LA

Lasaga AC (1979) The treatment of multi-component diffusion and ion pairs in diagenetic fluxes. Am J Sci 279:324-346

Lasaga AC (1981a) Influence of diffusion coupling on diagenetic concentration pofiles. Am J Sci 281, 553-575.

Lasaga AC (1981b) Transition state theory. In: Kinetics of Geochemical Processes. AC Lasaga, RJ Kirkpatrick (eds) Rev Mineral 8:135-169

Lovley DR, Klug MJ (1983) Sulfate reducers can outcompete methanogens at freshwater sulfate concentrations. Appl Environ Microbiol 45:187-192

Lovley DR, Klug MJ (1986) Model for the distribution of sulfate reduction and methanogenesis in freshwater sediments. Geochim Cosmochim Acta 50:11-18

Luther GW, Kostka JE, Church TM, Sulzberger B, Stumm W (1992) Seasonal cycling in the salt-marsh sedimentary environment: The importance of ligand complexes with Fe(II) and Fe(III) in the dissolution of Fe(III) minerals and pyrite, respectively. Marine Chem 40:81-103

Mackin JE (1987) Boron and silica behavior in salt-marsh sediments: Implications for paleoboron distributions and the early diagenesis of silica. Am J Sci 287:197-241

Mackin JE, Aller RC (1984) Dissolved Al in sediments and waters of the East China Sea: Implications for authigenic mineral formation. Geochim Cosmochim Acta 48:281-297

Mackin JE, Swider KT (1989) Organic matter decomposition pathways and oxygen consumption in coastal sediments. J Marine Res 47:681-716

Manheim FT (1970) The diffusion of ions in unconsolidated sediments. Earth Planet Sci Lett 9:307-309

Martin WR, Banta GT (1992) The measurement of sediment irrigation rate: A comparison of the Br tracer and $^{222}Rn/^{226}Ra$ disequilibrium techniques. J Marine Res 50:125-154

Matisoff G (1982) Mathematical models of bioturbation. In: Animal-Sediment Relations: The Biotic Alteration of Sediments. PL McCall, MJ Tevesz (eds) p 289-331. Plenum Press, New York

Matisoff G (1995) Effects of bioturbation on solute and particle transport in sediments. In: Metal Contaminated Aquatic Sediments. HE Allen (ed) p 201-272. Ann Arbor Press, Chelsea, Michigan

McCall PL, Fisher JB (1980) Effects of tubificid oligochaetes on physical and chemical properties of Lake Erie sediments. In: Aquatic Oligochaete Biology. RO Brinkhurts, DG Cook (eds) p 253-317. Plenum Press, New York

McDuff RE, Ellis RA (1979) Determining diffusion coefficients in marine sediments: A laboratory study of the validity of resistivity techniques. Am J Sci 279, 666-675

McNichol AP, Lee C, Druffel ERM (1988) Carbon cycling in coastal sediments: 1. A quantitative estimate of the remineralization of organic carbon in the sediments of Buzzards Bay, MA. Geochim Cosmochim Acta 52:1531-1543

Meyer-Reil L-A (1991) Ecological aspects of enzymatic activity in marine sediments. In: Microbial Enzymes in Aquatic Environments. RJ Chrost (ed) p 84-95. Springer-Verlag, Berlin

Middelburg JJ (1989) A simple rate model for organic matter decomposition in marine sediments. Geochim Cosmochim Acta 53:1577-1581

Morel FMM, Hering JG (1993) Principles and Applications of Aquatic Chemistry. Wiley, New York

Morgan JJ, Sung W, Stone A (1985) Chemistry of metal oxides in natural water: Catalysis of the oxidation of manganese(II) by γ-FeOOH and reductive dissolution of manganese(III) and (IV) oxides. In: Environmental Inorganic Chemistry. KJ Irgolic, AE Martell (eds) p 167-184. VCH Publ, Weinheim, Germany

Moses CO, Nordstrom DK, Herman JS, Mills AL (1986) Aqueous pyrite oxidation by dissolved oxygen and by ferric iron. Geochim Cosmochim Acta 51:1561-1571

Nealson KH, Myers CR (1992) Microbial reduction of manganese and iron: New approaches to carbon cycling. Appl Environ Microbiol 58:429-443

Nielsen LP, Christensen PB, Revsbech NP (1990) Denitrification and oxygen respiration in biofilms studied with a microsensor for nitrous oxide and oxygen. Microbiol Ecol 19:63-72

Novelli PC, Michelson AR, Scranton MI, Banta GT, Hobbie JE, Howarth RW (1988) Hydrogen and acetate cycling in two sulfate reducing sediments: Buzzards Bay and Town Cove, Mass. Geochim Cosmochim Acta 52:2477-2486

Poulsen LK, Ballard G, Stahl DA (1993) Use of rRNA fluorescence in situ hybridization for measuring the activity of single cells in young and established biofilms. Appl Environ Microbiol 59:1354-1360.

Postma D (1985) Concentration of Mn and separation from Fe in sediments: I. Kinetics and stoichiometry of the reaction between birnessite and dissolved Fe(II) at 10°C. Geochim Cosmochim Acta 49:1023-1033

Rabouille C, Gaillard J-F (1991) A coupled model representing the deep-sea organic carbon mineralization and oxygen consumption in surficial sediments. J Geophys Res 96:2761-2776

Rabouille C, Gaillard J-F, Tréguer P, Vincendeau M-A (1996) Biogenic silica recycling in surficial sedimnets across the Polar Front zone of the Southern Ocean (Indian Sector). Deep-Sea Res II, Antares I Special Vol (in press)

Raiswell R, Canfield DE, Berner, RA (1994) A comparison of iron extraction methods for the determination of degree of pyritisation and the recognition of iron-limited pyrite formation. Chem Geol 111:101-110

Ramsing N, Fossing H, Ferdelman TG, Andersen F, Thamdrup B (1996) Distribution of bacterial populations in a stratified fjord (Mariager Fjord, Denmark) quantified by in situ hybridization and related to chemical gradients in the water column. Appl Environ Microbiol 62:1391-1404

Revsbech NP, Jørgensen BB (1983) Photosynthesis of microbenyhos measured with high spatial resolution by the oxygen microprofile method. Limnol Oceanogr 28:749-756.

Revsbech NP, Jørgensen BB (1986) Microelectrodes: their use in microbial ecology. Adv Microb Ecol 9:293-352

Rhoads DC, Morse JW (1971) Evolutionary and ecologic significance of oxygen-deficient marine basins. Lethaia 4:413-428

Robbins JA, McCall PL, Fisher JB, Krezoski JR (1979) Effect of deposit-feeders on the migration of Cs-137 in lake sediments. Earth Planet Sci Lett 42:277-287

Robinson RA, Stokes RH (1959) Electrolyte Solutions. Butterworths, London

Ruttenberg KC (1992) Development of a sequential extraction method for different forms of phosphorus in marine sediments. Limnol Oceanogr 37:1460-1482

Simonin JP, Turq P, Soualhia E, Michard G, Gaillard J-F (1989) Transport coupling of ions: Influence of ion pairing and pH gradient - Application to the study of diagenetic fluxes. Chem Geol 78:343-356

Sposito G (1989) The Chemistry of Soils. Oxford University Press, New York

Stone AT (1987) Microbial metabolites and the reductive dissolution of manganese oxides: Oxalate and pyruvate. Geochim Cosmochim Acta, 51:919-925

Stumm W, Lee, GF (1961) Oxygenation of ferrous iron. Ind Eng Chem 53:143-146.

Stumm W, Morgan JJ (1996) Aquatic Chemistry. Wiley, New York

Sunda WG, Kieber DJ (1994) Oxidation of humic substances by manganese oxides yields low-molecular weight organic substrates. Nature 367:62-64

Taylor RJ (1987) Mn-Geochemistry in Galveston Bay Sediments. PhD Dissertation, Texas A & M Univ

Tebo BM, Emerson S (1985) Effect of oxygen tension, Mn(II) concentration, and temperature on the microbially catalyzed Mn(II) oxidation rate in a marine Fjord. Appl Environ Microbiol 50:1268-1273

Toth DJ, Lerman, A (1977) Organic matter reactivity and sedimentation rates in the ocean. Am J Sci 277:465-485

Tromp TK (1992) Global Carbon Cycles: A Coupled Atmosphere-Ocean-Sediment Model. PhD Dissertation, Princeton Univ, Princeton, New Jersey

Tromp TK, Van Cappellen P, Key RM (1995) A global model for the early diagenesis of organic carbon and organic phosphorus in marine sediments. Geochim Cosmochim Acta 59:1259-1284

Tyson RV, Pearson TH (1991) Modern and ancient continental shelf anoxia: An overview. In: Modern and Ancient Continental Shelf Anoxia. RV Tyson, TH Pearson (eds) Geol Soc Spec Publ 58:1-24, London

Ullman WJ, Aller RC (1982) Diffusion coefficients in nearshore sediments. Limnol Oceanogr 27552-556

Van Cappellen P (1996) Reactive surface area control on the dissolution kinetics of biogenic silica in deep-sea sediments. Chem Geol (in press)

Van Cappellen P, Wang Y (1995) Metal cycling in surface sediments. In: Metal Contaminated Aquatic Sediments. HE Allen (ed) p 21-64. Ann Arbor Press, Chelsea, Michigan

Van Cappellen P, Wang Y (1996) Cycling of iron and manganese in surface sediments: A general theory for the coupled transport and reaction of carbon, oxygen, nitrogen, sulfur, iron, and manganese. Am J Sci 296:197-243

Van Cappellen P, Qiu L (1996) Biogenic silica dissolution in sediments of the Southern Ocean. I. Solubility. Deep-Sea Res II, Antares I Special Vol (in press)

Van Cappellen P, Gaillard J-F, Rabouille C (1993) Biogeochemical transformations in sediments: Kinetic models of early diagenesis. In: Interactions of C, N, P and S Biogeochemical Cycles and Global Change. R Wollast, FT Mackenzie, L Chou (eds) p 401-445. Springer-Verlag, Berlin

Wang Y, Van Cappellen P (1996) A multicomponent reactive transport model of early diagenesis: Application to redox cycling in coastal marine sediments. Geochim Cosmochim Acta 60:2993-3014

Wehrli B (1990) Redox reactions of metal ions at mineral surfaces. In: Aquatic Chemical Kinetics. W Stumm (ed) p 311-336. Wiley, New York

Wehrli B, Sulzberger B, Stumm W (1989) Redox processes catalyzed by hydrous oxide surfaces. Chem Geol 78:167-179

Westrich JT, Berner RA (1984) The role of sedimentary organic matter in bacterial sulfate reduction: The G model tested. Limnol Oceanogr 29:236-249

Wheatcroft RA, Jumars PA, Smith CR, Nowell ARM (1990) A mechanistic view of the particulate biodiffusion coefficient: Step lengths, rest periods and transport directions. J Marine Res 48:177-207

Whitfield M (1973) A chemical model for the major electrolyte component of seawater based on the Brønsted-Guggenheim hypothesis. Marine Chem 1:251-266

Wollast R (1993) Interactions of carbon and nitrogen cycles in the coastal zone. In: Interactions of C, N, P and S: Biogeochemical Cycles and Global Change. R Wollast, FT Mackenzie, L Chou (eds) p 198-210. Springer-Verlag, Berlin

Chapter 9

REACTIVE TRANSPORT MODELING OF ACIDIC METAL-CONTAMINATED GROUND WATER AT A SITE WITH SPARSE SPATIAL INFORMATION

Pierre Glynn

U.S. Geological Survey
432 National Center,
Reston, Virginia 22091 U.S.A.

James Brown

U.S. Geological Survey
375 S. Euclid Avenue
Tucson, Arizona 85719 U.S.A.

INTRODUCTION

The construction of a multispecies reactive transport model used to predict the future evolution and movement of ground-water contaminants requires, at a minimum, three separate but related elements: (1) an understanding of the ground water flow system and its possible transients, (2) an understanding of the dispersive processes and other processes causing observed dilution or "mixing" of different water types, and (3) an understanding of the primary processes controlling the reactions of the various contaminants, not only across various phases but also within the ground water itself. The degree of understanding of all three of these elements, and perhaps more importantly an appreciation for the remaining knowledge gaps, will be essential in determining not only the usefulness of the constructed model but perhaps also its purpose. Indeed, even though a ground-water model may not adequately predict the future evolution of a contaminant plume, the construction of the model and its use may often result in an improved understanding of contaminant transport at the site.

Most ground-water contamination sites have less geochemical and hydrogeologic information known about them than may be desirable for predictive modeling of reactive contaminant transport. Detailed hydrogeologic and geochemical studies are usually much too expensive to consider*. The resulting lack of knowledge, on the operative chemical and hydrologic processes at a given site, means that investigations should try to use, as efficiently as possible, all tools and knowledge available. It is our belief that a combination of inverse and forward modeling of ground-water flow, inverse and forward modeling of advective/dispersive transport, and inverse and forward modeling of the geochemical evolution of the contaminated ground waters may often provide the greatest knowledge gains for the least amount of money and time. In particular, geochemical inverse modeling should be used first, prior to forward geochemical modeling, both to explain the currently observed ground-water chemistry in the aquifer system, and to make predictions on the future chemical evolution of the ground waters.

* Studies at the Cape Cod (LeBlanc, 1984) and Borden sites (Mackay et al., 1986) are examples of what we would consider detailed studies. On the order of 10^3 to 10^4 sampling points were installed to study plumes on the order of 10^2 meters to a few kilometers long. However, even at these sites many questions remain regarding the operative geochemical and hydrogeologic processes, and even after more than a decade, studies continue to refine and improve the existing knowledge.

0275-0279/96/0034-0009$05.00

This paper focuses on geochemical modeling and will show how both inverse and forward geochemical modeling approaches were used to better understand the evolution of acidic heavy-metal contaminated ground waters in the Pinal Creek basin, near Globe Arizona. The Pinal Creek basin is a site with sparse spatial information (30 wells distributed in a 15 km long and 10^2 to 10^3 m wide sulfate plume) and with significant temporal variations in both chemical and hydrological characteristics (water-table movements of more than 15 m during a three month period, ground-water velocities on the order of 3 to 5 m/day; Brown and Harvey, 1994). The Pinal Creek site is therefore eminently suited to test our modeling philosophy.

INVERSE GEOCHEMICAL MODELING: BASIC THEORY

Inverse geochemical modeling uses existing ground-water chemical and isotopic analyses, which are assumed to be representative of the chemical and isotopic evolution of a ground-water along a given ground-water flow path, and attempts to identify and quantify the reactions that may have been responsible for the chemical and isotopic evolution of the ground water along the flow path. Although an aqueous speciation code may be used to identify thermodynamically possible (or impossible) reactions and to determine the dissolved inorganic carbon content and the redox state (RS) of the ground waters, the inverse modeling approach does not require that reactions proceed to thermodynamic equilibrium. Indeed, mass-balance constraints and the judgment of the user concerning the possibly occurring reactions are the only constraints posed in the inverse modeling approach.

Mathematical formulation: Inverse modeling with the NETPATH computer code

The following discussion is based on Parkhurst and Plummer's (1993) excellent review of geochemical modeling in ground-water environments. Inverse geochemical modeling codes (BALANCE, Parkhurst et al.,. 1982; NETPATH, Plummer et al., 1991; PHREEQC, Parkhurst, 1995) solve a system of mass balance equations. For the simple case of chemical evolution between an initial and a final water, the mass balance equation for any component *i* can be written:

$$\Delta(i^{total}) = \sum_{p=1}^{P} b_{p,i}\alpha_p \qquad (1)$$

where $\Delta(i^{total})$ is the change in the total molality of element *i* between the initial and final waters, $b_{p,i}$ is the stoichiometric coefficient of element *i* in phase *p*; α_p is the mass transfer of phase *p* in moles per kilogram of pure water; and *P* is the total number of phases. The $\Delta(i^{total})$ concentration changes for each element are known or can be calculated; the reacting phases and stoichiometric coefficients $b_{p,i}$ are postulated by the user; the reaction mass-transfers α_p therefore constitute the unknowns to be solved for. The simulated reactions can include mineral dissolution/precipitation reactions, gas dissolution/exsolution reactions, ion exchange reactions or any other heterogeneous reaction (the NETPATH code, however, does not keep a mass balance on hydrogen or on oxygen, and therefore those elemental mass balances will be ignored if the user, for example defines an H^+/Na^+ exchange reaction).

The mass transfer amounts calculated for the postulated heterogeneous reactions in an inverse geochemical model represent the *net* mass transfer amounts between the initial and final waters chosen for that particular model. An inverse geochemical model does not calculate the specific reaction mass transfers that may be occurring at points along the flowpath in between the final and initial waters chosen for the model.

If the inverse modeling problem requires that several chemically or isotopically different waters be mixed together to form the observed final water, the mass balance equations will instead be stated as follows:

$$(i_{final}^{total}) = \sum_{j=1}^{J} \alpha_j (i_j^{total}) + \sum_{p=1}^{P} b_{p,i}\alpha_p \tag{2}$$

where J is the total number of initial waters to be mixed together to form the final water; and α_j are the fractions of each initial solution j. The α_j and α_p values are the unknowns to be solved for. The number of initial waters used in inverse geochemical modeling is usually less than 2.

Because total dissolved inorganic carbon (TDIC) is not usually measured in most ground-water sampling programs, TDIC concentrations must typically be calculated with a speciation program into which alkalinity and pH data have been entered along with other relevant analytical data.

The total element concentrations referred to in Equations (1) and (2) include all oxidation states of any given element in solution. Therefore, an additional "redox state" (RS) mass balance equation is required to ensure that electrons are conserved during any postulated redox processes. The redox state RS of a solution is defined as

$$RS = \sum_{k=1}^{K} v_k(k) \tag{3}$$

where K is the total number of redox species, and v_k is an operational valence assigned to each aqueous species k. The convention introduced by Parkhurst et al. (1982) defines the operational valence of a species as the charge on the species minus the number of hydrogen atoms in the species plus two times the number of oxygen atoms in the species. Plummer et al. (1983) used the additional convention that all redox-inactive species are assigned an operational valence of zero. Similarly to aqueous species, mineral and gas phases can also be assigned an operational valence μ_p. A few examples of operational valences for aqueous species and mineral and gas phases follow: Ca^{2+}, $\upsilon = 0$; Fe^{2+}, $\upsilon = 2$; SO_4^{2-}, $\upsilon = 6$, $H_2S_{(g)}$, $\mu_p = -2$; $FeSO_{4,\ (s)}$, $\mu_p = 8$. Consequently, the redox mass-balance equation that can be solved along with the element mass-balances represented by Equation (1) is (for a no-mixing problem):

$$\Delta RS = \sum_{p=1}^{P} \mu_{p,i}\alpha_p \tag{4}$$

where ΔRS is the change in redox state between the final and initial solutions.

Although the equations above relate to chemical mass-balances, isotopic mass-balances can also be used to further constrain the amounts and types of different reactions responsible for the chemical evolution of the ground water, as well as solve for the fractions of various initial waters involved in some "mixing" process that may also contribute to the observed chemical and isotopic composition of the final water.

Given a set of I mass-balance equations to solve, a set of P reactant phases can be postulated by the user. The number of reactant phases must necessarily be greater or equal to the number of mass-balance constraints imposed. If $P > I$, NETPATH and PHREEQC will test all possible subsets of I reactions (within the set of P reactions) and will determine whether or not a solution exists for each subset. A solution represents a set of mass-transfer amounts for each of the I reactions present in each subset of P. The user must use his judgment in postulating the possible reactions responsible for the chemical evolution of the ground water. This judgment may be based on knowledge of the mineralogy of the aquifer system, on speciation calculations predicting thermodynamically possible reactions for the initial and final ground-water compositions, and on a judgment of the kinetics of these reactions in the ground-water environment. More specifically the user must estimate the ground-water travel time between the initial and the final water sampling points and judge whether the travel time will be sufficient for any significant reaction mass-transfers.

NETPATH and PHREEQC will allow the user to place constraints on the direction of the reaction mass transfers, that is the user may specify whether a given phase should be allowed to only dissolve or only precipitate. The user may also force NETPATH to consider only solutions, i.e. mass-transfer models, that incorporate specific reactions [a capability absent in earlier versions of PHREEQC]. In so doing, the user shows his conviction that those specific reactions must be occurring and must be at least partially responsible for the chemical/isotopic evolution of the ground water. Placing such restrictions as well as limiting the total number of specified possible reactions is often essential to narrowing down the number of possible models, and also has the effect of forcing the user to think about the reaction processes that may be causing the chemical and isotopic evolution of the ground-water system. Ideally, the user will be able to narrow the number of possible models down to a single one, that quantifies and best represents, at least from the user's view, the various chemical reactions responsible for the chemical and isotopic evolution of the ground water.

Inverse modeling accounting for uncertainties, water and proton mass-balances: the PHREEQC code

In addition to solving the element and redox state (RS) mole-balance equations used in NETPATH (Eqns. 1-4), PHREEQC solves mole-balance equations for (1) the individual valence states of redox-active elements, (2) alkalinity, (3) water and also solves a charge balance equation for each aqueous solution. In its solution of this expanded set of equations, PHREEQC solves for the mixing fraction α_q of each aqueous solution, the aqueous mole transfers α_r between valence states of each redox element, the heterogeneous mole transfers α_p of minerals and gases and a set of analytical adjustments $\delta_{m,q}$ that account for uncertainties in the analytical data. Parkhurst (1995) provides a complete description of the equations solved by PHREEQC. The generalized mole-balance equation used in PHREEQC (Parkhurst, 1995, 1996) can be written:

$$\sum_{q}^{Q} c_q \alpha_q (T_{m,q} + \delta_{m,q}) + \sum_{r} c_{m,r} \alpha_r + \sum_{p} c_{m,p} \alpha_p = 0 \qquad (5)$$

where $T_{m,q}$ is the total number of moles of element or element valence state m in an initial aqueous solution q; Q is the final aqueous solution number; $q = 1$ to $q = Q - 1$ are any number of initial solutions; $\delta_{m,q}$ is the analytical error adjustment to the number of moles $T_{m,q}$ (this analytical adjustment is computed based on the maximum analytical uncertainties specified and the total charge imbalance of the solution); $c_{m,r}$ is the coefficient of the element or element valence state m in the redox reaction r; $c_{m,p}$ is the coefficient in the dissolution reaction p; α_r and α_p are the redox reaction and dissolution reaction mass transfer amounts and α_q is the mixing fraction of solution q. The c_q coefficients are defined so that $c_q = 1.0$ for $q < Q$ and $c_Q = -1.0$ for $q = Q$.

By optionally allowing mass-balances on individual element valence states, PHREEQC (in both its inverse and forward geochemical modeling modes) allows the user to specify multiple redox couples and to appropriately select which redox couples (or an Eh measurement) will control redox equilibria for given elements. For example, the user may have measured dissolved Fe(II) instead or in addition to total dissolved Fe, and dissolved sulfide in addition to dissolved sulfate, and may have made many other possible measurements of redox-active species. Several of those measurements will typically indicate redox disequilibrium in a given water. Those measurements represent valuable information that is often lost when an inverse modeling code like NETPATH or a forward modeling code like PHREEQE or PHREEQM are used. Indeed, those codes normally specify a single redox potential (which may be based on an Eh measurement or on a specific single redox couple, such as the SO_4^{2-}/S^{2-} couple) that applies to all redox-active elements.

The electron balance equation used in PHREEQC can be written:

$$\sum_r c_{e^-,r}\alpha_r + \sum_p c_{e^-,p}\alpha_p = 0 \tag{6}$$

where $c^-_{e,r}$ and $c^-_{e,p}$ represent the number of electrons released/consumed in each redox or phase dissolution reaction.

The alkalinity-balance equation is similar to the general mole-balance equation:

$$\sum_q^Q c_q\alpha_q(T_{Alk,q} + \delta_{Alk,q}) + \sum_r c_{Alk,r}\alpha_r + \sum_p c_{Alk,p}\alpha_p = 0 \tag{7}$$

When alkalinity has not been measured, PHREEQC will determine the alkalinity of a solution from the following equation:

$$T_{Alk,q} = \sum_i c_{Alk,i} m_{i,q} \tag{8}$$

where $c_{Alk,i}$ is the alkalinity contribution of aqueous species i and $m_{i,q}$ is the number of moles of species i in solution q. The alkalinity contribution values, $c_{Alk,m}$, for the master species (also known as basis specis or component species) are chosen such that the reference state for each element of element valence state is the predominant aqueous species at a pH of 4.5. The alkalinity of a solution is determined from a speciation calculation.

The charge imbalance $T_{z,q}$ in an aqueous-solution q is corrected by specifying the analytical error adjustments such that:

$$\sum_m z_m^{tot}\delta_{m,q} = -T_{z,q} \tag{9}$$

$T_{z,q}$ is determined from an aqueous-speciation calculation. For an element or element valence state, z_m^{tot} is the sum of the charge on the master species for that element or valence state plus the alkalinity assigned to the master species, $z_m^{tot} = z_m + Alk_m$. For alkalinity however, z_m^{tot} is -1.0.

The water mole-balance equation used by PHREEQC is:

$$\sum_q^Q \frac{W_{aq,q}}{gfw_{H_2O}} c_q\alpha_q + \sum_r c_{H_2O,r}\alpha_r + \sum_p c_{H_2O,p}\alpha_p + \delta_{H_2O,Q} = 0 \tag{10}$$

where gfw_{H2O} is the gram formula weight of water. $c_{H2O,r}$ and $c_{H2O,p}$ are the stoichiometric coefficients of water in the aqueous redox reactions and in the phase dissolution reactions. $\delta_{H2O,Q}$ is the error adjustment for the number of moles of water in the final aqueous solution Q, but actually accounts for all the uncertainty in the moles of water everywhere in the system.

The analytical error adjustments, $\delta_{m,q}$, computed by PHREEQC are constrained to be smaller than the specified uncertainties $u_{m,q}$:

$$|\delta_{m,q}| \leq u_{m,q} \tag{11}$$

Finally, the aqueous-solution mixing-fractions are also constrained to be positive:

$$\alpha_q \geq 0 \quad . (12)$$

In its search for possible mass-transfer solutions, PHREEQC solves all the mole-balance and charge-balance equations given above, subject to the inequalities given by Equations (11) and (12).

The most important difference between NETPATH and PHREEQC is that PHREEQC allows each analytical datum for each aqueous solution to be adjusted within an uncertainty range specified by the user. PHREEQC will determine sets of phase mass-transfers, solution mixing fractions, and adjustments to the analytical data that satisfy the mass-balance constraints, are consistent with the specified uncertainties, and minimize the sum of the adjustments to the analytical data. As an option, PHREEQC will also determine mass-transfer models that minimize the number of phases involved. The constraints used by PHREEQC inverse modeling are automatically specified by providing a list of the potentially reactive phases. For example, if tremolite ($Ca_2Mg_5Si_8O_{22}(OH)_2$) is identified as a potential reactant, PHREEQC will automatically use mass-balance constraints on Ca, Mg, Si, as well as on alkalinity and water. The water and alkalinity mole-balance equations are always solved, irrespective of postulated phases or reactions. In addition to the mass-balance constraints defined by specifying a list of potential reactants, PHREEQC also lets the user specify additional mass-balance constraints that may be used in determining the mixing fractions for two or more solutions that mix to form a final solution. Unlike NETPATH, PHREEQC includes a charge-balance constraint, which specifies that the sum of the deviations from the analytical data for a given solution must equal the charge imbalance present in that solution. PHREEQC also uses a water mass-balance constraint to account for mixing, water derived from mineral reactions, and water evaporation or dilution. PHREEQC will therefore account for water gained by the dissolution of hydrated minerals. The charge-balance and water mass-balance constraints used by PHREEQC are equivalent to including a mass balance on hydrogen or oxygen. During the inverse modeling simulation, PHREEQC will adjust not only the analytical element concentrations, it will also adjust the pH of the waters. The adjustment to total dissolved inorganic carbon is constrained to be consistent with the adjustments to pH and alkalinity.

Assumptions used in inverse modeling

A model is by definition a construct of assumptions that is meant to help understand some facet(s) of reality. Inverse geochemical modeling of ground waters requires the user to make many assumptions concerning (1) the types of geochemical reactions postulated to be present, (2) the rates of reaction relative to the movement of the water and its mobile constituents, (3) the present distribution of chemical constituents in the aquifer system studied and the prior evolution of this distribution. The last 2 sets of assumptions require that the user have some presumptive knowledge of the ground-water flow and transport system and of its prior evolution. The present discussion addresses some of the common assumptions that need to be made regarding flow and transport processes in inverse geochemical modeling of ground-water systems.

Knowledge of flowpaths and the assumption of a steady-state ground-water flow field. These are the most important and possibly the most tenuous assumptions used in inverse modeling of the chemical and isotopic evolution in a ground-water system. The user often does not have enough hydrologic knowledge to precisely determine the flow paths in a ground-water system. Furthermore, even if there is sufficient refined knowledge of the hydrogeologic system, already-existing wells must often be used. One rarely has the luxury of emplacing new sampling wells. When analyses are available from several wells, the spatial array of chemical and isotopic information may itself be used to decide the most likely flowpath. In most cases, the user will pick a direction that shows the least amount of dilution for the more conservative solutes characterizing the flowpath and the greatest chemical evolution for the more reactive solutes.

Most groundwater systems are likely to experience some seasonal and multi-year fluctuations in hydraulic heads. Therefore, flowlines and groundwater velocities are likely to change at least seasonally, and steady-state conditions may not apply during the time scale of interest. The time scale of interest will normally be the time required for ground-water flow between the wells used in the inverse modeling simulation. The user typically assumes a steady-state ground-water flow field over the time scale of interest, or at the least assumes that any fluctuations in ground-water flow directions and velocities can be averaged out to the ground-water flow field observed at the time of sampling.

The assumption of chemical steady-state. The ground-water analyses used in an inverse model usually represent samples taken concurrently or near-concurrently. The inverse modeling approach assumes that the parcel of water sampled from a final well (well B in Fig. 1) used to have the same composition as that of the water sampled concurrently at the initial well (well A in Fig. 1). This assumption will certainly be reasonable if the ground-water system has remained in chemical (and isotopic) steady-state at least during the travel time required for the water to move from the initial well to the final well. The assumption of chemical and isotopic steady-state simply states that although chemical and isotopic composition may vary spatially, they may not vary in time at any given point in the ground-water system. In ground-water systems with spatially varying chemical and isotopic compositions, the assumption of chemical steady-state will also imply a steady-state ground-water flow field, flow lines and ground-water velocities that have not varied in time at any given point in the system.

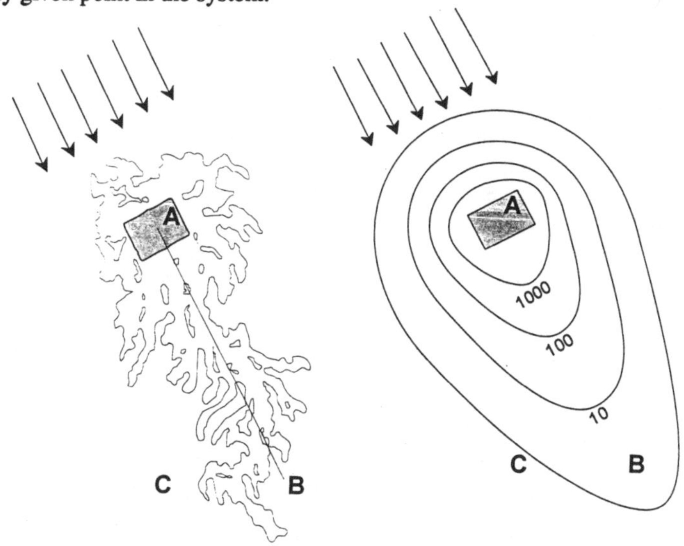

Figure 1. Two map views of a ground-water contaminant plume. Left: actual layout of the plume, drawn with a single concentration contour of concern. A is a well emplaced near or in the source of the contamination (stippled) and B and C are wells further downgradient. Right: Results of a transport model for the same ground-water contaminant plume based on a fit of concentration data obtained from several observation wells. Additional concentration contours are drawn. The large transverse and longitudinal dispersion of the modeled plume results not only from the mixing that actually occurs in the ground but also occurs during pumping at the observation wells, but is also caused by the inability to obtained a sufficiently detailed time-dependent representation of the contaminant plume and of the transient ground-water velocity field.

Most ground-water contamination cases involve dynamically evolving contaminant plumes, for which there can be no assumption of chemical steady-state throughout the ground-water system (steady-state plumes, in which the rate of diffusive/dispersive loss of solute balances the rate of solute influx, are possible but must considered the exception rather than the rule). Fortunately, although it may be desirable, the assumption of chemical and isotopic steady-state is not really required for the inverse geochemical modeling exercise to be meaningful in such situations. Indeed, the less stringent constraint is that the ground-water composition (chemical and isotopic) at the initial sampling point (well A on Fig. 1) must be the same at the time of sampling (t_s) as it was t years ago (at time t_0) when the parcel of water sampled at the final well was near the initial well. (This also assumes that there was a pathline responsible for the transport of water from well A to well B). This less stringent constraint allows the ground-water compositions at points in between the initial and final wells (A and B) to have varied with time, as long as the chemical composition of waters from the initial well has remained invariant. Strictly from a mass-balance point of view, it could be argued that the constraint could be reduced further to require only that the *changes in composition* between waters from well A (at time t_0) and B (at time t_s), rather than the actual compositions of waters from wells A (at time t_0) and B (at time t_s), should have remained constant. A uniform dilution or concentration of the waters sampled at the initial and final wells, however, could lead the user to conclude from his inspection of the mineral saturation indices and general speciation of the waters that some other set of reactions was responsible for the evolution of water A into water B.

How does "mixing" occur in ground-water systems? The U.S. Geological Survey inverse geochemical modeling codes (BALANCE, NETPATH, and the general geochemical code PHREEQC) have the capability of calculating the proportions of two or more initial solutions postulated to have "mixed" together and reacted with themselves and with the surrounding solid and gas phases in their chemical evolution towards the composition of the final water. Clearly in most ground-water environments, "mixing" of ground waters should really be modeled as a continuous process rather than as a discrete process where a small number of specified water compositions are mixed together. Unfortunately, the inverse geochemical codes presented here can not replicate a "continuous" mixing process. Forward transport modeling codes can replicate a continuous mixing process such as dispersion, but even then their results and the very basis of their conceptual models are usually fraught with uncertainty. In using a set of discrete initial water compositions, inverse geochemical models inherently assume that the initial waters chosen encompass the range of intermediate waters that are actually involved in the real, continuous mixing process.

Although the location and timing or sequencing of the mixing and reaction processes is of no mathematical significance in the solution of the mass-balance equations, the user should try to determine where, when and why such mixing processes may have occurred in the ground-water environment. The "mixing" of initial waters by dispersion for example may well have led to heterogeneous mass-transfers in areas that are not on directly on the flowpath between a principal initial water and the final water. The inverse models will, nevertheless, implicitely incorporate those mass transfers in their solution of ***net*** mass-transfer amounts.

The premise of inverse geochemical modeling is that the "final" and "initial" ground-waters used in a model should be related to each other. Ideally, they would represent very small volumes of water sampled from a unique flow-line or path-line. If it were indeed possible to do so (it is not), then the "final" ground water sampled could only have experienced "mixing" as the result of two different processes:

(1) diffusion of chemical and isotopic constituents (and possibly of water) to or

from the flow line (or path line) to neighboring flow lines or to stagnant water zones.

(2) sampling from the path line for more than an infinitesimally small amount of time, in the case of a system not in chemical steady-state. Although most ground-water analyses do not require large samples, and the samples are therefore typically collected over usually small time periods, this may be important in the case of contaminated ground-water evolution. If the system is not in chemical steady-state, the concentration of various constituents may be changing as a function of time at the final well and therefore a sample may in fact represent some ground-water composition averaged out in time and therefore in space.

From a more practical point of view sampling an instantaneous punctual ground-water composition is impossible and unwarranted considering the many sources of other uncertainties. Therefore "mixing" often results from:

(3) the sampling of multiple flow lines that have undergone different chemical and isotopic evolution. This will occur particularly in regions of converging flow, and also when sampling from wells screened across large and/or multiple intervals. Flow convergence may occur naturally or may be the direct result of pumping.

Using the "mixing" option in an inverse geochemical model may also of course be needed because the "initial" and "final" ground waters may not be truly related despite the belief of the user. Just as excessively high values of dispersivity are often used in ground-water transport modeling because of a lack of precise spatial and temporal information (Fig. 1), the use of the "mixing" option in NETPATH or in PHREEQC can often be the result of insufficient information on a ground-water system. For example in Figure 1 if the final well used in the inverse model (well C for example) was off to the side of the path line of heaviest contamination (on which wells A and B are) and if well A was used as the initial well, the inverse model defined by the user would probably require a significant contribution of "background" water to explain the extent of "dilution" between well A and the final well. Similarly pumping a large amount of water from the final well chosen (well B or C) and using the average composition of this water as the final water composition in the inverse model could also lead to a serious misrepresentation of the amount of mixing between the initial most heavily contaminated water (well A) and less contaminated, or even clean, "background" water used as the diluting water. An error in the mixing fractions of initial waters could result in significant errors in the amounts of reaction mass transfers calculated by the inverse model. Furthermore, using water compositions averaged out over a large volume by the sampling process could also mislead the modeler into thinking that certain reactions were thermodynamically impossible, when in fact proper sampling, and location, of the initial and final waters would have indicated that these reactions were in fact possible.

FORWARD GEOCHEMICAL MODELING: THE PHREEQM AND PHREEQC REACTIVE TRANSPORT CODES

Because the concepts of forward geochemical modeling are widely known (at least much more than those of inverse modeling), this discussion will limit itself to a brief description of the reactive transport capabilities of the PHREEQM and PHREEQC geochemical codes.

PHREEQM is a geochemical code developed by Appelo and Willemsen (1987). PHREEQM adds several subroutines to the U.S. Geological Survey geochemical speciation and mass-transfer code PHREEQE (Parkhurst et al., 1980). These subroutines allow PHREEQM to simulate the transport of aqueous-solutions by advection, dispersion and diffusion in a 1-dimensional column (made up of a sequential series of "cells") and to simulate the reaction of those solutions with minerals and surfaces inside the column. PHREEQM has all the chemical reaction simulation capabilities of PHREEQE (it can even be run in a non-transport mode), but in addition PHREEQM is also able to simulate ion exchange processes. Because it is based on the equilibrium mass transfer code PHREEQE, PHREEQM typically uses the local equilibrium assumption in its modeling of reactive transport. Irreversible, zero-order kinetic reactions can be specified, however, and the code can be easily modified to account for first-order element decay or production (Glynn, unpublished work).

The transport algorithm in PHREEQM uses an operator splitting technique. Advection is modeled by shifting cell contents from one cell to the next at every time step or "shift". Dispersion and/or diffusion is simulated by mixing the aqueous contents of each cell with that of its adjacent cells. This algorithm gives PHREEQM the advantage (over most typical finite-difference and finitie-element codes) of being able to simulate not only an advective-dispersive transport process or a diffusive transport process, but also a purely advective, albeit one-dimensional, transport process. The mixing factors f calculated are a function of both aquifer dispersivity α and molecular diffusivity D^*:

$$f_i = \frac{\alpha_i + \alpha_{i+1}}{l_i + l_{i+1}} + 4D^* \frac{\Delta t}{(l_i + l_{i+1})^2} \tag{13}$$

where i is a given cell number and Δt is the time step. Equation (13) can be derived from a finite difference approximation (ignoring advection) centered in space and forward in time. Because its simulation of dispersion is centered in space, PHREEQM shows no numerical dispersion error for conservative constituents when simulating advection-dispersion processes. Numerical dispersion does occur, however, in the case of non-conservative dispersing constituents and is dependent on the amount of retardation experienced by each constituent and on the cell lengths chosen (the maximum numerical dispersivity equals 1/2 the cell length). PHREEQM does not show any numerical dispersion in simulations with only diffusion as a transport process. The lack of sequential iterations between PHREEQM's solution of the chemical equilibrium equations and its simulation of the transport processes at every time step can theoretically generate some error, although our comparisons (Glynn et al., 1991, for example; see Figs. 9 and 10 discussed later) with the sequential iteration finite difference code MST1D (Engesgaard and Kipp, 1992) lead us to believe that the error is typically small as long as an appropriate discretization is used. Finally, operator splitting in itself can also generate error, although our comparisons of PHREEQM with the MST1D code and results by Steefel and MacQuarrie (this volume) suggest that this error is usually minor. A much more complete description of the PHREEQM code and its capabilities can be found in Appelo and Postma (1993).

The recently published geochemical code PHREEQC (Parkhurst, 1995) has the capability of doing both inverse and forward geochemical modeling. Its inverse modeling capabilities have been described in the previous section. PHREEQC also includes all the forward geochemical modeling capabilities of PHREEQE and adds the capability to simulate (1) ion-exchange reactions, (2) sorption processes (using Dzombak and Morel's (1990) diffuse double-layer surface-complexation model and associated thermodynamic data for hydrous ferric oxide), (3) gas bubble formation and (4) advection in a 1-dimensional column. Unlike PHREEQE, PHREEQC also keeps track of mineral amounts, an essential requirement for any reactive-transport simulation. A newer version of PHREEQC has been developed by Tony Appelo (written communication, December 1995) that incorporates all of the features of the PHREEQM transport simulation into PHREEQC, as well as several additional capabilities (including some reaction kinetics). This newer, still unpublished, version of PHREEQC uses essentially the same transport algorithm as PHREEQM, and includes dispersion and diffusion processes. In contrast, the recently published version of PHREEQC includes only pure advection, although dispersion and diffusion can be simulated to some extent through a cell-mixing option (Parkhurst, 1995; Brown, 1996). Unless mentioned otherwise, all PHREEQC simulations referred to in this paper were performed with the published version of the code.

The forward geochemical modeling approach (using the PHREEQE, PHREEQM or PHREEQC codes) is conceptually different from the inverse modeling approach. The inverse modeling approach uses *existing* aqueous-solution data and calculates the mass-transfer amounts for various reactions suspected of accounting for the chemical and isotopic evolution of an "initial" water into a "final" one. The user is responsible for determining that the "initial" and "final" waters are truly related. In contrast, forward modeling allows the *prediction* of aqueous-solution chemical composition given an initial solution and given certain postulated reactions, some of which are usually considered to go to thermodynamic equilibrium. Forward modeling is most suitable and most useful, when the amount of chemical and isotopic data available for a given ground-water system is limited, and also when the objective is to predict the future evolution of the system. Inverse modeling is most useful when abundant chemical, isotopic, hydrologic and mineralogic data are present and all that is desired is a possible explanation of the past chemical evolution of the ground-water system. Of course, just as understanding the past is a key to understanding the future, inverse modeling can also provide some understanding of the reactions that may control the future chemical evolution of a ground-water system.

Forward modeling codes can also be used for the purpose of inverse modeling in a series of trial and error simulations attempting to simulate some real observations (Van Cappellen and Gaillard, Steefel and MacQuarrie, this volume). Although this latest approach can be extremely time consuming, it does have the distinct advantage over simpler inverse geochemical codes of offering a potentially more accurate representation of ground-water mixing as a continuous (rather than discrete) process. This latest approach also does not require the assumption of chemical steady-state. The disadvantages of this approach over that of a non-transport-oriented inverse geochemical modeling approach are essentially the computer time requirements, the significantly greater number of adjustable parameters (flow and transport related) and the consequently greater number of possible solutions that may explain the actual observations. Further references in this paper to inverse modeling will generally not refer to the use of forward codes as part of an inverse modeling approach, although many of the statements made may apply equally well to this latter more sophisticated approach.

The Pinal Creek Toxics Program investigation site, a site of ground-water contamination by acidic metal-laden sulfate-rich wastewater near Globe, Arizona provides a good example of the improved understanding of the chemical reaction and transport processes that may be gained through the combined use of both inverse modeling and forward modeling approaches. The site is described below.

THE PINAL CREEK BASIN SITE: BRIEF DESCRIPTION

The Pinal Creek basin is located in central Arizona, about 100 km east of the Phoenix (Fig. 2). The surface drainage area of the basin occupies 516 km^2, of which 170 km^2 is covered by alluvium and basin fill, which form the regional aquifer. 27 km^2 are covered by mine tailings. Mining, mainly for copper, began in the late 1870s and has been the largest economic activity in the basin.

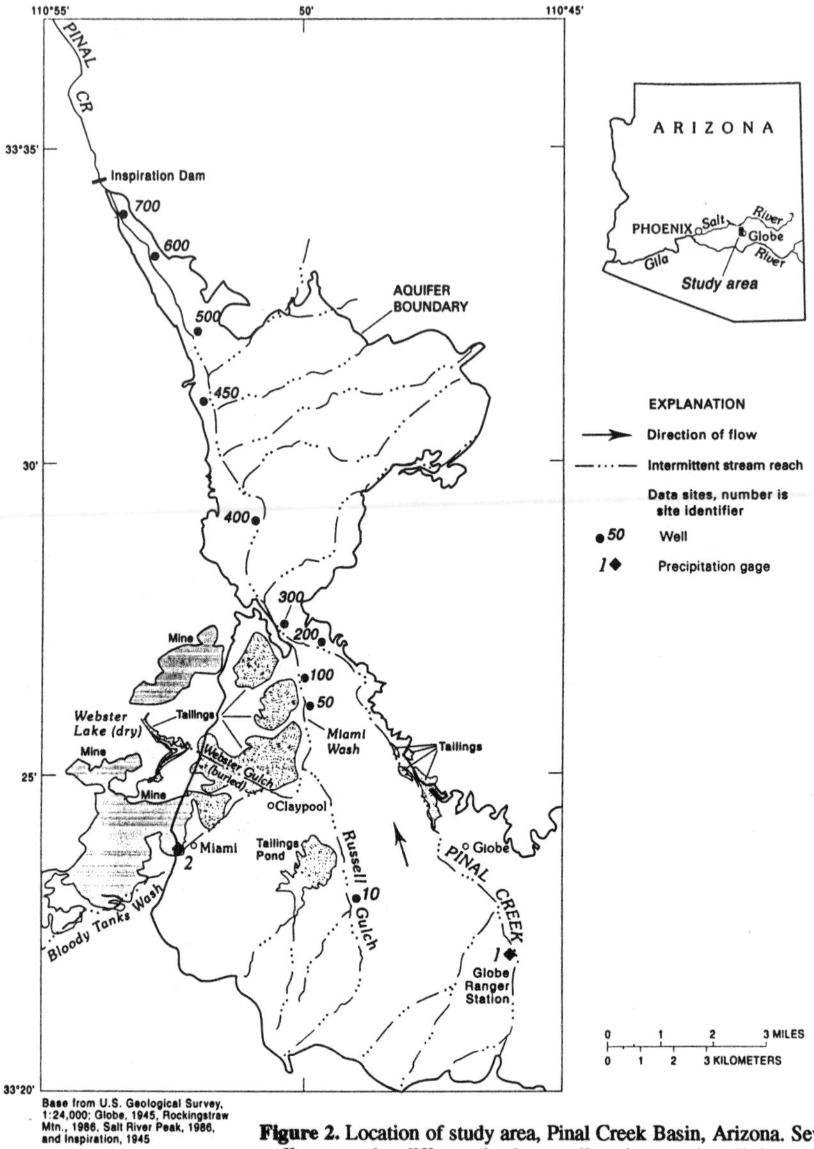

Figure 2. Location of study area, Pinal Creek Basin, Arizona. Several wells screened at different depths actually exist at each well site.

Because of the long history of mining in the basin, there are many potential sources of contamination to the regional aquifer (Eychaner, 1991; Brown and Harvey, 1994). Pyrite is the most abundant sulfide mineral in the tailings. Following significant rainfall, oxidation of pyrite and subsequent runoff into permeable streambeds could represent a significant source of acidity, iron, and sulfate to the regional aquifer. Unlined impoundments of water used in mine processing are a likely source of contamination. In mineralized areas, runoff and ground-water flow may be mineralized naturally, though the amount of this flow is small in relation to flow from unmineralized areas.

The largest suspected single source of contamination in the basin was Webster Lake, an unlined surface water impoundment that existed from 1940 to 1988, when it was drained at the order of the U.S. Environmental Protection Agency (Tolle and Arthur, 1991). Maximum volume of the lake was more than 7 million m^3. In 1988, a sample of water from the lake had a pH of 2.7. Concentrations of iron and sulfate were 6 g/L and 20 g/L, respectively. Aluminum, copper, cobalt, nickel, and zinc were present at concentrations greater than 20 mg/L.

The present U.S. Geological Survey investigation began in 1984 with the drilling of monitor wells along the length of the 15 km-long plume in the stream alluvium. In all, more than 30 wells have been drilled since 1984 to determine the location, chemistry, and movement of the contaminant plume. These wells have been sampled on a regular basis, usually biannually, since November 1984.

Geology

Peterson (1962) describes in detail the geology and mineral deposits of the area. Rocks in Pinal Creek basin range in age from Precambrian to Holocene. Rocks of Precambrian age include schist, diorite, granite, conglomerate, quartzite, limestone, and basalt, which are widely exposed in the hills and mountains throughout the study area. Rocks of Paleozoic age include quartzite, limestone, and shale. Rocks of Mesozoic and Cenozoic age are mainly intrusive and include granite, granodiorite, diabase, and monzonite, all of which are Cretaceous or Tertiary in age. Rocks of Mesozoic and Cenozoic age are exposed in the hills and mountains north of Globe and Miami.

The igneous and metamorphic rocks include a major body of copper porphyry ore. Chalcocite, chalcopyrite, and pyrite predominate in the deeper parts of the ore body, while chrysocolla, malachite, and azurite predominate in its upper, oxidized zone.

The present basin configuration was created by high-angle block faulting associated with basin subsidence that began 19 to 15 million years ago and continued until about 8 million years ago. Basin fill, which is derived from rocks of the surrounding mountains, is Tertiary in age. Lithology ranges "from completely unsorted and unconsolidated rubble of angular blocks as much as 4.5 m in diameter, to well-stratified deposits of firmly cemented sand, silt, and gravel containing well-rounded pebbles and cobbles" (Peterson (1962). Carbonate content of the basin fill is about 1.5 percent (Eychaner, 1989, p 570).

Unconsolidated stream alluvium, which is quaternary in age, overlies the basin fill along Miami Wash, Pinal Creek, and other major drainages. The alluvium is from 300 to 800 m wide and is less than 50 m thick. The alluvium contains cobble- to clay-sized material (Hydro Geo Chem, 1989) although sand- to gravel-sized material is most abundant. Drill cuttings from USGS monitor wells generally contained greater than 90-percent sand and gravel by weight; auger samples indicated the presence of silt or clay beds several inches thick. Sand-sized particles contained mainly quartz, feldspar, and lesser amounts of mica and a variety of rock fragments. Gravel-sized material consisted mainly of rock fragments of granite, volcanic rocks, and schist. Alluvium contains interbedded clays and lenticular clay layers that were as much as 12 m thick at Nugget Wash (Hydro Geo Chem, Inc., 1989). Particle-size analyses of drill cuttings indicate that these lenticular clay layers thin toward the center of the basin and extend an indeterminate distance in length parallel to the axis of the basin.

A sample of alluvium collected in 1985 (Eychaner and Stollenwerk, 1985) contained 0.34 percent calcite by weight. This is equivalent to 0.18 moles of carbonate material per liter of water, using the bulk density of 1.65 g/cc and porosity of 0.316 determined for alluvium used in a column experiment (Stollenwerk, 1994).

Estimates of the concentration of primary manganese oxide minerals were based on samples of alluvium not affected by acidic contamination. At well site 500, in the neutralized part of the plume, the depth-averaged content of manganese oxides was 0.079 mol/l, based on sequential extractions done by Ficklin and Others (1991). Stollenwerk (1994) estimated that 0.0449 mol/l of manganese could be available for reaction in a sample of alluvium obtained from a gravel quarry just east of well site 200.

Geohydrology

The geohydrology of the Pinal Creek basin is the result of past geologic events, the past and present climate, and human activities. Because the area climate is semiarid, most of the drainages in the basin are usually dry but convey large amounts of storm runoff during and after severe storms. Streams that drain the Pinal Mountains also flow during and following snowmelt in late winter and early spring. The amount and distribution of rainfall controls the size, frequency, and duration of streamflow, and the quantity and distribution of water that infiltrates permeable stream alluvium and recharges the regional aquifer.

Ground water in basin fill flows generally northward from the flanks of the Pinal Mountains and westward from the Apache Peak alluvial fan. Most ground water in the basin fill eventually flows upward into the alluvium and then moves generally north to the perennial reach of Pinal Creek. A greater quantity of water, however, recharges the alluvium directly and moves north, mixing with the water from the basin fill. In the northern part of the basin, the aquifer is constricted by impermeable rocks. This constriction forces ground water to the land surface, generating perennial flow from about 6 km above Inspiration Dam to the Salt River, which is a major source of drinking water for the Phoenix Metropolitan area.

INVERSE GEOCHEMICAL MODELING AT THE PINAL CREEK SITE

Like most sites of point-source ground-water contamination, the chemistry of ground waters in the Pinal Creek basin exhibits both spatial and temporal variations. The most heavily contaminated ground-waters are typically found near the base of the unconsolidated alluvial aquifer, where a zone of coarser (and possibly less carbonate-rich) material is suspected to be present. The wells with the most contaminated waters at each well site are 51, 101, 302, 402, 503, 601 and 702 in a down-gradient direction (Fig. 2). Although other wells at each site also show the presence of contaminated water, wells with the most contaminated waters (as measured by total dissolved solids, or chloride or any other relatively conservative constituent) present the most logical choice for inverse modeling. To further narrow the scope of the inverse-modeling study, we focus on the two wells which show the most significant change in the chemical characteristics of their waters, wells 402 and 503. The two water samples chosen represent an acidic contaminated water sampled from well 402 in January 1989 and a neutralized contaminated water sampled from well 503 in November 1991. The two wells are 5.6 kilometers apart. From the difference in sample times and from the distance, we calculate that a parcel of water leaving well 402 would have to travel at a linear ground-water velocity of about 5.2 m per day. This velocity is in the range of the 4.2 to 5.6 m/day velocities estimated by Brown (1995, 1996) using Darcy's law, assuming an effective porosity of 0.3 and a hydraulic conductivity of 200 m/d.

Inverse modeling with NETPATH

To start off our study, we will use the NETPATH inverse modeling code. Unlike the present version of PHREEQC, NETPATH has the advantage of being an interactive code and allows the user to quickly determine the primary issues of concern in an inverse geochemical modeling simulation. The NETPATH code will (1) help us identify some of the possible reaction mechanisms responsible for the chemical evolution of the ground waters between wells 402 and 503 and (2) will quantify some of the reaction mass-transfers involved.

Examination of end-member waters and their conservative constituents. The first step in an inverse modeling study is to examine the chemical composition and thermodynamic state of the waters that will be used in the model. The chemical analyses for the waters chosen for our study are given below in Table 1.

Table 1. Chemical composition of three ground waters from the Pinal Creek basin: an acidic contaminated water (well 402), a background uncontaminated water (well 504), and a neutralized contaminated water (well 503). Concentrations in mg/L. Concentration changes are expressed relative to well 402. ND: not determined. TDIC: total dissolved inorganic carbon. [1]Assumes chloride is conservative. The relative change expressed represents the relative difference in concentration between well 503 water and a mixture of waters from wells 402 and 504, determined on the basis of chloride concentrations. [2]Values were estimated by inspecting earlier and later analyses. [3]Average of two analyses.

	well 402 89/1/12	well 504 91/11/22	well 503 91/11/22	Change between wells 503 and 402	Change due to reaction only[1]
pH	4.13	7.05	5.59		
Eh (in mV)	420 est.[2]	350	410		
Temperature	18 °C	20.5 °C	18.2 °C	1.1 %	
Dissolved oxygen	0.3	6.64	< 0.1		
Calcium	502	44.6 [3]	634	26 %	57 %
Magnesium	161	15.6 [3]	200	24 %	54 %
Sodium	121	19.8 [3]	86	-29 %	-13 %
Potassium	7 est.[2]	2.1	5 est.[2]	-29 %	-16 %
Iron	591	0.004	< 0.1	-100 %	-100 %
Manganese	71.6	< 0.001	116	62 %	106 %
Aluminum	18.4	< 0.01	2.3 est.[2]	-88 %	-84 %
Strontium	2.29	0.335	2.7	18 %	44 %
Silica (as SiO_2)	85.6	27 [3]	91.8	7.2 %	26 %
Chloride	140	9.7 [3]	112	-20 %	0 %
TDIC (as C)	50	ND	ND	ND	ND
Alkalinity (as HCO_3)	ND	227	66	ND	ND
Sulfate	3260	14.2 [3]	2350	-28 %	-8 %
Fluoride	10 est.[2]	0.3	1.5 est.[2]	-85 %	-81 %

As can be seen, the most significant changes in the chemical evolution of the ground-water between wells 402 and 503 are: the increase in pH from 3.9 to 5.6, the 25% increase in calcium and magnesium, the complete removal of 590 mg/L of dissolved iron, the 90%

removal of 18.4 mg/L of dissolved aluminum, the 60% increase in manganese, the 30% decrease in sulfate, and the near-constant dissolved silica concentrations.

As is the case in any geochemical modeling analysis, however, conservative (i.e. non-reactive) constituents are perhaps the most important constituents to examine because they give some information on the physical flow and transport processes. Any ground-water sampling and analysis program should ensure the measurement of at least one, but preferably two or more, relatively non-reactive tracers, such as chloride, bromide, ^{18}O and ^{2}H contents. Sodium may also be relatively conservative although it may increase in solution due to ion exchange, feldspar dissolution, or evaporite dissolution processes. Sodium is rarely taken out of solution by reaction processes except in some instances by sorption and ion exchange processes. In the case of the Pinal Creek ground waters, the high Ca/Na ratio in the acidic waters (Ca/Na = 2.4 mol/mol in well 402) and the even higher ratio in the neutralized waters (Ca/Na = 4.2 mol/mol in well 503) suggests that removal of sodium by ion exchange is not a likely process.

The decrease in both Cl and Na between wells 402 and 503 suggests that a dilution process is occurring. This dilution may be caused either by longitudinal and transverse dispersion processes along the flow path or may be caused in part by the well-sampling process. It is also important to recognize that the ground water sample taken from well 503 in November 1991 was probably not exactly on the pathline originating from well 402 in January 1989. Well 503 may be further off the most contaminated pathline. In recognition of the difficulty in determining the causes and the exact proportions of the various ground waters responsible for the dilution of the well 503 water relative to the well 402 water, an uncontaminated water sampled in November 1991 from below the plume at well 504 was used as the source of diluting water in our inverse geochemical model.

Although chloride undergoes a 20% decrease between wells 402 and 503, sodium undergoes an even greater decrease of about 29%. If the decrease in chloride is used to calculate the fraction (0.2149) of water from well 504 diluting the water from well 402, the observed sodium concentration in well 503 is still 13% lower than the calculated diluted sodium concentration (Table 1, last column). This greater observed decrease in sodium may be at least partly due to a greater Cl/Na ratio in the average diluting water relative to that of the background water used (well 504) used in the calculation. Indeed, although the average Cl/Na ratio in the uncontaminated waters found below the plume (wells 404, 504) or upgradient (well 010) from the plume is 0.44 mg/mg (± 0.10), the average Cl/Na ratio for the most contaminated waters along the flow path is close to three times higher (well 51: 1.48 ± 0.74, well 101: 1.35 ± 0.44, well 302: 1.32 ± 0.36, well 402: 1.29 ± 0.44, well 503: 1.17 ± 0.43; all ratios in mg/mg). [Note the decreasing Cl/Na ratio with distance downgradient, i.e. with increasing neutralization and dilution of the contaminated waters.] Dilute, only slightly contaminated, ground waters sampled from wells on the side of the plume (wells 201, 202) also have a much higher average Cl/Na ratio (0.91 ± 0.30) than that of the uncontaminated ground waters. An argument can therefore be made that these slightly contaminated waters should have been used as the source of the diluting waters in the NETPATH modeling, instead of the uncontaminated water chosen here. The discrepancy in the chloride and sodium dilution factors can be used, however, as a measure of the uncertainty inherent in trying to model the ground-water mixing process with a simple inverse geochemical model.

The thermodynamic state of the end-member waters. After examining the conservative constituent concentrations of the ground waters, the next step is to examine the aqueous-speciation results, in particular the mineral saturation indices (Table 2) calculated for the three end-member waters chosen in our model. The speciation calculations were performed with the WATEQFP code incorporated in the database management code DB distributed with the NETPATH code. The thermodynamic database used in WATEQFP is a subset of the database described in Nordstrom et al. (1990) and is essentially similar to the thermodynamic databases used in all U.S. Geological Survey ion association codes.

Table 2. Saturation indices and carbon dioxide equilibrium partial pressures for an acidic ground water (well 402), an uncontaminated ground water (well 504) and a neutralized contaminated ground water (well 503) from the Pinal Creek alluvial and basin fill aquifers. NC: could not be calculated

Mineral	well 402 89/1/12	well 504 91/11/22	well 503 91/11/22
Calcite	-4.966	-0.448	-1.812
Dolomite	-10.170	-1.062	-3.867
Siderite	-2.602	-11.470	NC
Rhodochrosite	-3.253	NC	-0.003
Gypsum	0.010	-2.573	0.051
Fluorite	-3.200	-2.163	-2.511
SiO_2(am)	-0.063	-0.598	-0.038
Chalcedony	0.800	0.257	0.826
$Al(OH)_3$(am)	-4.032	NC	-0.849
Gibbsite	-1.277	NC	1.904
Kaolinite	0.745	NC	7.156
Alunite	1.817	NC	6.599
$Fe(OH)_3$(am)	-0.966	0.518	NC
Goethite	4.926	6.409	NC
K-Jarosite	0.033	-8.453	NC (< -3)
log pCO_2 (in atmospheres)	-0.992	-1.730	-0.921

The background water (well 504) is essentially a $Ca(Mg)HCO_3$ water typical of all the uncontaminated ground waters in the Pinal Creek basin. These waters are usually near saturation with calcite, dolomite and chalcedony, have near to slightly above neutral pH values and have equilibrium CO_2 partial pressures between $10^{-1.5}$ and $10^{-2.0}$. The uncontaminated ground waters are also typically rich in dissolved oxygen and other dissolved atmospheric gases (Glynn and Busenberg, 1994a, Robertson, 1991, Winograd and Robertson, 1982).

In comparison, the acidic water from well 402 is highly undersaturated with respect to calcite, dolomite, siderite ($FeCO_3$, SI: -2.60) and rhodochrosite ($MnCO_3$, SI: -3.25) and is near saturation with amorphous silica, kaolinite and gypsum. The water is also undersaturated with respect to amorphous $Fe(OH)_3$ but supersaturated with respect to goethite, and very highly undersaturated with respect to all manganese oxides (pyrolusite (MnO_2), hausmanite (Mn_3O_4), manganite (MnOOH) and pyrochroite ($Mn(OH)_2$)). These speciation results, based on the relatively high measured Eh (420 mV), are consistent with the high Fe and Mn contents of the water and the lack of any evidence of sulfate reduction. Surprisingly, the calculated equilibrium CO_2 partial pressure, $10^{-0.99}$ is very close to that of the neutralized water.

In comparison to the acidic water from well 402, the partially neutralized water from well 503 is not as highly undersaturated with respect to calcite and dolomite and remains close to saturation with respect to both gypsum and amorphous silica. Unlike its more acidic precursor, the water is highly supersaturated with respect to kaolinite and is instead near saturation with respect to an $Al(OH)_3$ phase. Although most of the dissolved iron has dropped out of the water, manganese has increased to near saturation with rhodochrosite. The water is still undersaturated with respect to several manganese oxides (pyrolusite SI: -9.14, hausmanite SI: -13.43, manganite SI: -4.67, pyrochroite SI: -7.22), although the uncertainty in these saturation indices is high, given the poor knowledge of manganese oxide thermodynamics and the dependence of the calculated saturation indices on the measured Eh. Indeed, lack of data on the vanishingly small dissolved Mn(IV) and Mn(III) concentrations makes any saturation index calculations for the Mn(IV) and Mn(III) minerals (pyrolusite, hausmanite, manganite) almost meaningless, because the calculations assume that the measured Eh values are representative of the Mn(IV)/Mn(II) and Mn(III)/Mn(II) aqueous activity ratios. Finally, the equilibrium CO_2 partial pressure ($10^{-0.94}$) is close to that of the acidic water from well 402, and is more than an order of magnitude higher than may be expected from equilibrium with unsaturated zone CO_2 partial pressures (Glynn and Busenberg, 1994b).

NETPATH inverse modeling: first simulation results. The first NETPATH simulation considered the following 11 mass balance constraints: Cl, Ca, Mg, Na, Al, Si, RS (redox state), Fe, Mn, C, S. The following 14 phases were considered (with additional limitations mentioned. Note: "forced inclusion" means that NETPATH was forced to consider only models that included the specific phase, or reaction):

calcite (forced inclusion dissolution only),
goethite (forced inclusion; precipitation only),
gypsum (forced inclusion; precipitation only),
kaolinite (precipitation only),
SiO_2,
dolomite (dissolution only),
MnO_2 (dissolution only),
rhodochrosite ($MnCO_3$),
anorthite ($CaAl_2Si_2O_8$; dissolution only),
gibbsite,
$Mn(OH)_3$ (precipitation only),
O_2 gas (dissolution only),
CO_2 gas (exsolution only),
a pure Na phase

This last phase was added simply to keep track of the Na imbalance. In this first simulation the mixing fractions of well 402 and well 504 waters were determined through the chloride concentrations, because no Cl phases were specified. As a result of the 11 element mass-balance constraints, and because of the mixing of the two initial waters (similar to having one forced phase mass-transfer), 10 out of 14 possible phases were present in each NETPATH model solution. Additional mass-transfer limitations were therefore necessary and were used to minimize the number of possible models found. NETPATH checked 330 models or possible solutions and actually found 12 that did not violate the limitations placed (whether a phase was forced to be included in all models, and whether it was allowed to dissolve only or to precipitate/exsolve only or both). Of the 12 models, 6 are given here (Table 3). They adequately represent the range of possible solutions given by the NETPATH code and will be further discussed. It should be remembered that linear combinations of any possible models also form possible solutions.

Table 3. NETPATH models. First simulation. Results in millimoles per kilogram of H_2O. Positive numbers indicate dissolution, negative numbers precipitation or degassing.

	Model 1 anorthite gibbsite	Model 2 gibbsite SiO_2	Model 3 kaolinite SiO_2	Model 4 e^- transfer	Model 5 O_2 gas	Model 6 O_2 gas rhodo. diss.
well 504 fraction	0.216	0.216	0.216	0.216	0.216	0.216
pure Na	-0.579	-0.579	-0.579	-0.579	-0.579	-0.579
dolomite +	2.899	2.899	2.899	2.899	2.899	2.899
gypsum **- F**	-2.219	-2.219	-2.219	-2.219	-2.219	-2.219
goethite **- F**	-8.339	-8.339	-8.339	-8.339	-8.339	-8.339
calcite **+ F**	4.929	5.086	5.086	4.929	4.929	4.929
anorthite	0.157			0.157	0.157	0.157
kaolinite			-0.226			
gibbsite	-0.766	-0.452		-0.766	-0.766	-0.766
SiO_2		0.314	0.766			
rhodochrosite	-2.972	-2.972	-2.972			1.092
MnO_2 +	4.064	4.064	4.064	7.036	1.092	
$Mn(OH)_3$ -				-5.944		
O_2 gas +					1.486	2.032
CO_2 gas -	-6.033	-6.190	-6.190	-9.005	-9.005	-10.097
net protons consumed	5.707	5.707	5.707	5.707	5.707	5.707

The phases in the simulation were chosen based on our knowledge of the mineralogy of the basin fill and alluvial aquifer materials and also on our examination of the saturation indices of the well 402 and well 503 water. Although gypsum is not present in the uncontaminated aquifer, the acidic and especially the neutralized contaminated waters are consistently slightly supersaturated with respect to gypsum. In fact samples brought back from the field precipitate gypsum over the course of several months. Calcite and dolomite are known to be present in the aquifer materials and were therefore included in the model. Similarly, there is no lack of manganese oxides in the alluvial materials. Manganese oxides form at the contact between the Mn(II) rich-ground waters and oxygenated ground-waters, and are also widely disseminated in the uncontaminated sand and gravel (Ficklin et al., 1991). Lind and Stollenwerk (1994) conducted an elution experiment reacting acidic iron and manganese-rich ground water from well 101 with alluvial material from well 601, downgradient from the manganese-contaminated ground waters. Based on X-ray diffraction results, Lind and Stollenwerk (1994) found that pyrolusite (β-MnO_2) and a solid resembling kutnahorite ($CaMn(CO_3)_2$) were present before, but not after, the elution of the alluvial materials. Although goethite was the Fe(III) phase chosen (for precipitation only), choosing any other Fe(III) oxide would have resulted in the same Fe mass transfer values. Thermodynamic stability is numerically irrelevant in NETPATH calculations. Similarly, we could have picked amorphous $Al(OH)_3$ instead of gibbsite. In fact, Fe and Al are most likely precipitating as fairly amorphous precipitates, that may recrystallize to more stable crystalline forms with time. The WATEQFP speciation results suggest that the waters near well 503 may be precipitating some $Al(OH)_3$ phase. The speciation results also suggest that kaolinite may be forming near well 402, but probably does not form very quickly near well 503 (as evidenced by the very high supersaturation with respect to kaolinite). The precipitation of amorphous forms of Al and Fe(III) minerals upon reaction of the alluvial sediments with acidic waters is also suggested by the elution experiments of Lind and Stollenwerk (1994) and by the selective extractions performed by Ficklin et al. (1991) on core materials from wells 107 (acidic), 451 (partially neutralized) and 505 (neutralized). Ficklin et al. (1991) also report no visible association between Al and SO_4 and argue therefore against the formation of an $AlOHSO_4$ phase. Stollenwerk and Eychaner (1987) had, however, earlier argued that this phase controlled aluminum concentrations in the acidic ground waters. Furthermore, in his column elution studies, Stollenwerk (1994) found that he could best simulate the behavior of dissolved aluminum by using amorphous $Al(OH)_3$ as the solubility-limiting phase at pH values above 4.7 and $AlOHSO_4$ at lower pH values. He did, however, change the solubility product of the $AlOHSO_4$ phase to best fit his experimental results (from log K = -3.23 to log K = -2.2). Considering the available evidence, we believe that the issue of $AlOHSO_4$ precipitation is not resolved and requires further research. The fact that the water from well 402 is close to saturation with kaolinite and the fact that kaolinite is known to form in acidic waters with high dissolved silica (Blair Jones, U.S. Geological Survey, pers. comm., 1996) leads us to prefer the hypothesis of Al control by kaolinite in the more acidic waters from the site. Nevertheless, we will also investigate the effect of using $AlOHSO_4$ in both our inverse and forward geochemical models.

Because well 503 water is close to saturation with respect to rhodochrosite, we chose this mineral as a possible Mn sink. We believe that reductive dissolution of MnO_2 is the primary process causing dissolved Fe(II) to oxidize and precipitate out of solution. The only problem with this reaction mechanism is that the increase in dissolved Mn(II) is too small relative to the decrease in Fe(II). Several other possible reactions could explain this fact. (1) Mn(II) may be precipitating out as rhodochrosite. (2) one mole of MnO_2 does not have to dissolve completely in order to oxidize two moles of Fe(II). An electron transfer process may be taking place during which the oxidation state of the Mn oxide simply decreases while only partially releasing Mn into the solution. (3) Mn(II) may be sorbing onto the freshly precipitated Fe-oxyhydroxides. (4) Oxygen is known to be diffusing through the unsaturated zone into the ground waters near the water table. Because of the depth of the well 402 and 503 waters below the water table, however, this last mechanism is not really considered to occur. Although O_2 ingassing was considered in our first NETPATH simulation, this reaction will be discarded in our second simulation.

For similar reasons, the possibility of CO_2 exsolution from a deep flow path can not be seriously entertained. Glynn and Busenberg (1994b) estimated, based on their measurements of dissolved gases in the Pinal Creek ground waters, that only waters down to ~2 m below the water table could possibly be exsolving dissolved gases and CO_2. Significant CO_2 exsolution would also cause exsolution of other dissolved gases such as N_2 and Ar. For example, exsolution of CH_4 and CO_2 from an hydrocarbon contaminant plume has been held responsible for the very low dissolved Ar and N_2 concentrations measured in ground waters from the U.S. Geological Survey Bemidji Toxics site (Revesz et al., 1995). Instead, ground waters from the Pinal Creek site show very high concentrations of both dissolved N_2 and Ar because of the large amounts (often above 20 mL/L) of excess air entrained during ground-water recharge at the site (Glynn and Busenberg, 1994b).

Interestingly enough the results of our first NETPATH simulation suggest that other

Ca and Mg sources (in addition to calcite and dolomite) are needed if CO_2 is disallowed as a sink for the excess carbon provided by the dissolution of the carbonates. We initially thought that rhodochrosite ($MnCO_3$) would provide an additional carbon sink but found out that given the Mn mass balance constraints, the rhodochrosite sink would not be a strong enough sink for the excess carbon. The presence of another Mn sink (such as Mn(II) sorption) instead or in addition to rhodochrosite precipitation would only exacerbate this problem. Therefore, because no other carbon sinks are likely to be present (the waters are undersaturated with respect to siderite), the next solution was to incorporate another Ca source, specifically a Ca silicate, so as to reduce the amount of carbon coming into solution. Although anorthite was chosen, it is likely that any silicate mineral dissolution accelerated by the acidic ground waters would also act as a source of Mg, Na and K (and probably Fe and Mn) to the solution. The dissolution of Ca-rich silicates and perhaps Mg-rich silicates, however, can be expected to be faster than that of the Na and K rich silicates. On the basis of their alluvium elution experiments, Lind and Stollenwerk (1994) suggest that tremolite ($Ca_2Mg_5Si_8O_{22}(OH)_2$) dissolution may be a source of both Ca and Mg to the Pinal Creek ground waters. Indeed, amphiboles, such as tremolite, and pyroxenes can be expected to have faster reaction rates than feldspar minerals, although their abundance in the alluvial materials is minor compared to that of the feldspar minerals. The presence of CO_2 degassing in all the models found by the the first NETPATH simulation suggests that some Mg-silicate phase (such as tremolite) must be included if models without CO_2 degassing are to be found.

The last row in Table 3 gives an estimate of the net number of millimoles of protons consumed by the various reaction models. Essentially, the number of protons consumed in each reaction model was calculated by estimating the number of protons consumed by the *dissolution* of one millimole of each solid or gaseous phase. The number of protons consumed is dependent on the degree of protonation or hydroxylation of the various aqueous species produced by the dissolution reactions. For example a calcite dissolution reaction will show consumption of two protons per mole of calcite dissolved if the reaction is written to produce $H_2CO_3^0$ (or equivalently aqueous CO_2; henceforth $H_2CO_3^0$ is meant to comprise the actual species and the much more dominant aqueous CO_2 species) but will show consumption of only one proton if the reaction is written to produce HCO_3^-. The proton consumption calculations shown here assume that the reaction byproducts are the dominant aqueous species determined from the speciation of the well 402 water, such as AlF^{2+}, AlF_2^+, $AlSO_4^+$, $Al(SO_4)_2^-$, Al^{+3} for Al, Mn^{2+}, $MnSO_4^0$ for Mn, Fe^{2+}, $FeSO_4^0$ for Fe, $H_2CO_3^0$ for TDIC (species listed are in order of decreasing predominance). Using this assumption, the number of moles of protons consumed per mole of phase dissolved are: 14 for tremolite, 8 for anorthite, 6 for kaolinite, 4 for dolomite, MnO_2 and O_2 gas, 3 for goethite (or $Fe(OH)_3$), gibbsite (or $Al(OH)_3$) and $Mn(OH)_3$, 2 for calcite and rhodochrosite, 1 for $AlOHSO_4$. All other phases mentioned in Tables 3 and 4 are assumed not to consume protons upon dissolution. The consumption of protons by the heterogeneous mass transfer reactions must necessarily be matched by an increase in solution pH and also by the release of protons from homogeneous deprotonation reactions (such as $H_2CO_3^0 \Rightarrow HCO_3^- + H^+$, and $HSO_4^- \Rightarrow SO_4^{2-} + H^+$). Given (1) that the increase in pH between wells 402 and 503 corresponds to approximately a 0.1 millimole decease in H^+ concentration, (2) that the difference in $H_2CO_3^0$ concentrations in well 402 and 503 waters, is less than 1 millimolal (and HCO_3^- is always at least 5 times lower than the $H_2CO_3^0$ concentration), (3) that the concentration of HSO_4^- in well 402 water is close to 0.1 millimolal, and (4) that there are no other major homogeneous deprotonation reactions, it appears that the 5.7 millimoles of proton consumption calculated for the various reaction models presented in Table 3 are at least 5 times too high. Unaccounted surface deprotonation or proton exchange reactions offer at least one possible reason for this discrepancy. Erroneous reaction models and analytical uncertainty in the basic data collected are other possible reasons.

The most interesting results of the first NETPATH inverse modeling simulation are the following. Gas exsolution or dissolution were found to be necessary in all models, even though anorthite dissolution and rhodochrosite precipitation were included. Of all the models found by the first NETPATH simulation, we prefer the 3 models that considered MnO_2 dissolution and rhodochrosite precipitation, rather than an electron-transfer mechanism ($MnO_2 \Rightarrow Mn(OH)_3$) or O_2 ingassing (with or without acccompanying rhodochrosite dissolution). Out of those three models, we also prefer the two models (models 1 and 2 in Table 3) that did not involve kaolinite precipitation. Although possible, the very high supersaturation of well 503 water with respect to kaolinite suggests that the mineral does not undergo very fast precipitation, at least at pH values >4. Instead, we favor aluminum control by $Al(OH)_3$ precipitation (with possible recrystallization to gibbsite).

The second NETPATH simulation. A second attempt to run the same NETPATH simulation using Na as the conservative constituent, instead of chloride, resulted only in "invalid" models that required the dissolution, rather than the precipitation, of 2.22 millimoles of gypsum per kg of water. Because both the acidic and neutralized ground waters at Pinal Creek are supersaturated with respect to gypsum, slightly but consistently, a model with gypsum dissolution was not plausible. A pure chloride source (0.484 millimoles) was used in this second simulation. The calculated mixing fraction of background water from well 504 was 0.347 (instead of 0.216).

The third NETPATH simulation. A third attempt to run the initial NETPATH simulation described above, using an intermediate mixing fraction of 0.281 (instead of 0.216 or 0.347) resulted in 12 models that were similar to those of the first NETPATH simulation, but had different mass-transfer amounts. Gypsum precipitation was very small (-0.003 millimoles). All models found required CO_2 mass transfer, but in somewhat smaller amounts (e.g. -5.4 instead of -6.0 millimoles for Model 1, Table 3).

The fourth NETPATH simulation. A fourth NETPATH simulation was used to explore the effects of including tremolite [$Ca_2Mg_5Si_8O_{22}(OH)_2$], biotite [$KMg_{1.5}Fe_{1.5}AlSi_3O_{10}(OH)_2$], forsterite [$Mg_2SiO_4$], a pure Mn sink (to simulate Mn II sorption) and $AlOHSO_4$, while excluding some of the reactions considered unrealistic in the previous simulations, namely CO_2 exsolution, O_2 dissolution and kaolinite precipitation. Forsterite was included for numerical reasons because it is a pure Mg-silicate with a high Mg/Si ratio. Tremolite and biotite were included because those minerals are commonly found in the Pinal Creek sediments. Although K is present in biotite no mass-balance for K was included in the NETPATH simulations because of the large uncertainties in our estimated K data. The NETPATH simulation resulted in 19 possible models. Of the 19 models, 6 included more than 1 Mg-silicate phase and are therefore not presented here (Table 4) for reasons of space and simplicity. Five models included tremolite as the only Mg-silicate phase, and differed from each other in their treatment of the Mn and Al mass-balances (Models 7, 10-13 in Table 4). Five other similar models included biotite instead of tremolite (Model 8 in Table 4) and 3 remaining models considered forsterite but did not include $AlOHSO_4$.

None of the models found included dolomite, a mineral that we definitely know is present in alluvium (we have seen dolomite rock fragments in the alluvium; dolomite formations are also present in the surrounding hills). Dolomite should certainly be as reactive with the acidic waters as some of the silicate minerals. The problem appears to be that to include dolomite dissolution as a primary contributor of dissolved magnesium a carbon sink must be found, in addition to rhodochrosite. This carbon sink remains enigmatic so far. In the meantime, of all the models presented in Table 4, we prefer the models (Models 7-9) that considered MnO_2 dissolution and rhodochrosite precipitation and did not include $AlOHSO_4$. The electron transfer models with any Mg-silicate phase are also believable. In reality, the reactions occurring and responsible for the evolution of well 402 water to well 503 water are likely to be some linear combination that may include, but will not be restricted to, the mass transfer models that we found using the NETPATH code. Many more models could have been found had we included other phases (silicates in particular), but their description and classification would have been pointless for the purposes of this manuscript.

Table 4 also gives the millimoles of protons consumed for each reaction, using the assumptions discussed earlier (see discussion of Table 3). Once again it appears that the millimoles of proton consumption calculated for the various reaction models presented in Table 4 are about 3 to 10 times too high. Unaccounted surface deprotonation and proton

Table 4. Fourth NETPATH simulation. Same phases included as in first three simulations, except for the following changes: (1) tremolite, biotite, forsterite, $AlOHSO_4$, a pure Mn sink included as possible phases; (2) kaolinite and gas mass transfers excluded. e^- trans.: electron transfer mechanism.

	Model 7 tremolite	Model 8 biotite	Model 9 forsterite	Model 10 tremolite e^- trans.	Model 11 tremolite Mn sink	Model 12 tremolite $AlOHSO_4$	Model 13 tremolite no anorthite $AlOHSO_4$
well 504 fraction	0.216	0.216	0.216	0.216	0.216	0.216	0.216
pure Na	-0.579	-0.579	-0.579	-0.579	-0.579	-0.579	-0.579
dolomite +							
gypsum **- F**	-2.219	-2.219	-2.219	-2.219	-2.219	-0.648	-0.089
goethite **- F**	-8.339	-11.238	-8.339	-8.339	-8.339	-8.339	-8.339
calcite **+ F**	4.696	6.146	4.696	1.724	1.724	4.696	4.696
anorthite	2.130	1.840	3.290	5.102	5.102	0.559	
$AlOHSO_4$ -						-1.571	-2.130
gibbsite	-4.712	-6.066	-7.032	-10.656	-10.656		1.678
SiO_2	-8.587	-9.166	-7.717	-14.530	-14.530	-5.445	-4.326
rhodochrosite	-2.972	-4.422	-2.972			-2.972	-2.972
MnO_2 +	4.064	5.514	4.064	7.036	4.064	4.064	4.064
$Mn(OH)_3$ -				-5.944			
Mn sink -					-2.972		
tremolite	0.580			0.580	0.580	0.580	0.580
biotite		1.933					
forsterite			1.450				
net protons consumed	5.711	7.642	5.711	5.711	11.655	5.708	5.711

exchange reactions and analytical data uncertainty offer possible reasons for the discrepancy. This problem will be circumvented in the PHREEQC inverse modeling demonstration discussed later on, because PHREEQC always includes alkalinity mass-balance and charge balance equations and also considers possible uncertainties in the analytical data.

The fifth and sixth NETPATH simulations. A fifth NETPATH simulation was conducted using Na, instead of Cl, as the conservative element dictating the mixing fractions of the waters from wells 402 and 504. Fifty-one models were found, but all dissolved more than 2 millimoles of gypsum. Similarly to the third NETPATH simulation conducted above, a sixth simulation was also conducted using an average mixing fraction of well 504 water (0.2814) midway between those of simulations 4 and 5. The 13 models found were similar to those of the fourth NETPATH simulation, but had different mass-transfer amounts. Gypsum precipitation was very small (-0.003 millimoles). None of the models found included $AlOHSO_4$ precipitation or dolomite dissolution. It is likely, however, that the dissolution of Ca- and Mg-silicates would also provide a source of Na. If that is the case, and if we assume that well 503 water (Nov. 1991) is indeed derived from well 402 water (Jan. 1989), then a Na sink remains to be found. Alternatively, we may consider that (1) the background water chosen does not have a Na composition representative of the diluting waters mixing with well 402 type water, or (2) analytical errors are present in the reported Na concentrations, which prevent the use of this element as a conservative constituent. Although the increase in Ca/Na ratios would normally argue against removal of dissolved Na by ion exchange, dissolution and accelerated weathering of silicate minerals by the acidic waters may be causing a significant increase in the cation exchange capacity of minerals exposed to the ground waters and may thus be responsible for a net removal of Na (and other cations) from solution (Blair Jones, U.S. Geological Survey, verbal communication).

Conclusions from the NETPATH simulations. Perhaps the most important conclusion provided by the NETPATH simulations above is that Ca- and Mg-silicate mineral dissolution must be a significant process. Many researchers at the Pinal Creek site originally believed that calcite and dolomite dissolution was responsible for the most of the acid neutralization. However, Glynn (1991) demonstrated that the increase in Sr concentrations between wells 51 and 402 must have been caused by silicate mineral dissolution, because the amount of Sr present in limestone and dolomite formations contributing to the carbonate content of the alluvial materials is too small relative to the amount of Sr that precipitates out as an impurity in gypsum. Sr is a significant impurity in Ca-silicate minerals and is released to the solution during their dissolution and weathering. Similarly, dissolved inorganic carbon $\delta^{13}C$ data (Glynn, Busenberg and Brown, unpublished data collected in June 1993) ranges from -9.15 to -12.90 per mil for the acidic or neutralized contaminated ground waters, and from -10.95 to -14.00 per mil for the uncontaminated waters, suggesting that neutralization of the acidic plume by silicate minerals must indeed be very important. [All $\delta^{13}C$ values are expressed relative to the Vienna PDB standard.] If the calcite and dolomite $\delta^{13}C$ values are near 0 per mil, as is reasonable for most marine carbonates, closed system dissolution of those carbonates, as may be caused by acid neutralization reactions deep within the aquifer, should have resulted in significantly higher $\delta^{13}C$ values. This theory awaits confirmation (or denial) from a determination of the actual solid carbonate $\delta^{13}C$ values.

Inverse geochemical modeling with PHREEQC. PHREEQC provides additional capabilities for modeling of the chemical evolution between the well 402 and well 503 waters, because it considers uncertainties associated with individual element analyses and also solves alkalinity-balance, water mass-balance and charge-balance equations. PHREEQC allows each analytical datum for each aqueous solution to be adjusted within an uncertainty range that is specified by the user. PHREEQC will determine sets of phase mass-transfers, solution mixing fractions, and adjustments to the analytical data that satisfy mass-balance constraints and are consistent with the specified uncertainties. As an option, PHREEQC will also determine mass-transfer models (later referred to as "minimal" models) that minimize the number of phases involved. The constraints used by PHREEQC inverse modeling are automatically specified by providing a list of the potentially reactive phases. For example, if tremolite is identified as a potential reactant, PHREEQC will automatically use mass-balance constraints on Ca, Mg, and Si (NETPATH does not do this). In addition to the mass-balance constraints defined by specifying a list of potential reactants, PHREEQC also lets the user specify additional mass-balance constraints that may be used in determining the mixing fractions for two or more solutions that mix to form a final solution. Unlike NETPATH, PHREEQC includes a charge-balance constraint, which specifies that the sum of the deviations from the analytical data for a given solution must equal the charge imbalance present in that solution. PHREEQC also uses a water mass-balance constraint to account for mixing, water derived from mineral reactions, and water evaporation or dilution (PHREEQC and NETPATH are not limited to ground-water problems). The charge-balance and water mass-balance constraints used by PHREEQC are equivalent to including a mass balance on hydrogen or oxygen. During the inverse modeling simulation, PHREEQC will adjust not only the analytical element concentrations, it will also adjust the pH of the waters. The adjustment to total dissolved inorganic carbon is constrained to be consistent with the adjustments to pH and alkalinity.

A PHREEQC simulation was constructed using all the phases used in the NETPATH simulations described above, except for kaolinite and CO_2 and O_2 gases which were excluded. Including biotite [$KMg_{1.5}Fe_{1.5}AlSi_3O_{10}(OH)_2$] in PHREEQC forced the code to account for K mass-balance. K, however, can not be expected to accumulate in the Pinal Creek ground waters and should also not be used to determine the mixing fractions of well 402 and well 504 waters. Therefore, a pure potassium-montmorillonite [$K_{0.33}Al_{2.33}Si_{3.67}O_{10}(OH)_2$] was also included in the simulations with biotite and only allowed to precipitate. Both sodium and chloride were also specified as mass balance constraints. Pure Na, pure Cl or pure Mn sinks were not specified, because PHREEQC does not allow the specification of charge-imbalanced phases. Adding charge imbalance would prevent PHREEQC from correctly adjusting the analytical data within the user-specified uncertainties. Adding Na as NaOH would also interfere with the PHREEQC results, because it would affect the alkalinity balance by causing a change in pH. A ±5% relative uncertainty was chosen for all elements, except for K (±20%) and for Cl for which an uncertainty of ±10% was initially chosen but later was reduced to ±5%. The lower uncertainty for Cl did not affect the number of possible mass-transfer models calculated by PHREEQC, minimal or otherwise, did not significantly affect the calculated mixing fractions of well 504 water, and did not result in any changes in the phases included in the mass-transfer models. The uncertainty in the pH of the three waters from wells 402, 504 and 503 was ±0.05 pH units. The "minimal" option was initially chosen to reduce the number of possible models to those minimizing the number of phases involved. Some additional precipitation-only and dissolution-only constraints were added on the phases chosen. The models shown in Table 5 represent a representative selection of all the models found by PHREEQC. Table 5 shows most of the tremolite-containing models, but all other models found (with biotite/K-montmorillonite, forsterite or tremolite) were essentially variations on the combination of reactions shown in Table 5. It should also be remembered that, similarly to the NETPATH models, linear combinations of PHREEQC inverse models also represent possible models. With the exception of Model 1 (compare with Model 2!), all models reported in Table 5 are "minimal" models, i.e. had the minimum number of phases.

PHREEQC also calculates, as an option, the minimum and maximum mass transfers associated with any given phase in any given model. These minimum and maximum mass-transfers are calculated from the minimum and maximum values associated with the various element and pH analyses. Due to space considerations, only the minimum and maximum mixing fractions of well 504 water are given in Table 5. The optimal mixing fractions of well 504 water reported by PHREEQC are very close to the minimum possible values and do not change significantly between the various models found. The mixing fraction of 0.258 reported in Table 5 is in between the values of 0.216 and 0.347 determined by NETPATH respectively assuming either conservative dissolved Cl or conservative dissolved Na.

The fact that gypsum precipitation is not present in any of the minimal or non-minimal models found is one of the most important results of the PHREEQC inverse modeling on the well 402 + well 504 ⇒ well 503 chemical evolution problem. PHREEQC revealed the fact that SO_4 could essentially be considered a conservative entity given the ±5% uncertainty associated with its analysis. Although most of the reasonable NETPATH mass-transfer models based on chloride conservation showed that 2 millimoles of gypsum should precipitate, this mass-transfer was essentially insignificant given the ±5% relative uncertainty on the SO_4 concentrations and the very high SO_4 concentrations in the well 402 and 503 waters. In contrast to the PHREEQC results, the mass-transfer models reported by NETPATH all included 10 phases. Given the uncertainties in the analytical data, PHREEQC shows that only 7 or 8 phases were really required.

The second major conclusion confirmed the earlier NETPATH simulation result that Ca- and Mg-silicate phases are definitely needed in addition to calcite and dolomite to explain the Ca, Mg and C mass balances. Another conclusion, not revealed by the previous NETPATH simulations, was the inclusion of dolomite in some of the minimal phase models. This conclusion was satisfying, because we do believe that dolomite is indeed reacting and it is definitely present in the aquifer.

The PHREEQC results presented in Table 5 appear to show a net production of close to 0.3 millimoles of protons. This was calculated, as previously done for the NETPATH model results (see discussion of Table 3), from the mass-transfer amounts given in the PHREEQC models. The amount of net proton consumption calculated for the PHREEQC models is an order of magnitude smaller than the amounts calculated for the NETPATH models presented earlier. Because of its solution of the alkalinity-balance and charge-balance equations, PHREEQC actually ensures that the net amount of protons released and consumed by heterogeneous and homogeneous reactions is consistent with the pH values of the initial and final waters, and with the uncertainties specified by the user. The proton consumption numbers can be checked against the change in alkalinity between the final well 503 water and the mixture of the well 402 and 504 waters. The alkalinity changes reported in Table 5 use the adjusted alkalinities used by PHREEQC in each calculated inverse model. The difference between the net proton consumption numbers and the net alkalinity changes is caused by the fact that the consumption of dissolved oxygen (6.64 mg/L in well 504) was not accounted for in our proton consumption calculations (David Parkhurst, pers. comm.). The difference of 0.49 millimoles is close to 4 times the difference between the oxygen content of the well 503 water and a mixture of the well 504 and well 402 waters ($4 \times 0.121 = 0.484$ millimoles). Essentially, the presence of dissolved O_2 in the well 504 water reduces the amount of MnO_2 which undergoes reductive dissolution. The fact that 4 protons are consumed rather than 2 for each mole of oxygen consumed appears to be an error in the PHREEQC code (David Parkhurst, pers. comm.). Because the MnO_2 mass transfer is 30 times greater than the oxygen mass-transfer, this error should not significantly affect our results.

Out of the 6 minimal models presented in Table 5, our favorite models are Models 7,

2, 4 and 3. These models do not involve $AlOHSO_4$ precipitation. The evidence for possible Al control by $AlOHSO_4$ precipitation during the evolution from well 402 water to well 503 water is weak, although the possibility cannot yet be rejected. Until proven wrong, we prefer to believe that amorphous $Al(OH)_3$ is the controlling Al phase. Tremolite is present in the alluvial materials and its reaction with acidic water from the Pinal Creek site has been documented by Lind and Stollenwerk (1994). Nevertheless, it probably contributes much less Ca, compared to Ca-rich plagioclase feldspars (such as anorthite and labradorite), during the neutralization of the acidic ground waters. The accompanying release of Na during feldspar dissolution could still remain a problem, however, because the above PHREEQC models consider Na as a conservative constituent, within the uncertainty of the analytical data. Too much Na dissolution would require finding a Na sink which remains elusive. As mentioned earlier, additional cation exchange capacity resulting from the trans-

Table 5. PHREEQC inverse-modeling simulation results. Amounts of mass transfer and net proton consumption are reported in millimoles per kilogram of H_2O. Only mass-transfer sets (models) with the minimum significant number of phases are shown. The sum of residuals gives an indication of the sum of the analytical data adjustments made by PHREEQC.

Abbreviations: 504 mf., well 504 mixing fraction; +, dissolution only; -, precipitation only; Alk., alkalinity change; non-min., not a "minimal" model; K-mont., K-montmorillonite.

	Model 1 tremolite non-min.	Model 2 tremolite	Model 3 tremolite no calcite	Model 4 tremolite e^- trans.	Model 5 tremolite $AlOHSO_4$	Model 6 tremolite $AlOHSO_4$ e^- trans.	Model 7 biotite
504 mf.	0.258	0.258	0.258	0.258	0.258	0.258	0.258
504 mf. min.	0.258	0.258	0.258	0.258	0.258	0.258	0.258
504 mf. max.	0.277	0.277	0.277	0.277	0.264	0.264	0.263
dolomite +	0.398		2.191		0.290		2.291
gypsum -							
goethite -	-8.292	-8.292	-8.292	-8.291	-8.292	-8.292	-9.074
calcite +	3.588	4.383		1.626	3.817	3.991	
anorthite +			2.512	2.202			2.691
$AlOHSO_4$ -					-0.423	-0.423	
gibbsite -	-0.423	-0.423	-5.448	-4.827			-2.167
SiO_2 -	-3.649	-3.649	-5.803	-8.687	-3.165	-3.629	
rhodochrosite	-2.757	-2.757	-2.756		-2.769	-2.364	-3.253
MnO_2 +	3.903	3.903	3.903	6.659	3.903	4.309	4.294
$Mn(OH)_3$ -				-5.512		-0.811	
tremolite +	0.500	0.500	0.142	0.580	0.440	0.498	
biotite +							0.521
K-mont. -							-1.796
Sum of resid.	6.180	6.611	6.182	6.184	7.226	7.227	10.830
Net protons consumed	-0.279	-0.281	-0.272	-0.266	-0.271	-0.270	-0.298
Alk. change	0.214	0.213	0.213	0.213	0.213	0.213	0.190
Difference	0.493	0.494	0.485	0.479	0.484	0.483	0.487

formation of the feldspars and other aluminosilicates into secondary clays minerals is the only Na sink that we can think of. Sorption on freshly precipitated iron oxyhydroxides could also represent a possible Na sink, although the affinity of Na for these precipitates is considerably weaker than the affinity of the other metals and cations present in the solution.

Two additional PHREEQC simulations were conducted allowing ion exchange of Na for Ca in one simulation and exchange of Na for H in another. The ion exchange reactions were simulated by adding phases NaX and CaX_2 or HX to the list of postulated phases (X is an ion exchange site in the PHREEQC database). Because element "X" is not specified in either the initial or final solutions, PHREEQC automatically sets its concentration (and its associated uncertainty) to 0 in all three solutions, and thereby ensures that any loss of Na by ion exchange is automatically matched by an equivalent release of Ca or H from CaX_2 or HX. For the sake of brevity, only a representative selection of the models found with the proton exchange reaction are presented here (Table 6). The mixing fraction of well 504 water determined in all the ion-exchange models found was close to, or significantly lower than 0.216, the fraction determined on the basis of conservative dissolved chloride in the NETPATH simulations. Interestingly, gypsum dissolution was included in three of the proton exchange models found that included biotite/K-montmorillonite. All three had unrealistically low mixing fractions of uncontaminated water from well 504 (2.6%). All models found that included tremolite, in either the Na/Ca exchange or the Na/H exchange simulations, also included $AlOHSO_4$ precipitation. The lower mixing fractions of uncontaminated water invariably resulted in lower dissolved-oxygen transfers. As discussed earlier and shown in Table 6, if the oxygen transfer (in millimoles of O) is multiplied by 4 and added to the calculated net proton consumption for the dissolution of the mineral phases, a very close match is obtained with the PHREEQC-calculated alkalinity change. We do not favor the ion-exchange models because we estimate that the cation exchange capacity of the Pinal Creek alluvial sediments is very low, probably lower than 1 meq/100 g, given the coarseness of the sediment and the less than 1% organic carbon content. Until more information becomes available concerning the possible ion-exchange reactions at the Pinal Creek site, Models 7, 4, 2 and 3 from the earlier PHREEQC simulation without ion exchange (Table 5) remain our preferred models. Simulating ion exchange in an inverse geochemical model presumes that the user has some knowledge of the thermodynamically preferred directions of exchange. Although we feel that Na replacement for Ca on exchange sites should generally not occur given the preferential dissolution of Ca over Na resulting from other mineral dissolution reactions, determining the direction of exchange for some other ion-exchange reactions is fraught with much greater uncertainty. Proton release from exchange sites during the neutralization of well 402 water is certainly conceivable, however. Our inverse geochemical modeling simulations point out the need for further experiments to determine more exactly actual cation exchange capacities and directions of exchange. The ability of inverse modeling to highlight knowledge gaps is perhaps one of its greatest benefits. As will be demonstrated in the next section, forward geochemical modeling may be able to provide greater insight into ion exchange reactions.

Although some mass transfer processes are likely to be occurring continuously throughout the flow path section used in inverse modeling, some mass transfer processes (such as ion exchange reactions) will be affecting the ground-water chemistry only in narrow portions of the flow system. In the case of continuous processes, an overall rate of reaction (expressed for example in moles per kg of H_2O per travel time and per traveled volume of aquifer) may be provided by the inverse modeling results. In the case of a non-continuous process, however, such rates will have little meaning. Unfortunately, inverse geochemical modeling can not provide information on the heterogeneous mass-transfer reactions occurring at specific points along a flow-path, but provides, instead, only the net

Table 6. PHREEQC inverse-modeling simulation with Na/H exchange. Amounts of mass transfer and net proton consumption reported in millimoles per kilogram of H_2O. Only mass-transfer sets (models) with the minimum significant number of phases are shown.

504 mf.: well 504 mixing fraction; + : dissolution only; – : precipitation only

	Model 1	Model 2	Model 3	Model 4	Model 5
504 mf.	0.205	0.219	0.026	0.135	0.142
504 mf. min.	0.190	0.206	0.026	0.259	0.103
504 mf. max.	0.274	0.254	0.091	0.195	0.220
dolomite +		2.242			0.825
gypsum -			-6.818		
goethite -	-8.457	-8.312	-11.640	-9.208	-10.480
calcite +	4.590		6.602	1.657	
anorthite +		3.872	3.798	1.420	3.958
$AlOHSO_4$ -	-0.459			-3.349	-2.733
gibbsite -		-8.194	-2.664		
SiO_2	-3.744	-7.428		-6.461	
rhodochrosite	-2.950	-2.848	-4.920		
MnO_2 +	4.028	3.944	5.763	7.932	9.183
$Mn(OH)_3$ -				-6.947	-8.188
tremolite	0.506			0.482	
NaX	-0.628	-0.568	-1.423	-0.941	-0.909
HX	0.628	0.568	1.423	0.941	0.909
Biotite			0.854		0.899
K-montmorillonite			-2.733		-2.826
Sum of residuals	2.580	3.996	4.531	3.003	3.179
Net protons consumed	0.018	-0.062	0.994	0.395	0.354
Alkalinity change	0.417	0.363	1.102	0.685	0.655
O-transfer	0.1	0.106	0.029	0.072	0.075
O-adjusted Difference	-0.002	0.002	-0.009	0.001	0.001

amounts of mass transfer between an initial and a final endpoint. Forward geochemical modeling can, however, provide insight on the possible evolution through time of chemical compositions at specific points along a postulated flow path. *Verification* of the forward modeling simulations may nevertheless also necessitate more spatial information than may be available.

REACTIVE TRANSPORT MODELING AT THE PINAL CREEK SITE

Inverse modeling is a valuable tool that can be used to gain an improved understanding of the geochemical processes that may be occurring, or have previously occurred, in an aquifer. By itself however, inverse modeling can not be used to make predictions on the future chemical evolution of a ground-water system, or in the case of a contaminated ground-water, on the movement of contaminants. Forward reactive transport modeling is needed to make any such predictions. Inverse modeling results, nevertheless, can be used to suggest the possible reactions that should be considered by a reactive transport model.

In metal-contaminated acidic ground waters such as those present at the Pinal Creek site, pH and Eh conditions are the primary chemical variables controlling the transport of metals and determining the quality of the ground waters. The partially-neutralized, Fe(II)-poor ground waters (such as the well 503 water used in our inverse modeling exercise) have significantly lower metal concentrations than the more acidic Fe(II)-rich waters (such as the well 402 water used above). The partially neutralized waters are still contaminated and have high SO_4, Ca, and Mn concentrations that make them unsuitable for normal human or domestic use, but nevertheless offer a significant improvement in water quality. Therefore, being able to predict the movement of the low-pH and Fe(II)-rich ground-water zones or at least their movement relative to that of unreactive conservative constituents, would be very desirable.

Inverse modeling can help identify the possible reactions affecting the neutralization and oxidation of the low-pH and Fe(II)-rich ground waters. Generally however, the movement of the low-pH and Fe(II)-rich waters will be controlled by the following factors: (1) the ground-water velocity field, (2) the dilution of the contaminated ground-waters by longitudinal and transverse dispersion, (3) the amounts of heterogeneous mass-transfer reactions affecting the pH and Fe(II) concentrations in the ground waters (causing the low-pH Fe(II)-rich waters to evolve into higher-pH Fe(II)-poor waters) and (4) the initial concentration and composition of phases present in the background or fully neutralized aquifer and responsible for the chemical evolution of the contaminated waters. The following sections briefly discuss ongoing research efforts aimed at a better understanding of the movement of acidic metal-rich ground waters at the Pinal Creek site. The research findings from the Pinal Creek site will hopefully provide information, not only on the processes controlling the spread of acidic metal-laden ground waters in semi-arid alluvial basins, but also on the most efficient techniques that can be used to characterize and model the spread of contaminated waters at similar sites with sparse spatial information.

The ground-water velocity field

Gound-water velocities can be determined through the construction of a ground-water flow model. Although calculations using Darcy's law on observed heads and estimated hydraulic conductivities have provided some estimates of ground-water velocities in the Pinal Creek basin (Brown, 1996; Neaville and Brown, 1993), efforts are currently under way to construct a general flow model for the alluvial and basin-fill aquifers, using not only observed heads and other hydrologic properties and characteristics, but also chemical and isotopic tracers (Glynn and Busenberg, 1994a) as evidence of ground-water provenance (O^{18}, H^2, dissolved Ar and N_2, Cl and conductivity) and ground-water residence times (chlorofluorocarbon, SF_6 and H^3/He^3 age-dating techniques). Because of the paucity of available hydrologic data, alternative sources of information, such as geochemical and isotopic information, are proving themselves invaluable at the Pinal Creek site. Inverse flow modeling with the MODFLOWP code (Hill, 1992) in conjunction with the MODPATH particle-tracking code (Pollock, 1989) will be used to estimate hydraulic conductivities and appropriate boundary conditions from observed hydraulic heads and from chemically and isotopically-estimated ground-water recharge dates and provenance. Having an adequate ground-water flow model is an essential prerequisite to understanding the movement of contaminated waters at the Pinal Creek site and the impact of anthropogenic or natural remediation processes.

Transport processes and contaminant dilution

The dilution of the acidic metal-contaminated ground waters is certainly one of the most important processes responsible for the downgradient decrease in dissolved-metal concentrations in the Pinal Creek basin. Although this dilution process was already evident in the dry to normal recharge years of the period 1984-1991, further dilution occurred as a result of the greater than normal recharge events that started in the spring of 1991, continued in 1992 and culminated in a 100-year-magnitude flooding event in spring 1993 (ground-water levels rose as much as 16 meters). Advanced modeling techniques are not needed to determine that such a dilution process exists; plots of metal concentrations as a function of chloride, or of other non-reactive constituents associated with the contaminated waters, clearly illustrate the process (Figs. 3 and 4). However, a 2- or 3-dimensional transient transport model of the site would clearly be of use in determining the effects and contributions of transverse dispersion, longitudinal dispersion, flow convergence and transient high-intensity recharge events in the dilution process. An appropriate ground-water flow model is required before such a useful transport model can be built; the construction of this transport model therefore also depends on the results of chemical and isotopic tracer and age-dating investigations currently being conducted at the site.

First simulation example:

The Brown (1996) 1-D reactive transport model for the Pinal Creek basin

Brown (1996) used the ground-water chemical data collected by the U.S. Geological Survey since 1984, to construct a 1-D reactive transport model of the site using the PHREEQC computer code. The author used measured chloride concentrations to back-fit the observed dilution of the most contaminated ground waters in the Pinal Creek basin as a function of time (from 1984 until 1994) and distance along a flow-path that extended from well 51, through wells 302, 402, 451, 503, 601, all the way to well 702 (Fig. 5). Only a few adjustments were made to the external dilution of the 1-D column after the initial fit of the 1984 chloride profile. Adjustments were made in March 1985 and in January 1988 just downgradient of well 101 (near km 1 in the profile), and more importantly in February 1993 in the profile section between wells 402 (km 5.8) and 503 (km 11.4). This later adjustment was necessitated by the very large flooding events that occurred during the spring of 1993. The PHREEQC code was used, because it allows simulation of advective transport, simulation of mixing processes and simulation of equilibrium geochemical reactions, including heterogeneous redox reactions and sorption with a diffuse double-layer surface-complexation model. The results of the modeling investigation showed that dilution, rather than sorption or other reaction processes, could account for the decrease of Cu, Zn, Ni and Co in the acidic ground-water upgradient from well 451. The results also suggested that reductive dissolution of Mn oxides by Fe(II) was taking place not only between wells 402 and 503 (where Fe(II) has completely disappeared), but also between wells 302 and 402, perhaps because of slow reaction kinetics. Indeed, the simulated Fe(II) concentrations are higher than the observed concentrations, and simulated Mn(II) concentrations are lower than the concentrations observed in this part of the flow path (Figs. 6 and 7). In the neutralized Fe(II)-poor ground-waters, dilution is a much less significant process and Brown (1996) suggested that sorption was an important removal process for Cu, Ni, Zn and Co. Brown (1996) also proposed that O_2 ingassing might be important in removing dissolved Mn(II) from the ground waters between wells 503 and 601. Rhodochrosite precipitation was not allowed in the Brown (1996) model (unlike most of the reactive transport simulations presented later in this paper). To better match the observed or calculated TDIC values (Fig. 8), the model also allowed CO_2 exsolution between wells 451 and 702 in equilibrium with a specified CO_2 partial pressure of $10^{-1.33}$. This pCO_2 value was determined from dissolved CO_2 measurements and considered appropriate for well 503 (Glynn and Busenberg, 1994a,b). The results of our inverse modeling work show, however, that CO_2 exsolution, between wells 402 and 503, does not need to be considered if Ca- and Mg-silicate mineral dissolution occurs.

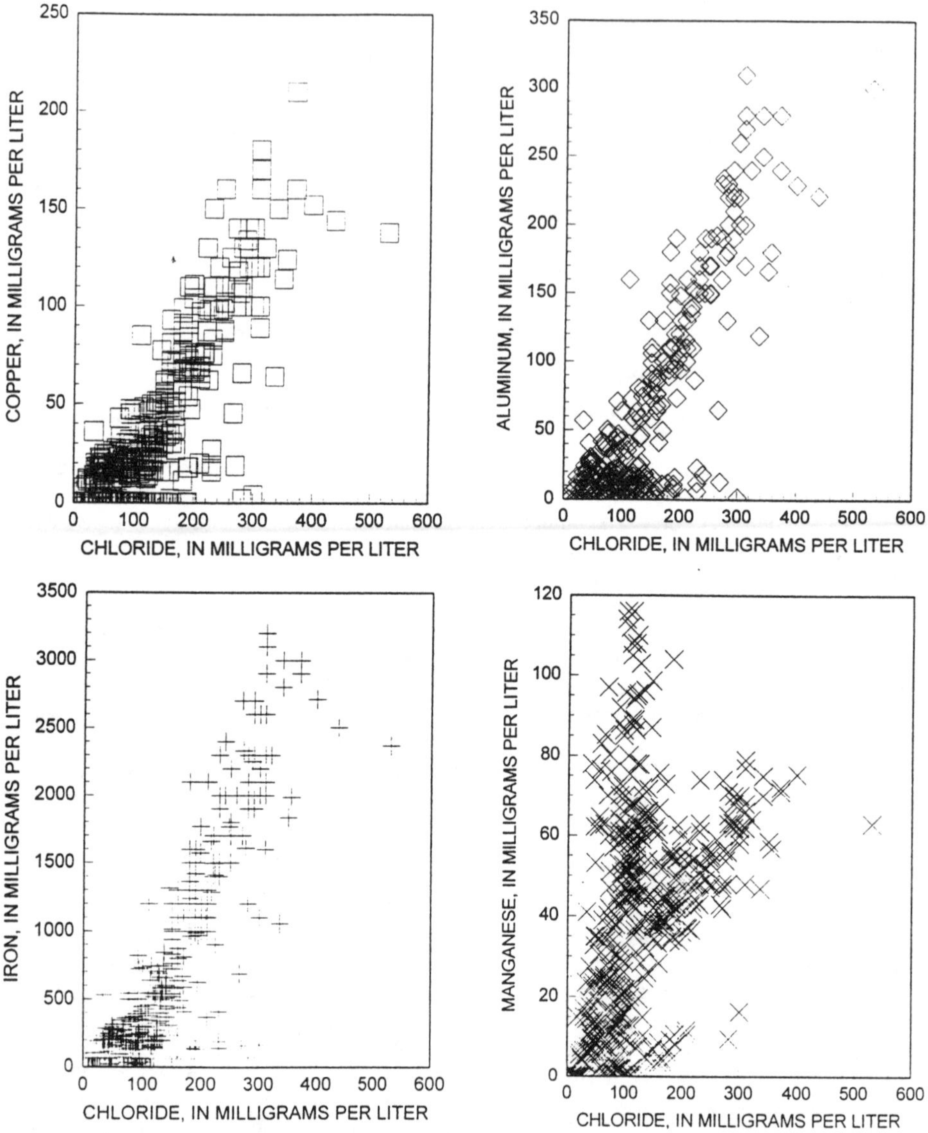

Figure 3. Copper, aluminum, iron and manganese concentrations in Pinal Creek ground waters as function of dissolved chloride. The linear decrease in metal concentrations as a function of chloride indicates the effects of dilution. Relatively low metal concentrations in waters with high chloride concentrations are the result of precipitation reactions. Waters with high manganese and relatively low chloride concentrations are typical of dilute partially-neutralized Fe(II)-poor contaminated ground waters.

Second simulation example:

The Glynn, Engesgaard and Kipp (1991) 1-D reactive transport model

In a comparison of the PHREEQM and MST1D (Engesgaard and Kipp, 1992) reactive transport codes, Glynn et al. (1991) simulated the movement of acidic water from well 51

(August 1987 sample) into a 1-dimensional column with the following minerals assumed to be initially present and distributed homogeneously throughout the column (contents expressed in mol/kg H_2O): calcite (4.2×10^{-2}), dolomite (2.1×10^{-2}), microcline (1.75×10^{-2}), birnessite (2.0×10^{-2}). Amorphous $Fe(OH)_3$, gypsum, silica gel and gibbsite were allowed to precipitate. Unlike most of the transport simulations presented later in this paper, the column was also left open to an equilibrium p_{CO_2} of 10^{-2} and rhodochrosite was not allowed to precipitate. The local equilibrium assumption was used for all simulated reactions.

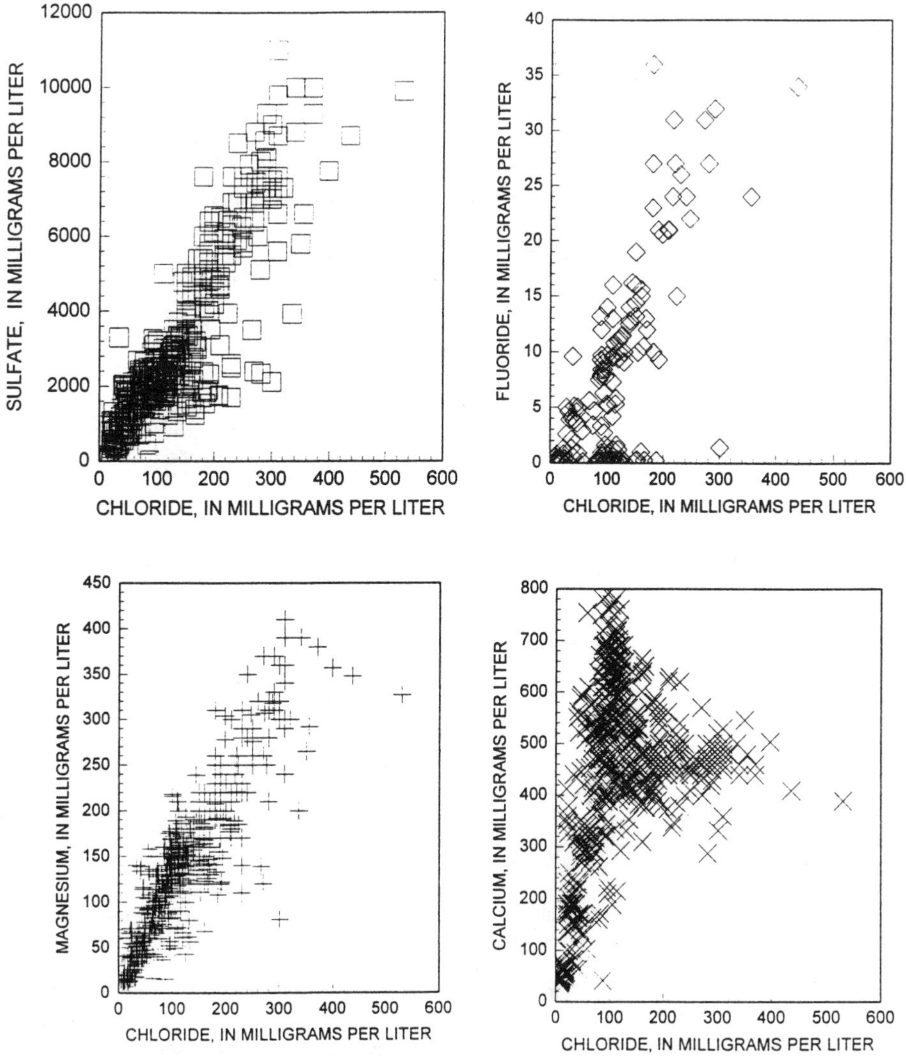

Figure 4. Sulfate, fluoride, magnesium and calcium concentrations in Pinal Creek ground waters as function of dissolved chloride. The chloride versus calcium plot indicates the effect of calcium mineral dissolution reactions in addition to dilution processes.

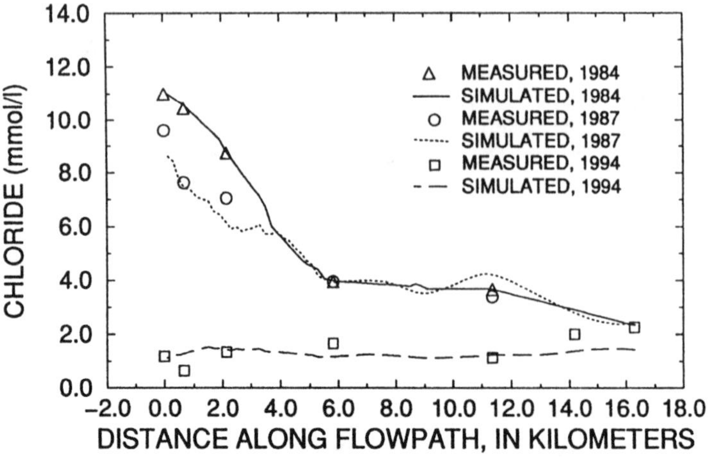

Figure 5. Measured and simulated chloride concentrations in Pinal Creek ground waters along an aquifer flowpath (modified from Brown, 1996). Observed data are for the following wells: 51 (km 0), 101 (km 0.7), 302 (km 2.1), 402 (km 5.8), 503 (km 11.4), 601 (km 14.2) and 702 (km 16.3).

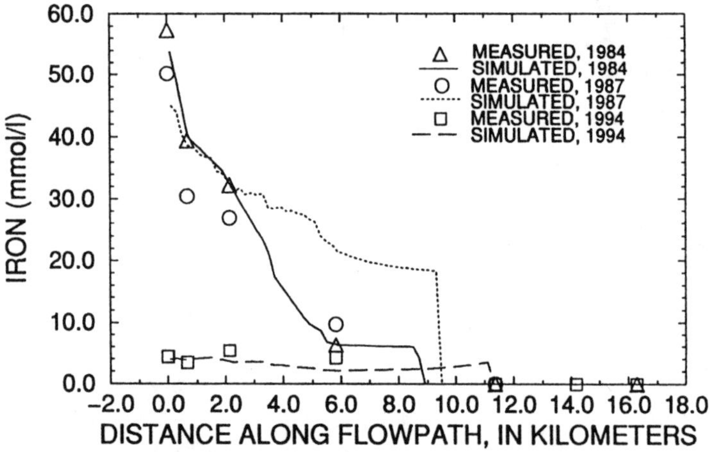

Figure 6. Measured and simulated iron concentrations in Pinal Creek ground waters along an aquifer flowpath (modified from Brown, 1996). See Figure 5 caption for well locations. Note the large decrease in dissolved iron concentrations between 1984 and 1994. This decrease (also seen in the Cl concentrations, Fig. 5) is attributable to the very large recharge events during 1992-1993 (and especially during the spring of 1993), to pumping of some of the more contaminated waters for remediation purposes and to the removal of one of the major sources of contamination (Webster Lake) during 1987.

The water from well 51 used in the simulation was significantly more contaminated than the well 402 water used in the inverse modeling example presented in this paper and used in reactive transport simulations presented later on. The concentration factors relative to the well 402 water used in this paper (Table 7) show that the well 51 water was approximately 2.4 times more concentrated in conservative consituents (Cl, SO_4 and Mg!), only 1.7 times more concentrated in Na (suggesting the existence of a Na source between wells 51 and 402), was significantly more concentrated in Fe and Al, and less concentrated

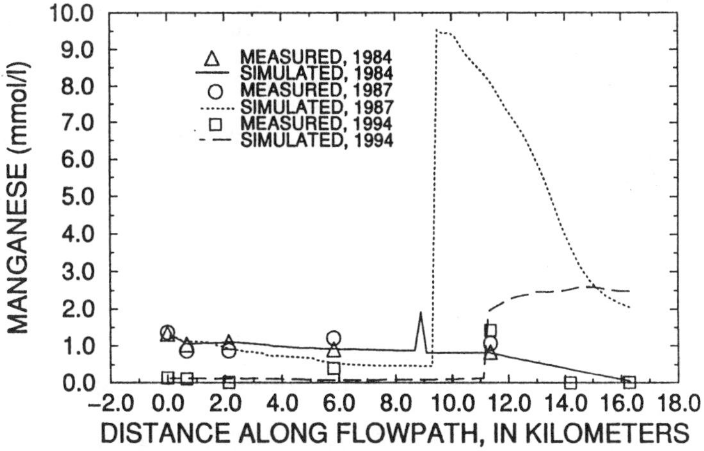

Figure 7. Measured and simulated manganese concentrations in Pinal Creek ground waters along an aquifer flowpath (modified from Brown, 1996). See Figure 5 caption for well locations.

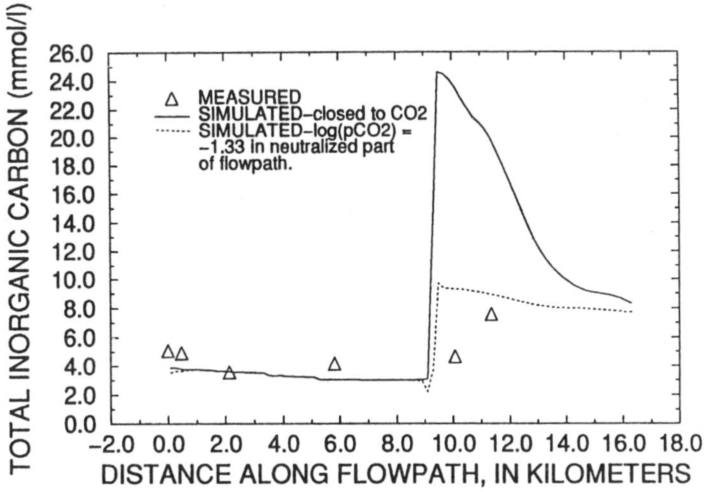

Figure 8. Measured and simulated dissolved inorganic carbon concentrations in Pinal Creek ground waters along an aquifer flowpath (modified from Brown, 1996). See Figure 5 caption for well locations.

ed in Ca and Sr, but had similar SiO_2, TDIC and Mn concentrations. The acidity represented by the potential oxidation and precipitation of the dissolved Fe(II), Mn(II) and Al in the water from well 51 was 1.31×10^{-1} moles of protons instead of 2.60×10^{-2} for the well 402 water, or about 5 times greater. Adding the acidity represented by TDIC does not significantly change the potential acidity of well 51 water, but does increase the potential acidity of well 402 water to a proton molality of 3.01×10^{-2}, a value still 4.5 times lower than that of well 51 water. [The potential acidity is calculated by assuming: (1) each mole of Fe^{2+} oxidation and precipitation (by reductive dissolution of MnO_2 or by reduction of dissolved O_2) can produce 2 moles of H^+, (2) oxidation of dissolved Mn^{2+} and precipitation as MnO_2 (or simply precipitation of $MnCO_3$) can produce 2 moles of H^+, (3) precipitation of Al^{3+} can produce 3 moles of protons, and (4) each mole of H_2CO_3 can produce one mole of H^+.]

Despite the difference in their algorithms, the PHREEQM and MST1D simulations gave very similar results (Figs. 9 and 10). [The MSTID code used a sequential iteration algorithm that iterated at each time step between the solution of the PHREEQE chemical equilibrium code and a solution of the advective-dispersive transport equations solved by an implicit finite difference approximation (based on the HST solute transport code).] The slight observed differences were most likely the result of numerical dispersion in the MST1D simulation. Both simulations showed that, given the mineral and aqueous concentrations used in the simulation, the velocity of the low-pH front would be retarded by a factor of 5 relative to that of the conservative constituents; the velocity of the Fe(II) front would be 1.8 times less than that of the conservative constituents.

Table 7. Chemical composition of the highly acidic ground water used as the infilling solution in the reactive transport simulation of Glynn et al. (1991). Concentrations are expressed as molalities. The 'concentration factors' represent the concentration of a constituent in well 51 water (87/8/18) divided by its concentration in well 402 water (89/1/12).

	Well 51 (87/8/18)	Concentration factors relative to well 402
pH	3.74	2.45 times H^+ activity
Calcium	1.10×10^{-2}	0.87
Magnesium	1.60×10^{-2}	2.40
Sodium	9.13×10^{-3}	1.73
Strontium	1.48×10^{-5}	0.56
Iron	5.01×10^{-2}	4.73
Manganese	1.37×10^{-3}	1.05
Aluminum	9.27×10^{-3}	13.5
Sulfate	9.16×10^{-2}	2.69
Chloride	9.59×10^{-3}	2.42
Silica	1.66×10^{-3}	1.16
TDIC	4.16×10^{-3}	0.99
potential Fe, Al, Mn and TDIC acidity (as proton molality)	1.35×10^{-1}	4.48

The Glynn et al. (1991) simulation of the reactive transport of the most acidic ground water (from well 51) found at the Pinal Creek site serves as a useful comparison for the reactive transport simulations discussed in the next section. These simulations will simulate the evolution of a less acidic water, water from well 402, and will be based on some of the inverse geochemical modeling results discussed earlier in this paper.

A 1-D reactive-transport sensitivity analysis on the movement of pH- and pe-controlling mineral fronts

Some of the reactions identified by inverse modeling of the chemical evolution of ground waters between wells 402 and 503 were used in 1-D PHREEQM and PHREEQC reactive-transport simulations to determine their effect on the movement of the low-pH and high-Fe(II) ground waters in the Pinal Creek basin. Many of results of this study will also be applicable at other ground-water contamination sites by sulfuric acid and heavy metals.

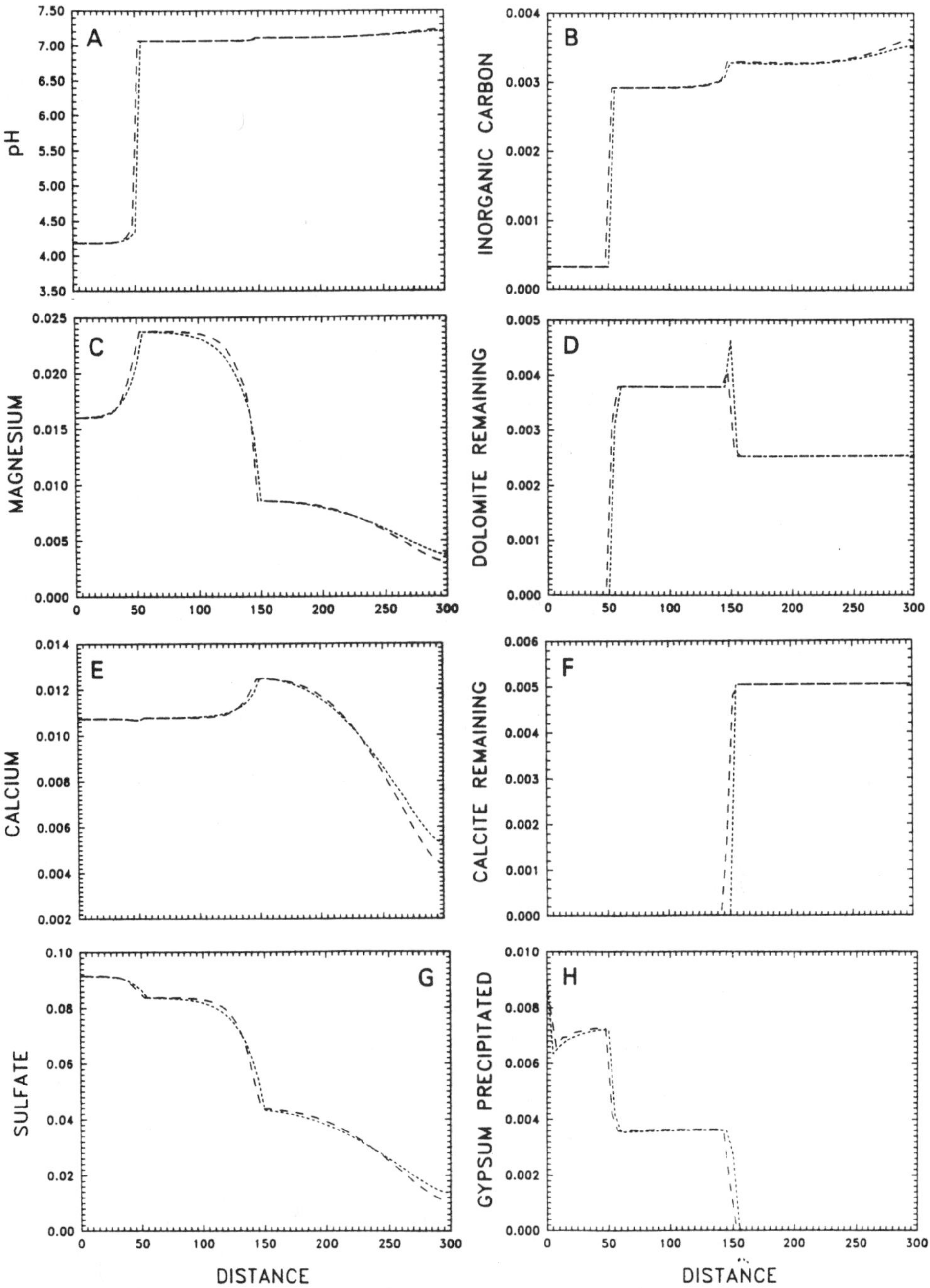

Figure 9. Profiles at 50 days of various pH-dependent aqueous and mineral constituents resulting from the PHREEQM (dashed lines) and MST1D (dotted lines) simulations of an acidic water (from well 51) from the Pinal Creek site intruding into a 1-dimensional model aquifer column with calcite, dolomite and MnO_2 initially present. The column is in equilibrium with a fixed pCO_2 of 10^{-2}. Rhodochrosite is not allowed to precipitate (Glynn et al., 1991).

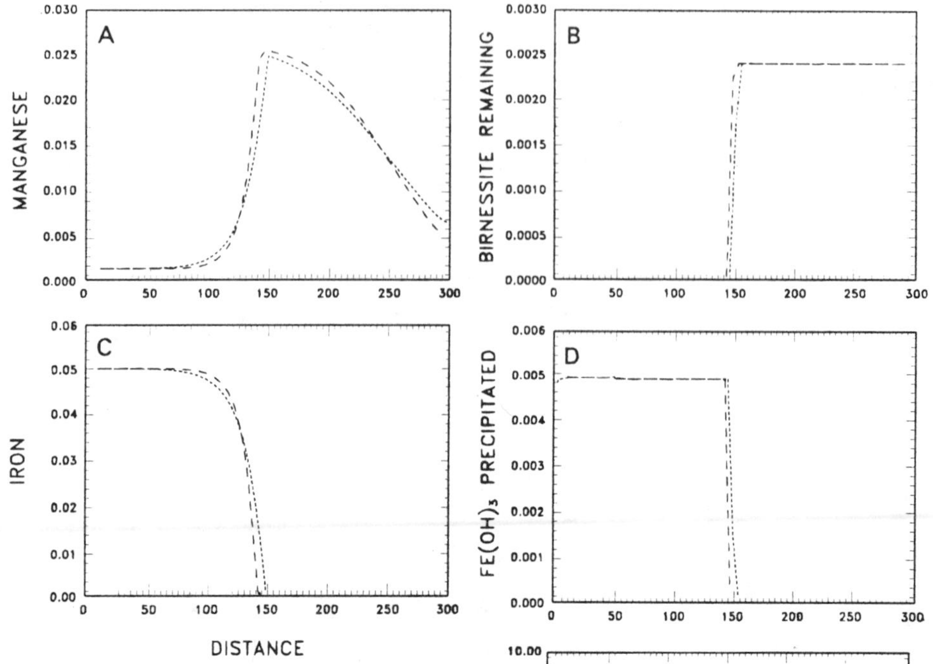

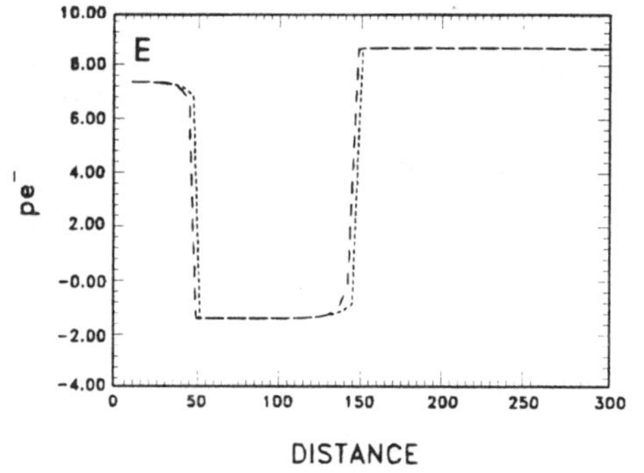

Figure 10. Profiles at 50 days of various pe-dependent aqueous and mineral constituents (Glynn et al. 1991). See Figure 9 caption for explanation.

Before discussing the results of the PHREEQM and PHREEQC transport simulations, however, we will show how the movement of a single mineral dissolution front, critical in controlling the redox state of the Pinal Creek ground waters, can be modeled without the use of a computer code.

A simple model for advective reactive transport of a dissolution front: the MnO_2 dissolution front. If O_2 ingassing is assumed not to affect the concentrations of Fe(II) at significant depths below the water table (11 m for well 402, 21 m for well 503), the movement of dissolved Fe(II) will probably be entirely determined by the reduction of manganese oxides such as exemplified by the following reaction:

$$2\ Fe^{2+} + MnO_{2,s} + 4\ H_2O \Leftrightarrow 2\ Fe(OH)_{3,am} + Mn^{2+} + 2\ H^+$$

The free energy change associated with the above reaction is so highly negative that for all practical purposes the reaction can be considered an irreversible reaction, regardless of which crystal structure is used either for $MnO_{2,s}$ (birnessite, pyrolusite) or for the precipitated iron oxyhydroxide. As a result, if the kinetics of the reductive dissolution reaction are fast relative to the movement of the water (5 m/day), the velocity of the dissolved Fe(II) front, V_{Fe}, can be related to the velocity, V_{H_2O}, of the water through an apparent retardation factor R

$$V_{Fe} = V_{H_2O}/R \tag{14}$$

where R is simply related to the amount, M_{ini}, of MnO_2 initially present in the aquifer (expressed in moles/kg H_2O) and to the amount, ΔM, of MnO_2 dissolved by a unit mass (1 kg of H_2O) of Fe(II)-rich water:

$$R = 1 + \frac{M_{ini}}{\Delta M} \tag{15}$$

Equation (15) can be related to the more general equation describing the retardation of sharp reaction fronts in systems with advective transport but no dispersive transport (Dria et al., 1987; similar expressions for the "traveling wave" approximation can also be found in Lichtner, 1988, 1985; Ortoleva et al., 1986):

$$R = 1 + \frac{\sum_{k=1}^{K} g_{ik}\left(m_k^D - m_k^U\right)}{\sum_{j=1}^{J} h_{ij}\left(c_j^D - c_j^U\right)} \tag{16}$$

where g_{ik} and h_{ij} are the stoichiometric coefficients of element i in minerals k and aqueous-species j. m_k and c_j are the mineral and aqueous-species concentrations, respectively. Superscripts D and U indicate downstream and upstream concentrations, respectively.

Determination of the initial MnO_2 and carbonate mineral concentrations. Although inverse modeling can be used to determine the mass-transfer amounts (ΔM) for various heterogeneous reactions, inverse modeling can not usually reveal the initial amounts (M_{ini}) of minerals present in an aquifer. Estimates of the average initial mineral contents must be made, either by using batch or column experiments on unaffected aquifer materials, or by judging what a reasonable amount may be on the basis of X-ray evidence or on some knowledge of the retardation observed in the field.

By determining the net mineral mass transfer (ΔM) experienced by a fluid packet between two points along a flow path, an inverse model may be used to set a lower bound on the initial concentration of that mineral ($M_{ini} \geq \Delta M$), but only if the mineral mass transfer occurs in a unique and localized part of the flow path, such as a single reaction front. Indeed, although commonly expressed in mol/kg of H_2O, the mineral mass-transfer determined by an inverse model really applies to the entire volume of aquifer traveled through by a unit packet of water between an initial and a final endpoint. In the case of a slow mass transfer reaction (relative to the movement of the water), the value of ΔM determined by an inverse model will change with increasing flow path length until some equilibrium or steady state is reached. The initial mineral concentration, M_{ini}, in Equation (15) is also expressed in terms of mol/kg H_2O, but really refers to a static mass of water and therefore a localized aquifer volume. In contrast, the value of ΔM determined by inverse modeling has a Lagrangian frame of reference and its units refer to a dynamic mass of H_2O that has traveled along a specific flow path length. In the case of a sharp reaction front responsible for the entire mass transfer, the difference in units may be moot, at least if the mineral concentration M_{ini} was initially uniform between the initial and final end points of the flow path. In any case, Equations (15) and (16) will be of limited use in describing the progression of a slow reaction, one that does not result in the development of a sharp reaction front*.

----

* Lichtner (1988) and Ortoleva et al. (1986) have shown that reaction kinetics will not affect the rate of front propagation *given enough time and distance*. The front may not be as sharp but it will still propagate at the same rate as a front resulting from a simulation that uses the Local Equilibrium Assumption.

We decided to use a value of 2×10^{-2} mol MnO_2 per kg H_2O in our PHREEQC and PHREEQM transport simulations, primarily to stay consistent with the simulations conducted by Glynn et al. (1991). A higher value would perhaps have been more appropriate. Indeed, on the basis of his column elution experiments, Stollenwerk (1994) suggests a value of 7.1 millimoles of MnO_2 per kg of sediment. Depending on the values of porosity and bulk density used, this is equivalent to a value between 3.2×10^{-2} (our estimate) to 4.49×10^{-2} mol/kg of H_2O (value calculated by Brown, 1996). Based on the selective extraction results of Ficklin et al. (1991), Brown (1996) used a value of 7.9×10^{-2} mol/kg H_2O in his 1-D simulation. The maximum and minimum amounts of MnO_2 dissolved between wells 402 and 503 in our inverse modeling simulations were 9.2×10^{-3} (Table 6) and 3.9×10^{-3} mol/kg H_2O (Table 5), respectively. These values may be considered lower bounds for the initial amount of MnO_2 present in the aquifer, if it is assumed that the reductive dissolution of MnO_2 is fast relative to the movement of the water, an assumption made in all the forward PHREEQM and PHREEQC simulations presented in this paper. The lack of accurate ground-water flow and transport models, the large distances between the well sites, and the suggestion that the reductive dissolution of MnO_2 may be slow compared to the ground-water velocities make it difficult to calculate with any accuracy what the exact Fe(II) retardation factor may be, and therefore what the appropriate initial MnO_2 content may be. In any case, the results of our simulations, and in particular the MnO_2 dissolution front retardation factors that we determine, can easily be extrapolated to other initial MnO_2 concentrations.

Determining appropriate initial carbonate (calcite and dolomite) concentrations to use in the reactive transport simulations was also a problem. Brown (1996) used a concentration of 0.18 mol/kg H_2O in his simulation, a value consistent with the carbonate content determined by Eychaner and Stollenwerk (1985) for a sample of alluvium collected in 1985. Brown (1996) noted that the buffering capacity measurements conducted by Hydro Geo Chem (1991) on alluvial samples collected from three different locations in the Pinal Creek basin could be translated into equivalent carbonate concentrations of between 0.12 mol/L for sand and gravel and 0.76 mol/L for calcareous clay. These measurements, however, did not correct for possible proton adsorption and silicate dissolution reactions. Based on a description of the Pinal Creek site by Eychaner (1989, and pers. comm.), Glynn et al. (1991) used an initial carbonate concentration of 0.084 mol/kg H_2O (0.048 mol/kg H_2O calcite and 0.021 mol/kg H_2O dolomite). The resulting pH front retardation factor of 5, for the acidic well 51 water used in the simulation, approximately matched the relative rate of advance of the low-pH waters at the Pinal Creek site over the last 50 years. The inverse modeling results discussed in the present paper show a net dissolution of usually 1.5×10^{-3} to 4.6×10^{-3} mol/kg H_2O of primary carbonate minerals (Table 5) between wells 402 and 503. If the questionable assumption is made that the carbonate mineral dissolution rates are relatively fast, resulting in localized dissolution, a value of 5×10^{-3} mol/kg H_2O may be considered a reasonable lower bound on the possible initial carbonate concentration representative for the Pinal Creek alluvial sediments. In any case, because the initial carbonate mineral concentration chosen for the simulations proved to be the most important adjustable parameter determining movement of the low-pH waters in our simulations (and is also probably the most important determining factor in Pinal Creek ground waters), a set of 8 different initial carbonate concentrations ranging from 5.25×10^{-3} to 3.32×10^{-1} mol/kg H_2O were used in different simulations to test other reaction-model modifications.

Setup of the 1-D reactive transport simulations. Unless specified otherwise, most of our simulations were conducted with the local equilibrium advective transport code PHREEQC. Dispersion was not usually simulated. The reactive transport simulations investigated the effect of the following model variations on the retardation of the low-pH and high Fe(II) fronts:

(1) changing solid-carbonate concentrations. The initial concentrations chosen were: 5.25×10^{-3}, 1.05×10^{-2}, 2.1×10^{-2}, 3.05×10^{-2}, 4.2×10^{-2}, 8.4×10^{-2}, 1.68×10^{-1}, and 3.32×10^{-1} mol/kg H_2O.

(2) using a longitudinal dispersivity of 560 m. This dispersivity represented 10% of the total simulation length, a rule of thumb often applied in ground-water transport modeling. Most simulations used a dispersivity of 0 m, primarily because of shorter execution times. The PHREEQM and PHREEQC v. 2 (unpublished) geochemical transport codes were used for all simulations with a dispersivity of 560 m.

(3) including or excluding dolomite. Most simulations did not include dolomite.

(4) allowing or disallowing rhodochrosite precipitation.

(5) allowing either $Al(OH)_3$, kaolinite, or $AlOHSO_4$ precipitation. Two different solubility products were used for $AlOHSO_4$ ($10^{-3.2}$ and $10^{-2.2}$). The higher solubility product is the value adopted by Stollenwerk (1994) in fitting the results of his laboratory column experiments, investigating the reaction of Pinal Creek acidic ground water with Pinal Creek sediments.

(6) allowing or disallowing equilibrium with an infinite reservoir of CO_2 at partial pressures of either $10^{-0.9865}$, a value based on unsaturated zone CO_2 gas measurements at the Pinal Creek site (Glynn and Busenberg, 1994b), or $10^{-1.33}$, the value used in Brown's (1996) simulation. This latter value was based on the dissolved CO_2 concentration at well 503 (91/11) reported by Glynn and Busenberg (1994a).

(7) allowing or disallowing cation exchange. Two different cation exchange capacities, 1 meq/100g and 10 meq/100g, were tested.

(8) allowing or disallowing sorption, using a diffuse double-layer surface-complexation model, based on Dzombak and Morel's (1990) thermodynamic data compilation for sorption onto hydrous ferric oxide.

(9) including or excluding the irreversible dissolution of Ca and Mg silicates to match the amounts calculated by two of our inverse modeling simulations (PHREEQC Models 2 and 7 in Table 5). These simulations all assumed a zero-order kinetic dissolution process for the silicate minerals with an inexhaustible supply of silicate minerals. Two of the simulations were also conducted assuming a zero-order kinetic dissolution process for MnO_2. In all cases, the zero-order kinetic dissolution processes were specified so that the acidic contaminated water would receive, during the course of its evolution through the transport column, exactly the silicate mineral mass-transfers (and possibly the MnO_2 mass-transfer) determined by inverse Models 2 and 7 in Table 5. The dissolution/precipitation of all other minerals was allowed to proceed to thermodynamic equilibrium at each time step and in each cell.

More than 160 reactive-transport simulations were conducted. All simulations used the water from well 402 (89/1/12) as the infilling solution. The water from well 504 (91/11/22) was used as the background water initially present in the 1-dimensional column. The 5.6 km-long column was subdivided into 10 cells of equal length and with initially homogenous mineral, surface, and aqueous concentrations. (Because mineral concentrations in PHREEQC and PHREEQM are expressed in terms of mol/kg H_2O, the porosity and bulk density of the sediments is implicitely ignored by the programs). A timestep of 112 days was used, thereby simulating an average linear ground-water velocity of 5 m/day (representative of the average ground-water velocity between wells 402 and 503). Up to a maximum of 5000 timesteps (1534 years) were simulated.

Amorphous $Fe(OH)_3$, and an aluminum phase (either amorphous $Al(OH)_3$, kaolinite or $AlOHSO_4$) were allowed to precipitate in all the simulations. The aluminum phase allowed was usually either amorphous $Al(OH)_3$ or $AlOHSO_4$ (with a solubility product of $10^{-2.2}$; Stollenwerk, 1994). Gypsum was allowed to precipitate in all the simulations except in some of the simulations that allowed irreversible, zero-order kinetic dissolution of Ca- and Mg-silicate minerals (according to Models 2 and 7 in Table 5). Rhodochrosite was allowed to precipitate in all except one set of simulations. An essentially infinite amount of chalcedony was present in all cells. An initial concentration of 2×10^{-2} mol/kg H_2O of MnO_2 was specified in all the simulations (except a few of the simulations that allowed irreversible dissolution of Ca and Mg silicates). Initial concentrations of carbonate minerals (calcite and dolomite) were specified in all the simulations and 8 different initial concentrations were specified within each set of simulations. Dolomite was included with calcite in only a few simulations (they all had an initial mineral carbonate concentration of 8.4×10^{-2} mol/kg H_2O), and was also the only carbonate mineral present in simulations that attempted to emulate Model 7 in Table 5. Although impossible to graphically present the results of all the simulations, a few representative pH-breakthrough curves are shown in Figures 11, 12 and 13 for the midpoint of the last cell (cell 10). We will attempt to summarize the essential findings of our study.

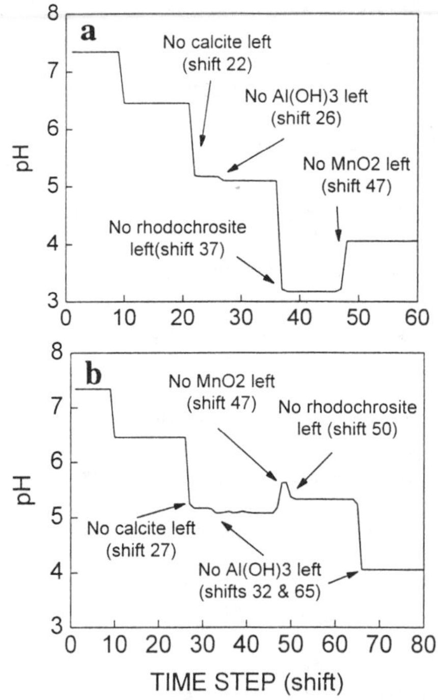

Figure 11. pH breakthrough curves for two PHREEQC advection/reaction simulations following our "basic reaction model" (BRM) with amorphous $Al(OH)_3$ as the Al-bearing phase allowed to precipitate. The BRM specifies fixed initial concentrations of calcite and of MnO_2, and an essentially infinite amount of chalcedony. In addition, the BRM allows secondary precipitation of the following minerals: gypsum, amorphous $Fe(OH)_3$, and an Al-bearing phase. An initial MnO_2 concentration of 2×10^{-2} mol/kg H_2O was used in both simulations shown, but the initial calcite concentrations differed. (a) *Initial carbonate concentration: 2.1 × 10^{-2} mol/kg H_2O.* (b) *Initial carbonate concentration: 3.0 x 10^{-2} mol/kg H_2O.*

Simulation results: movement of the Fe(II)-rich waters and of the MnO_2 dissolution front. The movement of the low-pe Fe(II)-rich ground-water zone was found to be primarily dependent on the amount of initial MnO_2 chosen for each simulation and was usually little affected by any other factors. A retardation factor (*R*) of 4.74 was determined for the simulations containing 2×10^{-2} mol MnO_2 per kg of H_2O. As mentioned earlier the movement of the MnO_2 dissolution front, and of the attendant Fe(II) and low-pe fronts, can easily be calculated by hand using Equation (15). Indeed, in that the

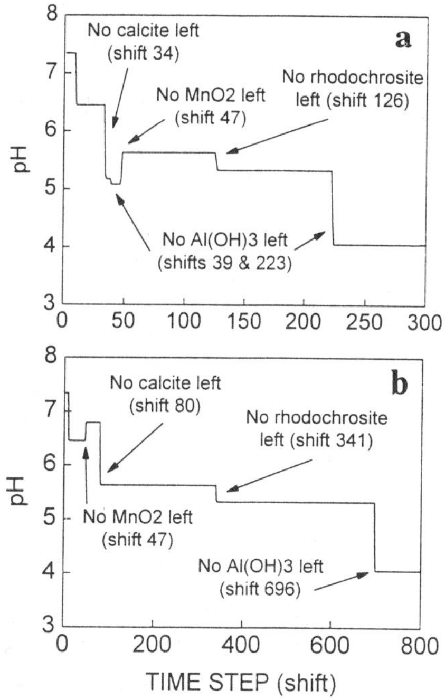

Figure 12. The pH breakthrough curves for two PHREEQC advection/reaction simulations following our "basic reaction model" with amorphous $Al(OH)_3$ as the Al-bearing phase allowed to precipitate. An initial MnO_2 concentration of 2×10^{-2} mol/kg H_2O was used in both simulations, but the initial calcite concentrations differed. (a) *Initial carbonate concentration: 4.2 × 10^{-2} mol/kg H_2O.* (b) *Initial carbonate concentration: 8.4 × 10^{-2} mol/kg H_2O.*

Fe(II) concentration of the infilling solution is 591 mg/L (Table 1), or 10.58 millimolar, and that it takes two moles of Fe(II) to dissolve one mole of MnO_2, the applicable ΔM will be close to 5.29×10^{-3} mol/L and if M_{ini} is 2×10^{-2} mol/kg H_2O, the calculated R is 4.78. The slight difference between our calculated result and the retardation factor determined from the PHREEQM simulations comes from the fact that we used molar concentration units for ΔM in our calculation instead of the correct molal units. The only processes that could conceivably affect the retardation of the MnO_2 dissolution front would be processes affecting the concentration of the reductant (dissolved Fe(II)) in the infilling water. Dispersion and cation exchange were the only two processes simulated that could potentially affect dissolved Fe(II) concentrations in the infilling water. And indeed, although the simulations that included a dispersivity of 560 m showed only a slight decrease on the retardation of the MnO_2 dissolution front (an R value of 4.6 for an MnO_2 concentration of 2×10^{-2}), the simulations that included cation exchange resulted in significantly greater MnO_2 retardation factors: 5.2 for the simulations with a cation exchange capacity (CEC) of 1 meq/100g and 9.2 for the simulations with a 10 meq/100g CEC (The lower CEC is more realistic for the Pinal Creek alluvial sediments). This effect is due to the retardation and lowering of the dissolved Fe(II) concentrations in the infilling water by cation exchange reactions, such as for example Fe/Ca exchange:

$$Fe^{2+} + CaX_2 \Leftrightarrow Ca^{2+} + FeX_2$$

Because Fe(II) sorption and Fe(II) surface complexation on hydrous ferric oxide was never simulated, the simulations that included surface-complexation sorption reactions did not result in increased retardation factors for the MnO_2 dissolution front.

Simulation results: evolution of the low-pH waters. Unlike the movement of the low-pe Fe(II)-rich front, the movement of the low-pH ground-waters is much more difficult to predict and it is not controlled by a unique mineral dissolution/precipitation

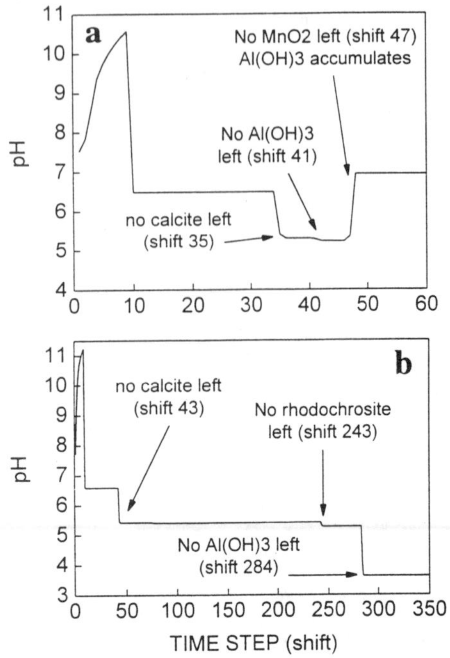

Figure 13. The pH breakthrough curves for two PHREEQC advection/reaction simulations allowing some of the reactions determined by inverse Model 2 in Table 5. The reactions allowed in the two simulations differ from our "basic reaction model" in two respects. First, the irreversible dissolution of tremolite (5.00×10^{-5} moles per cell per time step) was simulated as a continuous zero-order reaction process matching over the length of the simulation column the amount of tremolite mass-transfer specified in inverse Model 2. Secondly, gypsum was not allowed to precipitate. An initial concentration of 3.0×10^{-2} mol/kg H_2O of calcite was specified in both simulations shown. The two simulations shown differed in their consideration of MnO_2. (a) *an initial concentration of* MnO_2 *of* 2×10^{-2} *mol/kg* H_2O *was specified and allowed to react to equilibrium at each time step.* (b) MnO_2 *was added as a continuous irreversible dissolution process at the rate of* 3.903×10^{-4} *moles per cell per time step so as to match over the length of the simulation column the amount of tremolite mass-transfer specified in inverse Model 2.*

reaction. Indeed, the PHREEQM and PHREEQC simulation results show the development of several pH fronts, all controlled by mineral dissolution and precipitation fronts. The consumption of all the initial calcite present in the column (and of dolomite if initially present), is typically followed by the complete dissolution of secondary rhodochrosite and usually much later, if at all, by the dissolution of the secondary Al-bearing phase ($Al(OH)_3$, $AlOHSO_4$ or kaolinite) allowed to precipitate in the simulation. The complete dissolution of these minerals and the loss of their proton-consuming capacity leads to abrupt pH decreases. In contrast, the disappearance of MnO_2, and the ensuing halt of Fe(II) oxidation and $Fe(OH)_3$ precipitation results in a sharp pH increase.

The pH breakthrough curves shown in Figures 11 and 12 for the last of the 10 cells in the column are typical of the pH breakthrough curves seen for most of the simulations without silicate dissolution. pH remains high at 7.34, until the first pore volume (at shift 10) flushes through and neutralized infilling water with a pH of about 6.5 comes in. If MnO_2 is the first mineral consumed and $Al(OH)_3$ is allowed to precipitate, the pH will increase upon its disappearance from 6.5 to 6.8. The complete consumption of calcite will then result in a pH decrease from about 6.8 to about 5.6, and the ensuing consumption of rhodochrosite will result in another pH drop to near 5.3. Finally, the disappearance (with a long enough simulation time) of the secondary aluminum phase will eventually finally result in yet another pH drop to about 4.06, close to the initial infilling water pH of 4.13. Although the pH values reached during the simulations with silicate mineral dissolution differ somewhat from the pH values mentioned above, the general pattern of pH evolution in those simulations is also strongly dependent on the disappearance of calcite (or dolomite), MnO_2, secondary rhodochrosite, and the secondary aluminum phase (Fig. 13).

Many of the pH values obtained during the transport simulations can be simulated by equilibrating the infilling solution (well 402 water) with the various possible assemblages of coexisting mineral phases found during the transport simulations (Table 8). The resulting pH values usually closely agree with the pH values obtained during most of the simulations (and are even within half a pH unit of the values reached in the simulations that incorporated CO_2 exsolution and surface-complexation). The PHREEQC equilibration results clearly indicate that although the loss of calcite results in a pH decrease down to 5.1 to 5.7, the subsequent loss of rhodochrosite results in a further pH decrease, to values well below 5 if $Al(OH)_3$ is not present. Table 8 indicates the importance of the aluminum phase chosen for equilibration in determining the lowest pH values reached during the simulations, after the complete dissolution of rhodochrosite. Invariably, the simulations that allowed amorphous $Al(OH)_3$ precipitation resulted in higher pH values than those found in the simulations allowing kaolinite, or especially, $AlOHSO_4$ precipitation. In the presence of rhodochrosite (and possibly calcite), however, the choice of aluminum phase allowed to precipitate had little effect on the resulting pH values. Finally, Table 8 confirms our finding that the lowest pH values (near 3.1) were invariably attained when MnO_2 was still present but all other carbonate and Al-phases had dissolved away.

Table 8. pH values computed with the PHREEQC geochemical code by equilibrating well 402 water with various mineral phases, including chalcedony, gypsum and amorphous $Fe(OH)_3$ and one of four possible Al bearing phases (see table note). The last pH column indicates the pH value in the absence of any aluminum phase.

Presence of additional minerals?			Computed pH				
			Identity of Al bearing phase				
MnO_2	Calcite	Rhodochrosite	1	2	3	4	no Al phase
Yes	Yes	Yes	6.453	6.442	6.462	6.467	
Yes	**No**	Yes	5.174	5.071	5.095	5.090	
Yes	No	**No**	**5.011**	**3.810**	**3.364**	**3.186**	**3.175**
No	Yes	Yes	6.788	6.772	6.748	6.818	
No	**No**	Yes	5.633	5.576	5.666	5.659	
No	No	**No**	**5.330**	**4.190**	**4.554**	**3.946**	4.056

Note: the identity of the Al bearing phase is indexed as follows. (1) $Al(OH)_3$, (2) kaolinite, (3) $AlOHSO_4$ with log K_{sp} = -2.2, (4) $AlOHSO_4$ with log K_{sp} = -3.23).

The effect of the initial carbonate to initial MnO_2 ratio on the evolution of the low-pH waters. The set of simulations that allowed irreversible dissolution of Ca- and Mg-silicate minerals gave very different results from the set of simulations that did not incorporate Ca- and Mg-silicate dissolution. Unless mentioned otherwise, the following discussions apply only to the set of simulations that did not incorporate silicate mineral dissolution, although many of our conclusions will also be relevant to the simulations with Ca- and Mg-silicate dissolution.

The initial carbonate mineral concentration specified in each simulation offered the most important control on the movement of the calcite, rhodochrosite and Al-phase dissolution fronts (Tables 9, 10 and 11). The ratio of initial carbonate to MnO_2 (CMR) was the second most important determining factor affecting the retardation of the dissolution fronts for rhodochrosite and the Al-bearing phase (the rhodochrosite dissolution front is important because of the decrease in pH associated with the front, particularly in the absence of an $Al(OH)_3$ phase). Indeed, the timing of the complete dissolution of MnO_2 in relation to the disappearance of the carbonate phases, and therefore the CMR ratio specified in a given simulation, is important because the presence or absence of MnO_2 affects the pH values reached upon disappearance of the various carbonate phases. The lowest pH values reached in the transport simulations usually occurred when MnO_2 remained in a cell, but all the initial and secondary carbonate (rhodochrosite) phases had completely dissolved. (If additionally the Al-bearing phase had been completely dissolved, the pH was even lower). This transient situation (third row in Table 8) occurred in all simulations with a CMR ratio of less than 3/2 (5.25×10^{-3}, 1.05×10^{-2} and 2.1×10^{-2} initial carbonate concentrations and 2×10^{-2} initial MnO_2 concentration). The lower pH values generally reached in the presence of MnO_2 resulted in more aggressive waters and therefore in relatively lower retardation factors for the rhodochrosite and Al-phase dissolution fronts in simulations with an initial carbonate to MnO_2 ratio of less than 3/2 (Tables 10 and 11; Fig. 11a).

In simulations with CMR ratios near 3/2, the complete dissolution of MnO_2 was usually accompanied by the nearly simultaneous disappearance of precipitated rhodochrosite, but was preceded by the complete dissolution of calcite (Fig. 11b). For most of these simulations, therefore, the rhodochrosite dissolution front had a retardation factor (R) close to 4.7, that of the MnO_2 front. Disregarding the Ca and Mg silicate dissolution simulations, the highest rhodochrosite R values were obtained for the simulations that specified a high ion exchange capacity (CEC) or a surface complexation model or equilibrium with a low fixed pCO_2 value ($10^{-1.33}$). In the high CEC simulations, however, the R value of the rhodochrosite front was still lower than that of the MnO_2 dissolution front (R = 9.2), i.e. the rhodochrosite dissolution front preceded the MnO_2 dissolution front.

Finally, in the simulations with a CMR ratio greater than 3/2 (4.2×10^{-2}, 8.4×10^{-2}, 1.68×10^{-1} and 3.32×10^{-1} initial carbonate concentrations and 2×10^{-2} initial MnO_2 concentration), exhaustion of the initial MnO_2 generally proceeded faster than that of the secondary rhodochrosite (with the exception of the 2 high CEC simulations). In simulations with an initial carbonate concentration of 8.4×10^{-2} or greater, exhaustion of the initial MnO_2 also proceeded faster than that of the initial carbonate minerals. The faster movement of the MnO_2 dissolution front relative to the carbonate mineral dissolution fronts resulted in less aggressive waters contacting the carbonate minerals and Al-phase minerals and therefore in higher retardation factors.

Influence of the aluminum mineral allowed to precipitate on the evolution of the low-pH waters. The choice of Al-bearing phase allowed to precipitate was the third major determining factor controlling the relative retardation of the rhodochrosite and Al-phase dissolution fronts in the simulations without Ca and Mg silicate dissolution. The simulations that allowed $AlOHSO_4$ to precipitate, instead of amorphous $Al(OH)_3$, all exhibited significantly greater retardation factors for the rhodochrosite and Al-phase dissolution fronts. Remarkably, the simulations that used the lower solubility product for $AlOHSO_4$ ($10^{-3.23}$) resulted in much higher R values for the $AlOHSO_4$ dissolution front, but did not result in significantly different R values for the rhodochrosite dissolution front, when compared to the values obtained in simulations with the less stable $AlOHSO_4$ phase. More generally, significant differences in retardation factors (for a given initial carbonate concentration) between the $Al(OH)_3$ and the $AlOHSO_4$ simulations occurred only in simulations with CMR ratios above 3/2. The exact reason for this behavior is not clear to us at the present time, but is probably related at least in part to the effect of the MnO_2 dissolution front moving faster than the rhodochrosite dissolution front. In any case, the precipitation of $Al(OH)_3$ generates 3 times more protons than that of $AlOHSO_4$, and would therefore result in significantly more aggressive waters and faster carbonate mineral dissolution, explaining the generally smaller retardation factors found in the simulations with $Al(OH)_3$:

$$Al^{3+} + SO_4^{2-} + H_2O \Leftrightarrow AlOHSO_{4,s} + H^+$$
$$Al^{3+} + 3\,H_2O \Leftrightarrow Al(OH)_{3,am} + 3\,H^+$$

Table 9. Rhodochrosite dissolution-front retardation factors as a function of initial carbonate concentration (in mol/kg H_2O) and for various reaction models. All models except one had a dispersivity of 0 meters. The models all allowed precipitation of rhodochrosite and of one of four possible Al-bearing phases. The specified reaction models have an index that refers to the allowed Al-bearing phase. These indices and the meanings of the various reaction models are discussed in the table notes and in the text.

Reaction Model	Initial Carbonate Mineral Concentration ($x10^{-2}$)							
	0.525	1.05	2.1	3.0	4.2	8.4	16.8	33.2
Basic reaction model, (1)	1.6	2.3	3.7	5.0	12.6	34.1	65.6	128.9
Dispersivity, (1), PHREEQM	1.5	2.3	3.7	4.5	12.7	35.1	67.7	133.0
Dispersivity, (1), PHREEQC	1.5	2.2	3.6	4.5	12.5			
Low CEC, (1)	1.6	2.3	3.8	5.1	11.8	33.7	65.4	128.6
High CEC, (1)	0.0	3.5	5.1	6.5	8.8	31.3	63.0	126.3
CO_2, log pCO_2= -0.9865, (1)	1.4	2.1	3.5	5.0	12.2	32.5	56.5	104.5
CO_2, log pCO_2 = -1.33,(1)	1.5	2.2	3.7	5.9	15.2	37.5	68.8	131.4
Basic reaction model, (3)	1.6	2.3	3.6	4.9	14.7	42.3	81.4	159.4
Basic reaction model, (4)	1.6	2.3	3.6	4.8	14.5	41.3	80.9	160.2
Low CEC, (3)	1.6	2.3	3.7	5.0	13.6	41.8	80.9	158.9
High CEC, (3)	0.0	3.5	5.0	6.4	8.6	38.0	77.2	155.4
Surf. Complex., (3), no Cu,Zn,Co,Ni		3.4	4.7	12.3	22.5	47.4	86.8	165.4
Surf. Complex., (3), with Cu,Zn,Co,Ni	2.7	3.4	4.6	7.8	19.8	46.4	79.5	140.8
Tremolite, mod. 2, (1)	Rhodochrosite accumulates!							
Tremolite, MnO_2, mod. 2, (1)	7.3	10.9	18.2	24.3	32.6	61.6	119.4	235.3
Biotite/K-mont/An, mod. 7, (1)	Rhodochrosite accumulates!							
Biotite/K-mont/An, MnO_2, mod. 7, (1)	12.1	17.9	29.8	40.1	53.9	101.7	197.5	388.9
Biotite/K-mont/An., mod. 7, (3)	Rhodochrosite accumulates!							

Note: the identity of the Al bearing phase is indexed as follows for each reaction model. (1) $Al(OH)_3$, (2) kaolinite, (3) $AlOHSO_4$ with log K_{sp} = -2.2, (4) $AlOHSO_4$ with log K_{sp} = -3.2).

Guide to reaction model abbreviations and meanings.

Dispersivity: basic reaction model with dispersivity of 560 m;
CO_2: column open to a specified fixed pCO_2
Low CEC: low cation exchange capacity (1 meq/100g);
High CEC: high cation exchange capacity (10 meq/100g);
Surf. Complex.: surface complexation model
Tremolite, mod. 2: irreversible tremolite dissolution (model 2 in Table 5);
Tremolite, MnO_2, mod. 2: as above but with zero-order MnO_2 dissolution;
Biotite/K-mont/An, mod. 7: irreversible biotite and anorthite dissolution and K-montmorillonite precipitation (model 7 in Table 5);
Biotite/K-mont/An, MnO_2, mod. 7: same as above but with zero-order MnO_2 dissolution;

The precipitation of $Al(OH)_3$ (or of $AlOHSO_4$) generally occurred as a result of the increase in pH of the infilling water caused by the dissolution of the carbonate minerals. Reaction with the more acidic infilling waters eventually caused redissolution of the Al-bearing phases. In the case of the simulations with low initial carbonate concentrations and with $AlOHSO_4$ (solubility product: $10^{-2.2}$) or with $Al(OH)_3$ allowed to precipitate, complete consumption of the Al-bearing phase occurred at least twice during the simulations. The retardation factors for the Al-phase dissolution fronts given in Table 11 refer only to the last and presumably final dissolution front in the simulations.

Effects of the irreversible dissolution of Ca- and Mg-silicates on the evolution of low-pH Fe(II)-rich waters. Several simulations were conducted using some of the reactions specified in the PHREEQC inverse Models 2 and 7 (Table 5). In addition to allowing secondary $Al(OH)_3$, $Fe(OH)_3$, and rhodochrosite to precipitate, and in addition to specifying initial amounts of calcite (or dolomite) also allowed to react to equilibrium, these simulations forced the dissolution of a fixed number of moles of Ca- and Mg-silicate minerals into each cell and at each time step, thereby simulating a 0-order kinetic dissolution process that matched (over the length of the 10-cell 5.6 km column) the net amount of Ca- and Mg-silicate minerals dissolved between wells 402 and 503 according to PHREEQC inverse Models 2 and 7 in Table 5. Irreversible dissolution of tremolite was specified in the simulations following Model 2 (Fig. 13). Irreversible dissolution of biotite and anorthite and removal from solution of K-montmorillonite was specified in the simulations following Model 7. The silicate dissolution processes would certainly have been more correctly simulated with a pH-dependent rate law, but this capability was not present in the PHREEQC code used. An inexhaustible supply of silicate minerals was also assumed available to all cells throughout the course of the simulations. Despite the crudeness and fallacy of these assumptions, it was hoped that these simulations would nevertheless contribute some insights as to the importance of silicate mineral dissolution in governing the pH evolution of acidic contaminated waters such as those found at the Pinal Creek site. [Ongoing investigations on the occurrence and kinetics of silicate mineral dissolution (and of carbonate and Al, Mn and Fe oxyhydroxide reactions) at the Pinal Creek site will hopefully result in more "realistic" models in the future.]

The Ca- and Mg-silicate dissolution simulations specified an essentially infinite amount of initial chalcedony. This mineral was allowed to react to equilibrium. Following the results of Models 2 and 7 in Table 5, gypsum precipitation was excluded from the simulations. (a few simulations were run allowing gypsum precipitation but the results of those simulations were not significantly different and are not presented here). Although an initial fixed amount of MnO_2 of 2×10^{-2} mol/kg H_2O was specified in most simulations, two sets of simulations specified instead a zero-order dissolution process for MnO_2, according to the amounts determined by Models 2 and 7 in Table 5. Finally, one set of simulations conducted tried to follow Model 7 (Table 5) but allowed $AlOHSO_4$ to precipitate instead of $Al(OH)_3$.

All the simulations that specified a fixed initial MnO_2 concentration (of 2×10^{-2} mol/kg H_2O) invariably resulted in rhodochrosite precipitation and accumulation (Table 9, Fig. 13a). A rhodochrosite dissolution front never formed, presumably because the continuous proton consumption caused by the dissolution of the silicate minerals and the high initial Mn and TDIC concentrations in the infilling water (from well 402) were sufficient to ensure that the infilling water never became undersaturated with respect to rhodochrosite. In contrast, the simulations that allowed the irreversible zero-order dissolution of MnO_2 did produce rhodochrosite dissolution fronts (Fig. 13b), probably because of the constant generation of acidity caused by the oxidation of dissolved Fe(II) and consequent precipitation of $Fe(OH)_3$. The retardation factors for the rhodochrosite dissolution fronts increased almost linearly as a function of the initial carbonate mineral content, nearly doubling for each doubling of the initial carbonate concentration, but only in simulations with carbonate concentrations above 4.2×10^{-2} mol/kg H_2O (at lower initial carbonate concentrations rhodochrosite *R* values increased only by 50% or less).

The two sets of simulations with irreversible tremolite dissolution (with and without fixed initial MnO_2 concentrations) also generated calcite dissolution fronts (Table 10, Fig. 13). In contrast, out of the three sets of simulations that tried to emulate inverse Model 7, only the one that also allowed continuous zero-order dissolution of MnO_2 (and consequent production of acidity) also generated calcite dissolution fronts. Interestingly, the calcite *R* values increased nearly linearly with increasing initial carbonate concentration in the simulations that allowed continuous MnO_2 zero-order dissolution, but did not do so in the set of simulations (with tremolite dissolution) that specified a fixed initial MnO_2 amount. In that set of simulations the calcite *R* value increased from 4.3 to 97.7 (a factor of 23!) although the initial calcite amount specified increased only from 4.2×10^{-2} to 8.4×10^{-2} mol/kg H_2O. This probably occurred because in the simulation with the higher initial carbonate concentration the calcite dissolution front started moving more slowly than the MnO_2 dissolution front (*R* value = 4.7). Even higher initial carbonate concentrations resulted in a tripling of the calcite *R* value for each doubling of the initial calcite concentration. In the set of simulations that also used a fixed initial MnO_2 concentration (2×10^{-2} mol/kg H_2O) but attempted to emulate inverse Model 7, any initial dolomite dissolution was always followed eventually by dolomite precipitation and accumulation. Because of the additional contribution of Fe(II)-containing biotite, a retardation factor of 4.5 was obtained for the MnO_2 dissolution front in those simulations.

Most of the simulations with Ca- and Mg-silicate dissolution did not produce an Al-phase dissolution front (or produced only a temporary one). The proton consumption caused by the silicate dissolution reactions instead caused the continuous precipitation of $Al(OH)_3$ from the infilling water (Table 11). The set of simulations with irreversible tremolite dissolution and zero-order MnO_2 dissolution did however generate an $Al(OH)_3$ dissolution front, again probably because of the acidity generated by the continuous addition of MnO_2 and its reaction with dissolved Fe(II) in the infilling water. The set of simulations emulating Model 7, but allowing $AlOHSO_4$ to precipitate instead of $Al(OH)_3$ also produced a $AlOHSO_4$ dissolution front. Surprisingly, the *R* value for that front actually decreased as the initial carbonate concentration increased to 4.2×10^{-2} mol/kg H_2O and then increased with higher initial carbonate concentrations.

The pH values obtained during the evolution of the well 402 water during the simulations with irreversible Ca- and Mg-silicate dissolution (Fig. 13) were fairly close to the pH values predicted for the various mineral assemblages in Table 8, but only when carbonate minerals (rhodochrosite or calcite) were still present and when MnO_2 was either being added or was still present. The first 10 shifts of the simulations show a rather dramatic increase in the pH of cell 10 because of the constant forced dissolution of Ca- and Mg-silicates into the background water, which takes 10 shifts to flush out.

The effect of not allowing rhodochrosite precipitation. One set of simulations, without Ca- and Mg-silicate dissolution, was conducted using a low cation exchange capacity (1 meq/100 g), but disallowing rhodochrosite precipitation. Comparison with a similar set of simulations that did allow rhodochrosite precipitation shows that while the retardation factors (*R* values) for the calcite dissolution front were greater in the simulations without rhodochrosite precipitation, they were nevertheless up to two times smaller than the *R* values for the rhodochrosite dissolution front in the runs that allowed the mineral to precipitate. The pH values attained after complete dissolution of calcite in the simulations without rhodochrosite precipitation closely match the low-pH values attained after complete rhodochrosite dissolution in the simulations that allowed precipitation of that

Table 10. Calcite dissolution-front retardation factors. See Table 9 caption and notes for explanation of headings and abbreviations. One additional set of simulations reported here was performed with a CEC of 1 meq/100g but with no rhodochrosite allowed to precipitate (low CEC, no rhodo.).

*All simulations referring to Model 7 in Table 5 have initial dolomite instead of calcite.

Reaction Model	Initial Carbonate Mineral Concentrations ($x10^{-2}$)							
	0.525	1.05	2.1	3.0	4.2	8.4	16.8	33.2
Basic reaction model, (1)	1.2	1.5	2.2	2.7	3.4	8.0	22.7	52.0
Dispersivity, (1), PHREEQM	1.0	1.3	2.0	2.5	3.3	8.0	23.0	53.0
Dispersivity, (1), PHREEQC	1.0	1.3	2.0	2.5	3.3			
Low CEC, (1)	1.5	1.9	2.5	3.1	3.8	8.2	22.9	52.3
High CEC, (1)	2.5	3.5	4.9	5.9	6.9	10.7	25.4	54.8
CO_2, log pCO_2 = -0.9865, (1)	1.1	1.4	2.1	2.6	3.3	7.3	18.4	40.8
CO_2, log pCO_2 = -1.33, (1)	1.2	1.5	2.2	2.8	3.6	8.8	23.0	51.2
Low CEC, no rhodo., (1)	1.5	2.1	3.0	3.9	5.0	13.8	32.4	69.5
Basic reaction model, (3)	1.2	1.6	2.2	2.7	3.4	8.7	24.7	56.7
Basic reaction model, (4)	1.3	1.6	2.2	2.7	3.5	8.9	25.6	59.1
Low CEC, (3)	1.5	1.9	2.6	3.1	3.8	8.9	25.0	57.0
High CEC, (3)	2.5	3.5	4.9	5.9	7.0	11.4	27.4	59.5
Surf. Complex., (3), no Cu,Zn,Co,Ni		2.1	2.7	3.2	3.9	10.2	26.3	58.4
Surf. Complex., (3), with Cu,Zn,Co,Ni	1.7	2.0	2.6	3.1	3.8	8.9	22.2	48.2
Tremolite, mod. 2, (1)	1.5	1.9	2.8	3.5	4.5	**113.7**	**353.5**	**>500**
Tremolite, MnO_2, mod. 2, (1)	1.6	2.2	3.3	4.3	5.5	10.0	18.9	36.8
Biotite/K-mont, mod. 7, (1)	No calcite specified. Dolomite accumulates!							
Biotite/K-mont, MnO_2, mod. 7, (1)	1.6*	2.2*	3.2*	4.2*	5.4*	9.7*	18.4*	35.7*
Biotite/K-mont., mod. 7, (3)	No calcite specified. Dolomite accumulates!							

mineral (Table 8). Therefore for proper comparison, the *R* values for the calcite dissolution front in the simulations without rhodochrosite precipitation can only be compared to the *R* values for the rhodochrosite dissolution fronts in the other simulations. In simulations with low initial carbonate (2.1×10^{-2} mol/kg H_2O and lower), the faster movement of the $Al(OH)_3$ dissolution front relative to the rhodochrosite dissolution front meant that the complete dissolution of rhodochrosite led to a large pH decrease (Fig. 11a). In simulations with higher initial carbonate concentrations, the $Al(OH)_3$ final dissolution front traveled more slowly than the rhodochrosite front and therefore the disappearance of rhodochrosite resulted in a relatively small decrease in pH (Figs. 11b and 12). The *R* values for the $Al(OH)_3$ final dissolution-front were smaller (by as much as 1/3 in the runs with high initial carbonate concentrations) in the simulations without rhodochrosite precipitation relative to the $Al(OH)_3$ *R* values obtained in the simulations that allowed rhodochrosite to precipitate. For simulations with the three lowest initial carbonate concentrations (5.25×10^{-3}, 1.05×10^{-2} and 2.1×10^{-2}), however, differences in the *R* values between the simulations that did or did not include rhodochrosite precipitation were relatively small. It should also be remembered that the discretization error inherent in the lower *R* values is much greater than for the higher *R* values (resulting from higher initial carbonate concentrations). Indeed, for greater accuracy all the simulations with low *R* values (for any of the mineral fronts) should have been run with a greater number of cells (and consequently a greater number of time steps and a smaller cell length).

Table 11. Retardation factors for the final aluminum phase dissolution-front. See Table 9 caption and notes for explanation of headings and abbreviations. One additional set of simulations reported here was performed with a CEC of 1 meq/100g but with no rhodochrosite allowed to precipitate (low CEC, no rhodo.). In some cases, the aluminum phase was never exhausted during the 5000 time steps allowed for a simulation.

Reaction Model	Initial Carbonate Mineral Concentrations ($\times 10^{-2}$)							
	0.525	1.05	2.1	3.0	4.2	8.4	16.8	33.2
Basic reaction model, (1)	1.4	1.8	2.6	6.5	22.3	69.6	143.3	290.8
Dispersivity, (1), PHREEQM	1.3	1.8	2.7	8.4	25.3	78.9	162.7	330.3
Dispersivity, (1), PHREEQC	1.2	1.7	2.7	7.6	23.3			
Low CEC, (1)	1.7	2.2	3.2	3.9	20.6	68.8	142.6	290.1
High CEC, (1)	2.7	3.7	5.3	6.9	9.1	62.6	136.8	284.5
CO_2, log pCO_2 = -0.9865(1)	1.3	1.9	3.2	5.3	20.8	61.9	113.6	217.3
CO_2, log pCO_2 = -1.33, (1)	1.5	2.3	3.9	13.9	39.6	106.4	203.9	399.0
Low CEC, no rhodo., (1)	1.6	2.2	3.3	4.2	8.6	37.7	91.4	198.6
Basic reaction model, (3)	0.0	2.3	3.6	7.9	30.3	89.1	171.3	335.5
Basic reaction model, (4)	1.8	2.7	4.5	>500	>500	>500	>500	>500
Low CEC, (3)	1.6	2.3	3.7	5.0	27.5	87.6	169.9	334.1
High CEC, (3)	0.0	0.0	0.0	6.4	8.8	76.0	158.5	323.3
Surf. Complex., (3), no Cu,Zn,Co,Ni		3.5	7.4	24.7	45.9	96.8	176.6	335.8
Surf. Complex., (3), with Cu,Zn,Co,Ni	2.9	3.5	8.8	31.6	48.1	102.8	172.3	300.8
Tremolite, mod. 2, (1)	$Al(OH)_3$ accumulates!							
Tremolite, MnO_2, mod. 2, (1)	8.2	12.5	21.1	28.4	38.2	72.6	141.1	278.4
Biotite/K-mont, mod. 7, (1)	$Al(OH)_3$ accumulates!							
Biotite/K-mont, MnO_2, mod. 7, (1)	$Al(OH)_3$ accumulates!							
Biotite/K-mont, mod. 7, (3)	9.2	8.0	6.9	6.5	5.0	5.8	6.6	7.3

The CO_2 open system simulations. A few simulations were conducted assuming that the 1-dimensional column was allowed to equilibrate with an infinite gaseous CO_2 reservoir with a fixed partial pressure of either $10^{-1.33}$ (the pCO_2 value used by Brown, 1996, a value representative of equilibrium with a measured dissolved-CO_2 concentration for well 503) or $10^{-0.9865}$ (a pCO_2 value representative of the unsaturated zone pCO_2 at site 500 in Nov. 1991). Opening the system up to CO_2 did affect the retardation of the calcite, rhodochrosite and $Al(OH)_3$ dissolution fronts, particularly in the simulations with at least 3×10^{-2} mol/kg of initial calcite (i.e. with CMR ratios of 3/2 and above). The simulations conducted under the higher fixed pCO_2 value ($10^{-0.9865}$) generally resulted in lower R values for the calcite, rhodochrosite and $Al(OH)_3$ dissolution fronts, compared to the values determined in the basic reaction model simulations. In contrast, the simulations conducted under the lower fixed pCO_2 value ($10^{-1.33}$) resulted in higher R values for all three mineral dissolution fronts, compared to the values determined in the basic reaction model simulations. The relevance of these simulations to the migration of the acidic ground waters at Pinal Creek field site is questionable, although the magnitude of the CO_2 equilibration effect clearly warrants further study and confirmation that it is actually not significant at the field site.

The effect of longitudinal dispersion. Assuming a longitudinal dispersivity (α) of 560 m, 10% of the distance between wells 402 and 503, as opposed to a dispersivity of 0 m, did not significantly affect the retardation of the calcite or rhodochrosite dissolution fronts, except for the simulations with the lowest initial carbonate concentrations (3×10^{-2} mol/kg H_2O and lower, corresponding to CMR ratios below 3/2). Dispersion typically has little effect on the propagation of sharp fronts, caused by simple mineral dissolution reactions, especially if the initial mineral concentrations are high enough to significantly retard the propagation of the fronts. At the limit, dispersion would have absolutely no effect on the retardation of a mineral dissolution front for which an infinite initial mineral concentration had been specified. Of course, using a dispersivity of 560 m has a tremendous effect on the spreading of non-reacting solutes, and can also be expected to affect the propagation of mineral dissolution fronts moving close to the advective speed of the water. The spreading of a conservative solute can easily be calculated after a given travel time, or travel distance, using the following equation (Appelo and Postma, 1993):

$$\sigma_x = \sqrt{2\alpha vt} = \sqrt{2\alpha x} \tag{17}$$

or if $\alpha = 560$ m and $x = 5600$ m, the distance between wells 402 and 503, $\sigma_x = 2504$ m. σ_x represents a distance between two specific points in a 1-D homogeneous column. At those two specific points, a conservative tracer injected continuously at the beginning of the column (i.e. at well 402), with a relative concentration c/c_0 of 100%, would have achieved relative concentrations of 50% and 16%, respectively. Essentially, σ_x represents the spreading of the solute mass due to longitudinal dispersion. Longitudinal dispersion implies that portions of the tracer move both faster and slower than the average ground-water velocity. When the relative concentration of a conservative tracer reaches 50% at well 503, 5600 m downgradient from well 402, the 16% relative concentration level has already moved ahead by 2504 m. In contrast, the 84% relative concentration level would be 2504 m upgradient from well 503. The above analysis assumes steady-state flow through a porous medium with homogeneous physical properties and also assumes that there are no "dead-water" zones into which the conservative tracer can diffuse.

The simulations that included longitudinal dispersion were conducted with both the PHREEQM code and with an unpublished version of the PHREEQC v. 2 code (obtained courtesy of Tony Appelo and David Parkhurst). PHREEQC v. 2 allows the simulation of longitudinal dispersion, using an algorithm similar to the one used in PHREEQM. Although efforts were made to try to ensure that the two thermodynamic databases were identical, particularly as concerns Al speciation data, some minor differences may still remain, although we are not aware of any such differences at the present time. The database used in the PHREEQC v. 2 runs was identical to that used in all the other PHREEQC runs (those without dispersion). Unfortunately, we were unable to make PHREEQC v. 2 converge for the entire set of simulations.

The influence of ion exchange and surface-complexation sorption processes. The reader is referred to Appelo (this volume), Appelo and Postma (1993), Dzombak and Morel (1990) and Davis and Kent (1990) for excellent descriptions of the theories of ion exchange and surface-complexation sorption processes. The ion exchange conventions used in the PHREEQM and PHREEQC codes are described in Appelo and Postma (1993). PHREEQC's simulation of surface complexation sorption processes largely follows the diffuse double-layer surface-complexation model presented by Dzombak and Morel (1990) and is fully described in Parkhurst (1995).

Allowing ion exchange with a cation exchange capacity (CEC) of 1 meq/100g did not result in major changes in the rhodochrosite and calcite retardation factors (Tables 9, 10). The ions allowed to exchange were the following: Al^{3+}, Fe^{2+}, Mn^{2+}, Ca^{2+}, Mg^{2+}, Sr^{2+}, Ba^{2+}, Na^{+}, K^{+}. Proton exchange was also simulated. The selectivity coefficients used were the default values present in the PHREEQC thermodynamic database (close to a set of values given in Appelo and Postma, 1993). A CEC of 1 meq/100g [equivalent to 52.2 meq/kg H_2O given the porosity and bulk density of the Pinal Creek alluvial sediments] appears to be a reasonable order of magnitude estimate for the Pinal Creek sediments, given their low content of organic matter and their relative coarseness. The empirical formula given by Breeuwsma et al. (1986) relating the CEC to the <2μm clay fraction and to the organic carbon content is assumed applicable (cited in Appelo and Postma, 1993):

$$\text{CEC (meq/100g)} = 0.7\ (\%\ \text{clay}) + 3.5\ (\%\ \text{C}) \tag{18}$$

Direct measurements of the CEC of Pinal Creek sediments, and of the exchangeable ion composition of the sediments, would certainly be preferable to using an empirical formula. The purpose of our simulations, however, was to determine the effect of including ion exchange processes on the evolution of the low-pH and high-Fe(II) contaminated ground waters at the site. Therefore, simulations were also conducted using an unrealistically high CEC of 10 meq/100g, a CEC that would be applicable to sediments with more than 10% clay content. As discussed earlier, these high CEC simulations resulted in a doubling of the retardation of the MnO_2 dissolution front (from 4.7 to 9.2 for an initial MnO_2 concentration of 2×10^{-2} mol/kg H_2O). The high CEC ion-exchange simulations also generally increased the retardation factors for the calcite dissolution front but did not result in a uniform increase in the retardation of the rhodochrosite or of the Al-phase dissolution fronts (as a function of initial carbonate). At low initial carbonate concentrations, the high CEC simulations did not result in any rhodochrosite precipitation. Nevertheless, at initial carbonate concentrations of 2.1×10^{-2} and 3.0×10^{-2} mol/kg H_2O, the simulations that allowed $Al(OH)_3$ precipitation and a high CEC had higher retardation factors for the rhodochrosite dissolution front. The retardation factors were lower than those of the simulations with no ion exchange, however, at the higher initial carbonate concentrations. The retardation factors for the $Al(OH)_3$ dissolution front also exhibited this complex behavior with increasing initial carbonate concentration.

The PHREEQC simulations that included a diffuse double-layer surface-complexation sorption model resulted in retardation factors for the rhodochrosite, calcite and $AlOHSO_4$ dissolution fronts that were generally higher than the *R* values determined in simulations without sorption. Surface protonation and deprotonation reactions were incorporated into the sorption model, along with surface complexation of Mn, Ca, Mg, Cu, Zn, Co, Ni and SO_4 at weak and strong surface sites. Cu (36.2 mg/L), Zn (4.97 mg/L), Co (4.14 mg/L) and Ni (1.57 mg/L) concentrations measured in the well 402 water were added to one set of simulations. Another set of simulations was also conducted without those elements. This second set of simulations resulted in higher *R* values for the rhodochrosite and calcite fronts, but in slightly lower *R* values for the $AlOHSO_4$ front except in the simulations with the two highest initial carbonate concentrations.

The thermodynamic model was based on the compilation of intrinsic constants for hydrous ferric oxide published by Dzombak and Morel (1990) and used by default in PHREEQC. The number of weak and strong sorption sites used (1.5×10^{-2} and 3.8×10^{-4} respectively) and the amount of surface area per kg of H_2O used (4032 m^2) were identical to the values used by Brown (1996) in his 1-D simulation of reactive transport at the Pinal Creek site, and also in his PHREEQC simulation-fit of a column experiment conducted by Stollenwerk (1994). In his experiment, Stollenwerk (1994) eluted acidic contaminated ground water from the Pinal Creek site through an 80-cm column containing uncontaminated alluvial sediments also taken from the site. Using a higher number of sorption sites in our simulations would probably have resulted in much more significant effects on the retardation of the mineral dissolution fronts.

Other minor effects on the evolution of the low-pH waters. The effects of specifying an initial concentration of dolomite in addition to calcite, and separately of allowing kaolinite to precipitate instead of $Al(OH)_3$ or $AlOHSO_4$ was tested in several simulations. All the simulations specified the same initial carbonate mineral concentration (8.4×10^{-2} mol/kg H_2O) and initial MnO_2 concentration, concentrations identical to the those used in the simulation of Glynn et al. (1991). Including or excluding dolomite did not, in and of itself, affect the retardation of the low-pH waters, as can be seen from the resulting retardation factors for the rhodochrosite dissolution front (Table 12). The important variable was the total moles of solid carbonate, whether in the form of calcite or dolomite.

Table 12. Calcite, dolomite and rhodochrosite retardation factors (*R*) for various simulations with 8.4×10^{-2} mol/kg H_2O of initial carbonate. The simulations may include dolomite and may allow kaolinite precipitation instead of $Al(OH)_3$ or $AlOHSO_4$ precipitation. See Table 9 caption and notes for explanation of the row headings. "1/2 dolomite" means that half of the initial solid-phase carbonate was provided by dolomite.

Reaction Model	Calcite	Dolomite	Rhodochrosite	Al-phase
basic reaction model, (1)	8.0		34.1	69.6
basic reaction model, (2)	7.6		26.4	>500
basic reaction model, 1/2 dolomite, (1)	7.7	4.5	33.6	68.3
dispersivity, (1), PHREEQM	8.0		35.1	78.9
dispersivity, (2), PHREEQM	7.6		27.8	>34
dispersivity, 1/2 dolomite, (1), PHREEQM	7.8	4.3	34.6	77.7
Low CEC, (1)	8.2		33.7	68.8
Low CEC, (2)	7.9		26.2	>500
Low CEC, 1/2 dolomite, (1)	8.0	4.8	33.2	67.6
High CEC, (1)	10.7		31.3	62.6
High CEC, (2)	10.4		24.3	488.2
High CEC, 1/2 dolomite, (1)	10.5	0.0	30.9	61.6
High CEC, 1/2 dolomite, (2)	10.2	0.0	24.0	480.2
CO_2, log pCO_2 = -0.9865, (1)	7.3		32.5	61.9
CO_2, log pCO_2 = -0.9865, (2)	7.1		25.5	>500
CO_2, log pCO_2 = -0.9865, 1/2 dolomite, (1)	6.9	4.5	31.8	60.4
Low CEC, no rhodo., (1)	13.8			37.7
Low CEC, no rhodo., (2)	13.0			305.3
Low CEC, no rhodo.,1/2 dolomite, (1)	13.6	4.9		37.0
Low CEC, no rhodo.,1/2 dolomite, (2)	12.8	4.9		299.6
High CEC, (3)	11.4		38.0	76.0
High CEC, 1/2 dolomite, (3)	11.2	0.0	37.4	74.8
High CEC, 2/3 dolomite, (3)	11.1	6.5	37.3	74.6

Allowing kaolinite to precipitate instead of $Al(OH)_3$ resulted in slightly lower retardation factors for the calcite and rhodochrosite dissolution fronts (Table 12). In contrast, the *R* values obtained for the final kaolinite dissolution front were generally much greater than those obtained for the final $Al(OH)_3$ dissolution front. These effects were both due to the greater stability of kaolinite in the presence of acidic water.

Comparison of the reactive transport simulation results with observations at the Pinal Creek site. Although ground water samples at the Pinal Creek site have been collected and analysed by the U.S. Geological Survey since 1984 until now (1996), the large scale of the basin and of the contaminant plume, and limited financial resources, have prevented the emplacement of a large density of wells. The sparseness of the available spatial information have made it difficult to determine the exact location and especially the movements of the low-pH and high-Fe(II) ground waters (Fig. 14). The width of the fronts (and especially of the Fe(II) front), which might give valuable information on reaction kinetics, has also been difficult to determine exactly.

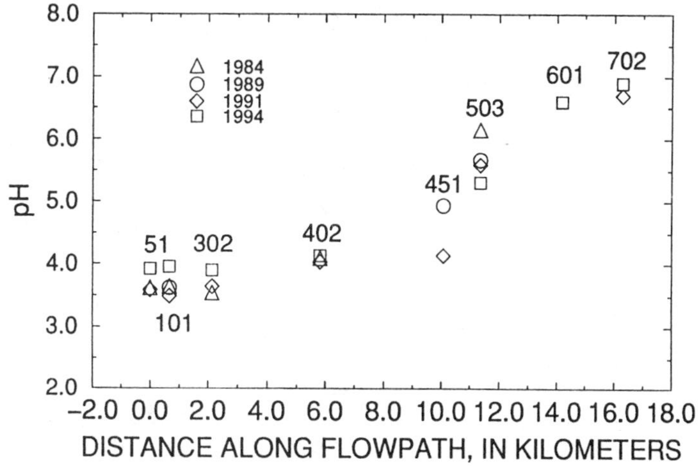

Figure 14. pH profiles along an aquifer flowpath based on the most contaminated wells in the Pinal Creek alluvial aquifer. Well 451 was emplaced in December 1988 and torn away during the floods of spring 1993, and therefore only January 1989 and November 1991 pH data are presented for that well. Note the large pH decrease at well 451 between those dates (actually, most of the decrease occurred between March and August 1989).

Nevertheless some estimates of the velocity of the low-pH front can be made, primarily because the breakthrough of low-pH waters (from a pH of 4.96 in March 1989 to a pH of 4.24 in August 1989) was observed at well 451 only a few months after its emplacement (December 1988). If we assume that the creation of Webster Lake in 1940, approximately 18.5 km upgradient from well 451 provided the principal source of acidic contaminated waters, then an effective velocity of about 1 m/day can be estimated for the low-pH front over that section of the aquifer. This velocity can be compared to an estimated average ground-water velocity of 8.4 m/day between Webster Lake and well 451, giving an estimated retardation factor of about 8.4 for the movement of the low-pH waters. This retardation factor estimate is thought to be a maximum estimate, because the applicable ground water velocity could actually be as low as 5 m/day. Eychaner (1991) and Glynn et al. (1991) estimated that a lower value of R of 5 would be reasonable for the movement of the pH front from Webster Lake to well 451.

Similarly, a low-pH (4.0 to 4.5) high-Fe(II) water was found to be already present during drilling (in February 1995) of a group of wells (LPC wells) emplaced by the Pinal Creek Group (a consortium of copper companies) slightly west of well 503 (on the other side of the creek bed). Given the distance of the LPC wells downgradient from well 451 (1.3 km), the minimum velocity of the low-pH front is estimated to be greater than 0.65

m/day. Comparing this velocity with an estimated average ground-water velocity of about 5 m/day (between 451 and 503) results in a maximum retardation factor of about 7.7 for the velocity of the low-pH front (between well 451 and the LPC well site). The acidity of the well 451 water in August 1989 was ~2 times lower than that of the well 402 water used in our simulations, although the potential acidity (see discussion of Table 7) of the well 451 water did increase with time until a maximum acidity, ~1/3 lower than that of well 402, was reached in November 1991.

In comparison to this field evidence, results of a column elution experiment conducted by Stollenwerk (1994), using water from well 51 (of composition similar to that presented in Table 7), indicated that the velocity of the low-pH front was about 2.5 times slower than that of the water. Many factors can explain the approximately 3 times lower retardation factor found in the column experiment, relative to the field-determined values. The pH front retardation factors estimated from the field evidence incorporate significant effects of dilution (Figs. 3 and 4, Table 7), which also are not accounted for in our transport simulations. The acidity of the initial solutions used in determining those retardation factors also needs to be taken into account if comparisons are to be made. The Webster Lake waters were certainly more acidic than those of well 51 used in Stollenwerk's column experiment. The potential acidity of the well 51 water was 4.5 times greater than that of the well 402 water used in our simulations. As a result, the retardation factor of 2.5 determined by Stollenwerk (1994) should translate to an retardation factor of approximately 7.8, $(2.5 - 1) \times 4.5 + 1 = 7.75$, had well 402 water been used in his column experiments. Equation (15) is used to crudely normalize these results, assuming that DM is proportional to the acidity of the low-pH solution and that the field and laboratory experiments both had identical homogeneous mineral concentrations and reactions. Similarly, considering that the water from well 451 in August 1989 was about 2 times less acidic than the water used in our simulations, the retardation factor of 7.7 determined between 451 and the LPC site would correspond to an R value of 4.35, if normalized to the acidity of the well 402 water (89/1/12). Comparing these normalized retardation factors (4.4 and 7.8) with the R values reported for the rhodochrosite dissolution front in Table 9, in the simulations without irreversible dissolution of Ca- and Mg-silicates, indicates that initial carbonate mineral concentrations between 2.1×10^{-2} and 4.2×10^{-2} mol/kg H_2O and certainly no lower than 1.05×10^{-2} mol/kg H_2O would give reasonable simulated retardation factors. This assumes that an MnO_2 concentration of 2×10^{-2} mol/kg H_2O was also reasonable. A lower MnO_2 concentration would result in a higher CMR ratio and therefore could increase the simulated retardation factors. Consequently, initial carbonate concentrations would have to be adjusted slightly downward in order to match the estimated lab and field retardation factors. A higher MnO_2 concentration would probably not, however, have any significant effect on the simulated pH front R values (for initial carbonate mineral concentrations between 1.05×10^{-2} and 3.0×10^{-2} mol/kg H_2O).

In the simulations with Ca- and Mg-silicate dissolution, the retardation factors for the rhodochrosite front are so high that an initial carbonate concentration of 5.25×10^{-3} or lower would have to be used to match the observed retardation of the pH front in Stollenwerk's experiments and in the field, if we assume that the disappearance of rhodochrosite controls the low-pH front. Such a low initial carbonate concentration is not really realistic given our knowledge of the alluvial sediments at the site, and we therefore conclude that the simulations with irreversible dissolution of Ca- and Mg-silicate should probably be conducted again using a finite (as opposed to an inexhaustible) source of these silicates. Further field information on the concentration, and on the rate of dissolution of silicate minerals at the Pinal Creek site, will however be needed. We note however, that the calcite R values obtained in our forward simulations of inverse Model 2 (Table 5), were reasonable but only for initial calcite concentrations below 4.2×10^{-2} mol/kg H_2O. The disappearance of calcite did not by itself, however, result in a sufficient pH decrease to match the pH 4.0 to 4.5 values (or lower) that we associate with the pH "front" in the field (Fig. 14).

The question of which secondary aluminum phase actually precipitates out, and its rate of redissolution, are also important considerations. Table 8 shows that the most acidic waters (below a pH of 5) can only be obtained if $Al(OH)_3$ is not present (or if its rate of dissolution is slow relative to the movement of the water). If we assume that it is present and that the Local Equilibrium Assumption is reasonable, then our simulations show that initial carbonate concentrations would have to be lower than 3.0×10^{-2} mol/kg H_2O to match the retardation of the pH front observed in the field and in Stollenwerk's laboratory columns.

There has been until now, to our knowledge, no direct observation of the breakthrough of the high-Fe(II) waters at the Pinal Creek site. Waters from wells 451 and the LPC site (near 503) were already found to have high Fe(II) concentrations during the emplacement of those wells (in November 1991 and February 1995 respectively). The fact that dissolved Fe(II) was present at well 451 before the low-pH breakthrough suggests that the Fe(II) front is moving faster than the low-pH front, or at least had progressed further than the low-pH front until that point in time. If we again use the creation of Webster Lake as an initial condition, retardation factors between 5 and 8.4 can be estimated for the high-Fe(II) front, between Webster Lake and well 451. Similarly, the evidence based on data from well 451 and from the LPC well site also suggests a maximum retardation factor of 7.7 over that section of the aquifer. The latter factor is equivalent to a maximum retardation factor of about 3.2 in our simulations, after having corrected for the three times higher Fe(II) concentration of well 402 relative to that of well 451 (in August 1989).

In his laboratory column experiments, Stollenwerk (1994) observed retardation factors of approximately 2 for the high-Mn(II) spike, and 2.5 for the high-Fe(II) front (Fig. 10 gives an example of a dispersion-affected Mn(II) spike and of the associated Fe(II) front). The lag in the Fe(II) front suggests that the rate of $Fe(OH)_3$ precipitation was delayed relative to the rate of MnO_2 dissolution. Normalizing those retardation factors to an infilling solution with 4.7 times less dissolved Fe(II) (for well 402) results in equivalent retardation factor of between 5.7 and 8.1.

Therefore, after normalization to the infilling water used in our simulations, the field and laboratory determined retardation factors for the high-Fe(II) front encompass a range of about 3.2 to 8.1. This range compares favorably with the retardation factor of 4.7 determined for most of the simulations presented here. Furthermore, although the MnO_2 dissolution front may be moving a little bit faster than that of the low-pH waters, according to field and laboratory observations, it probably is not moving much faster, and therefore the CMR ratio is probably close to 3/2.

The pH values determined for the various pH-plateaus in our simulations (Table 8, Figs. 11, 12, 13) are within the range of the values observed in the field (Fig. 14) and therefore the belief expressed in our simulations that the complete dissolution of secondary rhodochrosite occurs after the dissolution of calcite and is accompanied by a significant decrease in pH (in the absence of $Al(OH)_3$) is not inconsistent with the field evidence.

CONCLUSIONS

The basic theory and assumptions of inverse geochemical modeling were presented. Although much less commonly used than reactive transport modeling, particularly in investigations of ground-water contamination, inverse geochemical modeling can provide a powerful tool in such investigations by helping the user identify the possible reactions that may be affecting the chemical and isotopic evolution of contaminated ground waters. This information can then be incorporated into a geochemical transport model and used to conduct a sensitivity analysis on the transport of various reactive contaminants of concern, as a function of the reaction models identified earlier by inverse modeling. This approach was demonstrated for the case of a site of acidic heavy-metal ground-water contamination in the Pinal Creek Basin, Arizona. The interactive inverse geochemical modeling code NETPATH was first used to construct a series of inverse models that quantified observed differences in chemical composition between an initial acidic Fe(II)-rich water and an evolutionary, partially-neutralized Fe(II)-poor water, according to sets of postulated reactions. Each inverse model created was proposed and evaluated according to our knowledge of the geochemistry of the aquifer and according to our knowledge of the thermodynamic and kinetic feasibility of its reactions. Once an apparently suitable and irreducible set of inverse models had been identified, the PHREEQC inverse modeling code was used to further refine and evaluate our set of possible inverse models. Unlike NETPATH, PHREEQC takes account of uncertainties in the analytical data and additionally maintains alkalinity-balance, charge-balance and water-balance constraints in its solution of possible inverse models.

Inverse modeling with NETPATH and with PHREEQC allowed us to quantify the reaction processes responsible for the evolution of an acidic Fe(II)-rich ground water into a partially neutralized Fe(II)-poor water at the Pinal Creek site. The principal reaction processes appear to be the reductive dissolution of solid MnO_2 by aqueous Fe(II), the consequent precipitation of $Fe(OH)_3$, the dissolution of calcite and/or dolomite, the precipitation of an aluminum phase, $Al(OH)_3$ or $AlOHSO_4$, and the possible precipitation of chalcedony. Results of the inverse modeling simulations also led us to conclude that:

(1) Ca- and Mg-silicate dissolution must be an important process during the neutralization of the low-pH waters, given that CO_2 exsolution probably does not occur.

(2) Dilution of the acidic ground waters does occur and can be quantified using Cl, Na and SO_4 concentrations as conservative constituents, constituents that do not undergo significant heterogeneous mass-transfer reactions. The PHREEQC models revealed that SO_4 could be considered a conservative constituent given its associated analytical uncertainty.

(3) Rhodochrosite precipitation (or possibly an MnO_2-$Mn(OH)_3$ electron-transfer or Mn^{2+} sorption mechanism) must be responsible for the lower than expected increase in dissolved Mn(II) concentrations caused by the aqueous Fe(II) reduction of MnO_2 solids.

After a brief review of the status of reactive transport modeling at the Pinal Creek site, the results of some new forward simulations were presented. These geochemical transport simulations explored the effect of the reactions identified through the previous inverse modeling simulations, and also of reactions previously mentioned by other researchers, on the chemical evolution of an acidic Fe(II) rich water from the Pinal Creek site. The purpose of the PHREEQC and PHREEQM transport simulations was to determine the relative rates of movement of the Fe(II)-rich and low-pH ground waters and to determine how the retardation of these contaminated waters was affected by the various postulated reaction processes.

The only factors affecting the retardation of the Fe(II)-rich waters and the propagation of the MnO_2 dissolution front in our simulations were the initial specified concentration of MnO_2 and the concentration of Fe(II) in the inflowing contaminated water. As we demonstrated, the rate of movement of the MnO_2 front could have easily been calculated by hand, but only in the absence of ion-exchange or other processes affecting the Fe(II) concentration of the inflowing water. Simulations conducted with a high ion-exchange capacity of 10 meq/100g resulted in a doubling of the retardation factor for the MnO_2 dissolution front.

The propagation of the various pH fronts caused by the complete dissolution of carbonate and aluminum minerals could not, however, have been so easily predicted. The use of a geochemical transport code such as PHREEQM or PHREEQC does seem to be required to determine the retardation factors applicable to each mineral dissolution front. The initial amount of carbonate (calcite and/or dolomite) specified in each simulation was the primary factor determining the movement of the low-pH waters. Other important factors were: the ratio of initial carbonate to initial MnO_2, the type of Al phase allowed to precipitate, and whether or not rhodochrosite was allowed to precipitate. High initial carbonate to MnO_2 ratios, allowing rhodochrosite precipitation and allowing $AlOHSO_4$ precipitation instead of $Al(OH)_3$ precipitation resulted in higher retardation factors for the movement of the low-pH waters. Allowing the irreversible dissolution of Ca- and Mg-silicates so as to match the mass-transfer amounts determined in a few of our inverse models, resulted in unrealistically high retardation factors for the rhodochrosite and $Al(OH)_3$ dissolution fronts, although the retardation factors determined for the calcite dissolution front were reasonable. More field and laboratory information is required on the abundance and rates of reaction of these silicate minerals if more "realistic" transport simulations are to be conducted.

Inclusion of ion-exchange processes did not have a significant effect on the movement of the pH fronts at low cation exchange capacities (1 meq/100g) but did have a significant effect at higher cation exchange capacities (10 meq/100g). Because of surface-protonation, allowing surface-complexation reactions generally resulted in higher retardation factors for the carbonate and $AlOHSO_4$ dissolution fronts. Allowing equilibrium with a pCO_2 of $10^{-0.9865}$ resulted in lower retardation factors for the carbonate and $Al(OH)_3$ dissolution fronts particularly at initial carbonate concentrations greater than 3×10^{-2} mol/kg H_2O. Allowing equilibrium with a lower fixed pCO_2 of $10^{-1.33}$ resulted instead in generally higher retardation factors for the carbonate and $Al(OH)_3$ dissolution fronts. Simulation of longitudinal dispersion was not an important factor controlling the movement of the calcite and rhodochrosite disssolution fronts except at very low initial carbonate concentrations. Longitudinal dispersion would also have had an effect on the rate of movement of the Fe(II)-rich waters at very low initial MnO_2 concentrations had such simulations been conducted. Including dolomite in addition to calcite in the background aquifer, and allowing kaolinite to precipitate instead of $Al(OH)_3$ did not significantly affect the propagation of the low-pH fronts associated with the dissolution of calcite and rhodocrosite.

Identifying knowledge gaps and critical data needs, preventing us from more accurately determining the identity and importance of the reactions occurring at the Pinal Creek site, was one of the most important results of the inverse and reactive transport modeling simulations conducted in this paper.

A preliminary comparison of the retardation factors for the low-pH and high-Fe(II) fronts determined in our local equilibrium simulations with retardation factors estimated from field evidence and from Stollenwerk's (1994) laboratory column elution tests suggests that an initial carbonate concentration between 2.1×10^{-2} and 4.2×10^{-2} mol/kg H_2O may be reasonable and that the initial MnO_2 concentration of 2×10^{-2} mol/kg H_2O used in our simulations was also reasonable. Finally, pH values obtained during the course of our local equilbrium simulations are reasonable given the pH values observed in the field.

The retardation factors determined for the mineral dissolution fronts as a result of our various simulations will be of use not only in estimating the rate of movement of the low-pH and high-Fe(II) ground waters at the Pinal Creek site, but will also be useful in identifying the most important chemical parameters controlling the movement of these contaminated waters. The results of our simulations also may provide information on the possible rate of movement of acidic metal-rich waters at other similar ground-water contamination sites, after adjustments are made for the pH and dissolved metal (Al, Fe, Mn) concentrations in the contaminated waters and for the mineralogical characteristics of the particular site.

This paper presents and demonstrates an approach for the investigation of the evolution and movement of contaminated ground waters: the use of inverse geochemical modeling to identify important possible reaction processes, followed by geochemical transport simulations that incorporate the possible reactions previously determined and a range of possible aquifer characteristics. Such an approach results in an improved understanding of the processes that may control the future evolution of contaminated ground waters. This information may then lead to better predictability of the transport of highly reactive contaminants and may be used for more effective mitigation of contaminated ground waters at sites with sparse spatial information.

How to obtain U.S. Geological Survey computer codes and the PHREEQM code

The latest USGS gechemical codes can be obtained by anonymous ftp to the following internet site, brrcrftp.cr.usgs.gov, or may be obtained along with other USGS hydrologic modeling codes, via the World Wide Web (WWW) at URL http://h2o.usgs.gov/software, or by anonymous ftp to h2o.usgs.gov. Codes and documentation may also be ordered from U.S. Geological Survey, Branch of Information Services, Box 25286, Denver CO, 80225-0286 (*Telephone:* 303 202-4700, *Fax:* 303 202-4693). PHREEQM may be purchased for a modest fee from: A.A. Balkema Publishers, P.O. Box 1675, 3000 BR Rotterdam, The Netherlands (*Fax:* 31-10-4135947).

ACKNOWLEDGMENTS

We are deeply grateful to David Parkhurst for helping us understand the power and intricacies of inverse modeling with his latest very powerful geochemical code, PHREEQC. We also greatly appreciate the extensive help of Joseph Vrabel, who conducted many of the reactive transport simulations presented in this paper. We extend our sincere appreciation to Carl Steefel, Eric Oelkers and Don Thorstenson, who managed to review this paper on very short notice and still provided very significant comments that led to distinct improvements in the paper. Finally, I thank Paul Ribbe for his patience and skill in editing this manuscript.

REFERENCES

Appelo CAJ, Willemsen A (1987) Geochemical calculations and observations on salt water intrusions I. A combined geochemical/mixing cell model. J Hydrol 94:313-330

Appelo CAJ, Postma D (1993) Geochemistry, groundwater and pollution. A.A. Balkema, Rotterdam, The Netherlands, 536 p

Breeuwsma A, Wösten JHM, Vleeshouwer JJ, Van Slobbe AM, Bouma J (1986) Derivation of land qualities to assess environmental problems from soil surveys. Soil Sci Soc Am J 50:186-190

Brown JG (1996) Movement of metal contaminants in ground water in the Pinal Creek basin, Arizona: Model assessment and simulation of reactive transport, MSc Thesis, Dept Hydrology and Water Resources, Univ Arizona, Tucson, 236 p

Brown JG, Harvey JW (1994) Hydrologic and geochemical factors affecting metal-contaminant transport in Pinal Creek basin near Globe, Arizona. In: Morganwalp DW, Aronson DA (eds) U S Geol Surv Toxic Substances Hydrology Program—Proc Technical Mtg, Colorado Springs, CO, September 20-24, 1993: U. S Geol Surv Water Resources Invest Rpt 94-4015 (in press)

Davis JA, Kent DB (1990) Surface complexation modeling in aqueous geochemistry. In: Mineral-Water Interface Geochemistry, Hochella MF Jr, White AF (eds) Rev Mineral 23:177-260

Dria MA, Bryant SL, Schechter RS, Lake LW (1987) Interacting precipitation dissolution waves: the movement of inorganic contaminants in groundwater. Water Resources Res 23:2076-2090

Dzombak DA, Morel FMM (1990) Surface Complexation Modeling: Hydrous Ferric Oxide. Wiley & Sons, New York, 393 p

Engesgaard P, Kipp (1992) A geochemical transport model for redox-controlled movement of mineral fronts in groundwater flow systems: a case of nitrate removal by oxidation of pyrite. Water Resources Res 28:2829-2843

Eychaner JH (1989) Movement of inorganic contaminants in acidic water near Globe, Arizona, In: Mallard GE, Ragone SE (eds) U S Geol Surv Toxic Substances Hydrology Program—Proc Technical Mtg, Phoenix, Arizona, September 26-30, 1989: U S Geol Surv Water Resources Invest Rpt 89-4220:567-575

Eychaner JH (1991) The Globe, Arizona, research site—Contaminants related to copper mining in a hydrologically integrated environment. In: Mallard GE, Aronson DA (eds) U S Geol Surv Toxic Substances Hydrology Program—Proc Technical Mtg, Monterey, California, March 11-15, 1991: U S Geol Surv Water Resources Invest Rpt 91-4034:475-480

Eychaner JH, Stollenwerk KG (1985) Neutralization of acidic ground water near Globe, Arizona, In: Schmidt KD (ed) Groundwater Contamination and Reclamation. Proc of a Symposium, Tucson, Arizona, August 14-15, 1985. Bethesda, Maryland: Am Water Resources Assoc p 141-148

Ficklin WH, Love AH, Papp CSE (1991) Solid-phase variations in an aquifer as the aqueous solution changes, Globe, Arizona, In: Mallard GE, Aronson DA (eds), U S Geological Survey Toxic Substances Hydrology Program - Proc Technical Mtg, Monterey, California, March 11-15, 1991: U S Geol Surv Water Resources Invest Rpt 91-4034:475-480

Glynn PD (1991) Effect of impurities in gypsum on contaminant transport at Pinal Creek, Arizona In: Mallard GE, Aronson DA (eds), U S Geol Surv Toxic Substances Hydrology Program - Proc Technical Mtg, Monterey, California, March 11-15, 1991: U S Geol Surv Water Resources Invest Rpt 91-4034:466-474

Glynn PD, Engesgaard P, Kipp KL (1991) Use and limitations of two computer codes for simulating geochemical mass transport at the Pinal Creek toxic-waste site, In: Mallard GE, Aronson DA (eds) U S Geol Surv Toxic Substances Hydrology Program—Proc Technical Mtg, Monterey, California, March 11-15, 1991: U S Geol Surv Water Resources Invest Rpt 91-4034:454-460

Glynn PD, Busenberg E (1994a) Dissolved gas and chlorofluorocarbon content of ground waters in the Pinal Creek Basin, Arizona. In: Morganwalp DW, Aronson DA (eds), U S Geol Surv Toxic Substances Hydrology Program—Proc Technical Mtg, Colorado Springs, CO, September 20-24, 1993: U S Geol Surv Water Resources Invest Rpt 94-4015 (in press)

Glynn PD, Busenberg E (1994b) Unsaturated zone diffusion of carbon dioxide and oxygen in the Pinal Creek Basin, Arizona. In: Morganwalp DW, Aronson DA (eds) U S Geol Surv Toxic Substances Hydrology Program—Proc Technical Mtg, Colorado Springs, CO, September 20-24, 1993: U S Geol Surv Water Resources Invest Rpt 94-4015 (in press)

Hill M (1992) A computer program (MODFLOWP) for estimating parameters of a transient, three-dimensional, ground-water flow model using nonlinear regresssion.U S Geological Survey Open-File Report 91-484, 358 p

Hydro Geo Chem Inc. (1991) Investigation of acid water contamination along Miami Wash and Pinal Creek, Gila County, Arizona: Claypool, Arizona, Cyprus Miami Mining Corporation, 140 p

LeBlanc DR (1984) Sewage plume in a sand and gravel aquifer, Cape Cod, Massachusetts. U S Geological Survey Water Supply Paper 2218, 28 p

Lichtner PC (1985) Continuum model for simultaneous chemical reactions and mass transport in hydrothermal systems. Geochim Cosmochim Acta 49:779-800

Lichtner PC (1988) The quasi-stationary state approximation to coupled mass transport and fluid-rock interaction in a porous medium. Geochim Cosmochim Acta 52:143-165

Lind CJ, Stollenwerk KG (1994) Alteration of alluvium of Pinal Creek, Arizona by acidic ground water resulting from copper mining. In: Morganwalp DW, Aronson DA (eds), U.S. Geological Survey Toxic Substances Hydrology Program - Proc Technical Mtg, Colorado Springs, CO, September 20-24, 1993. U S Geol Surv Water Resources Invest Rpt 94-4015 (in press)

Mackay DM, Freyberg DL, Roberts PV, Cherry JA (1986) A natural gradient experiment on solute transport in a sand aquifer—1. Approach and overview of plume movement. Water Resources Res 22:2017-2029

Neaville CC, Brown JG (1993) Hydrogeology and hydrologic system of Pinal Creek Basin, Gila County, Arizona. U S Geol Surv Water Resources Invest Rpt 93-4212

Nordstrom DK, Plummer LN, Langmuir D, Busenberg E, May HM, Jones BF, Parkhurst DL (1990) Revised chemical equilibrium data for major water-mineral reactions and their limitations. In: Melchior DC, Bassett RL (eds), Chemical Modeling in Aqueous Systems II, Am Chem Soc Symp Series 416:398-413

Ortoleva P, Auchmuty G, Chadam J, Hettmer J, Merino E, Moore CH, Ripley E (1986) Redox front propagation and banding molalities. Physica 19D:334-354

Parkhurst DL, Plummer LN, Thorstenson DC (1982) BALANCE—A computer program for geochemical calculations. U S Geol Surv Water Resources Invest Rpt 82-14

Parkhurst DL, Plummer LN (1993) Geochemical models. In: Alley WM (ed) Regional Ground-Water Quality. Van Nostrand Reinhold, New York, Chapter 9:199-225

Parkhurst DL (1995) User's guide to PHREEQC—A computer program for speciation, reaction-path, advective-transport, and inverse geochemical calculations. U S Geol Surv Water Resources Invest Rpt 95-4227

Parkhurst DL (1996) Including uncertainties in geochemical mole-balance modeling. (In preparation)

Petersen NP (1962) Geology and ore deposits of the Globe-Miami district, Arizona. U S Geol Surv Prof Paper 342, 151 p

Pollock DW (1989) Documentation of computer programs to compute and display pathlines using results from the U.S.Geological Survey modular three-dimensional finite-difference ground-water flow model. U S Geol Surv Open File Report 89-381, 188 p

Plummer LN, Parkhurst DL, Thorstenson DC (1983) Development of reaction models for ground-water systems. Geochim Cosmochim Acta 47:665-686

Plummer LN, Prestemon EC, Parkhurst DL (1991) An interactive code (NETPATH) for modeling Net Geochemical Reactions along a flow path. U S Geol Surv Water Resources Invest Rpt 91-4078

Ransome FL (1903) Geology of the Globe copper district, Arizona. U S Geol Surv Prof Paper 12, 168 p

Revesz K, Coplen TB, Baedecker MJ, Glynn PD, Hult M (1995) Methane production and consumption monitored by stable H and C isotope ratios at a crude oil spill site, Bemidji, Minnesota. Applied Geochem 10:505-516

Robertson FN (1991) Geochemistry of ground water in alluvial basins of Arizona and adjacent parts of Nevada, New Mexico, and California. Regional Aquifer-System Analysis-Southwest Alluvial Basins, Arizona and Adjacent States. U S Geol Surv Prof Paper 1406-C.

Stollenwerk KG, Eychaner JH (1987) Acidic ground water contamination from copper mining near Globe, Arizona. In: Franks BJ (ed) U S Geol Surv Toxic Substances Hydrology Program—Proc of the 3rd technical meeting, Pensacola, Florida, March 23-27, 1987: U.S. Geol Surv Open File Report 87-109:D19-D24

Stollenwerk KG (1994) Geochemical interactions between constituents in acidic groundwater and alluvium in an aquifer near Globe, Arizona. Applied Geochem 9:353-369

Tolle S, Arthur GV (1991) Aquifer restoration under the clean water act. In: Mallard GE, Aronson DA (eds), U.S. Geological Survey Toxic Substances Hydrology Program—Proc Technical Mtg, Monterey, California, March 11-15, 1991: U S Geol Surv Water Resources Invest Rpt 91-4034:520-523

Winograd IJ, Robertson FN (1982) Deep oxygenated ground water: anomaly or common occurrence? Science 216:1227-1230